389.6 <del>(300)</del>

LIBRARY
No. B 6279 B

CAINS 26473

REFERENCE
COPY

CONVERSION TABLES FOR SI METRICATION

CONVERSION TABLES FOR SI METRICATION

WILLIAM J. SEMIOLI
and
PAUL B. SCHUBERT

INDUSTRIAL PRESS Inc., 200 Madison Avenue, New York, N. Y. 10016

Library of Congress Cataloging in Publication Data

Semioli, William J.   1938–

   Conversion tables for SI metrication.

   1. Metric system–Conversion tables.

   I. Schubert, Paul Bert, 1923–   joint author
   II. Title.

   QC94.S36   389'.152   74-1104

   ISBN 0-8311-1104-6

Tables 2 to 37A are photographically reproduced from computer printout to avoid typographical errors. Computer printouts were prepared by: Charles S. Davis, Technical Consultant, 154 E. 89th St., New York, N.Y. 10028.

CONVERSION TABLES FOR SI METRICATION

Copyright © 1974 by Industrial Press Inc., New York, N.Y. Printed in the United States of America. All rights reserved. This book, or parts thereof, may not be reproduced in any form without permission of the publishers.

Contents

---

*In Section C, the English—SI unit relations that are underlined can be converted by
exact values taken from the appropriate table, without decimal shifting or multiplying
by a power of ten.

Preface

This reference book is designed to assist engineers and those in related technical disciplines. It will enable them to convert numerical values of a wide variety of physical quantities in English units of measure to and from SI, The International System of Units.

For the reader's convenience, the contents have been arranged into four main sections:

Section A—An introduction to the SI System.

Section B—An extensive list of useful conversion factors.

Section C—Includes conversion tables Nos. 1 through 37, with special introductions to each, for more common SI and English units.

Section D—A refresher on using powers of ten.

Section A, the main introduction, is brief. Depending on the reader's familiarity with SI, it can serve as a primer or as a review of many essential features of this modern metric system. It also presents clear and succinct instructions for using the conversion tables in Section C.

For more information on SI practice, the editors refer the reader to the following publications:

Metric Practice Guide, published by the American Society for Testing and Materials, 1916 Race St., Philadelphia, Pa. 19103.

ISO International Standard 1000. This publication covers the rules for use of SI units, their multiples and submultiples. It can be obtained from the American National Standards Institute, 1430 Broadway, New York, N.Y. 10018.

The International System of Units, Special Publication 330 of the National Bureau of Standards—available from the Superintendent of Documents, U.S. Government Printing Office, Washington, D.C. 20402.

Section B, devoted to conversion factors, serves as a ready reference source for independent calculations in a wide variety of SI and non-SI units.

The tables in Section C permit the user to convert—without a calculator and to a high degree of accuracy—numerical quantities for more common English and SI units. These tables cover both simple and compound physical quantities. Each table value has been generated by computer, where permissible, to seven significant digits.

The values in Tables 2 to 37A, appearing in Section C, have been computed by use of the first conversion factor given in the special introductions to each table. Each calculation was carried to an answer of eleven significant figures and rounded to seven.

In this computer rounding procedure, if the eighth to eleventh digits of the answer were 5000, or greater, the seventh significant figure was raised by one in value. If the eighth to eleventh digits were 4999, or less, the seventh significant figure remained the same, for example:

$$12.941576235 \text{ would round to } 12.94158$$
$$2457.6835000 \text{ would round to } 2457.684$$
$$135.29144631 \text{ would round to } 135.2914$$

As an additional aid in using the tables, each has its own special introduction. They acquaint the reader with a variety of units which can be converted through the particular table. Also, the special introductions and headings within the tables provide guidelines for continued correct usage.

Section D presents a description of working with Powers of Ten Notation. If necessary, this section can provide the reader with a review of manipulating the powers of ten. This can prove helpful not only with applying the conversion tables given in the book, but also for various calculations in decimally based SI.

The editors feel that the approach taken in organizing this reference book has been particularly oriented to the use that will be made of it. Once the reader is familiar with its layout, he will be able to obtain, from any section, such information and values as are sufficient for desired accuracy. He will also be helped by not having to resort continually to extensive introductory reading.

SECTION A

Introduction

Introduction

The International System of Units (SI) is a modernized version of the tradi-tional MKSA Metric System (Metre, Kilogram, Second, Ampere). Like MKSA, the SI system is based on decimal arithmetic. For each physical quantity, units of different sizes are formed by multiplying the value of a base number by powers of 10 or by shifting decimal places.

SI Base Units

As a further simplification, the SI system is founded on seven base units. Figure A-1 presents: the physical quantity specified by each base unit, the name of the base unit, its symbol and definition.

PHYSICAL QUANTITY	NAME OF UNIT	UNIT SYMBOL	DEFINITION
Basic SI Units			
Length	metre	m	1,650,763.73 wavelengths in vacuo of the radiation corresponding to the transition between the energy levels $2p_{10}$ and $5d_5$ of the krypton — 86 atom.
Mass	kilogram	kg	Mass of the international prototype which is in the custody of the Bureau International des Poids et Mesures (BIPM) at Sèvres, near Paris.
Time	second	s	The duration of 9,192,631,770 periods of the radiation corresponding to the transition between the two hyperfine levels of the ground state of the cesium-133 atom.
Electric Current	ampere	A	The constant current which, if main-tained in two parallel rectilinear conductors of infinite length, of neg-ligible circular cross section, and placed at a distance of one metre apart in a vacuum, would produce between these conductors a force equal to 2×10^{-7} N/m length.
Thermodynamic Temperature	degree Kelvin	K	The fraction 1/273.16 of the ther-modynamic temperature of the triple point of water.
Amount of Substance	mole	mol	The amount of substance of a system which contains as many elementary entities as there are atoms in 0.012 kilogram of carbon 12.
Luminous Intensity	candela	cd	The luminous intensity, in the per-pendicular direction, of a surface of 1/600,000 square metre of a black body at the temperature of freezing platinum under a pressure of 101,325 newtons per square metre.

Fig. A-1

SI Supplementary Units

The two SI supplementary units can be considered base units devoted to plane and solid angles:

Physical Quantity	SI Supplementary Unit Name	Symbol	Definition
Plane angle	radian	rad	The plane angle between two radii of a circle which cut off, on the circumference, an arc equal in length to the radius.
Solid angle	steradian	sr	The steradian is the solid angle which, having its vertex in the center of a sphere, cuts off an area of the surface of the sphere equal to that of a square with sides equal in length to the radius of the sphere.

SI Derived Units

As their name suggests, SI Derived Units are designations of physical quantities obtained from algebraic combinations which can include base or supplementary SI units, or both.

For example, the SI derived unit for acceleration is m/s^2 or metre per second squared. For angular acceleration, the derived unit is rad/s^2 or radian per second squared.

Certain compound derived units are expressed by or contain special names. In Fig. A-2 it will be noted that the name of the SI unit is written in lower case, even though it may be drawn from a proper name. Symbols are written mostly in lower case letters, except where they derive from a proper name. In these instances, the first letter of the symbol is capitalized.

USING SI

Convenient Numerical Range

A user commonly selects a particular decimal multiple or submultiple of an SI unit which is most convenient (see Fig. A-3). For example, multiples or submultiples may be chosen so that numerical values are within a practical range of 0.1 to 1000. This range tends to eliminate awkward zeros, decimals, and exponential powers of 10.

SI Units Having Special Names			
Force	newton	$N = kg \cdot m/s^2$	That force which, when applied to a body having a mass of one kilogramme, gives it an acceleration of one metre per second squared.
Work, Energy, Quantity of Heat	joule	$J = N \cdot m$	The work done when the point of application of a force of one newton is displaced through a distance of one metre in the direction of the force.
Power	watt	$W = J/s$	One joule per second.
Electric Charge	coulomb	$C = A \cdot s$	The quantity of electricity transported in one second by a current of one ampere.
Electric Potential	volt	$V = W/A$	The difference of potential between two points of a conducting wire carrying a constant current of one ampere, when the power dissipated between these points is equal to one watt.
Electric Capacitance	farad	$F = C/V$	The capacitance of a capacitor between the plates of which there appears a difference of potential of one volt when it is charged by a quantity of electricity equal to one coulomb.

Fig. A-2

Examples: Where W = watt, MW = megawatt, m = metre, mm = millimetre, s = second, A = ampere, and nA = nanoampere.

$265,000 \text{ W} = 0.265(10^6)\text{W} = 0.265 \text{ MW}$

$0.0000035 \text{ m}^2 = (3.5 \times 10^{-6})(10^3 \text{ mm})^2 = (3.5 \times 10^{-6})$
$(10^6) \text{ mm}^2 = 3.5 \text{ mm}^2$

$71(10^{-6}) \text{ m}^2/\text{s} = (71 \times 10^{-6})(10^3 \text{ mm})^2/\text{s} = (71 \times 10^{-6})$
$(10^6)\text{mm}^2/\text{s} = 71 \text{ mm}^2/\text{s}$

$9.73 \times 10^{-7} \text{ A} = 973 \times 10^{-9} \text{ A} = 973 \text{ nA}$

Prefixes

Table A-3 presents the factors, prefixes and symbols for forming decimal multiples and submultiples of simple and compound SI units.

Factors and Prefixes for Forming Decimal Multiples and Sub-multiples of the SI Units

Factor by which the unit is multiplied	Prefix	Symbol	Factor by which the unit is multiplied	Prefix	Symbol
10^{12}	tera	T	10^{-2}	centi	c
10^9	giga	G	10^{-3}	milli	m
10^6	mega	M	10^{-6}	micro	μ
10^3	kilo	k	10^{-9}	nano	n
10^2	hecto	h	10^{-12}	pico	p
10	deka	da	10^{-15}	femto	f
10^{-1}	deci	d	10^{-18}	atto	a

Fig. A-3

At present, a preferred method in the selection of multiple and submultiple prefixes for any SI base unit employs factors which differ in orders of 1000 or 10^3.

Example: Applying this preference to the base unit for length, which is the *metre*:

*atto*metre (am), *femto*metre (fm), *pico*metre (pm), *nano*metre (nm), *micro*metre (μm), *milli*metre (mm), *metre* (m), *kilo*metre (km), *mega*metre (Mm), *giga*metre (Gm), *tera*metre (Tm).

Note that double prefixes and double prefix symbols are avoided.

Example: (10^{-12}) m = 1 pm, and *not* 1$\mu\mu$m.

Also one of the SI Base Units, the kilogram for mass, already contains the prefix "kilo-," as a multiple of the gram. However, all multiples and sub-multiples for mass units are formed by adding prefixes and their symbols to the "gram."

Example: (10^{-6}) kg = (10^{-3}) g = 1 mg, and *not* 1 μkg.

Units of Mass and Force

Unlike the traditional metric system, which uses the kilogram-force, the unit of force in SI is the *newton*.

In SI, the *kilogram*, and its multiples and submultiples are restricted to mass. Mass indicates the quantity of matter contained within a body. Technically, the term "weight," and therefore the newton, is used when the influence of gravi-tational force is taken into account. Since the newton is defined as "that force which, when applied to a body having a mass of one kilogram, gives it an acceleration of one metre per second squared," ($N = Kg \cdot m/s^2$), it is inde-pendent of gravitational acceleration "g."

As a result when using SI, the factor g disappears from a wide range of form-ulas in dynamics. In some formulas in statics, however, in which the weight of a body — rather than mass — is important, g does appear where it was formerly absent. The acceleration of free fall, g, actually varies with both time and loca-tion. However, the standard gravitational acceleration in SI is agreed upon as 9.80556 m/s^2.

SI with Non-SI Units

Generally, a combination of SI units with English or other non-SI units should be avoided. An example of this is "kg/gal." However, certain units outside SI are acceptable because of their convenience and importance in various scientific and technical fields.

Some examples of acceptable non-SI units are:

Time (SI unit = s) – minute (min), hour (h), day (d)

Plane angles (SI unit = rad) – degree (°), minute ('), second (")

Liquid volume (SI unit = m^3) – litre (l), where 1 l = 1 dm^3

Mass (SI unit = kg) – tonne or metric ton (t), where 1 t = 1000 kg

Fluid pressure (SI unit = Pa) – bar (bar), where 1 bar = 10^5 Pa.

Though the SI thermodynamic temperature is in kelvin (K), where necessary the Celsius (formerly "Centigrade") temperature scale can be used:

Kelvin temperature (K) = 273.15 + Celsius Temperature (°C).

When using the above equation for different temperature values it will be noted that the interval of change in Kelvin temperature is the same as the interval of change in Celsius.

For an SI temperature interval, either the K or °C designation may be used.

USE OF CONVERSION TABLES IN SECTION C

Directions specific to the use of Tables 1 and 1A (Common Fractional Inches to Millimetres) and Table 37B (Fahrenheit-Celsius) are given in the special introductions to these tables, on pages 22 and 447, respectively.

For all other tables, English or SI numerical values are converted by means of reference columns of whole-numbers (REF), located within each table. These REF numbers vary in increments of one, in ranges of 1 to 1000 or 1 to 100.

To convert a numerical value from English to SI units, find the corresponding number under the REF column in the appropriate table. Its SI equivalent will be on the same line in the column *at the left*. When the value being converted is in SI units, again refer to the REF column and the English equivalent will be found in the column *at the right*.

In order to make the following illustrations concerning the use of the tables easy to follow, the given examples have been confined to the use of Table 2 (Inch-Millimetre). Although Table 2 has a REF column range of 1 to 1000, the same techniques can be applied to those tables having a 1 to 100 REF range.

Example (a): Convert 732 inches to millimetres.
From Table 2, page 34 :

$$\underline{\text{mm}} \qquad\qquad \underline{\text{REF}}$$
$$18{,}592.80 \longleftarrow\!\!\!\text{———}\ 732$$

Example (b): Convert 732 millimetres to inches.
From the same location in REF column as Example (a):

$$\underline{\text{REF}} \qquad\qquad \underline{\text{in.}}$$
$$732 \ \text{———}\!\!\!\longrightarrow 28.81890$$

Values obtained can then be rounded off to a desired number of significant figures.*

Converting Any Decimal Value

As can be seen from Examples a and b above, all values given under the REF columns have an implied decimal point after the last digit. That is, REF 732 = 732.0000. All conversion equivalents given in the tables to the left and right of the REF columns correspond to this decimalization.

However, the table user can convert any decimal number through the REF column. Simply shift the decimal place of the obtained conversion equivalent. This decimal shift is in the same number of places and direction as the implied decimal point of the REF number would have to be moved to conform to the decimal number being converted.

Example (c): Convert 0.732 inches to millimetres.

From Example (a), this requires a decimal shift of 732 (REF) three places to the left. The same is done with the mm equivalent 18,592.80. Therefore:

$$0.732 \text{ inches} = 18.59280 \text{ mm} \approx 18.6 \text{ mm}$$

Example (d): Convert 7.32 millimetres to inches.

From Example (b), this requires a decimal shift of the REF and conversion equivalent values two places to the left. Therefore:

$$7.32 \text{ mm} = 0.2881890 \text{ in.} \approx 0.288 \text{ in.}$$

**Converting Values Having More Significant Figures Than
Given in the REF Columns**

Values having more significant figures than are given in an REF column in an appropriate table are also easily converted to the required accuracy. To

---

*See ASTM. *Metric Practice Guide,* E-380, for recommended rounding procedures.

accomplish this requires the separate addition of properly decimalized REF values and their conversion equivalents. The totals of these additions will correspond to the decimal number being converted and its English or SI equivalent.

Example (*e*): Convert 7.329 inches to millimetres.

From Table 2:

REF		mm
732	=	18,592.80
9	=	228.6000

Using appropriate decimal shifting and addition:

7.320	=	185.9280
0.009	=	0.2286
7.329	=	186.1566 ≈ 186.2 mm

A similar approach is used for converting SI units to English units:

Example (*f*): Convert 7.329 mm to inches.
From Table 2:

REF		in.
732	=	28.81890
9	=	0.3543307

Using appropriate decimal shifting and addition:

7.32	=	0.2881890
0.009	=	0.0003543307
7.329	=	0.28854333 ≈ 0.2885 in.

Using Special Introduction to Conversion Tables

The purpose of the special introduction given at the beginning of each conversion table in Section C is, in part, to acquaint the reader with methods for correctly using the table when working with SI unit multiples and submultiples.

Where a variety of English or SI units can be converted by a table, these units are indicated in the heading of the special introduction. Specific conversion relations within each heading have been underlined. Conversions of numerical values between the English-SI units so underlined can be made directly with the quantities taken from the individual given table.

To obtain the appropriate numerical values for the relations *not* underlined, the user is required to shift the decimal places of the numbers obtained from the table or multiply them by a power of ten. The decimal-shift diagrams given with each table, and explained later in this section, serve as a guide for this procedure.

All quantities within a table have been derived from the first of the two conversion factor equations given just below the English-SI units listed in the heading. The conversion factor given in the second equation is the reciprocal of that used in the first, and also has been rounded to seven significant figures. It is provided as a further convenience to the table user.

The examples in the text and the decimal-shift diagrams given in each special introduction, and at the top of each table page, are intended to guide the user in proper decimal transposition when dealing with SI multiples and submultiples. This aspect of decimal shifting reflects the similarity of conversion factors between a primary unit and its multiples and submultiples, which differ only by a power of 10. The following examples are based on the information obtained from Table 3 and its special introduction.

Since: 1 ft = 0.3048 m, it is obvious that:
$$1 \text{ ft} = 0.3048 \times 10^3 \text{ mm} = 304.8 \text{ mm}$$
$$1 \text{ ft} = 0.3048 \times 10^2 \text{ cm} = 30.48 \text{ cm}$$
$$1 \text{ ft} = 0.3048 \times 10^{-3} \text{ km} = 0.0003048 \text{ km}$$

Also to be seen in the same introduction is that Table 3 can conveniently be used to convert physical quantities other than linear — such as: velocity and acceleration. Any use of a table for more than one physical quantity, however, requires that consistency be maintained between English and SI units. Again, this consistency is such that the numerical value of the conversion factor for the new physical quantity remains the same as the original, or differs from it only by a power of 10. As may also be noted in the special introduction to Table 3, the consistency in this specific case depends upon the same time units being used in both the SI and English quantities.

Example: 1 ft = 0.3048 m (length)
$$1 \text{ ft/min} = 304.8 \text{ mm/min (velocity)}$$
$$1 \text{ ft/s}^2 = 30.48 \text{ cm/s}^2 \text{ (acceleration)}$$

Many of the special introductions to the tables will present the user with various parallelisms of this kind.

The decimal-shift diagrams used with the examples from the special introductions to the tables, and repeated in the table page-headings, serve as quick reference guides for multiple and submultiple decimal shifting. Referring to the introduction to Table 3, Section C, a portion of Diagram 3-A is taken as follows:

The directional arrow and the +2 digit shown between the given metric symbols indicate the number *and* direction of the decimal place shift for changing the primary unit to a multiple or submultiple. This portion of the diagram can be read as: "A numerical value in metres (m) is changed to centimetres (cm) by shifting its decimal two plaees to the right, or by multiplying it by 10 to the +2 power."

Example: 1.4 m = 140 cm, or 1.4×10^2 cm.

Diagram 3-B, from Table 3, Section C, assists in reversing this procedure. It aids in converting a multiple, or submultiple, back to the primary unit on which the calculation of Table 3 is based. A portion of Diagram 3-B follows:

$$m \;\vdash\!\!\overset{-3}{\xleftarrow{\hspace{2cm}}}\; mm$$

It can be read as: "A numerical value in millimetres (mm) is changed to metres (m) by shifting its decimal place three places to the left, or by multiplying it by 10 to the −3 power.

Example: 570 mm = 0.570 m, or 570×10^{-3} m.

As a user becomes more familiar with SI and the tables in this book, he will be able to shift decimals automatically without the use of such diagrams. However, where a particular table expresses compound units and is used only intermittently, the multiple or submultiple decimal shifting may be less obvious. In these instances, the diagrams will prove valuable in saving time, and possibly provide greater accuracy for the user.

Example: From the Decimal-shift Diagram 33-B in the special introduction to
          Table 33, Section C:

$$N/cm^2 \;\overset{-4}{\xleftarrow{\hspace{2cm}}}\!\dashv\; Pa$$

This segment of the Decimal-shift Diagram 33-B indicates that a numerical value in Pascals or Pa (N/m^2) is changed to newtons per centimetre squared (N/cm^2) by shifting its decimal four places to the left, or by multiplying it by 10 to the −4 power. Therefore:

$$7,500 \text{ Pa} = 0.750 \text{ N/cm}^2 \text{ or } 7,500 \times 10^{-4} \text{ N/cm}^2$$

At times, it may be desired to expand the use of this book to determine certain physical quantities beyond those specifically mentioned in the tables. A case may arise during the course of one's work for converting less common, or empirically developed, relations between the English and SI units. These relations may be important to a specific task and might be used continuously or intermittently. A conversion factor for such a less common or empirical relation may be the same as (or differ only by a power of ten from) one on which a table in the book is based. In this case the user should make note of it in the special introduction to the proper table. This table then, can be used consistently for converting any numerical value in this relation. Where necessary, a table user might even prepare his own decimal-shift diagram to help in working with this new relation. With this in mind, the individual owner of this book can make it considerably more useful and a continuously helpful and growing reference tool.

SECTION B

Metric Conversion Factors
Reference List

Metric Conversion Factors

(Symbols of SI units, multiples and submultiples are
given in parentheses in the right-hand column)

Multiply	By	To Obtain
LENGTH		
centimetre	0.03280840	foot
centimetre	0.3937008	inch
fathom	1.8288*	metre (m)
foot	0.3048*	metre (m)
foot	30.48*	centimetre (cm)
foot	304.8*	millimetre (mm)
inch	0.0254*	metre (m)
inch	2.54*	centimetre (cm)
inch	25.4*	millimetre (mm)
kilometre	0.6213712	mile [U. S. statute]
metre	39.37008	inch
metre	0.5468066	fathom
metre	3.280840	foot
metre	0.1988388	rod
metre	1.093613	yard
metre	0.0006213712	mile [U. S. statute]
microinch	0.0254*	micrometre [micron] (μm)
micrometre [micron]	39.37008	microinch
mile [U. S. statute]	1609.344*	metre (m)
mile [U. S. statute]	1.609344*	kilometre (km)
millimetre	0.003280840	foot
millimetre	0.03937008	inch
rod	5.0292*	metre (m)
yard	0.9144*	metre (m)
AREA		
acre	4046.856	metre^2 (m^2)
acre	0.4046856	hectare
centimetre^2	0.1550003	inch^2
centimetre^2	0.001076391	foot^2
foot^2	0.09290304*	metre^2 (m^2)
foot^2	929.0304*	centimetre^2 (cm^2)
foot^2	92,903.04*	millimetre^2 (mm^2)
hectare	2.471054	acre
inch^2	645.16*	millimetre^2 (mm^2)
inch^2	6.4516*	centimetre^2 (cm^2)
inch^2	0.00064516*	metre^2 (m^2)
metre^2	1550.003	inch^2
metre^2	10.763910	foot^2
metre^2	1.195990	yard^2
metre^2	0.0002471054	acre
millimetre^2	0.0001076387	foot^2
millimetre^2	0.001550003	inch^2
yard^2	0.8361274	metre^2 (m^2)

* Where an asterisk is shown, the figure is exact.

Metric Conversion Factors (*Continued*)

Multiply	By	To Obtain
VOLUME (including CAPACITY)		
centimetre^3	0.06102376	inch^3
foot^3	0.02831685	metre^3 (m^3)
foot^3	28.31685	litre
gallon [U. K. liquid]	0.004546092	metre^3 (m^3)
gallon [U. K. liquid]	4.546092	litre
gallon [U. S. liquid]	0.003785412	metre^3 (m^3)
gallon [U. S. liquid]	3.785412	litre
inch^3	16,387.06	millimetre^3 (mm^3)
inch^3	16.38706	centimetre^3 (cm^3)
inch^3	0.00001638706	metre^3 (m^3)
litre	0.001*	metre^3 (m^3)
litre	0.2199692	gallon [U. K. liquid]
litre	0.2641720	gallon [U. S. liquid]
litre	0.03531466	foot^3
metre^3	219.9692	gallon [U. K. liquid]
metre^3	264.1720	gallon [U. S. liquid]
metre^3	35.31466	foot^3
metre^3	1.307951	yard^3
metre^3	1000.*	litre
metre^3	61,023.76	inch^3
millimetre^3	0.00006102376	inch^3
yard^3	0.7645549	metre^3 (m^3)
VELOCITY, ACCELERATION, and FLOW		
centimetre/second	1.968504	foot/minute
centimetre/second	0.03280840	foot/second
centimetre/minute	0.3937008	inch/minute
foot/hour	0.00008466667	metre/second (m/s)
foot/hour	0.00508*	metre/minute
foot/hour	0.3048*	metre/hour
foot/minute	0.508*	centimetre/second
foot/minute	18.288*	metre/hour
foot/minute	0.3048*	metre/minute
foot/minute	0.00508*	metre/second (m/s)
foot/second	30.48*	centimetre/second
foot/second	18.288*	metre/minute
foot/second	0.3048*	metre/second (m/s)
foot/second^2	0.3048*	metre/second^2 (m/s^2)
foot^3/minute	28.31685	litre/minute
foot^3/minute	0.0004719474	metre^3/second (m^3/s)
gallon [U. S. liquid]/min.	0.003785412	metre^3/minute
gallon [U. S. liquid]/min.	0.00006309020	metre^3/second (m^3/s)
gallon [U. S. liquid]/min.	0.06309020	litre/second
gallon [U. S. liquid]/min.	3.785412	litre/minute
gallon [U. K. liquid]/min.	0.004546092	metre^3/minute
gallon [U. K. liquid]/min.	0.00007576820	metre^3/second (m^3/s)
inch/minute	25.4*	millimetre/minute
inch/minute	2.54*	centimetre/minute
inch/minute	0.0254*	metre/minute
inch/second^2	0.0254*	metre/second^2 (m/s^2)

* Where an asterisk is shown, the figure is exact.

Metric Conversion Factors (*Continued*)

Multiply	By	To Obtain
VELOCITY, ACCELERATION, and FLOW (*Continued*)		
kilometre/hour	0.6213712	mile/hour [U. S. statute]
litre/minute	0.03531466	foot^3/minute
litre/minute	0.2641720	gallon [U. S. liquid]/minute
litre/second	15.85032	gallon [U. S. liquid]/minute
mile/hour	1.609344*	kilometre/hour
millimetre/minute	0.03937008	inch/minute
metre/second	11,811.02	foot/hour
metre/second	196.8504	foot/minute
metre/second	3.280840	foot/second
metre/second^2	3.280840	foot/second^2
metre/second^2	39.37008	inch/second^2
metre/minute	3.280840	foot/minute
metre/minute	0.05468067	foot/second
metre/minute	39.37008	inch/minute
metre/hour	3.280840	foot/hour
metre/hour	0.05468067	foot/minute
metre^3/second	2118.880	foot^3/minute
metre^3/second	13,198.15	gallon [U. K. liquid]/minute
metre^3/second	15,850.32	gallon [U. S. liquid]/minute
metre^3/minute	219.9692	gallon [U. K. liquid]/minute
metre^3/minute	264.1720	gallon [U. S. liquid]/minute
MASS and DENSITY		
grain [1/7000 lb avoirdupois]	0.06479891	gram (g)
gram	15.43236	grain
gram	0.001*	kilogram (kg)
gram	0.03527397	ounce [avoirdupois]
gram	0.03215074	ounce [troy]
gram/centimetre^3	0.03612730	pound/inch^3
hundredweight [long]	50.80235	kilogram (kg)
hundredweight [short]	45.35924	kilogram (kg)
kilogram	1000.*	gram (g)
kilogram	35.27397	ounce [avoirdupois]
kilogram	32.15074	ounce [troy]
kilogram	2.204622	pound [avoirdupois]
kilogram	0.06852178	slug
kilogram	0.0009842064	ton [long]
kilogram	0.001102311	ton [short]
kilogram	0.001*	ton [metric]
kilogram	0.001*	tonne
kilogram	0.01968413	hundredweight [long]
kilogram	0.02204622	hundredweight [short]
kilogram/metre^3	0.06242797	pound/foot^3
kilogram/metre^3	0.008345406	pound/gallon [U. K. liquid]
kilogram/metre^3	0.01002242	pound/gallon [U. S. liquid]
ounce [avoirdupois]	28.34952	gram (g)
ounce [avoirdupois]	0.02834952	kilogram (kg)

* Where an asterisk is shown, the figure is exact.

Metric Conversion Factors (*Continued*)

Multiply	By	To Obtain
MASS and DENSITY (*Continued*)		
ounce [troy]	31.10348	gram (g)
ounce [troy]	0.03110348	kilogram (kg)
pound [avoirdupois]	0.4535924	kilogram (kg)
pound/foot³	16.01846	kilogram/metre³ (kg/m³)
pound/inch³	27.67990	gram/centimetre³ (g/cm³)
pound/gal [U. S. liquid]	119.8264	kilogram/metre³ (kg/m³)
pound/gal [U. K. liquid]	99.77633	kilogram/metre³ (kg/m³)
slug	14.59390	kilogram (kg)
ton [long 2240 lb]	1016.047	kilogram (kg)
ton [short 2000 lb]	907.1847	kilogram (kg)
ton [metric]	1000.*	kilogram (kg)
tonne	1000.*	kilogram (kg)
FORCE and FORCE/LENGTH		
dyne	0.00001*	newton (N)
kilogram-force	9.806650*	newton (N)
kilopond	9.806650*	newton (N)
newton	0.1019716	kilogram-force
newton	0.1019716	kilopond
newton	0.2248089	pound-force
newton	100,000.*	dyne
newton	7.23301	poundal
newton	3.596942	ounce-force
newton/metre	0.005710148	pound/inch
newton/metre	0.06852178	pound/foot
ounce-force	0.2780139	newton (N)
pound-force	4.448222	newton (N)
poundal	0.1382550	newton (N)
pound/inch	175.1268	newton/metre (N/m)
pound/foot	14.59390	newton/metre (N/m)
BENDING MOMENT or TORQUE		
dyne-centimetre	0.0000001*	newton-metre (N · m)
kilogram-metre	9.806650*	newton-metre (N · m)
ounce-inch	7.061552	newton-millimetre
ounce-inch	0.007061552	newton-metre (N · m)
newton-metre	0.7375621	pound-foot
newton-metre	10,000,000.*	dyne-centimetre
newton-metre	0.1019716	kilogram-metre
newton-metre	141.6119	ounce-inch
newton-millimetre	0.1416119	ounce-inch
pound-foot	1.355818	newton-metre (N · m)

* Where an asterisk is shown, the figure is exact.

Metric Conversion Factors (*Continued*)

Multiply	By	To Obtain
MOMENT OF INERTIA and SECTION MODULUS		
moment of inertia [kg · m²]	23.73036	pound-foot²
moment of inertia [kg · m²]	3417.171	pound-inch²
moment of inertia [lb · ft²]	0.04214011	kilogram-metre² (kg · m²)
moment of inertia [lb · inch²]	0.0002926397	kilogram-metre² (kg · m²)
moment of section [foot⁴]	0.008630975	metre⁴ (m⁴)
moment of section [inch⁴]	41.62314	centimetre⁴
moment of section [metre⁴]	115.8618	foot⁴
moment of section [centimetre⁴]	0.02402510	inch⁴
section modulus [foot³]	0.02831685	metre³ (m³)
section modulus [inch³]	0.00001638706	metre³ (m³)
section modulus [metre³]	35.31466	foot³
section modulus [metre³]	61,023.76	inch³
MOMENTUM		
kilogram-metre/second	7.233011	pound-foot/second
kilogram-metre/second	86.79614	pound-inch/second
pound-foot/second	0.1382550	kilogram-metre/second (kg · m/s)
pound-inch/second	0.01152125	kilogram-metre/second (kg · m/s)
PRESSURE and STRESS		
atmosphere [14.6959 lb/inch²]	101,325.	pascal (Pa)
bar	100,000.*	pascal (Pa)
bar	14.50377	pound/inch²
bar	100,000.*	newton/metre² (N/m²)
hectobar	0.6474898	ton [long]/inch²
kilogram/centimetre²	14.22334	pound/inch²
kilogram/metre²	9.806650*	newton/metre² (N/m²)
kilogram/metre²	9.806650*	pascal (Pa)
kilogram/metre²	0.2048161	pound/foot²
kilonewton/metre²	0.1450377	pound/inch²
newton/centimetre²	1.450377	pound/inch²
newton/metre²	0.00001*	bar
newton/metre²	1.0*	pascal (Pa)
newton/metre²	0.0001450377	pound/inch²
newton/metre²	0.1019716	kilogram/metre²
newton/millimetre²	145.0377	pound/inch²
pascal	0.00000986923	atmosphere
pascal	0.00001*	bar
pascal	0.1019716	kilogram/metre²
pascal	1.0*	newton/metre² (N/m²)
pascal	0.02088543	pound/foot²
pascal	0.0001450377	pound/inch²

* Where an asterisk is shown, the figure is exact.

Metric Conversion Factors (*Continued*)

Multiply	By	To Obtain
PRESSURE and STRESS (*Continued*)		
pound/foot²	4.882429	kilogram/metre²
pound/foot²	47.88026	pascal (Pa)
pound/inch²	0.06894757	bar
pound/inch²	0.07030697	kilogram/centimetre²
pound/inch²	0.6894757	newton/centimetre²
pound/inch²	6.894757	kilonewton/metre²
pound/inch²	6894.757	newton/metre² (N/m²)
pound/inch²	0.006894757	newton/millimetre² (N/mm²)
pound/inch²	6894.757	pascal (Pa)
ton [long]/inch²	1.544426	hectobar
ENERGY and WORK		
Btu [International Table]	1055.056	joule (J)
Btu [mean]	1055.87	joule (J)
calorie [mean]	4.19002	joule (J)
foot-pound	1.355818	joule (J)
foot-poundal	0.04214011	joule (J)
joule	0.0009478170	Btu [International Table]
joule	0.0009470863	Btu [mean]
joule	0.2386623	calorie [mean]
joule	0.7375621	foot-pound
joule	23.73036	foot-poundal
joule	0.9998180	joule [International U. S.]
joule	0.9999830	joule [U. S. legal, 1948]
joule [International U. S.]	1.000182	joule (J)
joule [U. S. legal, 1948]	1.000017	joule (J)
joule	.0002777778	watt-hour
watt-hour	3600.*	joule (J)
POWER		
Btu [International Table]/hour	0.2930711	watt (W)
foot-pound/hour	0.0003766161	watt (W)
foot-pound/minute	0.02259697	watt (W)
horsepower [550 ft-lb/s]	0.7456999	kilowatt (kW)
horsepower [550 ft-lb/s]	745.6999	watt (W)
horsepower [electric]	746.*	watt (W)
horsepower [metric]	735.499	watt (W)
horsepower [U. K.]	745.70	watt (W)
kilowatt	1.341022	horsepower [550 ft-lb/s]
watt	2655.224	foot-pound/hour
watt	44.25372	foot-pound/minute
watt	0.001341022	horsepower [550 ft-lb/s]
watt	0.001340483	horsepower [electric]
watt	0.001359621	horsepower [metric]
watt	0.001341022	horsepower [U. K.]
watt	3.412141	Btu [International Table]/hour

* Where an asterisk is shown, the figure is exact.

Metric Conversion Factors (*Continued*)

Multiply	By	To Obtain
VISCOSITY		
centipoise	0.001*	pascal-second (Pa · s)
centistoke	0.000001*	metre²/second (m²/s)
metre²/second	1,000,000.*	centistoke
metre²/second	10,000.*	stoke
pascal-second	1000.*	centipoise
pascal-second	10.*	poise
poise	0.1*	pascal-second (Pa · s)
stoke	0.0001*	metre²/second (m²/s)
TEMPERATURE		

To Convert From	To	Use Formula
temperature Celsius, t_C	temperature Kelvin, t_K	$t_K = t_C + 273.15$
temperature Fahrenheit, t_F	temperature Kelvin, t_K	$t_K = (t_F + 459.67)/1.8$
temperature Celsius, t_C	temperature Fahrenheit, t_F	$t_F = 1.8\,t_C + 32$
temperature Fahrenheit, t_F	temperature Celsius, t_C	$t_C = (t_F - 32)/1.8$
temperature Kelvin, t_K	temperature Celsius, t_C	$t_C = t_K - 273.15$
temperature Kelvin, t_K	temperature Fahrenheit, t_F	$t_F = 1.8\,t_K - 459.67$
temperature Kelvin, t_K	temperature Rankine, t_R	$t_R = 9/5\,t_K$
temperature Rankine, t_R	temperature Kelvin, t_K	$t_K = 5/9\,t_R$

* Where an asterisk is shown, the figure is exact.

SECTION   C

Conversion Tables

NOTE: Certain English-SI unit relations are underlined in the headings to the special introductions to the tables in this section. Conversions can be made between any of the underlined relations by the exact decimal values obtained from an appropriate table without use of pertinent decimal-shift diagrams.

$$\mathbf{1}^{\text{A}}_{\text{B}}$$

LENGTH

Common Fractional Inches to Millimetres
    (in. to mm)

Common Fractional Inches to Centimetres
    (in. to cm)

Common Fractional Inches to Metres
    (in. to m)

Conversion Factors:
      1 inch = 25.4 millimetres (mm) (exactly)
      (1 mm = 0.03937008 in.)

Tables 1A and 1B can be used together to convert to millimetres a range of values from 1/64 inch to 13 inches, in increments of 1/64 inch.

Example: Convert 9 29/32 inches to millimetres.

From Table 1A: 29/32 in. = 23.109 mm
From Table 1B: 9   in. = 228.6   mm
              9 29/32 in. = 251.619 mm*

Note that mm values given in Table 1B are exact. For convenience, the zeros after the first decimal place have been omitted. Full-inch values greater than 12 inches as well as decimal inch quantities can be converted through Table 2.

By simply shifting the decimal places in the mm conversions, Tables 1A and 1B can be used to obtain cm and m values.

From above example:

251.619 mm = 25.1619 cm* (a shift of one decimal place to the left)
        = 0.251619 m* (a shift of three decimal places to the left)

---

*All conversion values obtained can be rounded to desired number of significant figures.

The following Diagram 1-A, summarizing the above operations, will assist in changing mm to the other SI units.

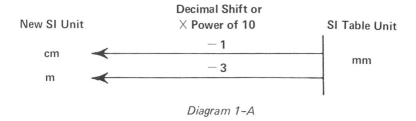

Diagram 1–A

Table 1-A

FRACTIONAL INCH TO MILLIMETRES							
in.	mm.	in.	mm.	in.	mm.	in.	mm.
1/64	0.397	17/64	6.747	33/64	13.097	49/64	19.447
1/32	0.794	9/32	7.144	17/32	13.494	25/32	19.844
3/64	1.191	19/64	7.541	35/64	13.891	51/64	20.241
1/16	1.588	5/16	7.938	9/16	14.288	13/16	20.638
5/64	1.984	21/64	8.334	37/64	14.684	53/64	21.034
3/32	2.381	11/32	8.731	19/32	15.081	27/32	21.431
7/64	2.778	23/64	9.128	39/64	15.478	55/64	21.828
1/8	3.175	3/8	9.525	5/8	15.875	7/8	22.225
9/64	3.572	25/64	9.922	41/64	16.272	57/64	22.622
5/32	3.969	13/32	10.319	21/32	16.669	29/32	23.019
11/64	4.366	27/64	10.716	43/64	17.066	59/64	23.416
3/16	4.762	7/16	11.112	11/16	17.462	15/16	23.812
13/64	5.159	29/64	11.509	45/64	17.859	61/64	24.209
7/32	5.556	15/32	11.906	23/32	18.256	31/32	24.606
15/64	5.953	31/64	12.303	47/64	18.653	63/64	25.003
1/4	6.350	1/2	12.700	3/4	19.050	1	25.400

Table 1-B

INCHES TO MILLIMETRES											
in.	mm.	in.	mm.	in.	mm.	in.	mm.	in.	mm.	in.	mm.
1	25.4	3	76.2	5	127.0	7	177.8	9	228.6	11	279.4
2	50.8	4	101.6	6	152.4	8	203.2	10	254.0	12	304.8

2

LENGTH

Inch — Millimetre
(in. — mm)

Inch — Centimetre
(in. — cm)

Inch — Metre
(in. — m)

VELOCITY

Inch/Second — Millimetre/Second
(in./s — mm/s)

Inch/Second — Centimetre/Second
(in./s — cm/s)

Inch/Second — Metre/Second
(in./s — m/s)

ACCELERATION

Inch / Second^2 — Millimetre/Second^2
(in./s^2 — mm/s^2)

Inch/Second^2 — Centimetre/Second^2
(in./s^2 — cm/s^2)

Inch/Second^2 — Metre/Second^2
(in./s^2 — m/s^2)

Conversion Factors:
1 inch (in.) = 25.4 millimetre (mm) (exactly)
(1 mm = 0.03937008 in.)

Velocity and Acceleration Values

Table 2 can be used to represent velocity and acceleration values, as long as the time units are the same in both English and SI equivalents:

Velocity Conversion Factor: 1 in./s = 25.4 mm/s (exactly)

Acceleration Conversion Factor: 1 in./s² = 25.4 mm/s² (exactly)

To convert mm values obtained from Table 2 to cm and m, Diagram 2-A will appear in the headings on alternate pages of this table for the user's convenience. This diagram can also be used for the velocity and acceleration values noted above. It is derived from: $1 \text{ mm} = (10^{-1}) \text{ cm} = (10^{-3}) \text{ m}$.

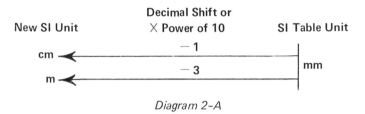

Diagram 2-A

Example (*a*): Convert 59 in. to mm, cm, m.

From table: 59 in. = 1,498.600 mm ≈ 1,500 mm

From Diagram 2-A: (a) $1,498.600 \text{ mm} = 1,498.600 \times 10^{-1}$ cm

= 149.8600 cm* ≈ 150 cm

(mm to cm by multiplying mm value by 10^{-1} or by shifting its decimal 1 place to the left.)

(b) $1,498.600 \text{ mm} = 1,498.600 \times 10^{-3}$ m

= 1.498600 m* ≈ 1.50 m

(mm to m by multiplying mm value by 10^{-3} or by shifting its decimal 3 places to the left.)

Diagram 2-B is also given at the top of alternate table pages:

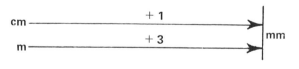

Diagram 2-B

It is the opposite of Diagram 2-A, and is used to convert cm and m values to in. using the table. The diagram guides the conversion of cm and m values to their mm equivalent. The mm equivalent is then applied to the REF column in

---

*All conversion values obtained can be rounded to desired number of significant figures.

the table to obtain in. values. This diagram can also be used for the velocity and acceleration values, as noted previously, as long as the time units are consistent.

Example (b): Convert 37.5 cm/s to in./s
From Diagram 2-B: 37.5 cm/s = 375 mm/s
From table: 375 mm/s = 14.76378 in./s* ≈ 14.8 in./s

Example (c): Convert 0.123 m/s² to in./s²
From Diagram 2-B: 0.123 m/s² = 123 mm/s²
Therefore, from table: 123 mm/s² = 4.842520 in./s² ≈ 4.84 in./s².

---

*All conversion values obtained can be rounded to desired number of significant figures.

cm ← ─1

m ← ─3

mm

mm mm/s mm/s²	REF	in in/s in/s²	mm mm/s mm/s²	REF	in in/s in/s²
25.40000	1	0.03937008	1295.400	51	2.007874
50.80000	2	0.07874016	1320.800	52	2.047244
76.20000	3	0.1181102	1346.200	53	2.086614
101.6000	4	0.1574803	1371.600	54	2.125984
127.0000	5	0.1968504	1397.000	55	2.165354
152.4000	6	0.2362205	1422.400	56	2.204724
177.8000	7	0.2755906	1447.800	57	2.244094
203.2000	8	0.3149606	1473.200	58	2.283465
228.6000	9	0.3543307	1498.600	59	2.322835
254.0000	10	0.3937008	1524.000	60	2.362205
279.4000	11	0.4330709	1549.400	61	2.401575
304.8000	12	0.4724409	1574.800	62	2.440945
330.2000	13	0.5118110	1600.200	63	2.480315
355.6000	14	0.5511811	1625.600	64	2.519685
381.0000	15	0.5905512	1651.000	65	2.559055
406.4000	16	0.6299213	1676.400	66	2.598425
431.8000	17	0.6692913	1701.800	67	2.637795
457.2000	18	0.7086614	1727.200	68	2.677165
482.6000	19	0.7480315	1752.600	69	2.716535
508.0000	20	0.7874016	1778.000	70	2.755906
533.4000	21	0.8267717	1803.400	71	2.795276
558.8000	22	0.8661417	1828.800	72	2.834646
584.2000	23	0.9055118	1854.200	73	2.874016
609.6000	24	0.9448819	1879.600	74	2.913386
635.0000	25	0.9842520	1905.000	75	2.952756
660.4000	26	1.023622	1930.400	76	2.992126
685.8000	27	1.062992	1955.800	77	3.031496
711.2000	28	1.102362	1981.200	78	3.070866
736.6000	29	1.141732	2006.600	79	3.110236
762.0000	30	1.181102	2032.000	80	3.149606
787.4000	31	1.220472	2057.400	81	3.188976
812.8000	32	1.259843	2082.800	82	3.228346
838.2000	33	1.299213	2108.200	83	3.267717
863.6000	34	1.338583	2133.600	84	3.307087
889.0000	35	1.377953	2159.000	85	3.346457
914.4000	36	1.417323	2184.400	86	3.385827
939.8000	37	1.456693	2209.800	87	3.425197
965.2000	38	1.496063	2235.200	88	3.464567
990.6000	39	1.535433	2260.600	89	3.503937
1016.000	40	1.574803	2286.000	90	3.543307
1041.400	41	1.614173	2311.400	91	3.582677
1066.800	42	1.653543	2336.800	92	3.622047
1092.200	43	1.692913	2362.200	93	3.661417
1117.600	44	1.732283	2387.600	94	3.700787
1143.000	45	1.771654	2413.000	95	3.740157
1168.400	46	1.811024	2438.400	96	3.779528
1193.800	47	1.850394	2463.800	97	3.818898
1219.200	48	1.889764	2489.200	98	3.858268
1244.600	49	1.929134	2514.600	99	3.897638
1270.000	50	1.968504	2540.000	100	3.937008

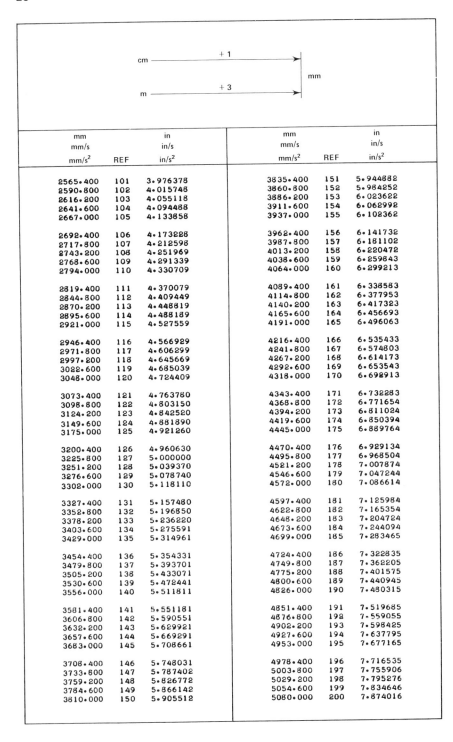

mm mm/s mm/s²	REF	in in/s in/s²	mm mm/s mm/s²	REF	in in/s in/s²
2565.400	101	3.976378	3835.400	151	5.944882
2590.800	102	4.015748	3860.800	152	5.984252
2616.200	103	4.055118	3886.200	153	6.023622
2641.600	104	4.094488	3911.600	154	6.062992
2667.000	105	4.133858	3937.000	155	6.102362
2692.400	106	4.173228	3962.400	156	6.141732
2717.800	107	4.212598	3987.800	157	6.181102
2743.200	108	4.251969	4013.200	158	6.220472
2768.600	109	4.291339	4038.600	159	6.259843
2794.000	110	4.330709	4064.000	160	6.299213
2819.400	111	4.370079	4089.400	161	6.338583
2844.800	112	4.409449	4114.800	162	6.377953
2870.200	113	4.448819	4140.200	163	6.417323
2895.600	114	4.488189	4165.600	164	6.456693
2921.000	115	4.527559	4191.000	165	6.496063
2946.400	116	4.566929	4216.400	166	6.535433
2971.800	117	4.606299	4241.800	167	6.574803
2997.200	118	4.645669	4267.200	168	6.614173
3022.600	119	4.685039	4292.600	169	6.653543
3048.000	120	4.724409	4318.000	170	6.692913
3073.400	121	4.763780	4343.400	171	6.732283
3098.800	122	4.803150	4368.800	172	6.771654
3124.200	123	4.842520	4394.200	173	6.811024
3149.600	124	4.881890	4419.600	174	6.850394
3175.000	125	4.921260	4445.000	175	6.889764
3200.400	126	4.960630	4470.400	176	6.929134
3225.800	127	5.000000	4495.800	177	6.968504
3251.200	128	5.039370	4521.200	178	7.007874
3276.600	129	5.078740	4546.600	179	7.047244
3302.000	130	5.118110	4572.000	180	7.086614
3327.400	131	5.157480	4597.400	181	7.125984
3352.800	132	5.196850	4622.800	182	7.165354
3378.200	133	5.236220	4648.200	183	7.204724
3403.600	134	5.275591	4673.600	184	7.244094
3429.000	135	5.314961	4699.000	185	7.283465
3454.400	136	5.354331	4724.400	186	7.322835
3479.800	137	5.393701	4749.800	187	7.362205
3505.200	138	5.433071	4775.200	188	7.401575
3530.600	139	5.472441	4800.600	189	7.440945
3556.000	140	5.511811	4826.000	190	7.480315
3581.400	141	5.551181	4851.400	191	7.519685
3606.800	142	5.590551	4876.800	192	7.559055
3632.200	143	5.629921	4902.200	193	7.598425
3657.600	144	5.669291	4927.600	194	7.637795
3683.000	145	5.708661	4953.000	195	7.677165
3708.400	146	5.748031	4978.400	196	7.716535
3733.800	147	5.787402	5003.800	197	7.755906
3759.200	148	5.826772	5029.200	198	7.795276
3784.600	149	5.866142	5054.600	199	7.834646
3810.000	150	5.905512	5080.000	200	7.874016

```
      cm  ←————— − 1 ——————
                              |
                              |  mm
      m   ←————— − 3 —————————|
```

mm mm/s mm/s^2	REF	in in/s in/s^2	mm mm/s mm/s^2	REF	in in/s in/s^2
5105.400	201	7.913386	6375.400	251	9.881890
5130.800	202	7.952756	6400.800	252	9.921260
5156.200	203	7.992126	6426.200	253	9.960630
5181.600	204	8.031496	6451.600	254	0.000000
5207.000	205	8.070866	6477.000	255	10.03937
5232.400	206	8.110236	6502.400	256	10.07874
5257.800	207	8.149606	6527.800	257	10.11811
5283.200	208	8.188976	6553.200	258	10.15748
5308.600	209	8.228346	6578.600	259	10.19685
5334.000	210	8.267717	6604.000	260	10.23622
5359.400	211	8.307087	6629.400	261	10.27559
5384.800	212	8.346457	6654.800	262	10.31496
5410.200	213	8.385827	6680.200	263	10.35433
5435.600	214	8.425197	6705.600	264	10.39370
5461.000	215	8.464567	6731.000	265	10.43307
5486.400	216	8.503937	6756.400	266	10.47244
5511.800	217	8.543307	6781.800	267	10.51181
5537.200	218	8.582677	6807.200	268	10.55118
5562.600	219	8.622047	6832.600	269	10.59055
5588.000	220	8.661417	6858.000	270	10.62992
5613.400	221	8.700787	6883.400	271	10.66929
5638.800	222	8.740157	6908.800	272	10.70866
5664.200	223	8.779528	6934.200	273	10.74803
5689.600	224	8.818898	6959.600	274	10.78740
5715.000	225	8.858268	6985.000	275	10.82677
5740.400	226	8.897638	7010.400	276	10.86614
5765.800	227	8.937008	7035.800	277	10.90551
5791.200	228	8.976378	7061.200	278	10.94488
5816.600	229	9.015748	7086.600	279	10.98425
5842.000	230	9.055118	7112.000	280	11.02362
5867.400	231	9.094488	7137.400	281	11.06299
5892.800	232	9.133858	7162.800	282	11.10236
5918.200	233	9.173228	7188.200	283	11.14173
5943.600	234	9.212598	7213.600	284	11.18110
5969.000	235	9.251969	7239.000	285	11.22047
5994.400	236	9.291339	7264.400	286	11.25984
6019.800	237	9.330709	7289.800	287	11.29921
6045.200	238	9.370079	7315.200	288	11.33858
6070.600	239	9.409449	7340.600	289	11.37795
6096.000	240	9.448819	7366.000	290	11.41732
6121.400	241	9.488189	7391.400	291	11.45669
6146.800	242	9.527559	7416.800	292	11.49606
6172.200	243	9.566929	7442.200	293	11.53543
6197.600	244	9.606299	7467.600	294	11.57480
6223.000	245	9.645669	7493.000	295	11.61417
6248.400	246	9.685039	7518.400	296	11.65354
6273.800	247	9.724409	7543.800	297	11.69291
6299.200	248	9.763780	7569.200	298	11.73228
6324.600	249	9.803150	7594.600	299	11.77165
6350.000	250	9.842520	7620.000	300	11.81102

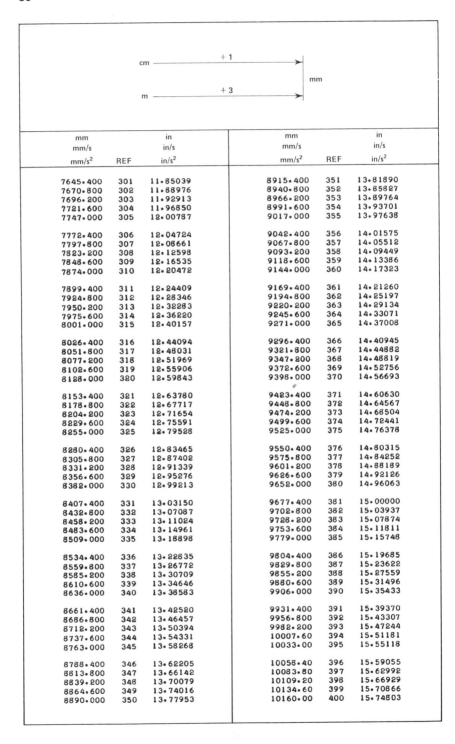

mm mm/s mm/s²	REF	in in/s in/s²	mm mm/s mm/s²	REF	in in/s in/s²
7645·400	301	11·85039	8915·400	351	13·81890
7670·800	302	11·88976	8940·800	352	13·85827
7696·200	303	11·92913	8966·200	353	13·89764
7721·600	304	11·96850	8991·600	354	13·93701
7747·000	305	12·00787	9017·000	355	13·97638
7772·400	306	12·04724	9042·400	356	14·01575
7797·800	307	12·08661	9067·800	357	14·05512
7823·200	308	12·12598	9093·200	358	14·09449
7848·600	309	12·16535	9118·600	359	14·13386
7874·000	310	12·20472	9144·000	360	14·17323
7899·400	311	12·24409	9169·400	361	14·21260
7924·800	312	12·28346	9194·800	362	14·25197
7950·200	313	12·32283	9220·200	363	14·29134
7975·600	314	12·36220	9245·600	364	14·33071
8001·000	315	12·40157	9271·000	365	14·37008
8026·400	316	12·44094	9296·400	366	14·40945
8051·800	317	12·48031	9321·800	367	14·44882
8077·200	318	12·51969	9347·200	368	14·48819
8102·600	319	12·55906	9372·600	369	14·52756
8128·000	320	12·59843	9398·000	370	14·56693
8153·400	321	12·63780	9423·400	371	14·60630
8178·800	322	12·67717	9448·800	372	14·64567
8204·200	323	12·71654	9474·200	373	14·68504
8229·600	324	12·75591	9499·600	374	14·72441
8255·000	325	12·79528	9525·000	375	14·76378
8280·400	326	12·83465	9550·400	376	14·80315
8305·800	327	12·87402	9575·800	377	14·84252
8331·200	328	12·91339	9601·200	378	14·88189
8356·600	329	12·95276	9626·600	379	14·92126
8382·000	330	12·99213	9652·000	380	14·96063
8407·400	331	13·03150	9677·400	381	15·00000
8432·800	332	13·07087	9702·800	382	15·03937
8458·200	333	13·11024	9728·200	383	15·07874
8483·600	334	13·14961	9753·600	384	15·11811
8509·000	335	13·18898	9779·000	385	15·15748
8534·400	336	13·22835	9804·400	386	15·19685
8559·800	337	13·26772	9829·800	387	15·23622
8585·200	338	13·30709	9855·200	388	15·27559
8610·600	339	13·34646	9880·600	389	15·31496
8636·000	340	13·38583	9906·000	390	15·35433
8661·400	341	13·42520	9931·400	391	15·39370
8686·800	342	13·46457	9956·800	392	15·43307
8712·200	343	13·50394	9982·200	393	15·47244
8737·600	344	13·54331	10007·60	394	15·51181
8763·000	345	13·58268	10033·00	395	15·55118
8788·400	346	13·62205	10058·40	396	15·59055
8813·800	347	13·66142	10083·80	397	15·62992
8839·200	348	13·70079	10109·20	398	15·66929
8864·600	349	13·74016	10134·60	399	15·70866
8890·000	350	13·77953	10160·00	400	15·74803

mm mm/s mm/s^2	REF	in in/s in/s^2		mm mm/s mm/s^2	REF	in in/s in/s^2
10185.40	401	15.78740		11455.40	451	17.75591
10210.80	402	15.82677		11480.80	452	17.79528
10236.20	403	15.86614		11506.20	453	17.83465
10261.60	404	15.90551		11531.60	454	17.87402
10287.00	405	15.94488		11557.00	455	17.91339
10312.40	406	15.98425		11582.40	456	17.95276
10337.80	407	16.02362		11607.80	457	17.99213
10363.20	408	16.06299		11633.20	458	18.03150
10388.60	409	16.10236		11658.60	459	18.07087
10414.00	410	16.14173		11684.00	460	18.11024
10439.40	411	16.18110		11709.40	461	18.14961
10464.80	412	16.22047		11734.80	462	18.18898
10490.20	413	16.25984		11760.20	463	18.22835
10515.60	414	16.29921		11785.60	464	18.26772
10541.00	415	16.33858		11811.00	465	18.30709
10566.40	416	16.37795		11836.40	466	18.34646
10591.80	417	16.41732		11861.80	467	18.38583
10617.20	418	16.45669		11887.20	468	18.42520
10642.60	419	16.49606		11912.60	469	18.46457
10668.00	420	16.53543		11938.00	470	18.50394
10693.40	421	16.57480		11963.40	471	18.54331
10718.80	422	16.61417		11988.80	472	18.58268
10744.20	423	16.65354		12014.20	473	18.62205
10769.60	424	16.69291		12039.60	474	18.66142
10795.00	425	16.73228		12065.00	475	18.70079
10820.40	426	16.77165		12090.40	476	18.74016
10845.80	427	16.81102		12115.80	477	18.77953
10871.20	428	16.85039		12141.20	478	18.81890
10896.60	429	16.88976		12166.60	479	18.85827
10922.00	430	16.92913		12192.00	480	18.89764
10947.40	431	16.96850		12217.40	481	18.93701
10972.80	432	17.00787		12242.80	482	18.97638
10998.20	433	17.04724		12268.20	483	19.01575
11023.60	434	17.08661		12293.60	484	19.05512
11049.00	435	17.12598		12319.00	485	19.09449
11074.40	436	17.16535		12344.40	486	19.13386
11099.80	437	17.20472		12369.80	487	19.17323
11125.20	438	17.24409		12395.20	488	19.21260
11150.60	439	17.28346		12420.60	489	19.25197
11176.00	440	17.32283		12446.00	490	19.29134
11201.40	441	17.36220		12471.40	491	19.33071
11226.80	442	17.40157		12496.80	492	19.37008
11252.20	443	17.44094		12522.20	493	19.40945
11277.60	444	17.48031		12547.60	494	19.44882
11303.00	445	17.51969		12573.00	495	19.48819
11328.40	446	17.55906		12598.40	496	19.52756
11353.80	447	17.59843		12623.80	497	19.56693
11379.20	448	17.63780		12649.20	498	19.60630
11404.60	449	17.67717		12674.60	499	19.64567
11430.00	450	17.71654		12700.00	500	19.68504

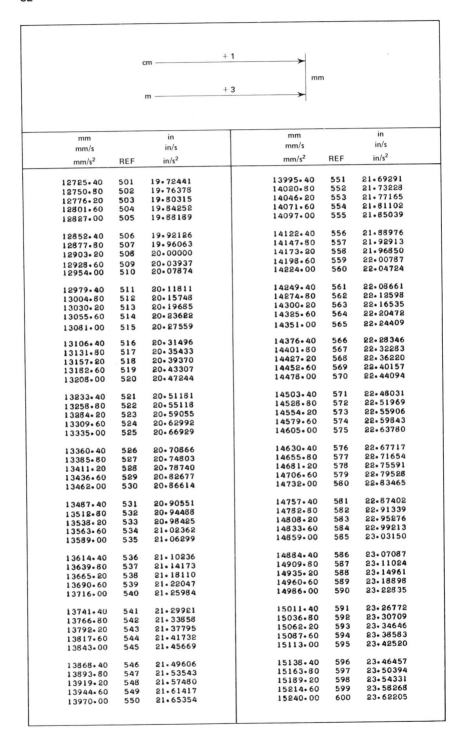

mm mm/s mm/s²	REF	in in/s in/s²	mm mm/s mm/s²	REF	in in/s in/s²
12725.40	501	19.72441	13995.40	551	21.69291
12750.80	502	19.76378	14020.80	552	21.73228
12776.20	503	19.80315	14046.20	553	21.77165
12801.60	504	19.84252	14071.60	554	21.81102
12827.00	505	19.88189	14097.00	555	21.85039
12852.40	506	19.92126	14122.40	556	21.88976
12877.80	507	19.96063	14147.80	557	21.92913
12903.20	508	20.00000	14173.20	558	21.96850
12928.60	509	20.03937	14198.60	559	22.00787
12954.00	510	20.07874	14224.00	560	22.04724
12979.40	511	20.11811	14249.40	561	22.08661
13004.80	512	20.15748	14274.80	562	22.12598
13030.20	513	20.19685	14300.20	563	22.16535
13055.60	514	20.23622	14325.60	564	22.20472
13081.00	515	20.27559	14351.00	565	22.24409
13106.40	516	20.31496	14376.40	566	22.28346
13131.80	517	20.35433	14401.80	567	22.32283
13157.20	518	20.39370	14427.20	568	22.36220
13182.60	519	20.43307	14452.60	569	22.40157
13208.00	520	20.47244	14478.00	570	22.44094
13233.40	521	20.51181	14503.40	571	22.48031
13258.80	522	20.55118	14528.80	572	22.51969
13284.20	523	20.59055	14554.20	573	22.55906
13309.60	524	20.62992	14579.60	574	22.59843
13335.00	525	20.66929	14605.00	575	22.63780
13360.40	526	20.70866	14630.40	576	22.67717
13385.80	527	20.74803	14655.80	577	22.71654
13411.20	528	20.78740	14681.20	578	22.75591
13436.60	529	20.82677	14706.60	579	22.79528
13462.00	530	20.86614	14732.00	580	22.83465
13487.40	531	20.90551	14757.40	581	22.87402
13512.80	532	20.94488	14782.80	582	22.91339
13538.20	533	20.98425	14808.20	583	22.95276
13563.60	534	21.02362	14833.60	584	22.99213
13589.00	535	21.06299	14859.00	585	23.03150
13614.40	536	21.10236	14884.40	586	23.07087
13639.80	537	21.14173	14909.80	587	23.11024
13665.20	538	21.18110	14935.20	588	23.14961
13690.60	539	21.22047	14960.60	589	23.18898
13716.00	540	21.25984	14986.00	590	23.22835
13741.40	541	21.29921	15011.40	591	23.26772
13766.80	542	21.33858	15036.80	592	23.30709
13792.20	543	21.37795	15062.20	593	23.34646
13817.60	544	21.41732	15087.60	594	23.38583
13843.00	545	21.45669	15113.00	595	23.42520
13868.40	546	21.49606	15138.40	596	23.46457
13893.80	547	21.53543	15163.80	597	23.50394
13919.20	548	21.57480	15189.20	598	23.54331
13944.60	549	21.61417	15214.60	599	23.58268
13970.00	550	21.65354	15240.00	600	23.62205

mm mm/s mm/s²	REF	in in/s in/s²	mm mm/s mm/s²	REF	in in/s in/s²
15265.40	601	23.66142	16535.40	651	25.62992
15290.80	602	23.70079	16560.80	652	25.66929
15316.20	603	23.74016	16586.20	653	25.70866
15341.60	604	23.77953	16611.60	654	25.74803
15367.00	605	23.81890	16637.00	655	25.78740
15392.40	606	23.85827	16662.40	656	25.82677
15417.80	607	23.89764	16687.80	657	25.86614
15443.20	608	23.93701	16713.20	658	25.90551
15468.60	609	23.97638	16738.60	659	25.94488
15494.00	610	24.01575	16764.00	660	25.98425
15519.40	611	24.05512	16789.40	661	26.02362
15544.80	612	24.09449	16814.80	662	26.06299
15570.20	613	24.13386	16840.20	663	26.10236
15595.60	614	24.17323	16865.60	664	26.14173
15621.00	615	24.21260	16891.00	665	26.18110
15646.40	616	24.25197	16916.40	666	26.22047
15671.80	617	24.29134	16941.80	667	26.25984
15697.20	618	24.33071	16967.20	668	26.29921
15722.60	619	24.37008	16992.60	669	26.33858
15748.00	620	24.40945	17018.00	670	26.37795
15773.40	621	24.44882	17043.40	671	26.41732
15798.80	622	24.48819	17068.80	672	26.45669
15824.20	623	24.52756	17094.20	673	26.49606
15849.60	624	24.56693	17119.60	674	26.53543
15875.00	625	24.60630	17145.00	675	26.57480
15900.40	626	24.64567	17170.40	676	26.61417
15925.80	627	24.68504	17195.80	677	26.65354
15951.20	628	24.72441	17221.20	678	26.69291
15976.60	629	24.76378	17246.60	679	26.73228
16002.00	630	24.80315	17272.00	680	26.77165
16027.40	631	24.84252	17297.40	681	26.81102
16052.80	632	24.88189	17322.80	682	26.85039
16078.20	633	24.92126	17348.20	683	26.88976
16103.60	634	24.96063	17373.60	684	26.92913
16129.00	635	25.00000	17399.00	685	26.96850
16154.40	636	25.03937	17424.40	686	27.00787
16179.80	637	25.07874	17449.80	687	27.04724
16205.20	638	25.11811	17475.20	688	27.08661
16230.60	639	25.15748	17500.60	689	27.12598
16256.00	640	25.19685	17526.00	690	27.16535
16281.40	641	25.23622	17551.40	691	27.20472
16306.80	642	25.27559	17576.80	692	27.24409
16332.20	643	25.31496	17602.20	693	27.28346
16357.60	644	25.35433	17627.60	694	27.32283
16383.00	645	25.39370	17653.00	695	27.36220
16408.40	646	25.43307	17678.40	696	27.40157
16433.80	647	25.47244	17703.80	697	27.44094
16459.20	648	25.51181	17729.20	698	27.48031
16484.60	649	25.55118	17754.60	699	27.51969
16510.00	650	25.59055	17780.00	700	27.55906

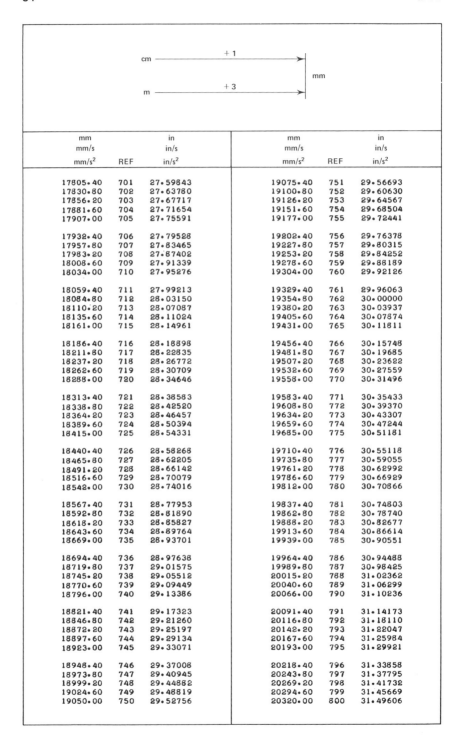

mm mm/s mm/s²	REF	in in/s in/s²	mm mm/s mm/s²	REF	in in/s in/s²
17805·40	701	27·59843	19075·40	751	29·56693
17830·80	702	27·63780	19100·80	752	29·60630
17856·20	703	27·67717	19126·20	753	29·64567
17881·60	704	27·71654	19151·60	754	29·68504
17907·00	705	27·75591	19177·00	755	29·72441
17932·40	706	27·79528	19202·40	756	29·76378
17957·80	707	27·83465	19227·80	757	29·80315
17983·20	708	27·87402	19253·20	758	29·84252
18008·60	709	27·91339	19278·60	759	29·88189
18034·00	710	27·95276	19304·00	760	29·92126
18059·40	711	27·99213	19329·40	761	29·96063
18084·80	712	28·03150	19354·80	762	30·00000
18110·20	713	28·07087	19380·20	763	30·03937
18135·60	714	28·11024	19405·60	764	30·07874
18161·00	715	28·14961	19431·00	765	30·11811
18186·40	716	28·18898	19456·40	766	30·15748
18211·80	717	28·22835	19481·80	767	30·19685
18237·20	718	28·26772	19507·20	768	30·23622
18262·60	719	28·30709	19532·60	769	30·27559
18288·00	720	28·34646	19558·00	770	30·31496
18313·40	721	28·38583	19583·40	771	30·35433
18338·80	722	28·42520	19608·80	772	30·39370
18364·20	723	28·46457	19634·20	773	30·43307
18389·60	724	28·50394	19659·60	774	30·47244
18415·00	725	28·54331	19685·00	775	30·51181
18440·40	726	28·58268	19710·40	776	30·55118
18465·80	727	28·62205	19735·80	777	30·59055
18491·20	728	28·66142	19761·20	778	30·62992
18516·60	729	28·70079	19786·60	779	30·66929
18542·00	730	28·74016	19812·00	780	30·70866
18567·40	731	28·77953	19837·40	781	30·74803
18592·80	732	28·81890	19862·80	782	30·78740
18618·20	733	28·85827	19888·20	783	30·82677
18643·60	734	28·89764	19913·60	784	30·86614
18669·00	735	28·93701	19939·00	785	30·90551
18694·40	736	28·97638	19964·40	786	30·94488
18719·80	737	29·01575	19989·80	787	30·98425
18745·20	738	29·05512	20015·20	788	31·02362
18770·60	739	29·09449	20040·60	789	31·06299
18796·00	740	29·13386	20066·00	790	31·10236
18821·40	741	29·17323	20091·40	791	31·14173
18846·80	742	29·21260	20116·80	792	31·18110
18872·20	743	29·25197	20142·20	793	31·22047
18897·60	744	29·29134	20167·60	794	31·25984
18923·00	745	29·33071	20193·00	795	31·29921
18948·40	746	29·37008	20218·40	796	31·33858
18973·80	747	29·40945	20243·80	797	31·37795
18999·20	748	29·44882	20269·20	798	31·41732
19024·60	749	29·48819	20294·60	799	31·45669
19050·00	750	29·52756	20320·00	800	31·49606

cm ← — 1

mm

m ← — 3

mm mm/s mm/s²	REF	in in/s in/s²	mm mm/s mm/s²	REF	in in/s in/s²
20345·40	801	31·53543	21615·40	851	33·50394
20370·80	802	31·57480	21640·80	852	33·54331
20396·20	803	31·61417	21666·20	853	33·58268
20421·60	804	31·65354	21691·60	854	33·62205
20447·00	805	31·69291	21717·00	855	33·66142
20472·40	806	31·73228	21742·40	856	33·70079
20497·80	807	31·77165	21767·80	857	33·74016
20523·20	808	31·81102	21793·20	858	33·77953
20548·60	809	31·85039	21818·60	859	33·81890
20574·00	810	31·88976	21844·00	860	33·85827
20599·40	811	31·92913	21869·40	861	33·89764
20624·80	812	31·96850	21894·80	862	33·93701
20650·20	813	32·00787	21920·20	863	33·97638
20675·60	814	32·04724	21945·60	864	34·01575
20701·00	815	32·08661	21971·00	865	34·05512
20726·40	816	32·12598	21996·40	866	34·09449
20751·80	817	32·16535	22021·80	867	34·13386
20777·20	818	32·20472	22047·20	868	34·17323
20802·60	819	32·24409	22072·60	869	34·21260
20828·00	820	32·28346	22098·00	870	34·25197
20853·40	821	32·32283	22123·40	871	34·29134
20878·80	822	32·36220	22148·80	872	34·33071
20904·20	823	32·40157	22174·20	873	34·37008
20929·60	824	32·44094	22199·60	874	34·40945
20955·00	825	32·48031	22225·00	875	34·44882
20980·40	826	32·51968	22250·40	876	34·48819
21005·80	827	32·55905	22275·80	877	34·52756
21031·20	828	32·59842	22301·20	878	34·56693
21056·60	829	32·63779	22326·60	879	34·60630
21082·00	830	32·67717	22352·00	880	34·64567
21107·40	831	32·71654	22377·40	881	34·68504
21132·80	832	32·75591	22402·80	882	34·72441
21158·20	833	32·79528	22428·20	883	34·76378
21183·60	834	32·83465	22453·60	884	34·80315
21209·00	835	32·87402	22479·00	885	34·84252
21234·40	836	32·91339	22504·40	886	34·88189
21259·80	837	32·95276	22529·80	887	34·92126
21285·20	838	32·99213	22555·20	888	34·96063
21310·60	839	33·03150	22580·60	889	35·00000
21336·00	840	33·07087	22606·00	890	35·03937
21361·40	841	33·11024	22631·40	891	35·07874
21386·80	842	33·14961	22656·80	892	35·11811
21412·20	843	33·18898	22682·20	893	35·15748
21437·60	844	33·22835	22707·60	894	35·19685
21463·00	845	33·26772	22733·00	895	35·23622
21488·40	846	33·30709	22758·40	896	35·27559
21513·80	847	33·34646	22783·80	897	35·31496
21539·20	848	33·38583	22809·20	898	35·35433
21564·60	849	33·42520	22834·60	899	35·39370
21590·00	850	33·46457	22860·00	900	35·43307

cm ——————————— +1 ———————→|
                                              mm
m ———————————— +3 ——————————→|

mm mm/s mm/s²	REF	in in/s in/s²	mm mm/s mm/s²	REF	in in/s in/s²
22885.40	901	35.47244	24155.40	951	37.44094
22910.80	902	35.51181	24180.80	952	37.48031
22936.20	903	35.55118	24206.20	953	37.51968
22961.60	904	35.59055	24231.60	954	37.55905
22987.00	905	35.62992	24257.00	955	37.59842
23012.40	906	35.66929	24282.40	956	37.63779
23037.80	907	35.70866	24307.80	957	37.67717
23063.20	908	35.74803	24333.20	958	37.71654
23088.60	909	35.78740	24358.60	959	37.75591
23114.00	910	35.82677	24384.00	960	37.79528
23139.40	911	35.86614	24409.40	961	37.83465
23164.80	912	35.90551	24434.80	962	37.87402
23190.20	913	35.94488	24460.20	963	37.91339
23215.60	914	35.98425	24485.60	964	37.95276
23241.00	915	36.02362	24511.00	965	37.99213
23266.40	916	36.06299	24536.40	966	38.03150
23291.80	917	36.10236	24561.80	967	38.07087
23317.20	918	36.14173	24587.20	968	38.11024
23342.60	919	36.18110	24612.60	969	38.14961
23368.00	920	36.22047	24638.00	970	38.18898
23393.40	921	36.25984	24663.40	971	38.22835
23418.80	922	36.29921	24688.80	972	38.26772
23444.20	923	36.33858	24714.20	973	38.30709
23469.60	924	36.37795	24739.60	974	38.34646
23495.00	925	36.41732	24765.00	975	38.38583
23520.40	926	36.45669	24790.40	976	38.42520
23545.80	927	36.49606	24815.80	977	38.46457
23571.20	928	36.53543	24841.20	978	38.50394
23596.60	929	36.57480	24866.60	979	38.54331
23622.00	930	36.61417	24892.00	980	38.58268
23647.40	931	36.65354	24917.40	981	38.62205
23672.80	932	36.69291	24942.80	982	38.66142
23698.20	933	36.73228	24968.20	983	38.70079
23723.60	934	36.77165	24993.60	984	38.74016
23749.00	935	36.81102	25019.00	985	38.77953
23774.40	936	36.85039	25044.40	986	38.81890
23799.80	937	36.88976	25069.80	987	38.85827
23825.20	938	36.92913	25095.20	988	38.89764
23850.60	939	36.96850	25120.60	989	38.93701
23876.00	940	37.00787	25146.00	990	38.97638
23901.40	941	37.04724	25171.40	991	39.01575
23926.80	942	37.08661	25196.80	992	39.05512
23952.20	943	37.12598	25222.20	993	39.09449
23977.60	944	37.16535	25247.60	994	39.13386
24003.00	945	37.20472	25273.00	995	39.17323
24028.40	946	37.24409	25298.40	996	39.21260
24053.80	947	37.28346	25323.80	997	39.25197
24079.20	948	37.32283	25349.20	998	39.29134
24104.60	949	37.36220	25374.60	999	39.33071
24130.00	950	37.40157	25400.00	1000	39.37008

3

LENGTH

Foot — Metre
  (ft — m)
Foot — Millimetre
  (ft — mm)
Foot — Centimetre
  (ft — cm)
Foot — Kilometre
  (ft — km)

VELOCITY

Foot/Second — Metre/Second
    (ft/s — m/s)
Foot/Second — Millimetre/Second
    (ft/s — mm/s)
Foot/Second — Centimetre/Second
    (ft/s — cm/s)
Foot/Second — Kilometre/Second
    (ft/s — km/s)

ACCELERATION

Foot/Second^2 — Metre/Second^2
    (ft/s^2 — m/s^2)
Foot/Second^2 — Millimetre/Second^2
    (ft/s^2 — mm/s^2)
Foot/Second^2 — Centimetre/Second^2
    (ft/s^2 — cm/s^2)
Foot/Second^2 — Kilometre/Second^2
    (ft/s^2 — km/s^2)

Conversion Factors:
$$1 \text{ foot (ft)} = 0.3048 \text{ metre (m) (exactly)}$$
$$(1 \text{ m} = 3.280840 \text{ ft})$$

Velocity and Acceleration Values

Table 3 can represent velocity and acceleration values, as long as the time units are the same in both English and SI equivalents. These consistent time units can be per second, per minute, per hour, per day, etc.

Velocity Conversion Factor: $1 \text{ ft/s} = 0.3048 \text{ m/s}$ (exactly)

Acceleration Conversion Factor: $1 \text{ ft/s}^2 = 0.3048 \text{ m/s}^2$ (exactly)

To convert m values obtained from tables to mm, cm, km, Diagram 3-A, which appears in the headings on alternate pages of this table, will assist the user. It can also be used to translate velocity and acceleration values noted above, and is derived from: $1 \text{ m} = (10^3) \text{ mm} = (10^2) \text{ cm} = (10^{-3}) \text{ km}$.

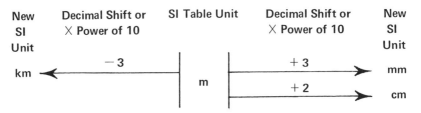

Diagram 3-A (See Introduction)

Example (a): Convert 0.125 ft to m, cm, mm.
  Since: $0.125 \text{ ft} = 125 \times 10^{-3} \text{ ft}$:
  From table: $125 \text{ ft} = 38.10000 \text{ m} = 38.1 \text{ m}$
  Therefore: $125 \times 10^{-3} \text{ ft} = 38.1 \times 10^{-3} \text{ m} = 0.0381 \text{ m}$.
  From Diagram 3-A: (a) $38.1 \times 10^{-3} \text{ m} = 38.1 \times 10^{-3} (10^2) \text{ cm}$
                                $= 38.1 \times 10^{-1} \text{ cm} = 3.81 \text{ cm}$
                                (m to cm by multiplying m value by 10^2 or by shifting its decimal 2 places to the right)
                            (b) $38.1 \times 10^{-3} \text{ m} = 38.1 \times 10^{-3} (10^3) \text{ mm} = 38.1 \text{ mm}$
                                (m to mm by multiplying m value by 10^3 or by shifting its decimal 3 places to the right)

Example (b): Convert 765 ft to km.
  From table: $765 \text{ ft} = 233.1720 \text{ m}*$
  From Diagram 3-A: $233.1720 \text{ m} = 233.1720 \times 10^{-3} \text{ km} = 0.233172 \text{ km}$
                          $\approx 0.233 \text{ km}$

---

*All conversion values obtained can be rounded to desired number of significant figures.

Diagram 3-B is also given at the top of alternate table pages:

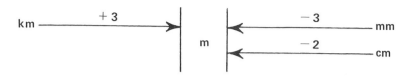

Diagram 3-B (See Introduction)

It is the opposite of Diagram 3-A, and is used to convert mm, cm, and km values to m. The m equivalent is then applied to the REF column in the table to obtain ft. Again, this diagram also refers to velocity and acceleration values as previously noted.

Example (*c*): Convert 951 mm/s to ft/s.
From Diagram 3-B: 951 mm/s = 951×10^{-3} m/s
From table: 951 m/s = 3120.079 ft/s.*
Therefore: 951×10^{-3} m/s = 3120.079×10^{-3} ft/s = 3.120079 ft/s
≈ 3.12 ft/s.

Example (*d*): Convert 0.855 km/min² to ft/min².
From Diagram 3-B: 0.855 km/min² = 0.855×10^3 m/min² = 855 m/min²
From table: 855 m/min² = 2805.118 ft/min²* $\approx$ 2810 ft/min².

---

*All conversion values obtained can be rounded to desired number of significant figures.

m m/s m/s^2	REF	ft ft/s ft/s^2	m m/s m/s^2	REF	ft ft/s ft/s^2
0.3048000	1	3.280840	15.54480	51	167.3228
0.6096000	2	6.561680	15.84960	52	170.6037
0.9144000	3	9.842520	16.15440	53	173.8845
1.219200	4	13.12336	16.45920	54	177.1654
1.524000	5	16.40420	16.76400	55	180.4462
1.828800	6	19.68504	17.06880	56	183.7270
2.133600	7	22.96588	17.37360	57	187.0079
2.438400	8	26.24672	17.67840	58	190.2887
2.743200	9	29.52756	17.98320	59	193.5696
3.048000	10	32.80840	18.28800	60	196.8504
3.352800	11	36.08924	18.59280	61	200.1312
3.657600	12	39.37008	18.89760	62	203.4121
3.962400	13	42.65092	19.20240	63	206.6929
4.267200	14	45.93176	19.50720	64	209.9738
4.572000	15	49.21260	19.81200	65	213.2546
4.876800	16	52.49344	20.11680	66	216.5354
5.181600	17	55.77428	20.42160	67	219.8163
5.486400	18	59.05512	20.72640	68	223.0971
5.791200	19	62.33596	21.03120	69	226.3780
6.096000	20	65.61680	21.33600	70	229.6588
6.400800	21	68.89764	21.64080	71	232.9396
6.705600	22	72.17848	21.94560	72	236.2205
7.010400	23	75.45932	22.25040	73	239.5013
7.315200	24	78.74016	22.55520	74	242.7822
7.620000	25	82.02100	22.86000	75	246.0630
7.924800	26	85.30184	23.16480	76	249.3438
8.229600	27	88.58268	23.46960	77	252.6247
8.534400	28	91.86352	23.77440	78	255.9055
8.839200	29	95.14436	24.07920	79	259.1864
9.144000	30	98.42520	24.38400	80	262.4672
9.448800	31	101.7060	24.68880	81	265.7480
9.753600	32	104.9869	24.99360	82	269.0289
10.05840	33	108.2677	25.29840	83	272.3097
10.36320	34	111.5486	25.60320	84	275.5905
10.66800	35	114.8294	25.90800	85	278.8714
10.97280	36	118.1102	26.21280	86	282.1522
11.27760	37	121.3911	26.51760	87	285.4331
11.58240	38	124.6719	26.82240	88	288.7139
11.88720	39	127.9528	27.12720	89	291.9948
12.19200	40	131.2336	27.43200	90	295.2756
12.49680	41	134.5144	27.73680	91	298.5564
12.80160	42	137.7953	28.04160	92	301.8373
13.10640	43	141.0761	28.34640	93	305.1181
13.41120	44	144.3570	28.65120	94	308.3989
13.71600	45	147.6378	28.95600	95	311.6798
14.02080	46	150.9186	29.26080	96	314.9606
14.32560	47	154.1995	29.56560	97	318.2415
14.63040	48	157.4803	29.87040	98	321.5223
14.93520	49	160.7612	30.17520	99	324.8032
15.24000	50	164.0420	30.48000	100	328.0840

```
km ————— +3 —————>           <————— -3 ——— mm
                    m
                             <————— -2 ——— cm
```

m m/s m/s²	REF	ft ft/s ft/s²	m m/s m/s²	REF	ft ft/s ft/s²
30.78480	101	331.3648	46.02480	151	495.4068
31.08960	102	334.6457	46.32960	152	498.6877
31.39440	103	337.9265	46.63440	153	501.9685
31.69920	104	341.2073	46.93920	154	505.2493
32.00400	105	344.4882	47.24400	155	508.5302
32.30880	106	347.7690	47.54880	156	511.8110
32.61360	107	351.0499	47.85360	157	515.0919
32.91840	108	354.3307	48.15840	158	518.3727
33.22320	109	357.6115	48.46320	159	521.6535
33.52800	110	360.8924	48.76800	160	524.9344
33.83280	111	364.1732	49.07280	161	528.2152
34.13760	112	367.4541	49.37760	162	531.4961
34.44240	113	370.7349	49.68240	163	534.7769
34.74720	114	374.0157	49.98720	164	538.0577
35.05200	115	377.2966	50.29200	165	541.3386
35.35680	116	380.5774	50.59680	166	544.6194
35.66160	117	383.8583	50.90160	167	547.9003
35.96640	118	387.1391	51.20640	168	551.1811
36.27120	119	390.4199	51.51120	169	554.4619
36.57600	120	393.7008	51.81600	170	557.7428
36.88080	121	396.9816	52.12080	171	561.0236
37.18560	122	400.2625	52.42560	172	564.3045
37.49040	123	403.5433	52.73040	173	567.5853
37.79520	124	406.8241	53.03520	174	570.8661
38.10000	125	410.1050	53.34000	175	574.1470
38.40480	126	413.3858	53.64480	176	577.4278
38.70960	127	416.6667	53.94960	177	580.7087
39.01440	128	419.9475	54.25440	178	583.9895
39.31920	129	423.2283	54.55920	179	587.2703
39.62400	130	426.5092	54.86400	180	590.5512
39.92880	131	429.7900	55.16880	181	593.8320
40.23360	132	433.0709	55.47360	182	597.1129
40.53840	133	436.3517	55.77840	183	600.3937
40.84320	134	439.6325	56.08320	184	603.6745
41.14800	135	442.9134	56.38800	185	606.9554
41.45280	136	446.1942	56.69280	186	610.2362
41.75760	137	449.4751	56.99760	187	613.5171
42.06240	138	452.7559	57.30240	188	616.7979
42.36720	139	456.0367	57.60720	189	620.0787
42.67200	140	459.3176	57.91200	190	623.3596
42.97680	141	462.5984	58.21680	191	626.6404
43.28160	142	465.8793	58.52160	192	629.9213
43.58640	143	469.1601	58.82640	193	633.2021
43.89120	144	472.4409	59.13120	194	636.4829
44.19600	145	475.7218	59.43600	195	639.7638
44.50080	146	479.0026	59.74080	196	643.0446
44.80560	147	482.2835	60.04560	197	646.3255
45.11040	148	485.5643	60.35040	198	649.6063
45.41520	149	488.8451	60.65520	199	652.8871
45.72000	150	492.1260	60.96000	200	656.1680

```
 km  ←——— -3 ———————|————— +3 ——————→ mm
                    m
                    |————— +2 ——————→ cm
```

m				ft	m			ft
m/s				ft/s	m/s			ft/s
m/s²	REF			ft/s²	m/s²	REF		ft/s²
61.26480	201			659.4488	76.50480	251		823.4908
61.56960	202			662.7297	76.80960	252		826.7717
61.87440	203			666.0105	77.11440	253		830.0525
62.17920	204			669.2913	77.41920	254		833.3333
62.48400	205			672.5722	77.72400	255		836.6142
62.78880	206			675.8530	78.02880	256		839.8950
63.09360	207			679.1339	78.33360	257		843.1758
63.39840	208			682.4147	78.63840	258		846.4567
63.70320	209			685.6955	78.94320	259		849.7375
64.00800	210			688.9764	79.24800	260		853.0184
64.31280	211			692.2572	79.55280	261		856.2992
64.61760	212			695.5381	79.85760	262		859.5800
64.92240	213			698.8189	80.16240	263		862.8609
65.22720	214			702.0997	80.46720	264		866.1417
65.53200	215			705.3806	80.77200	265		869.4226
65.83680	216			708.6614	81.07680	266		872.7034
66.14160	217			711.9423	81.38160	267		875.9843
66.44640	218			715.2231	81.68640	268		879.2651
66.75120	219			718.5039	81.99120	269		882.5459
67.05600	220			721.7848	82.29600	270		885.8268
67.36080	221			725.0656	82.60080	271		889.1076
67.66560	222			728.3465	82.90560	272		892.3885
67.97040	223			731.6273	83.21040	273		895.6693
68.27520	224			734.9081	83.51520	274		898.9501
68.58000	225			738.1890	83.82000	275		902.2310
68.88480	226			741.4698	84.12480	276		905.5118
69.18960	227			744.7507	84.42960	277		908.7926
69.49440	228			748.0315	84.73440	278		912.0735
69.79920	229			751.3123	85.03920	279		915.3543
70.10400	230			754.5932	85.34400	280		918.6352
70.40880	231			757.8740	85.64880	281		921.9160
70.71360	232			761.1549	85.95360	282		925.1968
71.01840	233			764.4357	86.25840	283		928.4777
71.32320	234			767.7165	86.56320	284		931.7585
71.62800	235			770.9974	86.86800	285		935.0394
71.93280	236			774.2782	87.17280	286		938.3202
72.23760	237			777.5591	87.47760	287		941.6011
72.54240	238			780.8399	87.78240	288		944.8819
72.84720	239			784.1207	88.08720	289		948.1627
73.15200	240			787.4016	88.39200	290		951.4436
73.45680	241			790.6824	88.69680	291		954.7244
73.76160	242			793.9633	89.00160	292		958.0052
74.06640	243			797.2441	89.30640	293		961.2861
74.37120	244			800.5249	89.61120	294		964.5669
74.67600	245			803.8058	89.91600	295		967.8478
74.98080	246			807.0866	90.22080	296		971.1286
75.28560	247			810.3675	90.52560	297		974.4094
75.59040	248			813.6483	90.83040	298		977.6903
75.89520	249			816.9291	91.13520	299		980.9711
76.20000	250			820.2100	91.44000	300		984.2520

| m | | | ft | | m | | | ft | |
| m/s | | | ft/s | | m/s | | | ft/s | |
m/s²	REF		ft/s²		m/s²	REF		ft/s²	
91.74480	301		987.5328		106.9848	351		1151.575	
92.04960	302		990.8136		107.2896	352		1154.856	
92.35440	303		994.0945		107.5944	353		1158.136	
92.65920	304		997.3753		107.8992	354		1161.417	
92.96400	305		1000.656		108.2040	355		1164.698	
93.26880	306		1003.937		108.5088	356		1167.979	
93.57360	307		1007.218		108.8136	357		1171.260	
93.87840	308		1010.499		109.1184	358		1174.541	
94.18320	309		1013.780		109.4232	359		1177.822	
94.48800	310		1017.060		109.7280	360		1181.102	
94.79280	311		1020.341		110.0328	361		1184.383	
95.09760	312		1023.622		110.3376	362		1187.664	
95.40240	313		1026.903		110.6424	363		1190.945	
95.70720	314		1030.184		110.9472	364		1194.226	
96.01200	315		1033.465		111.2520	365		1197.507	
96.31680	316		1036.745		111.5568	366		1200.787	
96.62160	317		1040.026		111.8616	367		1204.068	
96.92640	318		1043.307		112.1664	368		1207.349	
97.23120	319		1046.588		112.4712	369		1210.630	
97.53600	320		1049.869		112.7760	370		1213.911	
97.84080	321		1053.150		113.0808	371		1217.192	
98.14560	322		1056.430		113.3856	372		1220.472	
98.45040	323		1059.711		113.6904	373		1223.753	
98.75520	324		1062.992		113.9952	374		1227.034	
99.06000	325		1066.273		114.3000	375		1230.315	
99.36480	326		1069.554		114.6048	376		1233.596	
99.66960	327		1072.835		114.9096	377		1236.877	
99.97440	328		1076.115		115.2144	378		1240.157	
100.2792	329		1079.396		115.5192	379		1243.438	
100.5840	330		1082.677		115.8240	380		1246.719	
100.8888	331		1085.958		116.1288	381		1250.000	
101.1936	332		1089.239		116.4336	382		1253.281	
101.4984	333		1092.520		116.7384	383		1256.562	
101.8032	334		1095.801		117.0432	384		1259.843	
102.1080	335		1099.081		117.3480	385		1263.123	
102.4128	336		1102.362		117.6528	386		1266.404	
102.7176	337		1105.643		117.9576	387		1269.685	
103.0224	338		1108.924		118.2624	388		1272.966	
103.3272	339		1112.205		118.5672	389		1276.247	
103.6320	340		1115.486		118.8720	390		1279.528	
103.9368	341		1118.766		119.1768	391		1282.808	
104.2416	342		1122.047		119.4816	392		1286.089	
104.5464	343		1125.328		119.7864	393		1289.370	
104.8512	344		1128.609		120.0912	394		1292.651	
105.1560	345		1131.890		120.3960	395		1295.932	
105.4608	346		1135.171		120.7008	396		1299.213	
105.7656	347		1138.451		121.0056	397		1302.493	
106.0704	348		1141.732		121.3104	398		1305.774	
106.3752	349		1145.013		121.6152	399		1309.055	
106.6800	350		1148.294		121.9200	400		1312.336	

km ←   −3   m   +3 → mm

+2 → cm

m m/s m/s^2	REF	ft ft/s ft/s^2	m m/s m/s^2	REF	ft ft/s ft/s^2
122.2248	401	1315.617	137.4648	451	1479.659
122.5296	402	1318.898	137.7696	452	1482.940
122.8344	403	1322.178	138.0744	453	1486.220
123.1392	404	1325.459	138.3792	454	1489.501
123.4440	405	1328.740	138.6840	455	1492.782
123.7488	406	1332.021	138.9888	456	1496.063
124.0536	407	1335.302	139.2936	457	1499.344
124.3584	408	1338.583	139.5984	458	1502.625
124.6632	409	1341.864	139.9032	459	1505.906
124.9680	410	1345.144	140.2080	460	1509.186
125.2728	411	1348.425	140.5128	461	1512.467
125.5776	412	1351.706	140.8176	462	1515.748
125.8824	413	1354.987	141.1224	463	1519.029
126.1872	414	1358.268	141.4272	464	1522.310
126.4920	415	1361.549	141.7320	465	1525.591
126.7968	416	1364.829	142.0368	466	1528.871
127.1016	417	1368.110	142.3416	467	1532.152
127.4064	418	1371.391	142.6464	468	1535.433
127.7112	419	1374.672	142.9512	469	1538.714
128.0160	420	1377.953	143.2560	470	1541.995
128.3208	421	1381.234	143.5608	471	1545.276
128.6256	422	1384.514	143.8656	472	1548.556
128.9304	423	1387.795	144.1704	473	1551.837
129.2352	424	1391.076	144.4752	474	1555.118
129.5400	425	1394.357	144.7800	475	1558.399
129.8448	426	1397.638	145.0848	476	1561.680
130.1496	427	1400.919	145.3896	477	1564.961
130.4544	428	1404.199	145.6944	478	1568.241
130.7592	429	1407.480	145.9992	479	1571.522
131.0640	430	1410.761	146.3040	480	1574.803
131.3688	431	1414.042	146.6088	481	1578.084
131.6736	432	1417.323	146.9136	482	1581.365
131.9784	433	1420.604	147.2184	483	1584.646
132.2832	434	1423.885	147.5232	484	1587.927
132.5880	435	1427.165	147.8280	485	1591.207
132.8928	436	1430.446	148.1328	486	1594.488
133.1976	437	1433.727	148.4376	487	1597.769
133.5024	438	1437.008	148.7424	488	1601.050
133.8072	439	1440.289	149.0472	489	1604.331
134.1120	440	1443.570	149.3520	490	1607.612
134.4168	441	1446.850	149.6568	491	1610.892
134.7216	442	1450.131	149.9616	492	1614.173
135.0264	443	1453.412	150.2664	493	1617.454
135.3312	444	1456.693	150.5712	494	1620.735
135.6360	445	1459.974	150.8760	495	1624.016
135.9408	446	1463.255	151.1808	496	1627.297
136.2456	447	1466.535	151.4856	497	1630.577
136.5504	448	1469.816	151.7904	498	1633.858
136.8552	449	1473.097	152.0952	499	1637.139
137.1600	450	1476.378	152.4000	500	1640.420

m			m		
m/s		ft	m/s		ft
m/s²	REF	ft/s	m/s²	REF	ft/s
		ft/s²			ft/s²
152.7048	501	1643.701	167.9448	551	1807.743
153.0096	502	1646.982	168.2496	552	1811.024
153.3144	503	1650.262	168.5544	553	1814.304
153.6192	504	1653.543	168.8592	554	1817.585
153.9240	505	1656.824	169.1640	555	1820.866
154.2288	506	1660.105	169.4688	556	1824.147
154.5336	507	1663.386	169.7736	557	1827.428
154.8384	508	1666.667	170.0784	558	1830.709
155.1432	509	1669.948	170.3832	559	1833.990
155.4480	510	1673.228	170.6880	560	1837.270
155.7528	511	1676.509	170.9928	561	1840.551
156.0576	512	1679.790	171.2976	562	1843.832
156.3624	513	1683.071	171.6024	563	1847.113
156.6672	514	1686.352	171.9072	564	1850.394
156.9720	515	1689.633	172.2120	565	1853.675
157.2768	516	1692.913	172.5168	566	1856.955
157.5816	517	1696.194	172.8216	567	1860.236
157.8864	518	1699.475	173.1264	568	1863.517
158.1912	519	1702.756	173.4312	569	1866.798
158.4960	520	1706.037	173.7360	570	1870.079
158.8008	521	1709.318	174.0408	571	1873.360
159.1056	522	1712.598	174.3456	572	1876.640
159.4104	523	1715.879	174.6504	573	1879.921
159.7152	524	1719.160	174.9552	574	1883.202
160.0200	525	1722.441	175.2600	575	1886.483
160.3248	526	1725.722	175.5648	576	1889.764
160.6296	527	1729.003	175.8696	577	1893.045
160.9344	528	1732.283	176.1744	578	1896.325
161.2392	529	1735.564	176.4792	579	1899.606
161.5440	530	1738.845	176.7840	580	1902.887
161.8488	531	1742.126	177.0888	581	1906.168
162.1536	532	1745.407	177.3936	582	1909.449
162.4584	533	1748.688	177.6984	583	1912.730
162.7632	534	1751.969	178.0032	584	1916.010
163.0680	535	1755.249	178.3080	585	1919.291
163.3728	536	1758.530	178.6128	586	1922.572
163.6776	537	1761.811	178.9176	587	1925.853
163.9824	538	1765.092	179.2224	588	1929.134
164.2872	539	1768.373	179.5272	589	1932.415
164.5920	540	1771.654	179.8320	590	1935.696
164.8968	541	1774.934	180.1368	591	1938.976
165.2016	542	1778.215	180.4416	592	1942.257
165.5064	543	1781.496	180.7464	593	1945.538
165.8112	544	1784.777	181.0512	594	1948.819
166.1160	545	1788.058	181.3560	595	1952.100
166.4208	546	1791.339	181.6608	596	1955.381
166.7256	547	1794.619	181.9656	597	1958.661
167.0304	548	1797.900	182.2704	598	1961.942
167.3352	549	1801.181	182.5752	599	1965.223
167.6400	550	1804.462	182.8800	600	1968.504

| m | | ft | | m | | ft |
| m/s | | ft/s | | m/s | | ft/s |
m/s²	REF	ft/s²		m/s²	REF	ft/s²
183.1848	601	1971.785		198.4248	651	2135.827
183.4896	602	1975.066		198.7296	652	2139.108
183.7944	603	1978.346		199.0344	653	2142.388
184.0992	604	1981.627		199.3392	654	2145.669
184.4040	605	1984.908		199.6440	655	2148.950
184.7088	606	1988.189		199.9488	656	2152.231
185.0136	607	1991.470		200.2536	657	2155.512
185.3184	608	1994.751		200.5584	658	2158.793
185.6232	609	1998.031		200.8632	659	2162.073
185.9280	610	2001.312		201.1680	660	2165.354
186.2328	611	2004.593		201.4728	661	2168.635
186.5376	612	2007.874		201.7776	662	2171.916
186.8424	613	2011.155		202.0824	663	2175.197
187.1472	614	2014.436		202.3872	664	2178.478
187.4520	615	2017.717		202.6920	665	2181.759
187.7568	616	2020.997		202.9968	666	2185.039
188.0616	617	2024.278		203.3016	667	2188.320
188.3664	618	2027.559		203.6064	668	2191.601
188.6712	619	2030.840		203.9112	669	2194.882
188.9760	620	2034.121		204.2160	670	2198.163
189.2808	621	2037.402		204.5208	671	2201.444
189.5856	622	2040.682		204.8256	672	2204.724
189.8904	623	2043.963		205.1304	673	2208.005
190.1952	624	2047.244		205.4352	674	2211.286
190.5000	625	2050.525		205.7400	675	2214.567
190.8048	626	2053.806		206.0448	676	2217.848
191.1096	627	2057.087		206.3496	677	2221.129
191.4144	628	2060.367		206.6544	678	2224.409
191.7192	629	2063.648		206.9592	679	2227.690
192.0240	630	2066.929		207.2640	680	2230.971
192.3288	631	2070.210		207.5688	681	2234.252
192.6336	632	2073.491		207.8736	682	2237.533
192.9384	633	2076.772		208.1784	683	2240.814
193.2432	634	2080.052		208.4832	684	2244.094
193.5480	635	2083.333		208.7880	685	2247.375
193.8528	636	2086.614		209.0928	686	2250.656
194.1576	637	2089.895		209.3976	687	2253.937
194.4624	638	2093.176		209.7024	688	2257.218
194.7672	639	2096.457		210.0072	689	2260.499
195.0720	640	2099.738		210.3120	690	2263.780
195.3768	641	2103.018		210.6168	691	2267.060
195.6816	642	2106.299		210.9216	692	2270.341
195.9864	643	2109.580		211.2264	693	2273.622
196.2912	644	2112.861		211.5312	694	2276.903
196.5960	645	2116.142		211.8360	695	2280.184
196.9008	646	2119.423		212.1408	696	2283.465
197.2056	647	2122.703		212.4456	697	2286.745
197.5104	648	2125.984		212.7504	698	2290.026
197.8152	649	2129.265		213.0552	699	2293.307
198.1200	650	2132.546		213.3600	700	2296.588

m m/s m/s²	REF	ft ft/s ft/s²	m m/s m/s²	REF	ft ft/s ft/s²
213.6648	701	2299.869	228.9048	751	2463.911
213.9696	702	2303.150	229.2096	752	2467.192
214.2744	703	2306.430	229.5144	753	2470.472
214.5792	704	2309.711	229.8192	754	2473.753
214.8840	705	2312.992	230.1240	755	2477.034
215.1888	706	2316.273	230.4288	756	2480.315
215.4936	707	2319.554	230.7336	757	2483.596
215.7984	708	2322.835	231.0384	758	2486.877
216.1032	709	2326.115	231.3432	759	2490.157
216.4080	710	2329.396	231.6480	760	2493.438
216.7128	711	2332.677	231.9528	761	2496.719
217.0176	712	2335.958	232.2576	762	2500.000
217.3224	713	2339.239	232.5624	763	2503.281
217.6272	714	2342.520	232.8672	764	2506.562
217.9320	715	2345.801	233.1720	765	2509.843
218.2368	716	2349.081	233.4768	766	2513.123
218.5416	717	2352.362	233.7816	767	2516.404
218.8464	718	2355.643	234.0864	768	2519.685
219.1512	719	2358.924	234.3912	769	2522.966
219.4560	720	2362.205	234.6960	770	2526.247
219.7608	721	2365.486	235.0008	771	2529.528
220.0656	722	2368.766	235.3056	772	2532.808
220.3704	723	2372.047	235.6104	773	2536.089
220.6752	724	2375.328	235.9152	774	2539.370
220.9800	725	2378.609	236.2200	775	2542.651
221.2848	726	2381.890	236.5248	776	2545.932
221.5896	727	2385.171	236.8296	777	2549.213
221.8944	728	2388.451	237.1344	778	2552.493
222.1992	729	2391.732	237.4392	779	2555.774
222.5040	730	2395.013	237.7440	780	2559.055
222.8088	731	2398.294	238.0488	781	2562.336
223.1136	732	2401.575	238.3536	782	2565.617
223.4184	733	2404.856	238.6584	783	2568.898
223.7232	734	2408.136	238.9632	784	2572.178
224.0280	735	2411.417	239.2680	785	2575.459
224.3328	736	2414.698	239.5728	786	2578.740
224.6376	737	2417.979	239.8776	787	2582.021
224.9424	738	2421.260	240.1824	788	2585.302
225.2472	739	2424.541	240.4872	789	2588.583
225.5520	740	2427.822	240.7920	790	2591.864
225.8568	741	2431.102	241.0968	791	2595.144
226.1616	742	2434.383	241.4016	792	2598.425
226.4664	743	2437.664	241.7064	793	2601.706
226.7712	744	2440.945	242.0112	794	2604.987
227.0760	745	2444.226	242.3160	795	2608.268
227.3808	746	2447.507	242.6208	796	2611.549
227.6856	747	2450.787	242.9256	797	2614.829
227.9904	748	2454.068	243.2304	798	2618.110
228.2952	749	2457.349	243.5352	799	2621.391
228.6000	750	2460.630	243.8400	800	2624.672

km ← ——— −3 ——— | ——— +3 ——→ mm
m
| ——— +2 ——→ cm

m m/s m/s²	REF	ft ft/s ft/s²	m m/s m/s²	REF	ft ft/s ft/s²
244.1448	801	2627.953	259.3848	851	2791.995
244.4496	802	2631.234	259.6896	852	2795.276
244.7544	803	2634.514	259.9944	853	2798.556
245.0592	804	2637.795	260.2992	854	2801.837
245.3640	805	2641.076	260.6040	855	2805.118
245.6688	806	2644.357	260.9088	856	2808.399
245.9736	807	2647.638	261.2136	857	2811.680
246.2784	808	2650.919	261.5184	858	2814.961
246.5832	809	2654.199	261.8232	859	2818.241
246.8880	810	2657.480	262.1280	860	2821.522
247.1928	811	2660.761	262.4328	861	2824.803
247.4976	812	2664.042	262.7376	862	2828.084
247.8024	813	2667.323	263.0424	863	2831.365
248.1072	814	2670.604	263.3472	864	2834.646
248.4120	815	2673.885	263.6520	865	2837.927
248.7168	816	2677.165	263.9568	866	2841.207
249.0216	817	2680.446	264.2616	867	2844.488
249.3264	818	2683.727	264.5664	868	2847.769
249.6312	819	2687.008	264.8712	869	2851.050
249.9360	820	2690.289	265.1760	870	2854.331
250.2408	821	2693.570	265.4808	871	2857.612
250.5456	822	2696.850	265.7856	872	2860.892
250.8504	823	2700.131	266.0904	873	2864.173
251.1552	824	2703.412	266.3952	874	2867.454
251.4600	825	2706.693	266.7000	875	2870.735
251.7648	826	2709.974	267.0048	876	2874.016
252.0696	827	2713.255	267.3096	877	2877.297
252.3744	828	2716.535	267.6144	878	2880.577
252.6792	829	2719.816	267.9192	879	2883.858
252.9840	830	2723.097	268.2240	880	2887.139
253.2888	831	2726.378	268.5288	881	2890.420
253.5936	832	2729.659	268.8336	882	2893.701
253.8984	833	2732.940	269.1384	883	2896.982
254.2032	834	2736.220	269.4432	884	2900.262
254.5080	835	2739.501	269.7480	885	2903.543
254.8128	836	2742.782	270.0528	886	2906.824
255.1176	837	2746.063	270.3576	887	2910.105
255.4224	838	2749.344	270.6624	888	2913.386
255.7272	839	2752.625	270.9672	889	2916.667
256.0320	840	2755.906	271.2720	890	2919.948
256.3368	841	2759.186	271.5768	891	2923.228
256.6416	842	2762.467	271.8816	892	2926.509
256.9464	843	2765.748	272.1864	893	2929.790
257.2512	844	2769.029	272.4912	894	2933.071
257.5560	845	2772.310	272.7960	895	2936.352
257.8608	846	2775.591	273.1008	896	2939.633
258.1656	847	2778.871	273.4056	897	2942.913
258.4704	848	2782.152	273.7104	898	2946.194
258.7752	849	2785.433	274.0152	899	2949.475
259.0800	850	2788.714	274.3200	900	2952.756

km ———— +3 ————→		←———— -3 ———— mm
	m	
		←———— -2 ———— cm

m		ft		m		ft
m/s		ft/s		m/s		ft/s
m/s²	REF	ft/s²		m/s²	REF	ft/s²
274.6248	901	2956.037		289.8648	951	3120.079
274.9296	902	2959.318		290.1696	952	3123.360
275.2344	903	2962.598		290.4744	953	3126.640
275.5392	904	2965.879		290.7792	954	3129.921
275.8440	905	2969.160		291.0840	955	3133.202
276.1488	906	2972.441		291.3888	956	3136.483
276.4536	907	2975.722		291.6936	957	3139.764
276.7584	908	2979.003		291.9984	958	3143.045
277.0632	909	2982.283		292.3032	959	3146.325
277.3680	910	2985.564		292.6080	960	3149.606
277.6728	911	2988.845		292.9128	961	3152.887
277.9776	912	2992.126		293.2176	962	3156.168
278.2824	913	2995.407		293.5224	963	3159.449
278.5872	914	2998.688		293.8272	964	3162.730
278.8920	915	3001.969		294.1320	965	3166.010
279.1968	916	3005.249		294.4368	966	3169.291
279.5016	917	3008.530		294.7416	967	3172.572
279.8064	918	3011.811		295.0464	968	3175.853
280.1112	919	3015.092		295.3512	969	3179.134
280.4160	920	3018.373		295.6560	970	3182.415
280.7208	921	3021.654		295.9608	971	3185.696
281.0256	922	3024.934		296.2656	972	3188.976
281.3304	923	3028.215		296.5704	973	3192.257
281.6352	924	3031.496		296.8752	974	3195.538
281.9400	925	3034.777		297.1800	975	3198.819
282.2448	926	3038.058		297.4848	976	3202.100
282.5496	927	3041.339		297.7896	977	3205.381
282.8544	928	3044.619		298.0944	978	3208.661
283.1592	929	3047.900		298.3992	979	3211.942
283.4640	930	3051.181		298.7040	980	3215.223
283.7688	931	3054.462		299.0088	981	3218.504
284.0736	932	3057.743		299.3136	982	3221.785
284.3784	933	3061.024		299.6184	983	3225.066
284.6832	934	3064.304		299.9232	984	3228.346
284.9880	935	3067.585		300.2280	985	3231.627
285.2928	936	3070.866		300.5328	986	3234.908
285.5976	937	3074.147		300.8376	987	3238.189
285.9024	938	3077.428		301.1424	988	3241.470
286.2072	939	3080.709		301.4472	989	3244.751
286.5120	940	3083.990		301.7520	990	3248.031
286.8168	941	3087.270		302.0568	991	3251.312
287.1216	942	3090.551		302.3616	992	3254.593
287.4264	943	3093.832		302.6664	993	3257.874
287.7312	944	3097.113		302.9712	994	3261.155
288.0360	945	3100.394		303.2760	995	3264.436
288.3408	946	3103.675		303.5808	996	3267.717
288.6456	947	3106.955		303.8856	997	3270.997
288.9504	948	3110.236		304.1904	998	3274.278
289.2552	949	3113.517		304.4952	999	3277.559
289.5600	950	3116.798		304.8000	1000	3280.840

4

LENGTH

Yard — Metre
 (yd — m)
Yard — Centimetre
 (yd — cm)
Yard — Kilometre
 (yd — km)

VELOCITY

Yard/Second — Metre/Second
 (yd/s — m/s)
Yard/Second — Centimetre/Second
 (yd/s — cm/s)
Yard/Second — Kilometre/Second
 (yd/s — km/s)

ACCELERATION

Yard/Second^2 — Metre/Second^2
 (yd/s^2 — m/s^2)
Yard/Second^2 — Centimetre/Second^2
 (yd/s^2 — cm/s^2)
Yard/Second^2 — Kilometre/Second^2
 (yd/s^2 — km/s^2)

Conversion Factors:
 1 yard (yd) = 0.9144 metre (m) (exactly)
 (1 m = 1.093613 yd)

Velocity and Acceleration Values

Table 4 can represent velocity and acceleration values, as long as the time units are the same in both English and SI conversions. These consistent time units can be per second, per minute, per hour, etc.

Velocity Conversion Factor: 1 yd/s = 0.9144 m/s

Acceleration Conversion Factor: 1 yd/s² = 0.9144 m /s²

Though most tables in this reference book have a REF number range of 1 to 1000, the 1 to 100 range given in this table is sufficient for most yard-metre conversions.

To convert m values obtained from the table to cm, and km, Diagram 4-A will assist the user. This diagram can also be used to translate velocity and acceleration values noted above, and is derived from: $1 \text{ m} = (10^2) \text{ cm} = (10^{-3}) \text{ km}$.

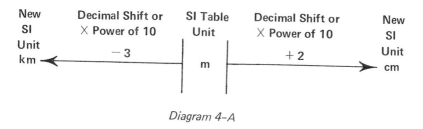

Diagram 4–A

Example (a): Convert 0.155 yd to cm.
  Since: 0.155 yd = (0.150 + 0.005) yd:
  From table: 15 yd = 13.716 m*
        5 yd = 4.572 m*
  By shifting decimals and adding:
    0.15 yd = 0.13716 m
    0.005 yd = 0.004572 m
    ‾‾‾‾‾‾‾‾‾‾‾‾‾‾‾‾‾‾‾‾‾‾
    0.155 yd = 0.141732 m
  From Diagram 4-A: $0.141732 \text{ m} = 0.141732 \times 10^2 \text{ cm} = 14.1732 \text{ cm}$
              ≈ 14.2 cm
              (m to cm by multiplying m value by 10^2 or by shifting
              its decimal 2 places to the right.)

Example (b): Convert 875 yd to km.
  Since: 875 yd = (870 + 5) yd:
  From table: 87 yd = 79.5528 m*
        5 yd = 4.572 m*

---

*All conversion values obtained can be rounded to desired number of significant figures.

By shifting decimals where necessary and adding:

870 yd = 795.528 m

5 yd =   4.572 m

875 yd = 800.100 m

From Diagram 4-A: 800.1 m = 800.1 × 10^{-3} km = 0.8001 km ≈ 0.8 km

(m to km by multiplying m value by 10^{-3} or by shifting its decimal 3 places to the left).

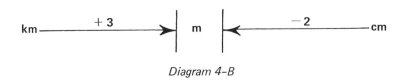

Diagram 4-B

Diagram 4-B is the opposite of the previous diagram, and is used to convert cm and km to yards through the table. This diagram guides the conversion of cm and km to their m equivalent. The m equivalent is then applied to the REF column in the table to obtain yards. Again, this diagram also refers to velocity and acceleration values as previously noted.

Example (c): Convert 98 cm/s to yd/s.

From Diagram 4-B: 98 cm/s = 98 × 10^{-2} m/s

From table: 98 m/s = 107.1741 yd/s*

Therefore: 98 × 10^{-2} m/s = 107.1741 × 10^{-2} yd/s = 1.071741 yd/s

≈ 1.07 yd/s .

Example (d): Convert 0.025 km to yd.

From Diagram 4-B: 0.025 km = 0.025 × 10^{3} m = 25 m

From table: 25 m = 27.34033 yd* ≈ 27 yd.

---

*All conversion values obtained can be rounded to desired number of significant figures.

m m/s m/s^2	REF	yd yd/s yd/s^2	m m/s m/s^2	REF	yd yd/s yd/s^2
0·9144000	1	1·093613	46·63440	51	55·77428
1·828800	2	2·187227	47·54880	52	56·86789
2·743200	3	3·280840	48·46320	53	57·96150
3·657600	4	4·374453	49·37760	54	59·05512
4·572000	5	5·468067	50·29200	55	60·14873
5·486400	6	6·561680	51·20640	56	61·24234
6·400800	7	7·655293	52·12080	57	62·33596
7·315200	8	8·748906	53·03520	58	63·42957
8·229600	9	9·842520	53·94960	59	64·52318
9·144000	10	10·93613	54·86400	60	65·61680
10·05840	11	12·02975	55·77840	61	66·71041
10·97280	12	13·12336	56·69280	62	67·80402
11·88720	13	14·21697	57·60720	63	68·89764
12·80160	14	15·31059	58·52160	64	69·99125
13·71600	15	16·40420	59·43600	65	71·08486
14·63040	16	17·49781	60·35040	66	72·17848
15·54480	17	18·59143	61·26480	67	73·27209
16·45920	18	19·68504	62·17920	68	74·36570
17·37360	19	20·77865	63·09360	69	75·45932
18·28800	20	21·87227	64·00800	70	76·55293
19·20240	21	22·96588	64·92240	71	77·64654
20·11680	22	24·05949	65·83680	72	78·74016
21·03120	23	25·15311	66·75120	73	79·83377
21·94560	24	26·24672	67·66560	74	80·92738
22·86000	25	27·34033	68·58000	75	82·02100
23·77440	26	28·43395	69·49440	76	83·11461
24·68880	27	29·52756	70·40880	77	84·20822
25·60320	28	30·62117	71·32320	78	85·30184
26·51760	29	31·71479	72·23760	79	86·39545
27·43200	30	32·80840	73·15200	80	87·48906
28·34640	31	33·90201	74·06640	81	88·58268
29·26080	32	34·99563	74·98080	82	89·67629
30·17520	33	36·08924	75·89520	83	90·76990
31·08960	34	37·18285	76·80960	84	91·86352
32·00400	35	38·27647	77·72400	85	92·95713
32·91840	36	39·37008	78·63840	86	94·05074
33·83280	37	40·46369	79·55280	87	95·14436
34·74720	38	41·55731	80·46720	88	96·23797
35·66160	39	42·65092	81·38160	89	97·33158
36·57600	40	43·74453	82·29600	90	98·42520
37·49040	41	44·83815	83·21040	91	99·51881
38·40480	42	45·93176	84·12480	92	100·6124
39·31920	43	47·02537	85·03920	93	101·7060
40·23360	44	48·11899	85·95360	94	102·7997
41·14800	45	49·21260	86·86800	95	103·8933
42·06240	46	50·30621	87·78240	96	104·9869
42·97680	47	51·39983	88·69680	97	106·0805
43·89120	48	52·49344	89·61120	98	107·1741
44·80560	49	53·58705	90·52560	99	108·2677
45·72000	50	54·68067	91·44000	100	109·3613

5

LENGTH

<u>Mile (statute) — Kilometre</u>
     (mi — km)

Mile — Metre
 (mi — m)

VELOCITY

<u>Mile/Hour — Kilometre/Hour</u>
   (mi/h — km/h)

ACCELERATION

<u>Mile/Hour^2 — Kilometre/Hour^2</u>
   (mi/h^2 — km/h^2)

Conversion Factors:
       1 mile (mi) = 1.609344 kilometres (km) (exactly)
       (1 km = 0.6213712 mi)

Velocity and Acceleration Values

Table 5 can represent velocity and acceleration values, as long as the time units are the same in both English and SI. These consistent time units can be per second, per minute, per hour, per day, etc.

Velocity Conversion Factor: 1 mi/h = 1.609344 km/h

Acceleration Conversion Factor: 1 mi/h^2 = 1.609344 km/h^2

Although most tables in this reference book have a REF number range of 1 to 1000, the 1 to 100 range given in this table is sufficient for most mile-kilometre conversions.

To convert km values obtained from the table to m, Diagram 5-A will assist the user, and is derived from: 1 km = (10^3) m. The diagram can also be used to translate velocity and acceleration values noted above.

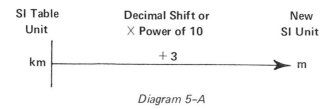

| SI Table
Unit | Decimal Shift or
X Power of 10 | New
SI Unit |

Diagram 5-A

Example (a): Convert 67 mi to m.
   From table: 67 mi = 107.8260 km
   From Diagram 5-A: 107.8260 km = 107.8260 X 10^3 m = 107,826.0 m
                ≈ 108,000 m
                (km to m by multiplying km value by 10^3 or by shifting
                its decimal 3 places to the right)

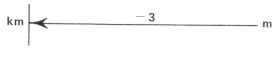

Diagram 5-B

Diagram 5-B is the opposite of Diagram 5-A, and is used to convert m to miles through the table. This diagram guides the conversion of a numerical value in m to km. The km equivalent is then applied to the REF column in the table to obtain mi. Again, this diagram also refers to velocity and acceleration values, as previously noted.

Example (b): Convert 999 m to mi.
   From Diagram 5-B: 999 m = 999 X 10^{-3} km = 0.999 km
   Since: 0.999 km = (0.990 + 0.009) km:
   From table: 99 km = 61.51575 mi*
              9 km = 5.592341 mi*
   By properly shifting decimals and adding:
       0.990 km = 0.6151575   mi
       0.009 km = 0.005592341 mi
       0.999 km = 0.6207498   mi ≈ 0.621 mi

---

*All conversion values obtained can be rounded to desired number of significant figures.

km km/h km/h²	REF	mi mi/h mi/h²	km km/h km/h²	REF	mi mi/h mi/h²
1·609344	1	0·6213712	82·07654	51	31·68993
3·218688	2	1·242742	83·68589	52	32·31130
4·828032	3	1·864114	85·29523	53	32·93267
6·437376	4	2·485485	86·90458	54	33·55404
8·046720	5	3·106856	88·51392	55	34·17542
9·656064	6	3·728227	90·12326	56	34·79679
11·26541	7	4·349598	91·73261	57	35·41816
12·87475	8	4·970969	93·34195	58	36·03953
14·48410	9	5·592341	94·95130	59	36·66090
16·09344	10	6·213712	96·56064	60	37·28227
17·70278	11	6·835083	98·16998	61	37·90364
19·31213	12	7·456454	99·77933	62	38·52501
20·92147	13	8·077825	101·3887	63	39·14638
22·53082	14	8·699197	102·9980	64	39·76776
24·14016	15	9·320568	104·6074	65	40·38913
25·74950	16	9·941939	106·2167	66	41·01050
27·35885	17	10·56331	107·8260	67	41·63187
28·96819	18	11·18468	109·4354	68	42·25324
30·57754	19	11·80605	111·0447	69	42·87461
32·18688	20	12·42742	112·6541	70	43·49598
33·79622	21	13·04879	114·2634	71	44·11735
35·40557	22	13·67017	115·8728	72	44·73873
37·01491	23	14·29154	117·4821	73	45·36010
38·62426	24	14·91291	119·0915	74	45·98147
40·23360	25	15·53428	120·7008	75	46·60284
41·84294	26	16·15565	122·3101	76	47·22421
43·45229	27	16·77702	123·9195	77	47·84558
45·06163	28	17·39839	125·5288	78	48·46695
46·67098	29	18·01976	127·1382	79	49·08832
48·28032	30	18·64114	128·7475	80	49·70969
49·88966	31	19·26251	130·3569	81	50·33107
51·49901	32	19·88388	131·9662	82	50·95244
53·10835	33	20·50525	133·5756	83	51·57381
54·71770	34	21·12662	135·1849	84	52·19518
56·32704	35	21·74799	136·7942	85	52·81655
57·93638	36	22·36936	138·4036	86	53·43792
59·54573	37	22·99073	140·0129	87	54·05929
61·15507	38	23·61211	141·6223	88	54·68066
62·76442	39	24·23348	143·2316	89	55·30204
64·37376	40	24·85485	144·8410	90	55·92341
65·98310	41	25·47622	146·4503	91	56·54478
67·59245	42	26·09759	148·0596	92	57·16615
69·20179	43	26·71896	149·6690	93	57·78752
70·81114	44	27·34033	151·2783	94	58·40889
72·42048	45	27·96170	152·8877	95	59·03026
74·02982	46	28·58307	154·4970	96	59·65163
75·63917	47	29·20445	156·1064	97	60·27301
77·24851	48	29·82582	157·7157	98	60·89438
78·85786	49	30·44719	159·3251	99	61·51575
80·46720	50	31·06856	160·9344	100	62·13712

6

AREA

Square-inch — Square-millimetre

$(\text{in.}^2 — \text{mm}^2)$

Square-inch — Square-centimetre

$(\text{in.}^2 — \text{cm}^2)$

Square-inch — Square-metre

$(\text{in.}^2 — \text{m}^2)$

KINEMATIC VISCOSITY

Square-inch/Second — Square-millimetre/Second

$(\text{in.}^2/\text{s} — \text{mm}^2/\text{s})$

Square-inch/Second — Square-metre/Second

$(\text{in.}^2/\text{s} — \text{m}^2/\text{s})$

Square-inch/Second — Stokes and Centistokes

$(\text{in.}^2/\text{s} — \text{St and cSt})$

Conversion Factors:

1 square-inch (in.^2) = 645.16 square-millimetres (mm^2) (exactly)

$(1 \text{ mm}^2 = 0.001550003 \text{ in.}^2)$

Kinematic Viscosity Values

Table 6 can be used for converting values in kinematic viscosity, which are represented in SI by the units m^2/s or mm^2/s. In applying kinematic viscosity conversions to this table, there must be a consistency in time units. This is shown above by the use of the per second in both English and SI. Also, conversions can be readily obtained between the 'cgs' (centimetre-gram-second) designation for kinematic viscosity, the stoke (St), and SI, by shifting decimal places. The cgs submultiple, centistoke (cSt) equals 1 mm^2/s. Therefore, centistoke values can be used with Table 6 in place of mm^2/s without any decimal shift.

Kinematic Viscosity Conversion Factors:

$$1 \text{ in.}^2/s = 645.16 \text{ mm}^2/s = 645.16 \text{ cSt}$$
$$1 \text{ St} = 1 \text{ cm}^2/s$$

In converting table values in mm^2 to cm^2 and m^2 quantities, as well as the non-SI units for kinematic viscosity, Diagram 6-A which appears in the headings on alternate pages of this table will assist the user. It is derived from:

$$1 \text{ mm}^2 = (10^{-2}) \text{ cm}^2 = (10^{-6}) \text{ m}^2 \text{ ; and } 1 \text{ mm}^2/s = (10^{-2}) \text{ cm}^2/s$$
$$= (10^{-2}) \text{ St} = 1 \text{ cSt.}$$

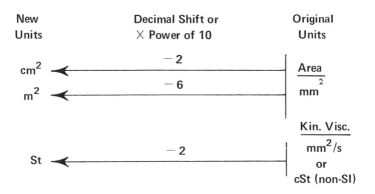

Diagram 6-A

Example (a): Convert 7.55 in.^{2} to mm^2, cm^2.
  Since: 7.55 in.^{2} = 755 × 10^{-2} in.^{2} :
  From table: 755 in.^{2} = 487095.8 mm^2 *
  Therefore: 755 × 10^{-2} in.^{2} = 487095.8 × 10^{-2} mm^2
          = 4870.958 mm^2 ≈ 4871 mm^2.
  From Diagram 6-A: 4870.958 mm^2 = 4870.958 × 10^{-2} cm^2
              = 48.70958 cm^2 ≈ 48.7 cm^2
              (mm^2 to cm^2 by multiplying mm^2 value by 10^{-2} or by shifting its decimal 2 places to the left).

Example (b): Convert 12 in.^{2}/s to St.
  From table: 12 in.^{2}/s = 7741.92 mm^2/s*
  From Diagram 6-B: 7741.92 mm^2/s = 7741.92 × 10^{-2} St
              = 77.4192 St ≈ 77 St.
              (mm^2/s to St by multiplying mm^2/s value by 10^{-2} or shifting its decimal 2 places to the left).

---

*All conversion values obtained can be rounded to desired number of significant figures.

Diagram 6-B is also given at the top of alternate table pages:

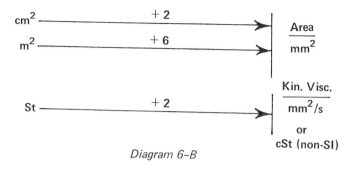

Diagram 6-B

This is the opposite of Diagram 6-A and is used to convert cm^2, m^2 values to their equivalent mm^2 quantities, and St to SI values, (mm^2/s), or their cgs submultiple (cSt). The mm^2 or mm^2/s equivalent is then applied to the REF column to obtain in.^{2} or in.$^2/s$.

Example (c): Convert 278 cm^2 to in.^{2}.
   From Diagram 6-B: 278 $cm^2 = 278 \times 10^2$ mm^2
   From table: 278 $mm^2 = 0.4309009$ in.^{2}.*
   Therefore: 278×10^2 $mm^2 = 0.4309009 \times 10^2$ in.$^2 = 43.09009$ in^2
       ≈ 43.1 in^2.

Example (d): Convert 95.5 St to in.$^2/s$.
   From Diagram 6-B: 95.5 St = 95.5×10^2 $mm^2/s = 955 \times 10$ mm^2/s
   From table: 955 $mm^2/s = 1.480253$ in.$^2/s$.*
   Therefore: 955×10 $mm^2/s = 1.480253 \times 10$ in.$^2/s$
       $= 14.80253$ in.$^2/s \approx 14.8$ in.$^2/s$.

---

*All conversion values obtained can be rounded to desired number of significant figures.

cm² ← —2

m² ← —6

St ← —2

Area / mm²

Kin. Vis. mm²/s or cSt (non-SI)

mm² mm²/s cSt	REF	in² in²/s	mm² mm²/s cSt	REF	in² in²/s
645.1600	1	0.001550003	32903.16	51	0.07905016
1290.320	2	0.003100006	33548.32	52	0.08060016
1935.480	3	0.004650009	34193.48	53	0.08215016
2580.640	4	0.006200012	34838.64	54	0.08370017
3225.800	5	0.007750015	35483.80	55	0.08525017
3870.960	6	0.009300019	36128.96	56	0.08680017
4516.120	7	0.01085002	36774.12	57	0.08835018
5161.280	8	0.01240002	37419.28	58	0.08990018
5806.440	9	0.01395003	38064.44	59	0.09145018
6451.600	10	0.01550003	38709.60	60	0.09300018
7096.760	11	0.01705003	39354.76	61	0.09455019
7741.920	12	0.01860004	39999.92	62	0.09610019
8387.080	13	0.02015004	40645.08	63	0.09765019
9032.240	14	0.02170004	41290.24	64	0.09920020
9677.400	15	0.02325005	41935.40	65	0.1007502
10322.56	16	0.02480005	42580.56	66	0.1023002
10967.72	17	0.02635005	43225.72	67	0.1038502
11612.88	18	0.02790006	43870.88	68	0.1054002
12258.04	19	0.02945006	44516.04	69	0.1069502
12903.20	20	0.03100006	45161.20	70	0.1085002
13548.36	21	0.03255006	45806.36	71	0.1100502
14193.52	22	0.03410007	46451.52	72	0.1116002
14838.68	23	0.03565007	47096.68	73	0.1131502
15483.84	24	0.03720007	47741.84	74	0.1147002
16129.00	25	0.03875008	48387.00	75	0.1162502
16774.16	26	0.04030008	49032.16	76	0.1178002
17419.32	27	0.04185008	49677.32	77	0.1193502
18064.48	28	0.04340009	50322.48	78	0.1209002
18709.64	29	0.04495009	50967.64	79	0.1224502
19354.80	30	0.04650009	51612.80	80	0.1240002
19999.96	31	0.04805010	52257.96	81	0.1255502
20645.12	32	0.04960010	52903.12	82	0.1271003
21290.28	33	0.05115010	53548.28	83	0.1286503
21935.44	34	0.05270011	54193.44	84	0.1302003
22580.60	35	0.05425011	54838.60	85	0.1317503
23225.76	36	0.05580011	55483.76	86	0.1333003
23870.92	37	0.05735011	56128.92	87	0.1348503
24516.08	38	0.05890012	56774.08	88	0.1364003
25161.24	39	0.06045012	57419.24	89	0.1379503
25806.40	40	0.06200012	58064.40	90	0.1395003
26451.56	41	0.06355013	58709.56	91	0.1410503
27096.72	42	0.06510013	59354.72	92	0.1426003
27741.88	43	0.06665013	59999.88	93	0.1441503
28387.04	44	0.06820014	60645.04	94	0.1457003
29032.20	45	0.06975014	61290.20	95	0.1472503
29677.36	46	0.07130014	61935.36	96	0.1488003
30322.52	47	0.07285015	62580.52	97	0.1503503
30967.68	48	0.07440015	63225.68	98	0.1519003
31612.84	49	0.07595015	63870.84	99	0.1534503
32258.00	50	0.07750015	64516.00	100	0.1550003

mm² mm²/s cSt	REF	in² in²/s	mm² mm²/s cSt	REF	in² in²/s
65161.16	101	0.1565503	97419.16	151	0.2340505
65806.32	102	0.1581003	98064.32	152	0.2356005
66451.48	103	0.1596503	98709.48	153	0.2371505
67096.64	104	0.1612003	99354.64	154	0.2387005
67741.80	105	0.1627503	99999.80	155	0.2402505
68386.96	106	0.1643003	100645.0	156	0.2418005
69032.12	107	0.1658503	101290.1	157	0.2433505
69677.28	108	0.1674003	101935.3	158	0.2449005
70322.44	109	0.1689503	102580.4	159	0.2464505
70967.60	110	0.1705003	103225.6	160	0.2480005
71612.76	111	0.1720503	103870.8	161	0.2495505
72257.92	112	0.1736003	104515.9	162	0.2511005
72903.08	113	0.1751503	105161.1	163	0.2526505
73548.24	114	0.1767004	105806.2	164	0.2542005
74193.40	115	0.1782504	106451.4	165	0.2557505
74838.56	116	0.1798004	107096.6	166	0.2573005
75483.72	117	0.1813504	107741.7	167	0.2588505
76128.88	118	0.1829004	108386.9	168	0.2604005
76774.04	119	0.1844504	109032.0	169	0.2619505
77419.20	120	0.1860004	109677.2	170	0.2635005
78064.36	121	0.1875504	110322.4	171	0.2650505
78709.52	122	0.1891004	110967.5	172	0.2666005
79354.68	123	0.1906504	111612.7	173	0.2681505
79999.84	124	0.1922004	112257.8	174	0.2697005
80645.00	125	0.1937504	112903.0	175	0.2712505
81290.16	126	0.1953004	113548.2	176	0.2728005
81935.32	127	0.1968504	114193.3	177	0.2743505
82580.48	128	0.1984004	114838.5	178	0.2759006
83225.64	129	0.1999504	115483.6	179	0.2774506
83870.80	130	0.2015004	116128.8	180	0.2790006
84515.96	131	0.2030504	116774.0	181	0.2805506
85161.12	132	0.2046004	117419.1	182	0.2821006
85806.28	133	0.2061504	118064.3	183	0.2836506
86451.44	134	0.2077004	118709.4	184	0.2852006
87096.60	135	0.2092504	119354.6	185	0.2867506
87741.76	136	0.2108004	119999.8	186	0.2883006
88386.92	137	0.2123504	120644.9	187	0.2898506
89032.08	138	0.2139004	121290.1	188	0.2914006
89677.24	139	0.2154504	121935.2	189	0.2929506
90322.40	140	0.2170004	122580.4	190	0.2945006
90967.56	141	0.2185504	123225.6	191	0.2960506
91612.72	142	0.2201004	123870.7	192	0.2976006
92257.88	143	0.2216504	124515.9	193	0.2991506
92903.04	144	0.2232004	125161.0	194	0.3007006
93548.20	145	0.2247504	125806.2	195	0.3022506
94193.36	146	0.2263005	126451.4	196	0.3038006
94838.52	147	0.2278505	127096.5	197	0.3053506
95483.68	148	0.2294005	127741.7	198	0.3069006
96128.84	149	0.2309505	128386.8	199	0.3084506
96774.00	150	0.2325005	129032.0	200	0.3100006

cm² ◄————— −2 ————— | Area
m² ◄——————— −6 ————— | mm²

Kin. Vis.
St ◄————————— −2 ————— | mm²/s
                                                     or
                                                cSt (non-SI)

mm²	in²	mm²	in²
mm²/s	in²/s	mm²/s	in²/s
cSt REF		cSt REF	
129677.2 201	0.3115506	161935.2 251	0.3890508
130322.3 202	0.3131006	162580.3 252	0.3906008
130967.5 203	0.3146506	163225.5 253	0.3921508
131612.6 204	0.3162006	163870.6 254	0.3937008
132257.8 205	0.3177506	164515.8 255	0.3952508
132903.0 206	0.3193006	165161.0 256	0.3968008
133548.1 207	0.3208506	165806.1 257	0.3983508
134193.3 208	0.3224006	166451.3 258	0.3999008
134838.4 209	0.3239506	167096.4 259	0.4014508
135483.6 210	0.3255006	167741.6 260	0.4030008
136128.8 211	0.3270507	168386.8 261	0.4045508
136773.9 212	0.3286007	169031.9 262	0.4061008
137419.1 213	0.3301507	169677.1 263	0.4076508
138064.2 214	0.3317007	170322.2 264	0.4092008
138709.4 215	0.3332507	170967.4 265	0.4107508
139354.6 216	0.3348007	171612.6 266	0.4123008
139999.7 217	0.3363507	172257.7 267	0.4138508
140644.9 218	0.3379007	172902.9 268	0.4154008
141290.0 219	0.3394507	173548.0 269	0.4169508
141935.2 220	0.3410007	174193.2 270	0.4185008
142580.4 221	0.3425507	174838.4 271	0.4200508
143225.5 222	0.3441007	175483.5 272	0.4216008
143870.7 223	0.3456507	176128.7 273	0.4231508
144515.8 224	0.3472007	176773.8 274	0.4247008
145161.0 225	0.3487507	177419.0 275	0.4262508
145806.2 226	0.3503007	178064.2 276	0.4278009
146451.3 227	0.3518507	178709.3 277	0.4293509
147096.5 228	0.3534007	179354.5 278	0.4309009
147741.6 229	0.3549507	179999.6 279	0.4324509
148386.8 230	0.3565007	180644.8 280	0.4340009
149032.0 231	0.3580507	181290.0 281	0.4355509
149677.1 232	0.3596007	181935.1 282	0.4371009
150322.3 233	0.3611507	182580.3 283	0.4386509
150967.4 234	0.3627007	183225.4 284	0.4402009
151612.6 235	0.3642507	183870.6 285	0.4417509
152257.8 236	0.3658007	184515.8 286	0.4433009
152902.9 237	0.3673507	185160.9 287	0.4448509
153548.1 238	0.3689007	185806.1 288	0.4464009
154193.2 239	0.3704507	186451.2 289	0.4479509
154838.4 240	0.3720007	187096.4 290	0.4495009
155483.6 241	0.3735507	187741.6 291	0.4510509
156128.7 242	0.3751007	188386.7 292	0.4526009
156773.9 243	0.3766508	189031.9 293	0.4541509
157419.0 244	0.3782008	189677.0 294	0.4557009
158064.2 245	0.3797508	190322.2 295	0.4572509
158709.4 246	0.3813008	190967.4 296	0.4588009
159354.5 247	0.3828508	191612.5 297	0.4603509
159999.7 248	0.3844008	192257.7 298	0.4619009
160644.8 249	0.3859508	192902.8 299	0.4634509
161290.0 250	0.3875008	193548.0 300	0.4650009

cm^2 —— + 2 —→ $\dfrac{\text{Area}}{\text{mm}^2}$

m^2 —— + 6 —→

St —— + 2 —→ $\dfrac{\text{Kin. Vis.}}{\text{mm}^2/\text{s}}$ or cSt (non-SI)

mm^2 mm^2/s cSt	REF	in^2 in^2/s	mm^2 mm^2/s cSt	REF	in^2 in^2/s
194193.2	301	0.4665509	226451.2	351	0.5440511
194838.3	302	0.4681009	227096.3	352	0.5456011
195483.5	303	0.4696509	227741.5	353	0.5471511
196128.6	304	0.4712009	228386.6	354	0.5487011
196773.8	305	0.4727509	229031.8	355	0.5502511
197419.0	306	0.4743009	229677.0	356	0.5518011
198064.1	307	0.4758509	230322.1	357	0.5533511
198709.3	308	0.4774010	230967.3	358	0.5549011
199354.4	309	0.4789510	231612.4	359	0.5564511
199999.6	310	0.4805010	232257.6	360	0.5580011
200644.8	311	0.4820510	232902.8	361	0.5595511
201289.9	312	0.4836010	233547.9	362	0.5611011
201935.1	313	0.4851510	234193.1	363	0.5626511
202580.2	314	0.4867010	234838.2	364	0.5642011
203225.4	315	0.4882510	235483.4	365	0.5657511
203870.6	316	0.4898010	236128.6	366	0.5673011
204515.7	317	0.4913510	236773.7	367	0.5688511
205160.9	318	0.4929010	237418.9	368	0.5704011
205806.0	319	0.4944510	238064.0	369	0.5719511
206451.2	320	0.4960010	238709.2	370	0.5735011
207096.4	321	0.4975510	239354.4	371	0.5750511
207741.5	322	0.4991010	239999.5	372	0.5766011
208386.7	323	0.5006510	240644.7	373	0.5781512
209031.8	324	0.5022010	241289.8	374	0.5797012
209677.0	325	0.5037510	241935.0	375	0.5812512
210322.2	326	0.5053010	242580.2	376	0.5828012
210967.3	327	0.5068510	243225.3	377	0.5843512
211612.5	328	0.5084010	243870.5	378	0.5859012
212257.6	329	0.5099510	244515.6	379	0.5874512
212902.8	330	0.5115010	245160.8	380	0.5890012
213548.0	331	0.5130510	245806.0	381	0.5905512
214193.1	332	0.5146010	246451.1	382	0.5921012
214838.3	333	0.5161510	247096.3	383	0.5936512
215483.4	334	0.5177010	247741.4	384	0.5952012
216128.6	335	0.5192510	248386.6	385	0.5967512
216773.8	336	0.5208010	249031.8	386	0.5983012
217418.9	337	0.5223510	249676.9	387	0.5998512
218064.1	338	0.5239010	250322.1	388	0.6014012
218709.2	339	0.5254510	250967.2	389	0.6029512
219354.4	340	0.5270010	251612.4	390	0.6045012
219999.6	341	0.5285510	252257.6	391	0.6060512
220644.7	342	0.5301011	252902.7	392	0.6076012
221289.9	343	0.5316511	253547.9	393	0.6091512
221935.0	344	0.5332011	254193.0	394	0.6107012
222580.2	345	0.5347511	254838.2	395	0.6122512
223225.4	346	0.5363011	255483.4	396	0.6138012
223870.5	347	0.5378511	256128.5	397	0.6153512
224515.7	348	0.5394011	256773.7	398	0.6169012
225160.8	349	0.5409511	257418.8	399	0.6184512
225806.0	350	0.5425011	258064.0	400	0.6200012

			Area mm²
cm² ←	−2		
m² ←	−6		Kin. Vis. mm²/s
St ←	−2		or cSt (non-SI)

mm² mm²/s cSt	REF	in² in²/s	mm² mm²/s cSt	REF	in² in²/s
258709.2	401	0.6215512	290967.2	451	0.6990514
259354.3	402	0.6231012	291612.3	452	0.7006014
259999.5	403	0.6246512	292257.5	453	0.7021514
260644.6	404	0.6262012	292902.6	454	0.7037014
261289.8	405	0.6277512	293547.8	455	0.7052514
261935.0	406	0.6293012	294193.0	456	0.7068014
262580.1	407	0.6308513	294838.1	457	0.7083514
263225.3	408	0.6324013	295483.3	458	0.7099014
263870.4	409	0.6339513	296128.4	459	0.7114514
264515.6	410	0.6355013	296773.6	460	0.7130014
265160.8	411	0.6370513	297418.8	461	0.7145514
265805.9	412	0.6386013	298063.9	462	0.7161014
266451.1	413	0.6401513	298709.1	463	0.7176514
267096.2	414	0.6417013	299354.2	464	0.7192014
267741.4	415	0.6432513	299999.4	465	0.7207514
268386.6	416	0.6448013	300644.6	466	0.7223014
269031.7	417	0.6463513	301289.7	467	0.7238514
269676.9	418	0.6479013	301934.9	468	0.7254014
270322.0	419	0.6494513	302580.0	469	0.7269515
270967.2	420	0.6510013	303225.2	470	0.7285015
271612.4	421	0.6525513	303870.4	471	0.7300515
272257.5	422	0.6541013	304515.5	472	0.7316015
272902.7	423	0.6556513	305160.7	473	0.7331515
273547.8	424	0.6572013	305805.8	474	0.7347015
274193.0	425	0.6587513	306451.0	475	0.7362515
274838.2	426	0.6603013	307096.2	476	0.7378015
275483.3	427	0.6618513	307741.3	477	0.7393515
276128.5	428	0.6634013	308386.5	478	0.7409015
276773.6	429	0.6649513	309031.6	479	0.7424515
277418.8	430	0.6665013	309676.8	480	0.7440015
278064.0	431	0.6680513	310322.0	481	0.7455515
278709.1	432	0.6696013	310967.1	482	0.7471015
279354.3	433	0.6711513	311612.3	483	0.7486515
279999.4	434	0.6727013	312257.4	484	0.7502015
280644.6	435	0.6742513	312902.6	485	0.7517515
281289.8	436	0.6758013	313547.8	486	0.7533015
281934.9	437	0.6773513	314192.9	487	0.7548515
282580.1	438	0.6789014	314838.1	488	0.7564015
283225.2	439	0.6804514	315483.2	489	0.7579515
283870.4	440	0.6820014	316128.4	490	0.7595015
284515.6	441	0.6835514	316773.6	491	0.7610515
285160.7	442	0.6851014	317418.7	492	0.7626015
285805.9	443	0.6866514	318063.9	493	0.7641515
286451.0	444	0.6882014	318709.0	494	0.7657015
287096.2	445	0.6897514	319354.2	495	0.7672515
287741.4	446	0.6913014	319999.4	496	0.7688015
288386.5	447	0.6928514	320644.5	497	0.7703515
289031.7	448	0.6944014	321289.7	498	0.7719015
289676.8	449	0.6959514	321934.8	499	0.7734515
290322.0	450	0.6975014	322580.0	500	0.7750015

mm²			mm²		
mm²/s		in²	mm²/s		in²
cSt	REF	in²/s	cSt	REF	in²/s
323225.2	501	0.7765515	355483.2	551	0.8540517
323870.3	502	0.7781015	356128.3	552	0.8556017
324515.5	503	0.7796516	356773.5	553	0.8571517
325160.6	504	0.7812016	357418.6	554	0.8587017
325805.8	505	0.7827516	358063.8	555	0.8602517
326451.0	506	0.7843016	358709.0	556	0.8618017
327096.1	507	0.7858516	359354.1	557	0.8633517
327741.3	508	0.7874016	359999.3	558	0.8649017
328386.4	509	0.7889516	360644.4	559	0.8664517
329031.6	510	0.7905016	361289.6	560	0.8680017
329676.8	511	0.7920516	361934.8	561	0.8695517
330321.9	512	0.7936016	362579.9	562	0.8711017
330967.1	513	0.7951516	363225.1	563	0.8726517
331612.2	514	0.7967016	363870.2	564	0.8742017
332257.4	515	0.7982516	364515.4	565	0.8757517
332902.6	516	0.7998016	365160.6	566	0.8773017
333547.7	517	0.8013516	365805.7	567	0.8788517
334192.9	518	0.8029016	366450.9	568	0.8804018
334838.0	519	0.8044516	367096.0	569	0.8819518
335483.2	520	0.8060016	367741.2	570	0.8835018
336128.4	521	0.8075516	368386.4	571	0.8850518
336773.5	522	0.8091016	369031.5	572	0.8866018
337418.7	523	0.8106516	369676.7	573	0.8881518
338063.8	524	0.8122016	370321.8	574	0.8897018
338709.0	525	0.8137516	370967.0	575	0.8912518
339354.2	526	0.8153016	371612.2	576	0.8928018
339999.3	527	0.8168516	372257.3	577	0.8943518
340644.5	528	0.8184016	372902.5	578	0.8959018
341289.6	529	0.8199516	373547.6	579	0.8974518
341934.8	530	0.8215016	374192.8	580	0.8990018
342580.0	531	0.8230516	374838.0	581	0.9005518
343225.1	532	0.8246016	375483.1	582	0.9021018
343870.3	533	0.8261516	376128.3	583	0.9036518
344515.4	534	0.8277017	376773.4	584	0.9052018
345160.6	535	0.8292517	377418.6	585	0.9067518
345805.8	536	0.8308017	378063.8	586	0.9083018
346450.9	537	0.8323517	378708.9	587	0.9098518
347096.1	538	0.8339017	379354.1	588	0.9114018
347741.2	539	0.8354517	379999.2	589	0.9129518
348386.4	540	0.8370017	380644.4	590	0.9145018
349031.6	541	0.8385517	381289.6	591	0.9160518
349676.7	542	0.8401017	381934.7	592	0.9176018
350321.9	543	0.8416517	382579.9	593	0.9191518
350967.0	544	0.8432017	383225.0	594	0.9207018
351612.2	545	0.8447517	383870.2	595	0.9222518
352257.4	546	0.8463017	384515.4	596	0.9238018
352902.5	547	0.8478517	385160.5	597	0.9253518
353547.7	548	0.8494017	385805.7	598	0.9269018
354192.8	549	0.8509517	386450.8	599	0.9284519
354838.0	550	0.8525017	387096.0	600	0.9300019

mm²	in²	mm²	in²
mm²/s cSt	REF	in²/s	mm²/s cSt	REF	in²/s

mm²/s cSt	REF	in²/s	mm²/s cSt	REF	in²/s
387741.2	601	0.9315519	419999.2	651	1.009052
388386.3	602	0.9331019	420644.3	652	1.010602
389031.5	603	0.9346519	421289.5	653	1.012152
389676.6	604	0.9362019	421934.6	654	1.013702
390321.8	605	0.9377519	422579.8	655	1.015252
390967.0	606	0.9393019	423225.0	656	1.016802
391612.1	607	0.9408519	423870.1	657	1.018352
392257.3	608	0.9424019	424515.3	658	1.019902
392902.4	609	0.9439519	425160.4	659	1.021452
393547.6	610	0.9455019	425805.6	660	1.023002
394192.8	611	0.9470519	426450.8	661	1.024552
394837.9	612	0.9486019	427095.9	662	1.026102
395483.1	613	0.9501519	427741.1	663	1.027652
396128.2	614	0.9517019	428386.2	664	1.029202
396773.4	615	0.9532519	429031.4	665	1.030752
397418.6	616	0.9548019	429676.6	666	1.032302
398063.7	617	0.9563519	430321.7	667	1.033852
398708.9	618	0.9579019	430966.9	668	1.035402
399354.0	619	0.9594519	431612.0	669	1.036952
399999.2	620	0.9610019	432257.2	670	1.038502
400644.4	621	0.9625519	432902.4	671	1.040052
401289.5	622	0.9641019	433547.5	672	1.041602
401934.7	623	0.9656519	434192.7	673	1.043152
402579.8	624	0.9672019	434837.8	674	1.044702
403225.0	625	0.9687519	435483.0	675	1.046252
403870.2	626	0.9703019	436128.2	676	1.047802
404515.3	627	0.9718519	436773.3	677	1.049352
405160.5	628	0.9734019	437418.5	678	1.050902
405805.6	629	0.9749519	438063.6	679	1.052452
406450.8	630	0.9765019	438708.8	680	1.054002
407096.0	631	0.9780519	439354.0	681	1.055552
407741.1	632	0.9796019	439999.1	682	1.057102
408386.3	633	0.9811520	440644.3	683	1.058652
409031.4	634	0.9827020	441289.4	684	1.060202
409676.6	635	0.9842520	441934.6	685	1.061752
410321.8	636	0.9858020	442579.8	686	1.063302
410966.9	637	0.9873520	443224.9	687	1.064852
411612.1	638	0.9889020	443870.1	688	1.066402
412257.2	639	0.9904520	444515.2	689	1.067952
412902.4	640	0.9920020	445160.4	690	1.069502
413547.6	641	0.9935520	445805.6	691	1.071052
414192.7	642	0.9951020	446450.7	692	1.072602
414837.9	643	0.9966520	447095.9	693	1.074152
415483.0	644	0.9982020	447741.0	694	1.075702
416128.2	645	0.9997520	448386.2	695	1.077252
416773.4	646	1.001302	449031.4	696	1.078802
417418.5	647	1.002852	449676.5	697	1.080352
418063.7	648	1.004402	450321.7	698	1.081902
418708.8	649	1.005952	450966.8	699	1.083452
419354.0	650	1.007502	451612.0	700	1.085002

cm² ———— + 2 ————→ Area
mm²

m² ———— + 6 ————→

Kin. Vis.

St ———— + 2 ————→ mm²/s
or
cSt (non-SI)

mm² mm²/s cSt	REF	in² in²/s	mm² mm²/s cSt	REF	in² in²/s
452257·2	701	1·086552	484515·2	751	1·164052
452902·3	702	1·088102	485160·3	752	1·165602
453547·5	703	1·089652	485805·5	753	1·167152
454192·6	704	1·091202	486450·6	754	1·168702
454837·8	705	1·092752	487095·8	755	1·170252
455483·0	706	1·094302	487741·0	756	1·171802
456128·1	707	1·095852	488386·1	757	1·173352
456773·3	708	1·097402	489031·3	758	1·174902
457418·4	709	1·098952	489676·4	759	1·176452
458063·6	710	1·100502	490321·6	760	1·178002
458708·8	711	1·102052	490966·8	761	1·179552
459353·9	712	1·103602	491611·9	762	1·181102
459999·1	713	1·105152	492257·1	763	1·182652
460644·2	714	1·106702	492902·2	764	1·184202
461289·4	715	1·108252	493547·4	765	1·185752
461934·6	716	1·109802	494192·6	766	1·187302
462579·7	717	1·111352	494837·7	767	1·188852
463224·9	718	1·112902	495482·9	768	1·190402
463870·0	719	1·114452	496128·0	769	1·191952
464515·2	720	1·116002	496773·2	770	1·193502
465160·4	721	1·117552	497418·4	771	1·195052
465805·5	722	1·119102	498063·5	772	1·196602
466450·7	723	1·120652	498708·7	773	1·198152
467095·8	724	1·122202	499353·8	774	1·199702
467741·0	725	1·123752	499999·0	775	1·201252
468386·2	726	1·125302	500644·2	776	1·202802
469031·3	727	1·126852	501289·3	777	1·204352
469676·5	728	1·128402	501934·5	778	1·205902
470321·6	729	1·129952	502579·6	779	1·207452
470966·8	730	1·131502	503224·8	780	1·209002
471612·0	731	1·133052	503870·0	781	1·210552
472257·1	732	1·134602	504515·1	782	1·212102
472902·3	733	1·136152	505160·3	783	1·213652
473547·4	734	1·137702	505805·4	784	1·215202
474192·6	735	1·139252	506450·6	785	1·216752
474837·8	736	1·140802	507095·8	786	1·218302
475482·9	737	1·142352	507740·9	787	1·219852
476128·1	738	1·143902	508386·1	788	1·221402
476773·2	739	1·145452	509031·2	789	1·222952
477418·4	740	1·147002	509676·4	790	1·224502
478063·6	741	1·148552	510321·6	791	1·226052
478708·7	742	1·150102	510966·7	792	1·227602
479353·9	743	1·151652	511611·9	793	1·229152
479999·0	744	1·153202	512257·0	794	1·230702
480644·2	745	1·154752	512902·2	795	1·232252
481289·4	746	1·156302	513547·4	796	1·233802
481934·5	747	1·157852	514192·5	797	1·235352
482579·7	748	1·159402	514837·7	798	1·236902
483224·8	749	1·160952	515482·8	799	1·238452
483870·0	750	1·162502	516128·0	800	1·240002

cm² ←———— −2 ————┐
                          ┌─────────
m² ←———— −6 ————┤   Area
                          │   mm²
                          │ Kin. Vis.
St ←———— −2 ————┘   mm²/s
                              or
                          cSt (non-SI)

mm² mm²/s cSt	REF	in² in²/s	mm² mm²/s cSt	REF	in² in²/s
516773.2	801	1.241552	549031.2	851	1.319053
517418.3	802	1.243102	549676.3	852	1.320603
518063.5	803	1.244652	550321.5	853	1.322153
518708.6	804	1.246202	550966.6	854	1.323703
519353.8	805	1.247752	551611.8	855	1.325253
519999.0	806	1.249302	552257.0	856	1.326803
520644.1	807	1.250852	552902.1	857	1.328353
521289.3	808	1.252402	553547.3	858	1.329903
521934.4	809	1.253953	554192.4	859	1.331453
522579.6	810	1.255502	554837.6	860	1.333003
523224.8	811	1.257052	555482.8	861	1.334553
523869.9	812	1.258602	556127.9	862	1.336103
524515.1	813	1.260153	556773.1	863	1.337653
525160.2	814	1.261703	557418.2	864	1.339203
525805.4	815	1.263253	558063.4	865	1.340753
526450.6	816	1.264803	558708.6	866	1.342303
527095.7	817	1.266353	559353.7	867	1.343853
527740.9	818	1.267903	559998.9	868	1.345403
528386.0	819	1.269453	560644.0	869	1.346953
529031.2	820	1.271003	561289.2	870	1.348503
529676.4	821	1.272553	561934.4	871	1.350053
530321.5	822	1.274103	562579.5	872	1.351603
530966.7	823	1.275653	563224.7	873	1.353153
531611.8	824	1.277203	563869.8	874	1.354703
532257.0	825	1.278753	564515.0	875	1.356253
532902.2	826	1.280303	565160.2	876	1.357803
533547.3	827	1.281853	565805.3	877	1.359353
534192.5	828	1.283403	566450.5	878	1.360903
534837.6	829	1.284953	567095.6	879	1.362453
535482.8	830	1.286503	567740.8	880	1.364003
536128.0	831	1.288053	568386.0	881	1.365553
536773.1	832	1.289603	569031.1	882	1.367103
537418.3	833	1.291153	569676.3	883	1.368653
538063.4	834	1.292703	570321.4	884	1.370203
538708.6	835	1.294253	570966.6	885	1.371753
539353.8	836	1.295803	571611.8	886	1.373303
539998.9	837	1.297353	572256.9	887	1.374853
540644.1	838	1.298903	572902.1	888	1.376403
541289.2	839	1.300453	573547.2	889	1.377953
541934.4	840	1.302003	574192.4	890	1.379503
542579.6	841	1.303553	574837.6	891	1.381053
543224.7	842	1.305103	575482.7	892	1.382603
543869.9	843	1.306653	576127.9	893	1.384153
544515.0	844	1.308203	576773.0	894	1.385703
545160.2	845	1.309753	577418.2	895	1.387253
545805.4	846	1.311303	578063.4	896	1.388803
546450.5	847	1.312853	578708.5	897	1.390353
547095.7	848	1.314403	579353.7	898	1.391903
547740.8	849	1.315953	579998.8	899	1.393453
548386.0	850	1.317503	580644.0	900	1.395003

		+ 2		
cm^2		→		**Area**
m^2		+ 6 →		**mm^2**
				Kin. Vis.
St		+ 2 →		**mm^2/s** or cSt (non-SI)

mm^2		in^2	mm^2		in^2
mm^2/s cSt	REF	in^2/s	mm^2/s cSt	REF	in^2/s
581289.2	901	1.396553	613547.2	951	1.474053
581934.3	902	1.398103	614192.3	952	1.475603
582579.5	903	1.399653	614837.5	953	1.477153
583224.6	904	1.401203	615482.6	954	1.478703
583869.8	905	1.402753	616127.8	955	1.480253
584515.0	906	1.404303	616773.0	956	1.481803
585160.1	907	1.405853	617418.1	957	1.483353
585805.3	908	1.407403	618063.3	958	1.484903
586450.4	909	1.408953	618708.4	959	1.486453
587095.6	910	1.410503	619353.6	960	1.488003
587740.8	911	1.412053	619998.8	961	1.489553
588385.9	912	1.413603	620643.9	962	1.491103
589031.1	913	1.415153	621289.1	963	1.492653
589676.2	914	1.416703	621934.2	964	1.494203
590321.4	915	1.418253	622579.4	965	1.495753
590966.6	916	1.419803	623224.6	966	1.497303
591611.7	917	1.421353	623869.7	967	1.498853
592256.9	918	1.422903	624514.9	968	1.500403
592902.0	919	1.424453	625160.0	969	1.501953
593547.2	920	1.426003	625805.2	970	1.503503
594192.4	921	1.427553	626450.4	971	1.505053
594837.5	922	1.429103	627095.5	972	1.506603
595482.7	923	1.430653	627740.7	973	1.508153
596127.8	924	1.432203	628385.8	974	1.509703
596773.0	925	1.433753	629031.0	975	1.511253
597418.2	926	1.435303	629676.2	976	1.512803
598063.3	927	1.436853	630321.3	977	1.514353
598708.5	928	1.438403	630966.5	978	1.515903
599353.6	929	1.439953	631611.6	979	1.517453
599998.8	930	1.441503	632256.8	980	1.519003
600644.0	931	1.443053	632902.0	981	1.520553
601289.1	932	1.444603	633547.1	982	1.522103
601934.3	933	1.446153	634192.3	983	1.523653
602579.4	934	1.447703	634837.4	984	1.525203
603224.6	935	1.449253	635482.6	985	1.526753
603869.8	936	1.450803	636127.8	986	1.528303
604514.9	937	1.452353	636772.9	987	1.529853
605160.1	938	1.453903	637418.1	988	1.531403
605805.2	939	1.455453	638063.2	989	1.532953
606450.4	940	1.457003	638708.4	990	1.534503
607095.6	941	1.458553	639353.6	991	1.536053
607740.7	942	1.460103	639998.7	992	1.537603
608385.9	943	1.461653	640643.9	993	1.539153
609031.0	944	1.463203	641289.0	994	1.540703
609676.2	945	1.464753	641934.2	995	1.542253
610321.4	946	1.466303	642579.4	996	1.543803
610966.5	947	1.467853	643224.5	997	1.545353
611611.7	948	1.469403	643869.7	998	1.546903
612256.8	949	1.470953	644514.8	999	1.548453
612902.0	950	1.472503	645160.0	1000	1.550003

AREA

Square-foot — Square-metre
$$(\text{ft}^2 - \text{m}^2)$$

Square-foot — Square-kilometre
$$(\text{ft}^2 - \text{km}^2)$$

Square-foot — Square-centimetre
$$(\text{ft}^2 - \text{cm}^2)$$

Square-foot — Square-millimetre
$$(\text{ft}^2 - \text{mm}^2)$$

Kinematic Viscosity

Square-foot/Second — Square-metre/Second
$$(\text{ft}^2/\text{s} - \text{m}^2/\text{s})$$

Square-foot/Second — Square-millimetre/Second
$$(\text{ft}^2/\text{s} - \text{mm}^2/\text{s})$$

Square-foot/Second — Stoke and Centistoke
$$(\text{ft}^2/\text{s} - \text{St and cSt})$$

Conversion Factors:
$$1 \text{ square-foot (ft}^2) = 0.09290304 \text{ square-metre (m}^2) \text{ (exactly)}$$
$$(1 \text{ m}^2 = 10.76391 \text{ ft}^2)$$

Kinematic Viscosity Values

Table 7 can be used for converting values in kinematic viscosity, which are represented in SI by the units m^2/s or mm^2/s. In applying kinematic viscosity conversions to this table, there must be a consistency in time units. This is shown above by the use of the per second in both English and SI. Also, conversions can be readily obtained between the 'cgs' (centimetre-gram-second) designa-

70

tion for kinematic viscosity, the stoke (St), and SI by shifting decimal places. The cgs submultiple, centistoke (cSt) equals 1 mm²/s, or (10^{-6}) m²/s.

Kinematic Viscosity Conversion Factors:
$$1 \text{ ft}^2/\text{s} = 0.09290304 \text{ m}^2/\text{s}$$
$$1 \text{ St} = 1 \text{ cm}^2/\text{s} = 0.0001 \text{ m}^2/\text{s}$$

In converting m² values obtained from the table to km², cm², mm², quantities, as well as to non-SI unit values for kinematic viscosity, Diagram 7-A, which appears in the headings on alternate pages of this table, will assist the user. It is derived from: $1 \text{ m}^2 = (10^{-6}) \text{ km}^2 = (10^4) \text{ cm}^2 = (10^6) \text{ mm}^2$; and $1 \text{ m}^2/\text{s} = (10^4) \text{ cm}^2/\text{s} = (10^6) \text{ mm}^2/\text{s} = (10^6) \text{ cSt} = (10^4) \text{ St}$.

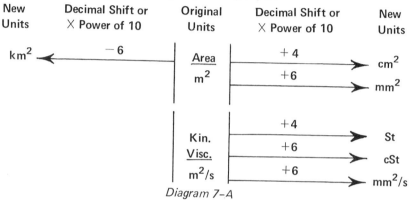

Diagram 7-A

Example (a): Convert 4,590,000 ft² to m² ; km² .
  Since: 4,590,000 ft² = 459 × 10⁴ ft² :
  From table: 459 ft² = 42.64250 m² *
  Therefore: 459 × 10⁴ ft² = 42.64250 × 10⁴ m² = 426,425.0 m²
       ≈ 426,000 m² .
  From Diagram 7-A: 426,425.0 m² = 426,425.0 (10^{-6}) km² = 0.426425 km²
       ≈ 0.426 km²
       (m² to km² by multiplying m² value by 10^{-6} or by shifting its decimal 6 places to the left)

Example (b): Convert 0.0459 ft² to cm² ; mm² .
  Since: 0.0459 ft² = 459 × 10⁻⁴ ft² :
  Again from table: 459 ft² = 42.64250 m² *
  Therefore: 459 × 10⁻⁴ ft² = 42.64250 × 10⁻⁴ m² = 0.004264250 m². *
  From Diagram 7-A: a) 0.004264250 m² = 0.004264250 (10^4)
       = 42.64250 cm² ≈ 42.6 cm²
       (m² to cm² by multiplying m² value by (10^4) or by shifting its decimal 4 places to the right)

---

*All conversion values obtained can be rounded to desired number of significant figures.

b) $0.004264250 \text{ m}^2 = 0.004264250 \, (10^6) \text{ mm}^2$
$= 4,264.250 \text{ mm}^2 \approx 4,260 \text{ mm}^2$
(m^2 to mm^2 by multiplying m^2 value by 10^6 or by shifting its decimal 6 places to the right)

Example (*c*): Convert $9.5 \times 10^{-5} \text{ ft}^2/\text{s}$ to m^2/s; St; cSt.
  Since: $9.5 \times 10^{-5} \text{ ft}^2/\text{s} = 95 \times 10^{-6} \text{ ft}^2/\text{s}$:
  From table: $95 \text{ ft}^2/\text{sec} = 8.825789 \text{ m}^2/\text{s}*$
  Therefore: $9 \, 5 \times 10^{-6} \text{ ft}^2/\text{sec} = 8.825789 \times 10^{-6} \text{ m}^2/\text{s} \approx 8.8 \times 10^{-6} \text{ m}^2/\text{s}$.
  From Diagram 7-A: a) $8.825789 \times 10^{-6} \text{ m}^2/\text{s} = 8.825789 \times 10^{-6} \, (10^4) \text{ St}$
$= 8.825789 \times 10^{-2} \text{ St or}$
$0.08825789 \text{ St} \approx 0.088 \text{ St}$

b) $8.825789 \times 10^{-6} \text{ m}^2/\text{s} = 8.825789 \times 10^{-6} \, (10^6) \text{ cSt}$
$= 8.825789 \text{ cSt} \approx 8.8 \text{ cSt}$

Diagram 7-B is also given at the top of alternate table pages:

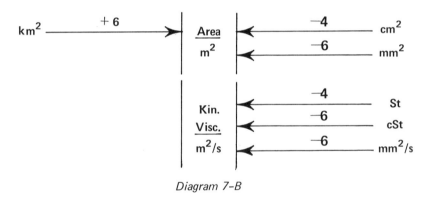

Diagram 7-B

This is the opposite of Diagram 7-A and is used to convert values containing km^2, cm^2, mm^2, to those with m^2, and the St and cSt values to m^2/s. The m^2 or m^2/s equivalent is then applied to the REF column of the table to obtain ft^2 or ft^2/s.

Example (*d*): Convert 0.333 km^2 to ft^2.
  From Diagram 7-B: $0.333 \text{ km}^2 = 0.333 \times 10^6 \text{ m}^2 = 333 \times 10^3 \text{ m}^2$
$= 333,000 \text{ m}^2$
  From table: $333 \text{ m}^2 = 3584.382 \text{ ft}^2 *$
  Therefore: $333 \times 10^3 \text{ m}^2 = 3,584.382 \times 10^3 \text{ ft}^2 = 3,584,382 \text{ ft}^2$
$\approx 3.58 \times 10^6 \text{ ft}^2$.

---

*All conversion values obtained can be rounded to desired number of significant figures.

Example (*e*): Convert 899 cm² to ft².
From Diagram 7-B: $899 \text{ cm}^2 = 899 \times 10^{-4} \text{ m}^2$
From table: $899 \text{ m}^2 = 9{,}676.755 \text{ ft}^2$ *
Therefore: $899 \times 10^{-4} \text{ m}^2 = 9{,}676.755 \times 10^{-4} \text{ ft}^2 = 0.9676755 \text{ ft}^2$
$\approx 0.968 \text{ ft}^2$.

Example (*f*): Convert 6.5 cSt to ft²/s.
From Diagram 7-B: $6.5 \text{ cSt} = 6.5 \times 10^{-6} \text{ m}^2/\text{s} = 65 \times 10^{-7} \text{ m}^2/\text{s}$
From table: $65 \text{ m}^2/\text{s} = 699.6542 \text{ ft}^2/\text{s}$*
Therefore: $65 \times 10^{-7} \text{ m}^2/\text{s} = 699.6542 \times 10^{-7} \text{ ft}^2/\text{s} \approx 7 \times 10^{-5} \text{ ft}^2/\text{s}$.

---

*All conversion values obtained can be rounded to desired number of significant figures.

	-6		Area $\dfrac{}{m^2}$	$+4$	cm^2
$km^2 \longleftarrow$				$+6$	mm^2

Kin. Vis. $\dfrac{}{m^2/s}$	$+4$	St
	$+6$	cSt
	$+6$	mm^2/s

m^2 m^2/s	REF	ft^2 ft^2/s	m^2 m^2/s	REF	ft^2 ft^2/s
0.09290304	1	10.76391	4.738055	51	548.9594
0.1858061	2	21.52782	4.830958	52	559.7233
0.2787091	3	32.29173	4.923861	53	570.4873
0.3716122	4	43.05564	5.016764	54	581.2512
0.4645152	5	53.81955	5.109667	55	592.0151
0.5574182	6	64.58346	5.202570	56	602.7790
0.6503213	7	75.34737	5.295473	57	613.5429
0.7432243	8	86.11128	5.388376	58	624.3068
0.8361274	9	96.87519	5.481279	59	635.0707
0.9290304	10	107.6391	5.574182	60	645.8346
1.021933	11	118.4030	5.667085	61	656.5985
1.114836	12	129.1669	5.759988	62	667.3624
1.207740	13	139.9308	5.852892	63	678.1264
1.300643	14	150.6947	5.945795	64	688.8903
1.393546	15	161.4587	6.038698	65	699.6542
1.486449	16	172.2226	6.131601	66	710.4181
1.579352	17	182.9865	6.224504	67	721.1820
1.672255	18	193.7504	6.317407	68	731.9459
1.765158	19	204.5143	6.410310	69	742.7098
1.858061	20	215.2782	6.503213	70	753.4737
1.950964	21	226.0421	6.596116	71	764.2376
2.043867	22	236.8060	6.689019	72	775.0015
2.136770	23	247.5699	6.781922	73	785.7655
2.229673	24	258.3339	6.874825	74	796.5294
2.322576	25	269.0978	6.967728	75	807.2933
2.415479	26	279.8617	7.060631	76	818.0572
2.508382	27	290.6256	7.153534	77	828.8211
2.601285	28	301.3895	7.246437	78	839.5850
2.694188	29	312.1534	7.339340	79	850.3489
2.787091	30	322.9173	7.432243	80	861.1128
2.879994	31	333.6812	7.525146	81	871.8767
2.972897	32	344.4451	7.618049	82	882.6407
3.065800	33	355.2090	7.710952	83	893.4046
3.158703	34	365.9730	7.803855	84	904.1685
3.251606	35	376.7369	7.896758	85	914.9324
3.344509	36	387.5008	7.989661	86	925.6963
3.437413	37	398.2647	8.082564	87	936.4602
3.530316	38	409.0286	8.175468	88	947.2241
3.623219	39	419.7925	8.268371	89	957.9880
3.716122	40	430.5564	8.361274	90	968.7519
3.809025	41	441.3203	8.454177	91	979.5158
3.901928	42	452.0842	8.547080	92	990.2798
3.994831	43	462.8481	8.639983	93	1001.044
4.087734	44	473.6121	8.732886	94	1011.808
4.180637	45	484.3760	8.825789	95	1022.571
4.273540	46	495.1399	8.918692	96	1033.335
4.366443	47	505.9038	9.011595	97	1044.099
4.459346	48	516.6677	9.104498	98	1054.863
4.552249	49	527.4316	9.197401	99	1065.627
4.645152	50	538.1955	9.290304	100	1076.391

m^2 m^2/s	REF	ft^2 ft^2/s	m^2 m^2/s	REF	ft^2 ft^2/s
9.383207	101	1087.155	14.02836	151	1625.350
9.476110	102	1097.919	14.12126	152	1636.114
9.569013	103	1108.683	14.21417	153	1646.878
9.661916	104	1119.447	14.30707	154	1657.642
9.754819	105	1130.211	14.39997	155	1668.406
9.847722	106	1140.975	14.49287	156	1679.170
9.940625	107	1151.738	14.58578	157	1689.934
10.03353	108	1162.502	14.67868	158	1700.698
10.12643	109	1173.266	14.77158	159	1711.462
10.21933	110	1184.030	14.86449	160	1722.226
10.31224	111	1194.794	14.95739	161	1732.990
10.40514	112	1205.558	15.05029	162	1743.753
10.49804	113	1216.322	15.14320	163	1754.517
10.59095	114	1227.086	15.23610	164	1765.281
10.68385	115	1237.850	15.32900	165	1776.045
10.77675	116	1248.614	15.42190	166	1786.809
10.86966	117	1259.378	15.51481	167	1797.573
10.96256	118	1270.141	15.60771	168	1808.337
11.05546	119	1280.905	15.70061	169	1819.101
11.14836	120	1291.669	15.79352	170	1829.865
11.24127	121	1302.433	15.88642	171	1840.629
11.33417	122	1313.197	15.97932	172	1851.393
11.42707	123	1323.961	16.07223	173	1862.156
11.51998	124	1334.725	16.16513	174	1872.920
11.61288	125	1345.489	16.25803	175	1883.684
11.70578	126	1356.253	16.35094	176	1894.448
11.79869	127	1367.017	16.44384	177	1905.212
11.89159	128	1377.781	16.53674	178	1915.976
11.98449	129	1388.544	16.62964	179	1926.740
12.07740	130	1399.308	16.72255	180	1937.504
12.17030	131	1410.072	16.81545	181	1948.268
12.26320	132	1420.836	16.90835	182	1959.032
12.35610	133	1431.600	17.00126	183	1969.796
12.44901	134	1442.364	17.09416	184	1980.560
12.54191	135	1453.128	17.18706	185	1991.323
12.63481	136	1463.892	17.27997	186	2002.087
12.72772	137	1474.656	17.37287	187	2012.851
12.82062	138	1485.420	17.46577	188	2023.615
12.91352	139	1496.184	17.55867	189	2034.379
13.00643	140	1506.947	17.65158	190	2045.143
13.09933	141	1517.711	17.74448	191	2055.907
13.19223	142	1528.475	17.83738	192	2066.671
13.28513	143	1539.239	17.93029	193	2077.435
13.37804	144	1550.003	18.02319	194	2088.199
13.47094	145	1560.767	18.11609	195	2098.963
13.56384	146	1571.531	18.20900	196	2109.726
13.65675	147	1582.295	18.30190	197	2120.490
13.74965	148	1593.059	18.39480	198	2131.254
13.84255	149	1603.823	18.48770	199	2142.018
13.93546	150	1614.587	18.58061	200	2152.782

km² ←	— 6	Area $\frac{m^2}{m^2}$	+ 4	cm²
			+ 6	mm²
		Kin. Vis. $\frac{m^2}{m^2/s}$	+ 4	St
			+ 6	cSt
			+ 6	mm²/s

m² m²/s	REF	ft² ft²/s	m² m²/s	REF	ft² ft²/s
18·67351	201	2163·546	23·31866	251	2701·742
18·76641	202	2174·310	23·41157	252	2712·505
18·85932	203	2185·074	23·50447	253	2723·269
18·95222	204	2195·838	23·59737	254	2734·033
19·04512	205	2206·602	23·69028	255	2744·797
19·13803	206	2217·366	23·78318	256	2755·561
19·23093	207	2228·129	23·87608	257	2766·325
19·32383	208	2238·893	23·96898	258	2777·089
19·41674	209	2249·657	24·06189	259	2787·853
19·50964	210	2260·421	24·15479	260	2798·617
19·60254	211	2271·185	24·24769	261	2809·381
19·69544	212	2281·949	24·34060	262	2820·145
19·78835	213	2292·713	24·43350	263	2830·908
19·88125	214	2303·477	24·52640	264	2841·672
19·97415	215	2314·241	24·61931	265	2852·436
20·06706	216	2325·005	24·71221	266	2863·200
20·15996	217	2335·769	24·80511	267	2873·964
20·25286	218	2346·532	24·89801	268	2884·728
20·34577	219	2357·296	24·99092	269	2895·492
20·43867	220	2368·060	25·08382	270	2906·256
20·53157	221	2378·824	25·17672	271	2917·020
20·62448	222	2389·588	25·26963	272	2927·784
20·71738	223	2400·352	25·36253	273	2938·548
20·81028	224	2411·116	25·45543	274	2949·311
20·90318	225	2421·880	25·54834	275	2960·075
20·99609	226	2432·644	25·64124	276	2970·839
21·08899	227	2443·408	25·73414	277	2981·603
21·18189	228	2454·172	25·82705	278	2992·367
21·27480	229	2464·935	25·91995	279	3003·131
21·36770	230	2475·699	26·01285	280	3013·895
21·46060	231	2486·463	26·10575	281	3024·659
21·55351	232	2497·227	26·19866	282	3035·423
21·64641	233	2507·991	26·29156	283	3046·187
21·73931	234	2518·755	26·38446	284	3056·951
21·83221	235	2529·519	26·47737	285	3067·714
21·92512	236	2540·283	26·57027	286	3078·478
22·01802	237	2551·047	26·66317	287	3089·242
22·11092	238	2561·811	26·75608	288	3100·006
22·20383	239	2572·575	26·84898	289	3110·770
22·29673	240	2583·339	26·94188	290	3121·534
22·38963	241	2594·102	27·03478	291	3132·298
22·48254	242	2604·866	27·12769	292	3143·062
22·57544	243	2615·630	27·22059	293	3153·826
22·66834	244	2626·394	27·31349	294	3164·590
22·76124	245	2637·158	27·40640	295	3175·354
22·85415	246	2647·922	27·49930	296	3186·117
22·94705	247	2658·686	27·59220	297	3196·881
23·03995	248	2669·450	27·68511	298	3207·645
23·13286	249	2680·214	27·77801	299	3218·409
23·22576	250	2690·978	27·87091	300	3229·173

km²		+6		Area m²			−4 −6		cm² mm²
				Kin. Vis. m²/s			−4 −6 −6		St cSt mm²/s

m² m²/s	REF	ft² ft²/s		m² m²/s	REF	ft² ft²/s
27.96382	301	3239.937		32.60897	351	3778.133
28.05672	302	3250.701		32.70187	352	3788.896
28.14962	303	3261.465		32.79477	353	3799.660
28.24252	304	3272.229		32.88768	354	3810.424
28.33543	305	3282.993		32.98058	355	3821.188
28.42833	306	3293.757		33.07348	356	3831.952
28.52123	307	3304.521		33.16639	357	3842.716
28.61414	308	3315.284		33.25929	358	3853.480
28.70704	309	3326.048		33.35219	359	3864.244
28.79994	310	3336.812		33.44509	360	3875.008
28.89285	311	3347.576		33.53800	361	3885.772
28.98575	312	3358.340		33.63090	362	3896.536
29.07865	313	3369.104		33.72380	363	3907.299
29.17155	314	3379.868		33.81671	364	3918.063
29.26446	315	3390.632		33.90961	365	3928.827
29.35736	316	3401.396		34.00251	366	3939.591
29.45026	317	3412.160		34.09542	367	3950.355
29.54317	318	3422.924		34.18832	368	3961.119
29.63607	319	3433.687		34.28122	369	3971.883
29.72897	320	3444.451		34.37413	370	3982.647
29.82188	321	3455.215		34.46703	371	3993.411
29.91478	322	3465.979		34.55993	372	4004.175
30.00768	323	3476.743		34.65283	373	4014.939
30.10058	324	3487.507		34.74574	374	4025.702
30.19349	325	3498.271		34.83864	375	4036.466
30.28639	326	3509.035		34.93154	376	4047.230
30.37929	327	3519.799		35.02445	377	4057.994
30.47220	328	3530.563		35.11735	378	4068.758
30.56510	329	3541.327		35.21025	379	4079.522
30.65800	330	3552.090		35.30316	380	4090.286
30.75091	331	3562.854		35.39606	381	4101.050
30.84381	332	3573.618		35.48896	382	4111.814
30.93671	333	3584.382		35.58186	383	4122.578
31.02962	334	3595.146		35.67477	384	4133.342
31.12252	335	3605.910		35.76767	385	4144.106
31.21542	336	3616.674		35.86057	386	4154.869
31.30832	337	3627.438		35.95348	387	4165.633
31.40123	338	3638.202		36.04638	388	4176.397
31.49413	339	3648.966		36.13928	389	4187.161
31.58703	340	3659.730		36.23219	390	4197.925
31.67994	341	3670.493		36.32509	391	4208.689
31.77284	342	3681.257		36.41799	392	4219.453
31.86574	343	3692.021		36.51089	393	4230.217
31.95865	344	3702.785		36.60380	394	4240.981
32.05155	345	3713.549		36.69670	395	4251.745
32.14445	346	3724.313		36.78960	396	4262.509
32.23736	347	3735.077		36.88251	397	4273.272
32.33026	348	3745.841		36.97541	398	4284.036
32.42316	349	3756.605		37.06831	399	4294.800
32.51606	350	3767.369		37.16122	400	4305.564

$$km^2 \xleftarrow{\quad -6 \quad}$$

Area $\overline{m^2}$	$\xrightarrow{+4}$	cm^2
	$\xrightarrow{+6}$	mm^2

Kin. Vis. m^2/s	$\xrightarrow{+4}$	St
	$\xrightarrow{+6}$	cSt
	$\xrightarrow{+6}$	mm^2/s

m^2 m^2/s	REF	ft^2 ft^2/s	m^2 m^2/s	REF	ft^2 ft^2/s
37.25412	401	4316.328	41.89927	451	4854.524
37.34702	402	4327.092	41.99217	452	4865.288
37.43993	403	4337.856	42.08508	453	4876.051
37.53283	404	4348.620	42.17798	454	4886.815
37.62573	405	4359.384	42.27088	455	4897.579
37.71863	406	4370.148	42.36379	456	4908.343
37.81154	407	4380.912	42.45669	457	4919.107
37.90444	408	4391.675	42.54959	458	4929.871
37.99734	409	4402.439	42.64250	459	4940.635
38.09025	410	4413.203	42.73540	460	4951.399
38.18315	411	4423.967	42.82830	461	4962.163
38.27605	412	4434.731	42.92120	462	4972.927
38.36896	413	4445.495	43.01411	463	4983.690
38.46186	414	4456.259	43.10701	464	4994.454
38.55476	415	4467.023	43.19991	465	5005.218
38.64767	416	4477.787	43.29282	466	5015.982
38.74057	417	4488.551	43.38572	467	5026.746
38.83347	418	4499.315	43.47862	468	5037.510
38.92637	419	4510.078	43.57153	469	5048.274
39.01928	420	4520.842	43.66443	470	5059.038
39.11218	421	4531.606	43.75733	471	5069.802
39.20508	422	4542.370	43.85023	472	5080.566
39.29799	423	4553.134	43.94314	473	5091.330
39.39089	424	4563.898	44.03604	474	5102.094
39.48379	425	4574.662	44.12894	475	5112.857
39.57669	426	4585.426	44.22185	476	5123.621
39.66960	427	4596.190	44.31475	477	5134.385
39.76250	428	4606.954	44.40765	478	5145.149
39.85540	429	4617.718	44.50056	479	5155.913
39.94831	430	4628.482	44.59346	480	5166.677
40.04121	431	4639.245	44.68636	481	5177.441
40.13411	432	4650.009	44.77927	482	5188.205
40.22702	433	4660.773	44.87217	483	5198.969
40.31992	434	4671.537	44.96507	484	5209.733
40.41282	435	4682.301	45.05797	485	5220.497
40.50573	436	4693.065	45.15088	486	5231.260
40.59863	437	4703.829	45.24378	487	5242.024
40.69153	438	4714.593	45.33668	488	5252.788
40.78443	439	4725.357	45.42959	489	5263.552
40.87734	440	4736.121	45.52249	490	5274.316
40.97024	441	4746.885	45.61539	491	5285.080
41.06314	442	4757.648	45.70830	492	5295.844
41.15605	443	4768.412	45.80120	493	5306.608
41.24895	444	4779.176	45.89410	494	5317.372
41.34185	445	4789.940	45.98700	495	5328.136
41.43476	446	4800.704	46.07991	496	5338.900
41.52766	447	4811.468	46.17281	497	5349.663
41.62056	448	4822.232	46.26571	498	5360.427
41.71347	449	4832.996	46.35862	499	5371.191
41.80637	450	4843.760	46.45152	500	5381.955

km²	+ 6	Area / m²	-4	cm²
			-6	mm²
		Kin. Vis. / m²/s	-4	St
			-6	cSt
			-6	mm²/s

m² / m²/s	REF	ft² / ft²/s	m² / m²/s	REF	ft² / ft²/s
46.54442	501	5392.719	51.18958	551	5930.915
46.63733	502	5403.483	51.28248	552	5941.679
46.73023	503	5414.247	51.37538	553	5952.442
46.82313	504	5425.011	51.46828	554	5963.206
46.91604	505	5435.775	51.56119	555	5973.970
47.00894	506	5446.539	51.65409	556	5984.734
47.10184	507	5457.303	51.74699	557	5995.498
47.19474	508	5468.066	51.83990	558	6006.262
47.28765	509	5478.830	51.93280	559	6017.026
47.38055	510	5489.594	52.02570	560	6027.790
47.47345	511	5500.358	52.11861	561	6038.554
47.56636	512	5511.122	52.21151	562	6049.318
47.65926	513	5521.886	52.30441	563	6060.082
47.75216	514	5532.650	52.39731	564	6070.845
47.84507	515	5543.414	52.49022	565	6081.609
47.93797	516	5554.178	52.58312	566	6092.373
48.03087	517	5564.942	52.67602	567	6103.137
48.12378	518	5575.706	52.76893	568	6113.901
48.21668	519	5586.469	52.86183	569	6124.665
48.30958	520	5597.233	52.95473	570	6135.429
48.40248	521	5607.997	53.04764	571	6146.193
48.49539	522	5618.761	53.14054	572	6156.957
48.58829	523	5629.525	53.23344	573	6167.721
48.68119	524	5640.289	53.32634	574	6178.485
48.77410	525	5651.053	53.41925	575	6189.248
48.86700	526	5661.817	53.51215	576	6200.012
48.95990	527	5672.581	53.60505	577	6210.776
49.05281	528	5683.345	53.69796	578	6221.540
49.14571	529	5694.109	53.79086	579	6232.304
49.23861	530	5704.872	53.88376	580	6243.068
49.33151	531	5715.636	53.97667	581	6253.832
49.42442	532	5726.400	54.06957	582	6264.596
49.51732	533	5737.164	54.16247	583	6275.360
49.61022	534	5747.928	54.25538	584	6286.124
49.70313	535	5758.692	54.34828	585	6296.888
49.79603	536	5769.456	54.44118	586	6307.651
49.88893	537	5780.220	54.53408	587	6318.415
49.98184	538	5790.984	54.62699	588	6329.179
50.07474	539	5801.748	54.71989	589	6339.943
50.16764	540	5812.512	54.81279	590	6350.707
50.26054	541	5823.276	54.90570	591	6361.471
50.35345	542	5834.039	54.99860	592	6372.235
50.44635	543	5844.803	55.09150	593	6382.999
50.53925	544	5855.567	55.18441	594	6393.763
50.63216	545	5866.331	55.27731	595	6404.527
50.72506	546	5877.095	55.37021	596	6415.291
50.81796	547	5887.859	55.46312	597	6426.055
50.91087	548	5898.623	55.55602	598	6436.818
51.00377	549	5909.387	55.64892	599	6447.582
51.09667	550	5920.151	55.74182	600	6458.346

m² m²/s	REF	ft² ft²/s		m² m²/s	REF	ft² ft²/s
55.83473	601	6469.110		60.47988	651	7007.306
55.92763	602	6479.874		60.57278	652	7018.070
56.02053	603	6490.638		60.66569	653	7028.833
56.11344	604	6501.402		60.75859	654	7039.597
56.20634	605	6512.166		60.85149	655	7050.361
56.29924	606	6522.930		60.94439	656	7061.125
56.39215	607	6533.694		61.03730	657	7071.889
56.48505	608	6544.458		61.13020	658	7082.653
56.57795	609	6555.221		61.22310	659	7093.417
56.67085	610	6565.985		61.31601	660	7104.181
56.76376	611	6576.749		61.40891	661	7114.945
56.85666	612	6587.513		61.50181	662	7125.709
56.94956	613	6598.277		61.59472	663	7136.473
57.04247	614	6609.041		61.68762	664	7147.237
57.13537	615	6619.805		61.78052	665	7158.000
57.22827	616	6630.569		61.87343	666	7168.764
57.32118	617	6641.333		61.96633	667	7179.528
57.41408	618	6652.097		62.05923	668	7190.292
57.50698	619	6662.861		62.15213	669	7201.056
57.59988	620	6673.624		62.24504	670	7211.820
57.69279	621	6684.388		62.33794	671	7222.584
57.78569	622	6695.152		62.43084	672	7233.348
57.87859	623	6705.916		62.52375	673	7244.112
57.97150	624	6716.680		62.61665	674	7254.876
58.06440	625	6727.444		62.70955	675	7265.640
58.15730	626	6738.208		62.80246	676	7276.403
58.25021	627	6748.972		62.89536	677	7287.167
58.34311	628	6759.736		62.98826	678	7297.931
58.43601	629	6770.500		63.08116	679	7308.695
58.52892	630	6781.264		63.17407	680	7319.459
58.62182	631	6792.027		63.26697	681	7330.223
58.71472	632	6802.791		63.35987	682	7340.987
58.80762	633	6813.555		63.45278	683	7351.751
58.90053	634	6824.319		63.54568	684	7362.515
58.99343	635	6835.083		63.63858	685	7373.279
59.08633	636	6845.847		63.73149	686	7384.043
59.17924	637	6856.611		63.82439	687	7394.806
59.27214	638	6867.375		63.91729	688	7405.570
59.36504	639	6878.139		64.01019	689	7416.334
59.45795	640	6888.903		64.10310	690	7427.098
59.55085	641	6899.667		64.19600	691	7437.862
59.64375	642	6910.430		64.28890	692	7448.626
59.73665	643	6921.194		64.38181	693	7459.390
59.82956	644	6931.958		64.47471	694	7470.154
59.92246	645	6942.722		64.56761	695	7480.918
60.01536	646	6953.486		64.66052	696	7491.682
60.10827	647	6964.250		64.75342	697	7502.446
60.20117	648	6975.014		64.84632	698	7513.209
60.29407	649	6985.778		64.93923	699	7523.973
60.38698	650	6996.542		65.03213	700	7534.737

km² ——————— + 6 ——————→	Area / m²	← −4 ——— cm²
		← −6 ——— mm²
	Kin. Vis. / m²/s	← −4 ——— St
		← −6 ——— cSt
		← −6 ——— mm²/s

m² m²/s	REF	ft² ft²/s	m² m²/s	REF	ft² ft²/s
65.12503	701	7545.501	69.77018	751	8083.697
65.21793	702	7556.265	69.86309	752	8094.461
65.31084	703	7567.029	69.95599	753	8105.225
65.40374	704	7577.793	70.04889	754	8115.988
65.49664	705	7588.557	70.14180	755	8126.752
65.58955	706	7599.321	70.23470	756	8137.516
65.68245	707	7610.085	70.32760	757	8148.280
65.77535	708	7620.849	70.42050	758	8159.044
65.86826	709	7631.612	70.51341	759	8169.808
65.96116	710	7642.376	70.60631	760	8180.572
66.05406	711	7653.140	70.69921	761	8191.336
66.14697	712	7663.904	70.79212	762	8202.100
66.23987	713	7674.668	70.88502	763	8212.864
66.33277	714	7685.432	70.97792	764	8223.628
66.42567	715	7696.196	71.07083	765	8234.391
66.51858	716	7706.960	71.16373	766	8245.155
66.61148	717	7717.724	71.25663	767	8255.919
66.70438	718	7728.488	71.34953	768	8266.683
66.79729	719	7739.252	71.44244	769	8277.447
66.89019	720	7750.016	71.53534	770	8288.211
66.98309	721	7760.779	71.62824	771	8298.975
67.07600	722	7771.543	71.72115	772	8309.739
67.16890	723	7782.307	71.81405	773	8320.503
67.26180	724	7793.071	71.90695	774	8331.267
67.35470	725	7803.835	71.99986	775	8342.031
67.44761	726	7814.599	72.09276	776	8352.794
67.54051	727	7825.363	72.18566	777	8363.558
67.63341	728	7836.127	72.27857	778	8374.322
67.72632	729	7846.891	72.37147	779	8385.086
67.81922	730	7857.655	72.46437	780	8395.850
67.91212	731	7868.419	72.55727	781	8406.614
68.00503	732	7879.182	72.65018	782	8417.378
68.09793	733	7889.946	72.74308	783	8428.142
68.19083	734	7900.710	72.83598	784	8438.906
68.28373	735	7911.474	72.92889	785	8449.670
68.37664	736	7922.238	73.02179	786	8460.434
68.46954	737	7933.002	73.11469	787	8471.198
68.56244	738	7943.766	73.20760	788	8481.961
68.65535	739	7954.530	73.30050	789	8492.725
68.74825	740	7965.294	73.39340	790	8503.489
68.84115	741	7976.058	73.48631	791	8514.253
68.93406	742	7986.822	73.57921	792	8525.017
69.02696	743	7997.585	73.67211	793	8535.781
69.11986	744	8008.349	73.76501	794	8546.545
69.21276	745	8019.113	73.85792	795	8557.309
69.30567	746	8029.877	73.95082	796	8568.073
69.39857	747	8040.641	74.04372	797	8578.837
69.49147	748	8051.405	74.13663	798	8589.600
69.58438	749	8062.169	74.22953	799	8600.364
69.67728	750	8072.933	74.32243	800	8611.128

m² (m²/s)	REF	ft² (ft²/s)	m² (m²/s)	REF	ft² (ft²/s)
74.41534	801	8621.892	79.06049	851	9160.088
74.50824	802	8632.656	79.15339	852	9170.852
74.60114	803	8643.420	79.24629	853	9181.616
74.69404	804	8654.184	79.33920	854	9192.380
74.78695	805	8664.948	79.43210	855	9203.143
74.87985	806	8675.712	79.52500	856	9213.907
74.97275	807	8686.476	79.61791	857	9224.671
75.06566	808	8697.240	79.71081	858	9235.435
75.15856	809	8708.004	79.80371	859	9246.199
75.25146	810	8718.767	79.89662	860	9256.963
75.34437	811	8729.531	79.98952	861	9267.727
75.43727	812	8740.295	80.08242	862	9278.491
75.53017	813	8751.059	80.17532	863	9289.255
75.62307	814	8761.823	80.26823	864	9300.019
75.71598	815	8772.587	80.36113	865	9310.782
75.80888	816	8783.351	80.45403	866	9321.546
75.90178	817	8794.115	80.54694	867	9332.310
75.99469	818	8804.879	80.63984	868	9343.074
76.08759	819	8815.643	80.73274	869	9353.838
76.18049	820	8826.406	80.82565	870	9364.602
76.27340	821	8837.170	80.91855	871	9375.366
76.36630	822	8847.934	81.01145	872	9386.130
76.45920	823	8858.698	81.10435	873	9396.894
76.55210	824	8869.462	81.19726	874	9407.658
76.64501	825	8880.226	81.29016	875	9418.422
76.73791	826	8890.990	81.38306	876	9429.186
76.83081	827	8901.754	81.47597	877	9439.949
76.92372	828	8912.518	81.56887	878	9450.713
77.01662	829	8923.282	81.66177	879	9461.477
77.10952	830	8934.046	81.75468	880	9472.241
77.20243	831	8944.810	81.84758	881	9483.005
77.29533	832	8955.573	81.94048	882	9493.769
77.38823	833	8966.337	82.03338	883	9504.533
77.48114	834	8977.101	82.12629	884	9515.297
77.57404	835	8987.865	82.21919	885	9526.061
77.66694	836	8998.629	82.31209	886	9536.825
77.75984	837	9009.393	82.40500	887	9547.589
77.85275	838	9020.157	82.49790	888	9558.352
77.94565	839	9030.921	82.59080	889	9569.116
78.03855	840	9041.685	82.68371	890	9579.880
78.13146	841	9052.449	82.77661	891	9590.644
78.22436	842	9063.213	82.86951	892	9601.408
78.31726	843	9073.976	82.96241	893	9612.172
78.41017	844	9084.740	83.05532	894	9622.936
78.50307	845	9095.504	83.14822	895	9633.700
78.59597	846	9106.268	83.24112	896	9644.464
78.68888	847	9117.032	83.33403	897	9655.228
78.78178	848	9127.796	83.42693	898	9665.992
78.87468	849	9138.560	83.51983	899	9676.755
78.96758	850	9149.324	83.61274	900	9687.519

km² ——————— + 6 ——————→	Area / m²	←———— − 4 ———— cm²
		←———— − 6 ———— mm²
	Kin. Vis. / m²/s	←———— − 4 ———— St
		←———— − 6 ———— cSt
		←———— − 6 ———— mm²/s

m² m²/s	REF	ft² ft²/s	m² m²/s	REF	ft² ft²/s
83.70564	901	9698.283	88.35079	951	10236.48
83.79854	902	9709.047	88.44369	952	10247.24
83.89145	903	9719.811	88.53660	953	10258.01
83.98435	904	9730.575	88.62950	954	10268.77
84.07725	905	9741.339	88.72240	955	10279.53
84.17015	906	9752.103	88.81531	956	10290.30
84.26306	907	9762.867	88.90821	957	10301.06
84.35596	908	9773.631	89.00111	958	10311.83
84.44886	909	9784.395	89.09402	959	10322.59
84.54177	910	9795.158	89.18692	960	10333.35
84.63467	911	9805.922	89.27982	961	10344.12
84.72757	912	9816.686	89.37272	962	10354.88
84.82048	913	9827.450	89.46563	963	10365.65
84.91338	914	9838.214	89.55853	964	10376.41
85.00628	915	9848.978	89.65143	965	10387.17
85.09918	916	9859.742	89.74434	966	10397.94
85.19209	917	9870.506	89.83724	967	10408.70
85.28499	918	9881.270	89.93014	968	10419.47
85.37789	919	9892.034	90.02305	969	10430.23
85.47080	920	9902.798	90.11595	970	10440.99
85.56370	921	9913.562	90.20885	971	10451.76
85.65660	922	9924.325	90.30175	972	10462.52
85.74951	923	9935.089	90.39466	973	10473.28
85.84241	924	9945.853	90.48756	974	10484.05
85.93531	925	9956.617	90.58046	975	10494.81
86.02822	926	9967.381	90.67337	976	10505.58
86.12112	927	9978.145	90.76627	977	10516.34
86.21402	928	9988.909	90.85917	978	10527.10
86.30692	929	9999.673	90.95208	979	10537.87
86.39983	930	10010.44	91.04498	980	10548.63
86.49273	931	10021.20	91.13788	981	10559.40
86.58563	932	10031.96	91.23079	982	10570.16
86.67854	933	10042.73	91.32369	983	10580.92
86.77144	934	10053.49	91.41659	984	10591.69
86.86434	935	10064.26	91.50949	985	10602.45
86.95725	936	10075.02	91.60240	986	10613.22
87.05015	937	10085.78	91.69530	987	10623.98
87.14305	938	10096.55	91.78820	988	10634.74
87.23596	939	10107.31	91.88111	989	10645.51
87.32886	940	10118.08	91.97401	990	10656.27
87.42176	941	10128.84	92.06691	991	10667.04
87.51466	942	10139.60	92.15982	992	10677.80
87.60757	943	10150.37	92.25272	993	10688.56
87.70047	944	10161.13	92.34562	994	10699.33
87.79337	945	10171.90	92.43853	995	10710.09
87.88628	946	10182.66	92.53143	996	10720.85
87.97918	947	10193.42	92.62433	997	10731.62
88.07208	948	10204.19	92.71723	998	10742.38
88.16499	949	10214.95	92.81014	999	10753.15
88.25789	950	10225.71	92.90304	1000	10763.91

8

AREA

Square-mile (statute) — Square-kilometre
$$(\text{mi}^2 - \text{km}^2)$$

Square-mile (statute) — Square-metre
$$(\text{mi}^2 - \text{m}^2)$$

Conversion Factors:
$$1 \text{ square-mile (mi}^2) = 2.589988 \text{ square-kilometres (km}^2)$$
$$(1 \text{ km}^2 = 0.3861022 \text{ mi}^2)$$

Although most tables in this reference book have a REF number range of 1 to 1000, the 1 to 100 range given in this table is sufficient for most square-mile—square-kilometre conversions.

To convert km^2 values obtained from the table to m^2, Diagram 8-A will assist the user. It is derived from: $1 \text{ km}^2 = (10^6) \text{ m}^2$.

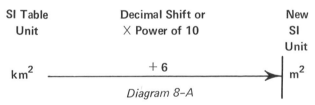

SI Table Unit	Decimal Shift or $\times$ Power of 10	New SI Unit
km^2	$+6$	m^2

Diagram 8–A

Example (a): Convert 9.15 mi^2 to km^2, m^2.
Since: 9.15 $\text{mi}^2 = (9.00 + 0.15) \text{ mi}^2$:
From table: 9 $\text{mi}^2 = 23.30989 \text{ km}^2$ *
           15 $\text{mi}^2 = 38.84982 \text{ km}^2$ *
By proper decimal shifting and addition:
   9.00 $\text{mi}^2 = 23.30989 \text{ km}^2$
   <u>0.15 mi^2 =   0.3884982 km^2</u>
   9.15 $\text{mi}^2 = 23.6983882 \text{ km}^2$ or $\approx 23.7 \text{ km}^2$

---

*All conversion values obtained can be rounded to desired number of significant figures.

From Diagram 8-A: $23.69839 \text{ km}^2 = 23.69839 \times 10^6 \text{ m}^2 \, {}^* \approx 23.7 \times 10^6 \text{ m}^2$

(km^2 to m^2 by multiplying km^2 value by 10^6)

Diagram 8-B

Diagram 8-B is the opposite of Diagram 8-A, and is used to convert m^2 to km^2. The km^2 equivalent is then applied to the REF column in the table to obtain mi^2.

Example (b): Convert 3,750,000 m^2 to mi^2.

From Diagram 8-B: $3,750,000 \text{ m}^2 = 3,750,000 \times 10^{-6} \text{ km}^2 = 3.75 \text{ km}^2$

Since: $3.75 \text{ km}^2 = (3.00 + 0.75) \text{ km}^2$:

From table: $30 \text{ km}^2 = 11.58307 \text{ mi}^2 \, {}^*$

$\qquad\qquad\quad 75 \text{ km}^2 = 28.95766 \text{ mi}^2 \, {}^*$

Therefore, by shifting decimals where necessary and adding:

$3.00 \text{ km}^2 = 1.158307 \;\; \text{mi}^2$

$\underline{0.75 \text{ km}^2 = 0.2895766 \text{ mi}^2}$

$3.75 \text{ km}^2 \quad 1.4478836 \text{ mi}^2 \text{ or} \approx 1.45 \text{ mi}^2.$

---

*All conversion values obtained can be rounded to desired number of significant figures.

km²	REF	mi²	km²	REF	mi²
2.589988	1	0.3861022	132.0894	51	19.69121
5.179976	2	0.7722044	134.6794	52	20.07731
7.769964	3	1.158307	137.2694	53	20.46342
10.35995	4	1.544409	139.8594	54	20.84952
12.94994	5	1.930511	142.4493	55	21.23562
15.53993	6	2.316613	145.0393	56	21.62172
18.12992	7	2.702715	147.6293	57	22.00782
20.71990	8	3.088817	150.2193	58	22.39393
23.30989	9	3.474920	152.8093	59	22.78003
25.89988	10	3.861022	155.3993	60	23.16613
28.48987	11	4.247124	157.9893	61	23.55223
31.07986	12	4.633226	160.5793	62	23.93833
33.66984	13	5.019328	163.1692	63	24.32444
36.25983	14	5.405430	165.7592	64	24.71054
38.84982	15	5.791533	168.3492	65	25.09664
41.43981	16	6.177635	170.9392	66	25.48274
44.02980	17	6.563737	173.5292	67	25.86885
46.61978	18	6.949839	176.1192	68	26.25495
49.20977	19	7.335941	178.7092	69	26.64105
51.79976	20	7.722044	181.2992	70	27.02715
54.38975	21	8.108146	183.8891	71	27.41325
56.97974	22	8.494248	186.4791	72	27.79936
59.56972	23	8.880350	189.0691	73	28.18546
62.15971	24	9.266452	191.6591	74	28.57156
64.74970	25	9.652554	194.2491	75	28.95766
67.33969	26	10.03866	196.8391	76	29.34377
69.92968	27	10.42476	199.4291	77	29.72987
72.51966	28	10.81086	202.0191	78	30.11597
75.10965	29	11.19696	204.6091	79	30.50207
77.69964	30	11.58307	207.1990	80	30.88817
80.28963	31	11.96917	209.7890	81	31.27428
82.87962	32	12.35527	212.3790	82	31.66038
85.46960	33	12.74137	214.9690	83	32.04648
88.05959	34	13.12747	217.5590	84	32.43258
90.64958	35	13.51358	220.1490	85	32.81869
93.23957	36	13.89968	222.7390	86	33.20479
95.82956	37	14.28578	225.3290	87	33.59089
98.41954	38	14.67188	227.9189	88	33.97699
101.0095	39	15.05798	230.5089	89	34.36309
103.5995	40	15.44409	233.0989	90	34.74920
106.1895	41	15.83019	235.6889	91	35.13530
108.7795	42	16.21629	238.2789	92	35.52140
111.3695	43	16.60239	240.8689	93	35.90750
113.9595	44	16.98850	243.4589	94	36.29360
116.5495	45	17.37460	246.0489	95	36.67971
119.1394	46	17.76070	248.6388	96	37.06581
121.7294	47	18.14680	251.2288	97	37.45191
124.3194	48	18.53290	253.8188	98	37.83801
126.9094	49	18.91901	256.4088	99	38.22412
129.4994	50	19.30511	258.9988	100	38.61022

AREA

Square-yard — Square-metre
   $(yd^2 - m^2)$

Square-yard — Square-kilometre
   $(yd^2 - km^2)$

Conversion Factors:
$$1 \text{ square-yard } (yd^2) = 0.8361274 \text{ square-metre } (m^2)$$
$$(1 \text{ m}^2 = 1.195990 \text{ yd}^2)$$

Although most tables in this reference book have an REF number range of 1 to 1000, the 1 to 100 range given in this table is sufficient for most square-yard—square-metre conversions.

To convert m^2 values from the table to km^2, Diagram 9-A will assist the user. It is derived from: $1 \text{ m}^2 = (10^{-6}) \text{ km}^2$.

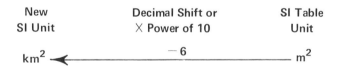

New SI Unit	Decimal Shift or × Power of 10	SI Table Unit
km^2 ◄	—6	m^2

Diagram 9–A

Example (a): Convert 39.5 yd^2 to m^2.
   Since: 39.5 yd^2 = (39.0 + 0.05) yd:
   From table: 39 yd^2 = 32.60897 m^2 *
                  5 yd^2 = 4.180637 m^2 *

*All conversion values obtained can be rounded to desired number of significant figures.

By shifting decimals where necessary and adding:

$$39.0 \text{ yd}^2 = 32.60897$$
$$0.5 \text{ yd}^2 = \underline{\ 0.4180637}$$
$$39.5 \text{ yd}^2 = 33.0270337 \text{ m}^2 \text{ or} \approx 33.03 \text{ m}^2$$

Example (*b*): Convert 551,000 yd^2 to km^2
   Since: 551,000 $\text{yd}^2 = (550 + 1) \times 10^3 \text{ yd}$:
   From table: 55 $\text{yd}^2 = 45.98701 \text{ m}^2$ *
                    1 $\text{yd}^2 = 0.8361274 \text{ m}^2$ *

By shifting decimals where necessary and adding:

$$550 \text{ yd}^2 = 459.8701 \quad \text{m}^2$$
$$1 \text{ yd}^2 = \underline{\ \ 0.8361274 \text{ m}^2}$$
$$551 \text{ yd}^2 = 460.7062 \quad \text{m}^2$$

   Therefore: $551 \times 10^3 \text{ yd}^2 = 460.7062 \times 10^3 \text{ m}^2 = 460,706.2 \text{ m}^2$.
   From Diagram 9-A: $460,706.2 \text{ m}^2 = 460,706.2 \, (10^{-6}) \text{ km}^2$
                              $= 0.4607062 \text{ km}^2 \approx 0.461 \text{ km}^2$
                              (m^2 to km^2 by multiplying m^2 value by 10^{-6} or by shifting decimal 6 places to the left)

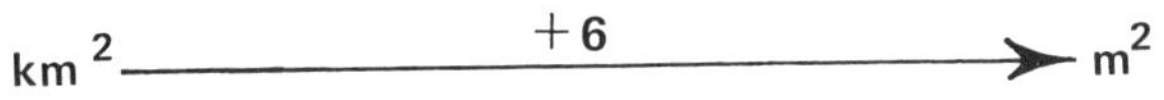

$$\text{km}^2 \xrightarrow{\hspace{2cm} +6 \hspace{2cm}} \text{m}^2$$

Diagram 9-B

Diagram 9-B is the opposite of Diagram 9-A, and is used to convert km^2 to m^2. The m^2 equivalent is then applied to the REF column in the table to obtain yd^2.

Example (*c*): Convert 0.55 km^2 to yd^2.
   From Diagram 9-B: $0.55 \text{ km}^2 = 0.55 \times 10^6 \text{ m}^2 = 55 \times 10^4 \text{ m}^2$
   From table: 55 $\text{m}^2 = 65.77945 \text{ yd}^2$ *
   Therefore: $55 \times 10^4 \text{ m}^2 = 65.77945 \times 10^4 \text{ yd}^2 = 657,794.5 \text{ yd}^2$
                    $\approx 658,000 \text{ yd}^2$.

---

*All conversion values obtained can be rounded to desired number of significant figures.

m²	REF	yd²		m²	REF	yd²
0.8361274	1	1.195990		42.64250	51	60.99549
1.672255	2	2.391980		43.47862	52	62.19148
2.508382	3	3.587970		44.31475	53	63.38747
3.344510	4	4.783960		45.15088	54	64.58346
4.180637	5	5.979950		45.98701	55	65.77945
5.016764	6	7.175940		46.82313	56	66.97544
5.852892	7	8.371930		47.65926	57	68.17143
6.689019	8	9.567920		48.49539	58	69.36742
7.525147	9	10.76391		49.33152	59	70.56341
8.361274	10	11.95990		50.16764	60	71.75940
9.197401	11	13.15589		51.00377	61	72.95539
10.03353	12	14.35188		51.83990	62	74.15138
10.86966	13	15.54787		52.67603	63	75.34737
11.70578	14	16.74386		53.51215	64	76.54336
12.54191	15	17.93985		54.34828	65	77.73935
13.37804	16	19.13584		55.18441	66	78.93534
14.21417	17	20.33183		56.02054	67	80.13133
15.05029	18	21.52782		56.85666	68	81.32732
15.88642	19	22.72381		57.69279	69	82.52331
16.72255	20	23.91980		58.52892	70	83.71930
17.55868	21	25.11579		59.36505	71	84.91529
18.39480	22	26.31178		60.20117	72	86.11128
19.23093	23	27.50777		61.03730	73	87.30727
20.06706	24	28.70376		61.87343	74	88.50326
20.90319	25	29.89975		62.70956	75	89.69925
21.73931	26	31.09574		63.54568	76	90.89524
22.57544	27	32.29173		64.38181	77	92.09123
23.41157	28	33.48772		65.21794	78	93.28722
24.24769	29	34.68371		66.05406	79	94.48321
25.08382	30	35.87970		66.89019	80	95.67920
25.91995	31	37.07569		67.72632	81	96.87519
26.75608	32	38.27168		68.56245	82	98.07118
27.59220	33	39.46767		69.39857	83	99.26717
28.42833	34	40.66366		70.23470	84	100.4632
29.26446	35	41.85965		71.07083	85	101.6591
30.10059	36	43.05564		71.90696	86	102.8551
30.93671	37	44.25163		72.74308	87	104.0511
31.77284	38	45.44762		73.57921	88	105.2471
32.60897	39	46.64361		74.41534	89	106.4431
33.44510	40	47.83960		75.25147	90	107.6391
34.28122	41	49.03559		76.08759	91	108.8351
35.11735	42	50.23158		76.92372	92	110.0311
35.95348	43	51.42757		77.75985	93	111.2271
36.78961	44	52.62356		78.59598	94	112.4231
37.62573	45	53.81955		79.43210	95	113.6190
38.46186	46	55.01554		80.26823	96	114.8150
39.29799	47	56.21153		81.10436	97	116.0110
40.13412	48	57.40752		81.94049	98	117.2070
40.97024	49	58.60351		82.77661	99	118.4030
41.80637	50	59.79950		83.61274	100	119.5990

10

VOLUME OR SECTION MODULUS

Cubic-inch — Cubic-millimetre
    $(in.^3 — mm^3)$
Cubic-inch — Cubic-centimetre
    $(in.^3 — cm^3)$
Cubic-inch — Cubic-metre
    $(in.^3 — m^3)$

VOLUME

Cubic-inches — Litres
    $(in.^2 — l)$

VOLUME FLOW RATE

Cubic-inch/Second — Cubic-millimetre/Second
    $(in.^3/s — mm^3/s)$
Cubic-inch/Second — Cubic-centimetre/Second
    $(in.^3/s — cm^3/s)$
Cubic-inch/Second — Cubic-metre/Second
    $(in.^3/s — m^3/s)$
Cubic-inch/Second — Litre/Second
    $(in.^3/s — l/s)$

Conversion Factors:
    1 cubic-inch $(in.^3)$ = 16,387.06 cubic-millimetres (mm^3)
    $(1\ mm^3 = 0.00006102376\ in.^3)$

Section Modulus and Volume Flow Values

Section modulus quantities are related in the same compound units as volume, in this instance in.^{3}. Therefore, Table 10 can be used for the conversion of these values.

Volume flow rate is also represented by this table, as long as the time units are the same for both English and SI conversion quantities. Besides per second shown above, these consistent time units can also be in per minute, per hour, etc.

Volume Flow Rate Conversion Factor:

1 in.^{3}/s (or in.^{3}/min, etc.) = 16,387.06 mm^3/s (or mm^3/min. etc.)

In converting values containing mm^3 units to expressions containing cm^3, m^3, l (litres), Diagram 10-A, which appears in the headings on alternate pages of this table, will assist the user. It is derived from:

1 mm^3 = (10^{-3}) cm^3 = (10^{-9}) m^3 = (10^{-6}) 1 (litre); 1 litre = 1 dm^3.

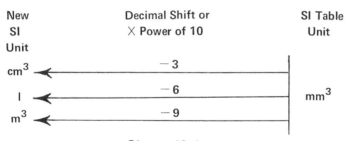

Diagram 10-A

Example (a): 0.135 in.^{3} to mm^3, cm^3.
   Since: 0.135 in.^{3} = 135 × 10^{-3} in.^{3}.
   From table: 135 in.^{3} = 2,212,253 mm^3 *
   Therefore: 135 × 10^{-3} in.^{3} = 2,212,253 × 10^{-3} mm^3 = 2,212.253 mm^3
      ≈ 2,210 mm^3.
   From Diagram 10-A: 2,212.253 mm^3 = 2,212.253 × 10^{-3} cm^3
                = 2.212253 cm^3 ≈ 2.21 cm^3
                (mm^3 to cm^3 by multiplying the mm^3 value by 10^{-3} or
                by shifting its decimal 3 places to the left)

Example (b): Convert 107 in.^{3}/s to l/s.
   From table: 107 in.^{3}/s = 1,753,415 mm^3/s*
   From Diagram 10-A: 1,753,415 mm^3/s = 1,753,415 × 10^{-6} l/s
                = 1.753415 l/s ≈ 1.75 l/s
                (mm^3/s to l/s by multiplying mm^3/s value by 10^{-6} or
                by shifting decimal 6 places to the left)

---

*All conversion values obtained can be rounded to desired number of significant figures.

Diagram 10-B is also given at the top of alternate table pages:

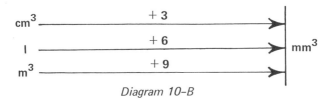

Diagram 10–B

This diagram is the opposite of Diagram 10-A and is used to convert values containing cm^3, m^3, and l to those with mm^3. The mm^3 equivalent is then applied to the REF column in the table to obtain in.^{3} quantities.

Example (c): Convert 358×10^{-9} m^3 to in.^{3}.
   From Diagram 10-B: 358×10^{-9} $m^3 = 358 \times 10^{-9}$ (10^9) $mm^3 = 358$ mm^3
   From table: 358 $mm^3 = 0.02184651$ in.$^{3*} \approx 0.0218$ in.^{3}

Example (d): Convert 245.5 l/sec to in.^{3}/s.
   From Diagram 10-B: 245.5 l/s $= 245.5 \times 10^6$ mm^3/s
   Since: 245.5 mm^3/s $= (240 + 5.5)$ mm^3/s:
   From table: 240 mm^3/s $= 0.01464570$ in.^{3}/s*
               55 mm^3/s $= 0.003356307$ in.^{3}/s*

By shifting decimals where necessary and adding:
   240.0 mm^3/s $= 0.01464570$ in.^{3}/s
   $\underline{\quad 5.5\ mm^3/s = 0.0003356307\ in.^3/s}$
   245.5 mm^3/s $= 0.01500133$ in.^{3}/s

Therefore: 245.5×10^6 mm^3/s $= 0.01500133$ in.^{3}/s $\times 10^6$ in.^{3}/s
   $= 15{,}001.33$ in.^{3}/s or $\approx 15{,}000$ in.^{3}/s.

---

*All conversion values obtained can be rounded to desired number of significant figures.

mm^3 mm^3/s	REF	in^3 in^3/s	mm^3 mm^3/s	REF	in^3 in^3/s
16387.06	1	0.0000610238	835740.1	51	0.003112212
32774.12	2	0.0001220475	852127.1	52	0.003173235
49161.18	3	0.0001830713	868514.2	53	0.003234259
65548.24	4	0.0002440950	884901.2	54	0.003295283
81935.30	5	0.0003051188	901288.3	55	0.003356307
98322.36	6	0.0003661426	917675.4	56	0.003417331
114709.4	7	0.0004271663	934062.4	57	0.003478354
131096.5	8	0.0004881901	950449.5	58	0.003539378
147483.5	9	0.0005492138	966836.5	59	0.003600402
163870.6	10	0.0006102376	983223.6	60	0.003661426
180257.7	11	0.0006712613	999610.7	61	0.003722449
196644.7	12	0.0007322851	1015998.	62	0.003783473
213031.8	13	0.0007933089	1032385.	63	0.003844497
229418.8	14	0.0008543326	1048772.	64	0.003905521
245805.9	15	0.0009153564	1065159.	65	0.003966544
262193.0	16	0.0009763801	1081546.	66	0.004027568
278580.0	17	0.001037404	1097933.	67	0.004088592
294967.1	18	0.001098428	1114320.	68	0.004149616
311354.1	19	0.001159451	1130707.	69	0.004210639
327741.2	20	0.001220475	1147094.	70	0.004271663
344128.3	21	0.001281499	1163481.	71	0.004332687
360515.3	22	0.001342523	1179868.	72	0.004393711
376902.4	23	0.001403546	1196255.	73	0.004454734
393289.4	24	0.001464570	1212642.	74	0.004515758
409676.5	25	0.001525594	1229030.	75	0.004576782
426063.6	26	0.001586618	1245417.	76	0.004637806
442450.6	27	0.001647641	1261804.	77	0.004698829
458837.7	28	0.001708665	1278191.	78	0.004759853
475224.7	29	0.001769689	1294578.	79	0.004820877
491611.8	30	0.001830713	1310965.	80	0.004881901
507998.9	31	0.001891737	1327352.	81	0.004942924
524385.9	32	0.001952760	1343739.	82	0.005003948
540773.0	33	0.002013784	1360126.	83	0.005064972
557160.0	34	0.002074808	1376513.	84	0.005125996
573547.1	35	0.002135832	1392900.	85	0.005187020
589934.2	36	0.002196855	1409287.	86	0.005248043
606321.2	37	0.002257879	1425674.	87	0.005309067
622708.3	38	0.002318903	1442061.	88	0.005370091
639095.3	39	0.002379927	1458448.	89	0.005431115
655482.4	40	0.002440950	1474835.	90	0.005492138
671869.5	41	0.002501974	1491222.	91	0.005553162
688256.5	42	0.002562998	1507610.	92	0.005614186
704643.6	43	0.002624022	1523997.	93	0.005675210
721030.6	44	0.002685045	1540384.	94	0.005736233
737417.7	45	0.002746069	1556771.	95	0.005797257
753804.8	46	0.002807093	1573158.	96	0.005858281
770191.8	47	0.002868117	1589545.	97	0.005919305
786578.9	48	0.002929140	1605932.	98	0.005980328
802965.9	49	0.002990164	1622319.	99	0.006041352
819353.0	50	0.003051188	1638706.	100	0.006102376

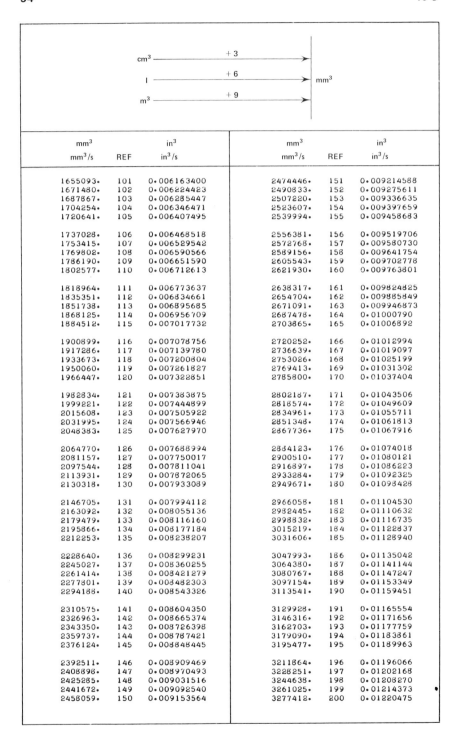

mm³ mm³/s	REF	in³ in³/s	mm³ mm³/s	REF	in³ in³/s
1655093.	101	0.006163400	2474446.	151	0.009214588
1671480.	102	0.006224423	2490833.	152	0.009275611
1687867.	103	0.006285447	2507220.	153	0.009336635
1704254.	104	0.006346471	2523607.	154	0.009397659
1720641.	105	0.006407495	2539994.	155	0.009458683
1737028.	106	0.006468518	2556381.	156	0.009519706
1753415.	107	0.006529542	2572768.	157	0.009580730
1769802.	108	0.006590566	2589156.	158	0.009641754
1786190.	109	0.006651590	2605543.	159	0.009702778
1802577.	110	0.006712613	2621930.	160	0.009763801
1818964.	111	0.006773637	2638317.	161	0.009824825
1835351.	112	0.006834661	2654704.	162	0.009885849
1851738.	113	0.006895685	2671091.	163	0.009946873
1868125.	114	0.006956709	2687478.	164	0.01000790
1884512.	115	0.007017732	2703865.	165	0.01006892
1900899.	116	0.007078756	2720252.	166	0.01012994
1917286.	117	0.007139780	2736639.	167	0.01019097
1933673.	118	0.007200804	2753026.	168	0.01025199
1950060.	119	0.007261827	2769413.	169	0.01031302
1966447.	120	0.007322851	2785800.	170	0.01037404
1982834.	121	0.007383875	2802187.	171	0.01043506
1999221.	122	0.007444899	2818574.	172	0.01049609
2015608.	123	0.007505922	2834961.	173	0.01055711
2031995.	124	0.007566946	2851348.	174	0.01061813
2048383.	125	0.007627970	2867736.	175	0.01067916
2064770.	126	0.007688994	2884123.	176	0.01074018
2081157.	127	0.007750017	2900510.	177	0.01080121
2097544.	128	0.007811041	2916897.	178	0.01086223
2113931.	129	0.007872065	2933284.	179	0.01092325
2130318.	130	0.007933089	2949671.	180	0.01098428
2146705.	131	0.007994112	2966058.	181	0.01104530
2163092.	132	0.008055136	2982445.	182	0.01110632
2179479.	133	0.008116160	2998832.	183	0.01116735
2195866.	134	0.008177184	3015219.	184	0.01122837
2212253.	135	0.008238207	3031606.	185	0.01128940
2228640.	136	0.008299231	3047993.	186	0.01135042
2245027.	137	0.008360255	3064380.	187	0.01141144
2261414.	138	0.008421279	3080767.	188	0.01147247
2277801.	139	0.008482303	3097154.	189	0.01153349
2294188.	140	0.008543326	3113541.	190	0.01159451
2310575.	141	0.008604350	3129928.	191	0.01165554
2326963.	142	0.008665374	3146316.	192	0.01171656
2343350.	143	0.008726398	3162703.	193	0.01177759
2359737.	144	0.008787421	3179090.	194	0.01183861
2376124.	145	0.008848445	3195477.	195	0.01189963
2392511.	146	0.008909469	3211864.	196	0.01196066
2408898.	147	0.008970493	3228251.	197	0.01202168
2425285.	148	0.009031516	3244638.	198	0.01208270
2441672.	149	0.009092540	3261025.	199	0.01214373
2458059.	150	0.009153564	3277412.	200	0.01220475

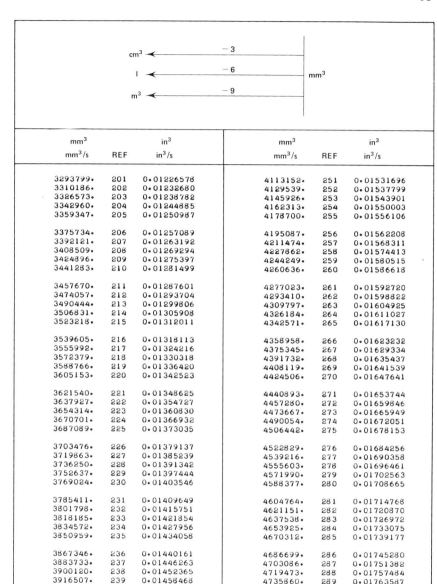

| mm³ | | in³ | mm³ | | in³ |
mm³/s	REF	in³/s	mm³/s	REF	in³/s
3293799.	201	0.01226578	4113152.	251	0.01531696
3310186.	202	0.01232680	4129539.	252	0.01537799
3326573.	203	0.01238782	4145926.	253	0.01543901
3342960.	204	0.01244885	4162313.	254	0.01550003
3359347.	205	0.01250987	4178700.	255	0.01556106
3375734.	206	0.01257089	4195087.	256	0.01562208
3392121.	207	0.01263192	4211474.	257	0.01568311
3408509.	208	0.01269294	4227862.	258	0.01574413
3424896.	209	0.01275397	4244249.	259	0.01580515
3441283.	210	0.01281499	4260636.	260	0.01586618
3457670.	211	0.01287601	4277023.	261	0.01592720
3474057.	212	0.01293704	4293410.	262	0.01598822
3490444.	213	0.01299806	4309797.	263	0.01604925
3506831.	214	0.01305908	4326184.	264	0.01611027
3523218.	215	0.01312011	4342571.	265	0.01617130
3539605.	216	0.01318113	4358958.	266	0.01623232
3555992.	217	0.01324216	4375345.	267	0.01629334
3572379.	218	0.01330318	4391732.	268	0.01635437
3588766.	219	0.01336420	4408119.	269	0.01641539
3605153.	220	0.01342523	4424506.	270	0.01647641
3621540.	221	0.01348625	4440893.	271	0.01653744
3637927.	222	0.01354727	4457280.	272	0.01659846
3654314.	223	0.01360830	4473667.	273	0.01665949
3670701.	224	0.01366932	4490054.	274	0.01672051
3687089.	225	0.01373035	4506442.	275	0.01678153
3703476.	226	0.01379137	4522829.	276	0.01684256
3719863.	227	0.01385239	4539216.	277	0.01690358
3736250.	228	0.01391342	4555603.	278	0.01696461
3752637.	229	0.01397444	4571990.	279	0.01702563
3769024.	230	0.01403546	4588377.	280	0.01708665
3785411.	231	0.01409649	4604764.	281	0.01714768
3801798.	232	0.01415751	4621151.	282	0.01720870
3818185.	233	0.01421854	4637538.	283	0.01726972
3834572.	234	0.01427956	4653925.	284	0.01733075
3850959.	235	0.01434058	4670312.	285	0.01739177
3867346.	236	0.01440161	4686699.	286	0.01745280
3883733.	237	0.01446263	4703086.	287	0.01751382
3900120.	238	0.01452365	4719473.	288	0.01757484
3916507.	239	0.01458468	4735860.	289	0.01763587
3932894.	240	0.01464570	4752247.	290	0.01769689
3949281.	241	0.01470673	4768635.	291	0.01775791
3965669.	242	0.01476775	4785022.	292	0.01781894
3982056.	243	0.01482877	4801409.	293	0.01787996
3998443.	244	0.01488980	4817796.	294	0.01794099
4014830.	245	0.01495082	4834183.	295	0.01800201
4031217.	246	0.01501184	4850570.	296	0.01806303
4047604.	247	0.01507287	4866957.	297	0.01812406
4063991.	248	0.01513389	4883344.	298	0.01818508
4080378.	249	0.01519492	4899731.	299	0.01824610
4096765.	250	0.01525594	4916118.	300	0.01830713

```
cm³ ————————— + 3 ——————————→
  l ————————— + 6 ——————————→  mm³
 m³ ————————— + 9 ——————————→
```

mm³ / mm³/s	REF	in³ / in³/s	mm³ / mm³/s	REF	in³ / in³/s
4932505.	301	0.01836815	5751858.	351	0.02141934
4948892.	302	0.01842918	5768245.	352	0.02148036
4965279.	303	0.01849020	5784632.	353	0.02154139
4981666.	304	0.01855122	5801019.	354	0.02160241
4998053.	305	0.01861225	5817406.	355	0.02166343
5014440.	306	0.01867327	5833793.	356	0.02172446
5030827.	307	0.01873429	5850180.	357	0.02178548
5047215.	308	0.01879532	5866568.	358	0.02184651
5063602.	309	0.01885634	5882955.	359	0.02190753
5079989.	310	0.01891737	5899342.	360	0.02196855
5096376.	311	0.01897839	5915729.	361	0.02202958
5112763.	312	0.01903941	5932116.	362	0.02209060
5129150.	313	0.01910044	5948503.	363	0.02215162
5145537.	314	0.01916146	5964890.	364	0.02221265
5161924.	315	0.01922248	5981277.	365	0.02227367
5178311.	316	0.01928351	5997664.	366	0.02233470
5194698.	317	0.01934453	6014051.	367	0.02239572
5211085.	318	0.01940556	6030438.	368	0.02245674
5227472.	319	0.01946658	6046825.	369	0.02251777
5243859.	320	0.01952760	6063212.	370	0.02257879
5260246.	321	0.01958863	6079599.	371	0.02263981
5276633.	322	0.01964965	6095986.	372	0.02270084
5293020.	323	0.01971067	6112373.	373	0.02276186
5309407.	324	0.01977170	6128760.	374	0.02282289
5325795.	325	0.01983272	6145148.	375	0.02288391
5342182.	326	0.01989375	6161535.	376	0.02294493
5358569.	327	0.01995477	6177922.	377	0.02300596
5374956.	328	0.02001579	6194309.	378	0.02306698
5391343.	329	0.02007682	6210696.	379	0.02312800
5407730.	330	0.02013784	6227083.	380	0.02318903
5424117.	331	0.02019886	6243470.	381	0.02325005
5440504.	332	0.02025989	6259857.	382	0.02331108
5456891.	333	0.02032091	6276244.	383	0.02337210
5473278.	334	0.02038194	6292631.	384	0.02343312
5489665.	335	0.02044296	6309018.	385	0.02349415
5506052.	336	0.02050398	6325405.	386	0.02355517
5522439.	337	0.02056501	6341792.	387	0.02361619
5538826.	338	0.02062603	6358179.	388	0.02367722
5555213.	339	0.02068705	6374566.	389	0.02373824
5571600.	340	0.02074808	6390953.	390	0.02379927
5587988.	341	0.02080910	6407341.	391	0.02386029
5604375.	342	0.02087013	6423728.	392	0.02392131
5620762.	343	0.02093115	6440115.	393	0.02398234
5637149.	344	0.02099217	6456502.	394	0.02404336
5653536.	345	0.02105320	6472889.	395	0.02410438
5669923.	346	0.02111422	6489276.	396	0.02416541
5686310.	347	0.02117524	6505663.	397	0.02422643
5702697.	348	0.02123627	6522050.	398	0.02428746
5719084.	349	0.02129729	6538437.	399	0.02434848
5735471.	350	0.02135832	6554824.	400	0.02440950

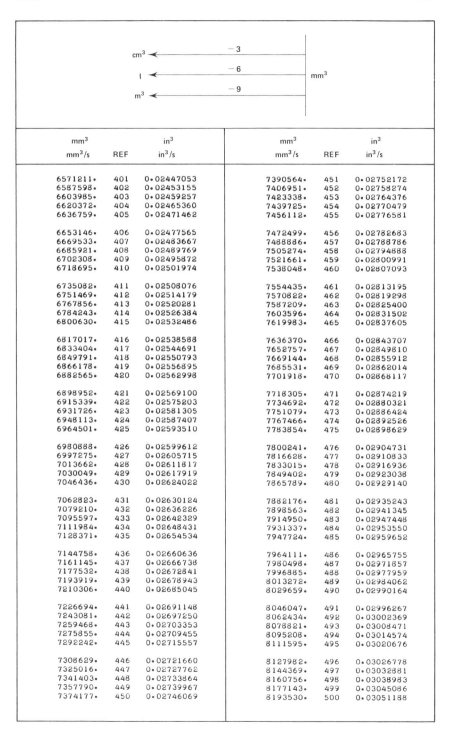

mm³		in³	mm³		in³
mm³/s	REF	in³/s	mm³/s	REF	in³/s
6571211.	401	0.02447053	7390564.	451	0.02752172
6587598.	402	0.02453155	7406951.	452	0.02758274
6603985.	403	0.02459257	7423338.	453	0.02764376
6620372.	404	0.02465360	7439725.	454	0.02770479
6636759.	405	0.02471462	7456112.	455	0.02776581
6653146.	406	0.02477565	7472499.	456	0.02782683
6669533.	407	0.02483667	7488886.	457	0.02788786
6685921.	408	0.02489769	7505274.	458	0.02794888
6702308.	409	0.02495872	7521661.	459	0.02800991
6718695.	410	0.02501974	7538048.	460	0.02807093
6735082.	411	0.02508076	7554435.	461	0.02813195
6751469.	412	0.02514179	7570822.	462	0.02819298
6767856.	413	0.02520281	7587209.	463	0.02825400
6784243.	414	0.02526384	7603596.	464	0.02831502
6800630.	415	0.02532486	7619983.	465	0.02837605
6817017.	416	0.02538588	7636370.	466	0.02843707
6833404.	417	0.02544691	7652757.	467	0.02849810
6849791.	418	0.02550793	7669144.	468	0.02855912
6866178.	419	0.02556895	7685531.	469	0.02862014
6882565.	420	0.02562998	7701918.	470	0.02868117
6898952.	421	0.02569100	7718305.	471	0.02874219
6915339.	422	0.02575203	7734692.	472	0.02880321
6931726.	423	0.02581305	7751079.	473	0.02886424
6948113.	424	0.02587407	7767466.	474	0.02892526
6964501.	425	0.02593510	7783854.	475	0.02898629
6980888.	426	0.02599612	7800241.	476	0.02904731
6997275.	427	0.02605715	7816628.	477	0.02910833
7013662.	428	0.02611817	7833015.	478	0.02916936
7030049.	429	0.02617919	7849402.	479	0.02923038
7046436.	430	0.02624022	7865789.	480	0.02929140
7062823.	431	0.02630124	7882176.	481	0.02935243
7079210.	432	0.02636226	7898563.	482	0.02941345
7095597.	433	0.02642329	7914950.	483	0.02947448
7111984.	434	0.02648431	7931337.	484	0.02953550
7128371.	435	0.02654534	7947724.	485	0.02959652
7144758.	436	0.02660636	7964111.	486	0.02965755
7161145.	437	0.02666738	7980498.	487	0.02971857
7177532.	438	0.02672841	7996885.	488	0.02977959
7193919.	439	0.02678943	8013272.	489	0.02984062
7210306.	440	0.02685045	8029659.	490	0.02990164
7226694.	441	0.02691148	8046047.	491	0.02996267
7243081.	442	0.02697250	8062434.	492	0.03002369
7259468.	443	0.02703353	8078821.	493	0.03008471
7275855.	444	0.02709455	8095208.	494	0.03014574
7292242.	445	0.02715557	8111595.	495	0.03020676
7308629.	446	0.02721660	8127982.	496	0.03026778
7325016.	447	0.02727762	8144369.	497	0.03032881
7341403.	448	0.02733864	8160756.	498	0.03038983
7357790.	449	0.02739967	8177143.	499	0.03045086
7374177.	450	0.02746069	8193530.	500	0.03051188

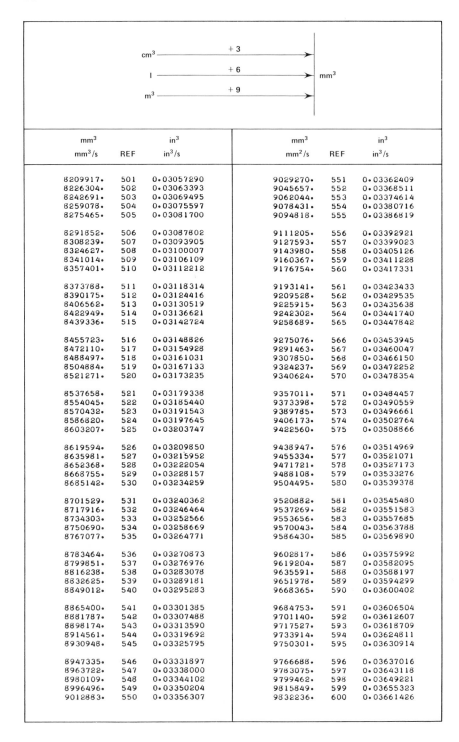

mm³ (mm³/s)	REF	in³ (in³/s)	mm³ (mm²/s)	REF	in³ (in³/s)
8209917.	501	0.03057290	9029270.	551	0.03362409
8226304.	502	0.03063393	9045657.	552	0.03368511
8242691.	503	0.03069495	9062044.	553	0.03374614
8259078.	504	0.03075597	9078431.	554	0.03380716
8275465.	505	0.03081700	9094818.	555	0.03386819
8291852.	506	0.03087802	9111205.	556	0.03392921
8308239.	507	0.03093905	9127593.	557	0.03399023
8324627.	508	0.03100007	9143980.	558	0.03405126
8341014.	509	0.03106109	9160367.	559	0.03411228
8357401.	510	0.03112212	9176754.	560	0.03417331
8373788.	511	0.03118314	9193141.	561	0.03423433
8390175.	512	0.03124416	9209528.	562	0.03429535
8406562.	513	0.03130519	9225915.	563	0.03435638
8422949.	514	0.03136621	9242302.	564	0.03441740
8439336.	515	0.03142724	9258689.	565	0.03447842
8455723.	516	0.03148826	9275076.	566	0.03453945
8472110.	517	0.03154928	9291463.	567	0.03460047
8488497.	518	0.03161031	9307850.	568	0.03466150
8504884.	519	0.03167133	9324237.	569	0.03472252
8521271.	520	0.03173235	9340624.	570	0.03478354
8537658.	521	0.03179338	9357011.	571	0.03484457
8554045.	522	0.03185440	9373398.	572	0.03490559
8570432.	523	0.03191543	9389785.	573	0.03496661
8586820.	524	0.03197645	9406173.	574	0.03502764
8603207.	525	0.03203747	9422560.	575	0.03508866
8619594.	526	0.03209850	9438947.	576	0.03514969
8635981.	527	0.03215952	9455334.	577	0.03521071
8652368.	528	0.03222054	9471721.	578	0.03527173
8668755.	529	0.03228157	9488108.	579	0.03533276
8685142.	530	0.03234259	9504495.	580	0.03539378
8701529.	531	0.03240362	9520882.	581	0.03545480
8717916.	532	0.03246464	9537269.	582	0.03551583
8734303.	533	0.03252566	9553656.	583	0.03557685
8750690.	534	0.03258669	9570043.	584	0.03563788
8767077.	535	0.03264771	9586430.	585	0.03569890
8783464.	536	0.03270873	9602817.	586	0.03575992
8799851.	537	0.03276976	9619204.	587	0.03582095
8816238.	538	0.03283078	9635591.	588	0.03588197
8832625.	539	0.03289181	9651978.	589	0.03594299
8849012.	540	0.03295283	9668365.	590	0.03600402
8865400.	541	0.03301385	9684753.	591	0.03606504
8881787.	542	0.03307488	9701140.	592	0.03612607
8898174.	543	0.03313590	9717527.	593	0.03618709
8914561.	544	0.03319692	9733914.	594	0.03624811
8930948.	545	0.03325795	9750301.	595	0.03630914
8947335.	546	0.03331897	9766688.	596	0.03637016
8963722.	547	0.03338000	9783075.	597	0.03643118
8980109.	548	0.03344102	9799462.	598	0.03649221
8996496.	549	0.03350204	9815849.	599	0.03655323
9012883.	550	0.03356307	9832236.	600	0.03661426

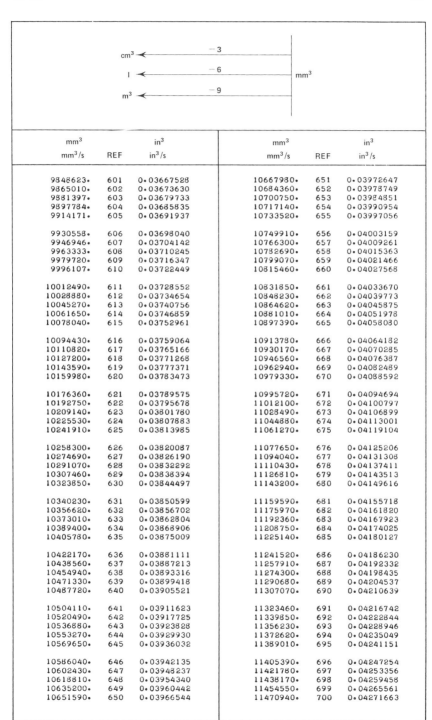

mm³			mm³		
mm³/s	REF	in³/s	mm³/s	REF	in³/s
9848623.	601	0.03667528	10667980.	651	0.03972647
9865010.	602	0.03673630	10684360.	652	0.03978749
9881397.	603	0.03679733	10700750.	653	0.03984851
9897784.	604	0.03685835	10717140.	654	0.03990954
9914171.	605	0.03691937	10733520.	655	0.03997056
9930558.	606	0.03698040	10749910.	656	0.04003159
9946946.	607	0.03704142	10766300.	657	0.04009261
9963333.	608	0.03710245	10782690.	658	0.04015363
9979720.	609	0.03716347	10799070.	659	0.04021466
9996107.	610	0.03722449	10815460.	660	0.04027568
10012490.	611	0.03728552	10831850.	661	0.04033670
10028880.	612	0.03734654	10848230.	662	0.04039773
10045270.	613	0.03740756	10864620.	663	0.04045875
10061650.	614	0.03746859	10881010.	664	0.04051978
10078040.	615	0.03752961	10897390.	665	0.04058030
10094430.	616	0.03759064	10913780.	666	0.04064132
10110820.	617	0.03765166	10930170.	667	0.04070285
10127200.	618	0.03771268	10946560.	668	0.04076387
10143590.	619	0.03777371	10962940.	669	0.04082489
10159980.	620	0.03783473	10979330.	670	0.04088592
10176360.	621	0.03789575	10995720.	671	0.04094694
10192750.	622	0.03795678	11012100.	672	0.04100797
10209140.	623	0.03801780	11028490.	673	0.04106899
10225530.	624	0.03807883	11044880.	674	0.04113001
10241910.	625	0.03813985	11061270.	675	0.04119104
10258300.	626	0.03820087	11077650.	676	0.04125206
10274690.	627	0.03826190	11094040.	677	0.04131308
10291070.	628	0.03832292	11110430.	678	0.04137411
10307460.	629	0.03838394	11126810.	679	0.04143513
10323850.	630	0.03844497	11143200.	680	0.04149616
10340230.	631	0.03850599	11159590.	681	0.04155718
10356620.	632	0.03856702	11175970.	682	0.04161820
10373010.	633	0.03862804	11192360.	683	0.04167923
10389400.	634	0.03868906	11208750.	684	0.04174025
10405780.	635	0.03875009	11225140.	685	0.04180127
10422170.	636	0.03881111	11241520.	686	0.04186230
10438560.	637	0.03887213	11257910.	687	0.04192332
10454940.	638	0.03893316	11274300.	688	0.04198435
10471330.	639	0.03899418	11290680.	689	0.04204537
10487720.	640	0.03905521	11307070.	690	0.04210639
10504110.	641	0.03911623	11323460.	691	0.04216742
10520490.	642	0.03917725	11339850.	692	0.04222844
10536880.	643	0.03923828	11356230.	693	0.04228946
10553270.	644	0.03929930	11372620.	694	0.04235049
10569650.	645	0.03936032	11389010.	695	0.04241151
10586040.	646	0.03942135	11405390.	696	0.04247254
10602430.	647	0.03948237	11421780.	697	0.04253356
10618810.	648	0.03954340	11438170.	698	0.04259458
10635200.	649	0.03960442	11454550.	699	0.04265561
10651590.	650	0.03966544	11470940.	700	0.04271663

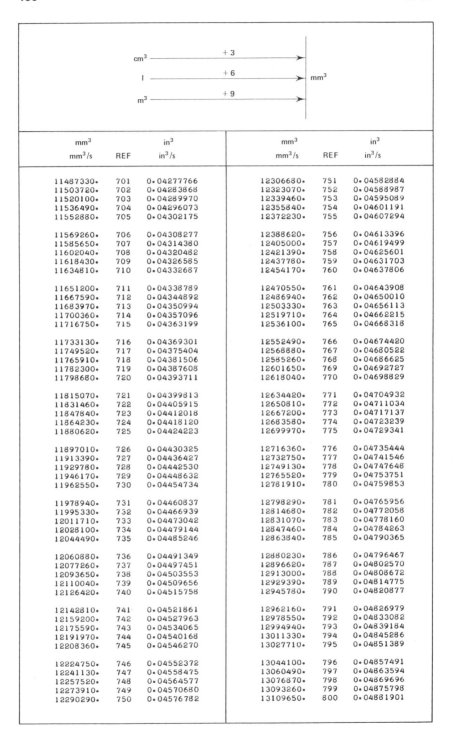

| mm³ | | in³ | mm³ | | in³ |
mm³/s	REF	in³/s	mm³/s	REF	in³/s
11487330.	701	0.04277766	12306680.	751	0.04582884
11503720.	702	0.04283868	12323070.	752	0.04588987
11520100.	703	0.04289970	12339460.	753	0.04595089
11536490.	704	0.04296073	12355840.	754	0.04601191
11552880.	705	0.04302175	12372230.	755	0.04607294
11569260.	706	0.04308277	12388620.	756	0.04613396
11585650.	707	0.04314380	12405000.	757	0.04619499
11602040.	708	0.04320482	12421390.	758	0.04625601
11618430.	709	0.04326585	12437780.	759	0.04631703
11634810.	710	0.04332687	12454170.	760	0.04637806
11651200.	711	0.04338789	12470550.	761	0.04643908
11667590.	712	0.04344892	12486940.	762	0.04650010
11683970.	713	0.04350994	12503330.	763	0.04656113
11700360.	714	0.04357096	12519710.	764	0.04662215
11716750.	715	0.04363199	12536100.	765	0.04668318
11733130.	716	0.04369301	12552490.	766	0.04674420
11749520.	717	0.04375404	12568880.	767	0.04680522
11765910.	718	0.04381506	12585290.	768	0.04686625
11782300.	719	0.04387608	12601650.	769	0.04692727
11798680.	720	0.04393711	12618040.	770	0.04698829
11815070.	721	0.04399813	12634420.	771	0.04704932
11831460.	722	0.04405915	12650810.	772	0.04711034
11847840.	723	0.04412018	12667200.	773	0.04717137
11864230.	724	0.04418120	12683580.	774	0.04723239
11880620.	725	0.04424223	12699970.	775	0.04729341
11897010.	726	0.04430325	12716360.	776	0.04735444
11913390.	727	0.04436427	12732750.	777	0.04741546
11929780.	728	0.04442530	12749130.	778	0.04747648
11946170.	729	0.04448632	12765520.	779	0.04753751
11962550.	730	0.04454734	12781910.	780	0.04759853
11978940.	731	0.04460837	12798290.	781	0.04765956
11995330.	732	0.04466939	12814680.	782	0.04772058
12011710.	733	0.04473042	12831070.	783	0.04778160
12028100.	734	0.04479144	12847460.	784	0.04784263
12044490.	735	0.04485246	12863840.	785	0.04790365
12060880.	736	0.04491349	12880230.	786	0.04796467
12077260.	737	0.04497451	12896620.	787	0.04802570
12093650.	738	0.04503553	12913000.	788	0.04808672
12110040.	739	0.04509656	12929390.	789	0.04814775
12126420.	740	0.04515758	12945780.	790	0.04820877
12142810.	741	0.04521861	12962160.	791	0.04826979
12159200.	742	0.04527963	12978550.	792	0.04833082
12175590.	743	0.04534065	12994940.	793	0.04839184
12191970.	744	0.04540168	13011330.	794	0.04845286
12208360.	745	0.04546270	13027710.	795	0.04851389
12224750.	746	0.04552372	13044100.	796	0.04857491
12241130.	747	0.04558475	13060490.	797	0.04863594
12257520.	748	0.04564577	13076870.	798	0.04869696
12273910.	749	0.04570680	13093260.	799	0.04875798
12290290.	750	0.04576782	13109650.	800	0.04881901

| mm³ | | | mm³ | | |
mm³/s	REF	in³/s	mm³/s	REF	in³/s
13126040.	801	0.04888003	13945390.	851	0.05193122
13142420.	802	0.04894105	13961780.	852	0.05199224
13158810.	803	0.04900208	13978160.	853	0.05205327
13175200.	804	0.04906310	13994550.	854	0.05211429
13191580.	805	0.04912413	14010940.	855	0.05217531
13207970.	806	0.04918515	14027320.	856	0.05223634
13224360.	807	0.04924617	14043710.	857	0.05229736
13240740.	808	0.04930720	14060100.	858	0.05235839
13257130.	809	0.04936822	14076480.	859	0.05241941
13273520.	810	0.04942924	14092870.	860	0.05248043
13289910.	811	0.04949027	14109260.	861	0.05254146
13306290.	812	0.04955129	14125650.	862	0.05260248
13322680.	813	0.04961232	14142030.	863	0.05266350
13339070.	814	0.04967334	14158420.	864	0.05272453
13355450.	815	0.04973436	14174810.	865	0.05278555
13371840.	816	0.04979539	14191190.	866	0.05284658
13388230.	817	0.04985641	14207580.	867	0.05290760
13404620.	818	0.04991743	14223970.	868	0.05296862
13421000.	819	0.04997846	14240360.	869	0.05302965
13437390.	820	0.05003948	14256740.	870	0.05309067
13453780.	821	0.05010051	14273130.	871	0.05315169
13470160.	822	0.05016153	14289520.	872	0.05321272
13486550.	823	0.05022255	14305900.	873	0.05327374
13502940.	824	0.05028358	14322290.	874	0.05333477
13519320.	825	0.05034460	14338680.	875	0.05339579
13535710.	826	0.05040562	14355060.	876	0.05345681
13552100.	827	0.05046665	14371450.	877	0.05351784
13568490.	828	0.05052767	14387840.	878	0.05357886
13584870.	829	0.05058870	14404230.	879	0.05363988
13601260.	830	0.05064972	14420610.	880	0.05370091
13617650.	831	0.05071074	14437000.	881	0.05376193
13634030.	832	0.05077177	14453390.	882	0.05382296
13650420.	833	0.05083279	14469770.	883	0.05388398
13666810.	834	0.05089381	14486160.	884	0.05394500
13683200.	835	0.05095484	14502550.	885	0.05400603
13699580.	836	0.05101586	14518940.	886	0.05406705
13715970.	837	0.05107689	14535320.	887	0.05412807
13732360.	838	0.05113791	14551710.	888	0.05418910
13748740.	839	0.05119893	14568100.	889	0.05425012
13765130.	840	0.05125996	14584480.	890	0.05431115
13781520.	841	0.05132098	14600870.	891	0.05437217
13797900.	842	0.05138201	14617260.	892	0.05443319
13814290.	843	0.05144303	14633640.	893	0.05449422
13830680.	844	0.05150405	14650030.	894	0.05455524
13847070.	845	0.05156508	14666420.	895	0.05461626
13863450.	846	0.05162610	14682810.	896	0.05467729
13879840.	847	0.05168712	14699190.	897	0.05473831
13896230.	848	0.05174815	14715580.	898	0.05479934
13912610.	849	0.05180917	14731970.	899	0.05486036
13929000.	850	0.05187020	14748350.	900	0.05492138

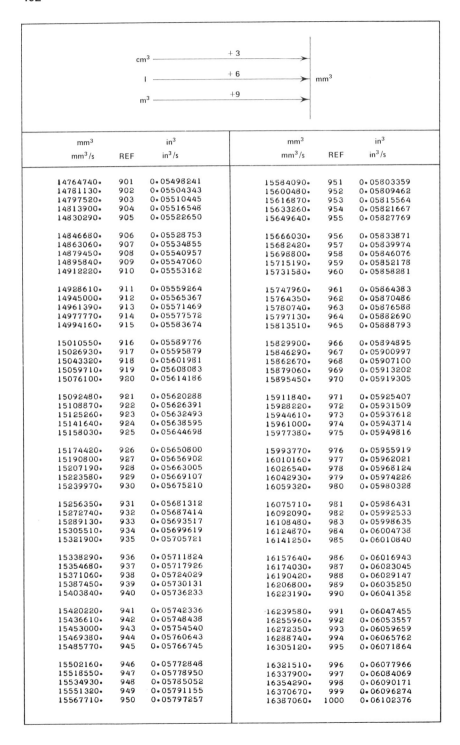

mm³		in³	mm³		in³
mm³/s	REF	in³/s	mm³/s	REF	in³/s
14764740.	901	0.05498241	15584090.	951	0.05803359
14781130.	902	0.05504343	15600480.	952	0.05809462
14797520.	903	0.05510445	15616870.	953	0.05815564
14813900.	904	0.05516548	15633260.	954	0.05821667
14830290.	905	0.05522650	15649640.	955	0.05827769
14846680.	906	0.05528753	15666030.	956	0.05833871
14863060.	907	0.05534855	15682420.	957	0.05839974
14879450.	908	0.05540957	15698800.	958	0.05846076
14895840.	909	0.05547060	15715190.	959	0.05852178
14912220.	910	0.05553162	15731580.	960	0.05858281
14928610.	911	0.05559264	15747960.	961	0.05864383
14945000.	912	0.05565367	15764350.	962	0.05870486
14961390.	913	0.05571469	15780740.	963	0.05876588
14977770.	914	0.05577572	15797130.	964	0.05882690
14994160.	915	0.05583674	15813510.	965	0.05888793
15010550.	916	0.05589776	15829900.	966	0.05894895
15026930.	917	0.05595879	15846290.	967	0.05900997
15043320.	918	0.05601981	15862670.	968	0.05907100
15059710.	919	0.05608083	15879060.	969	0.05913202
15076100.	920	0.05614186	15895450.	970	0.05919305
15092480.	921	0.05620288	15911840.	971	0.05925407
15108870.	922	0.05626391	15928220.	972	0.05931509
15125260.	923	0.05632493	15944610.	973	0.05937612
15141640.	924	0.05638595	15961000.	974	0.05943714
15158030.	925	0.05644698	15977380.	975	0.05949816
15174420.	926	0.05650800	15993770.	976	0.05955919
15190800.	927	0.05656902	16010160.	977	0.05962021
15207190.	928	0.05663005	16026540.	978	0.05968124
15223580.	929	0.05669107	16042930.	979	0.05974226
15239970.	930	0.05675210	16059320.	980	0.05980328
15256350.	931	0.05681312	16075710.	981	0.05986431
15272740.	932	0.05687414	16092090.	982	0.05992533
15289130.	933	0.05693517	16108480.	983	0.05998635
15305510.	934	0.05699619	16124870.	984	0.06004738
15321900.	935	0.05705721	16141250.	985	0.06010840
15338290.	936	0.05711824	16157640.	986	0.06016943
15354680.	937	0.05717926	16174030.	987	0.06023045
15371060.	938	0.05724029	16190420.	988	0.06029147
15387450.	939	0.05730131	16206800.	989	0.06035250
15403840.	940	0.05736233	16223190.	990	0.06041352
15420220.	941	0.05742336	16239580.	991	0.06047455
15436610.	942	0.05748438	16255960.	992	0.06053557
15453000.	943	0.05754540	16272350.	993	0.06059659
15469380.	944	0.05760643	16288740.	994	0.06065762
15485770.	945	0.05766745	16305120.	995	0.06071864
15502160.	946	0.05772848	16321510.	996	0.06077966
15518550.	947	0.05778950	16337900.	997	0.06084069
15534930.	948	0.05785052	16354290.	998	0.06090171
15551320.	949	0.05791155	16370670.	999	0.06096274
15567710.	950	0.05797257	16387060.	1000	0.06102376

11

VOLUME OR SECTION MODULUS

Cubic-foot — Cubic-metre
     $(ft^3 — m^3)$
Cubic-foot — Cubic-centimetre
     $(ft^3 — cm^3)$
Cubic-foot — Cubic-kilometre
     $(ft^3 — km^3)$

VOLUME

Cubic-foot — Litre
     $(ft^3 — l)$
Register-ton — Cubic-metre
     $(ton [r] — m^3)$

VOLUME FLOW RATE

Cubic-foot/Second — Cubic-metre/Second
     $(ft^3/s — m^3/s)$
Cubic-foot/Second — Cubic-centimetre/Second
     $(ft^3/s — cm^3/s)$
Cubic-foot/Second — Litre/Second
     $(ft^3/s — l/s)$

Conversion Factors:
     1 cubic-foot $(ft^3) = 0.02831685$ cubic-metre (m^3)
     $(1 \ m^3 = 35.31466 \ ft^3)$

Section Modulus and Volume Flow Rate Values

Section modulus quantities are expressed in the same compound units as volume, in this instance ft^3. Therefore Table 11 can be used for the conversion of these values.

Volume flow rate is also represented by this table, as long as the time units are the same for both English and SI conversion quantities. Besides per second shown above, these consistent time units can also be in per minute, per hour, per day, etc.

Volume Flow Rate Conversion Factor:

1 ft^3/s (or ft^3/min, etc.) = 0.02831685 m^3/s (or m^3/min, etc.)

Register Ton Values

The Register Ton, normally used in the maritime trade, is a measure of volume where:

1 ton (register) = 100 cubic feet
Therefore, 1 ton (r) = 2.831685 m^3.

This conversion factor differs from the above only by a power of 10. Table 11 can then be used for conversion of register tons to and from SI units by properly shifting through two places the decimal point of the conversion values obtained from the table. The Decimal-shift Diagram 11-A will help the table user.

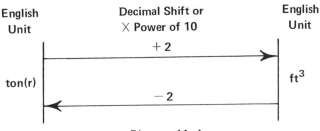

Diagram 11-A

Example (a): Convert 392 register tons to m^3.
From Diagram 11-A: 392 ton (r) = 392 × 10^2 ft^3
From table: 392 ft^3 = 11.10021 m^3 *
Therefore: 392 ton (r) $-$ 11.10021 × 10^2 m^3 = 1,110.021 m^3
$\approx$ 1,110 m^3.

---

*All conversion values obtained can be rounded to desired number of significant figures.

Example (b): Convert 750 m³ to register tons.
   From table: 750 m³ = 26,486.00 ft³*
   From Diagram 11-A: 26,486 ft³ = 26,486 × 10⁻² or 264.86 ton (r)
                              ≈ 265 ton (r)
                              (ft³ to ton (r) by shifting decimal 2 places to the left.)

In converting the quantities mentioned above to units containing cm³, km³, and l (litres), Diagram 11-B, which also appears in the headings on alternate pages of this table will assist the user. It is derived from: 1 m³ = (10⁶) cm³ = (10³) l (litres) = (10⁻⁹) km³; 1 litre = 1 dm³

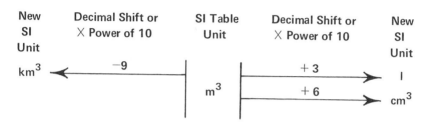

New SI Unit	Decimal Shift or × Power of 10	SI Table Unit	Decimal Shift or × Power of 10	New SI Unit
km³	← −9			
		m³	+3 →	l
			+6 →	cm³

Diagram 11-B

Example (c): Convert 879 ft³/min to m³/min; cm³/min; l/min.
   From table: 879 ft³/min = 24.89051 m³/min*
   From Diagram 11-A: 1) 24.89051 m³/min = 24.89051 × 10⁶ cm³/min
                              = 24,890,510 cm³/min* ≈ 24.9 × 10⁶ cm³/min
                              (m³ to cm³ by multiplying m³ value by 10⁶ or by
                              shifting its decimal 6 places to the right.)

                          2) 24.89051 m³/min = 24.89051 × 10³ l/min
                              = 24,890.51 l/min ≈ 24,900 l/min
                              (m³ to l by multiplying m³ value by 10³ or by
                              shifting its decimal 3 places to the right.)

Example (d): Convert 619,000,000 ft³ to km³.
   Since: 619,000,000 ft³ = 619 × 10⁶ ft³:
   From table: 619 ft³ = 17.52813 × 10⁶ m³*
   Therefore: 619 × 10⁶ ft³ = 17.52813 × 10⁶ m³.
   From Diagram 11-B: 17.52813 × 10⁶ m³ = 17.52813 × 10⁶ (10⁻⁹) km³
                              = 17.52813 × 10⁻³ or 0.01752813 km³
                              ≈ 1.75 × 10⁻² km³

Diagram 11-C is also given at the top of alternate table pages:

_____
*All conversion values obtained can be rounded to desired number of significant figures.

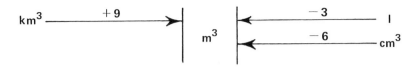

Diagram 11-C

It is the opposite of Diagram 11-B and is used to convert km^3, cm^3, and l to m^3. The m^3 equivalent is then applied to the REF column in the table to obtain ft^3 quantities.

Example: Convert 57 kilolitres per day to cubic feet per day.
   Since: 57 kl/d = 57 × 10^3 l/d
   From Diagram 11-C: 57 × 10^3 l/d = 57 × 10^3 (10^{-3}) m^3/d = 57 m^3/d
   From table: 57 m^3/d = 2,012.936 ft^3/d* ≈ 2,000 ft^3/d

---

*All conversion values obtained can be rounded to desired number of significant figures.

m^3		ft^3	m^3		ft^3
m^3/s	REF	ft^3/s	m^3/s	REF	ft^3/s
0.02831685	1	35.31466	1.444159	51	1801.048
0.05663370	2	70.62932	1.472476	52	1836.362
0.08495055	3	105.9440	1.500793	53	1871.677
0.1132674	4	141.2586	1.529110	54	1906.992
0.1415843	5	176.5733	1.557427	55	1942.306
0.1699011	6	211.8880	1.585744	56	1977.621
0.1982180	7	247.2026	1.614060	57	2012.936
0.2265348	8	282.5173	1.642377	58	2048.250
0.2548517	9	317.8320	1.670694	59	2083.565
0.2831685	10	353.1466	1.699011	60	2118.880
0.3114854	11	388.4613	1.727328	61	2154.194
0.3398022	12	423.7760	1.755645	62	2189.509
0.3681190	13	459.0906	1.783962	63	2224.824
0.3964359	14	494.4053	1.812278	64	2260.138
0.4247527	15	529.7199	1.840595	65	2295.453
0.4530696	16	565.0346	1.868912	66	2330.768
0.4813865	17	600.3493	1.897229	67	2366.082
0.5097033	18	635.6639	1.925546	68	2401.397
0.5380201	19	670.9786	1.953863	69	2436.712
0.5663370	20	706.2933	1.982180	70	2472.026
0.5946539	21	741.6079	2.010496	71	2507.341
0.6229707	22	776.9226	2.038813	72	2542.656
0.6512875	23	812.2372	2.067130	73	2577.970
0.6796044	24	847.5519	2.095447	74	2613.285
0.7079213	25	882.8666	2.123764	75	2648.600
0.7362381	26	918.1812	2.152081	76	2683.914
0.7645550	27	953.4959	2.180397	77	2719.229
0.7928718	28	988.8105	2.208714	78	2754.544
0.8211887	29	1024.125	2.237031	79	2789.858
0.8495055	30	1059.440	2.265348	80	2825.173
0.8778224	31	1094.755	2.293665	81	2860.488
0.9061392	32	1130.069	2.321982	82	2895.802
0.9344561	33	1165.384	2.350299	83	2931.117
0.9627729	34	1200.699	2.373615	84	2966.432
0.9910898	35	1236.013	2.406932	85	3001.746
1.019407	36	1271.328	2.435249	86	3037.061
1.047723	37	1306.643	2.463566	87	3072.376
1.076040	38	1341.957	2.491883	88	3107.690
1.104357	39	1377.272	2.520200	89	3143.005
1.132674	40	1412.587	2.548517	90	3178.320
1.160991	41	1447.901	2.576833	91	3213.634
1.189308	42	1483.216	2.605150	92	3248.949
1.217625	43	1518.530	2.633467	93	3284.264
1.245941	44	1553.845	2.661784	94	3319.578
1.274258	45	1589.160	2.690101	95	3354.893
1.302575	46	1624.474	2.718418	96	3390.208
1.330892	47	1659.789	2.746734	97	3425.522
1.359209	48	1695.104	2.775051	98	3460.837
1.387526	49	1730.418	2.803368	99	3496.152
1.415843	50	1765.733	2.831685	100	3531.466

$km^3 \xrightarrow{\quad +9 \quad} m^3 \xleftarrow{\quad -3 \quad} I$

$m^3 \xleftarrow{\quad -6 \quad} cm^3$

m^3 m^3/s	REF	ft^3 ft^3/s	m^3 m^3/s	REF	ft^3 ft^3/s
2.860002	101	3566.781	4.275844	151	5332.514
2.888319	102	3602.096	4.304161	152	5367.829
2.916636	103	3637.410	4.332478	153	5403.143
2.944952	104	3672.725	4.360795	154	5438.458
2.973269	105	3708.040	4.389112	155	5473.773
3.001586	106	3743.354	4.417429	156	5509.087
3.029903	107	3778.669	4.445745	157	5544.402
3.058220	108	3813.984	4.474062	158	5579.717
3.086537	109	3849.298	4.502379	159	5615.031
3.114854	110	3884.613	4.530696	160	5650.346
3.143170	111	3919.928	4.559013	161	5685.661
3.171487	112	3955.242	4.587330	162	5720.975
3.199804	113	3990.557	4.615647	163	5756.290
3.228121	114	4025.872	4.643963	164	5791.605
3.256438	115	4061.186	4.672280	165	5826.919
3.284755	116	4096.501	4.700597	166	5862.234
3.313071	117	4131.815	4.728914	167	5897.549
3.341388	118	4167.130	4.757231	168	5932.863
3.369705	119	4202.445	4.785548	169	5968.178
3.398022	120	4237.760	4.813865	170	6003.493
3.426339	121	4273.074	4.842181	171	6038.807
3.454656	122	4308.389	4.870498	172	6074.122
3.482973	123	4343.703	4.898815	173	6109.437
3.511289	124	4379.018	4.927132	174	6144.751
3.539606	125	4414.333	4.955449	175	6180.066
3.567923	126	4449.647	4.983766	176	6215.381
3.596240	127	4484.962	5.012082	177	6250.695
3.624557	128	4520.277	5.040399	178	6286.010
3.652874	129	4555.591	5.068716	179	6321.325
3.681191	130	4590.906	5.097033	180	6356.639
3.709507	131	4626.221	5.125350	181	6391.954
3.737824	132	4661.535	5.153667	182	6427.269
3.766141	133	4696.850	5.181984	183	6462.583
3.794458	134	4732.165	5.210300	184	6497.898
3.822775	135	4767.479	5.238617	185	6533.213
3.851092	136	4802.794	5.266934	186	6568.527
3.879408	137	4838.109	5.295251	187	6603.842
3.907725	138	4873.423	5.323568	188	6639.157
3.936042	139	4908.738	5.351885	189	6674.471
3.964359	140	4944.053	5.380202	190	6709.786
3.992676	141	4979.367	5.408518	191	6745.101
4.020993	142	5014.682	5.436835	192	6780.415
4.049310	143	5049.997	5.465152	193	6815.730
4.077626	144	5085.311	5.493469	194	6851.044
4.105943	145	5120.626	5.521786	195	6886.359
4.134260	146	5155.941	5.550103	196	6921.674
4.162577	147	5191.255	5.578419	197	6956.989
4.190894	148	5226.570	5.606736	198	6992.303
4.219211	149	5261.885	5.635053	199	7027.618
4.247528	150	5297.199	5.663370	200	7062.932

km³ ← − 9 | m³ | + 3 → l
m³ | + 6 → cm³

m³		ft³	m³		ft³
m³/s	REF	m³/s	m³/s	REF	ft³/s
5.691687	201	7098.247	7.107529	251	8863.980
5.720004	202	7133.562	7.135846	252	8899.295
5.748321	203	7168.876	7.164163	253	8934.610
5.776637	204	7204.191	7.192480	254	8969.924
5.804954	205	7239.506	7.220797	255	9005.239
5.833271	206	7274.820	7.249114	256	9040.554
5.861588	207	7310.135	7.277430	257	9075.868
5.889905	208	7345.450	7.305747	258	9111.183
5.918222	209	7380.764	7.334064	259	9146.498
5.946539	210	7416.079	7.362381	260	9181.812
5.974855	211	7451.394	7.390698	261	9217.127
6.003172	212	7486.708	7.419015	262	9252.442
6.031489	213	7522.023	7.447332	263	9287.756
6.059806	214	7557.338	7.475648	264	9323.071
6.088123	215	7592.652	7.503965	265	9358.385
6.116440	216	7627.967	7.532282	266	9393.700
6.144756	217	7663.282	7.560599	267	9429.015
6.173073	218	7698.596	7.588916	268	9464.330
6.201390	219	7733.911	7.617233	269	9499.644
6.229707	220	7769.226	7.645550	270	9534.959
6.258024	221	7804.540	7.673866	271	9570.274
6.286341	222	7839.855	7.702183	272	9605.588
6.314658	223	7875.170	7.730500	273	9640.903
6.342974	224	7910.484	7.758817	274	9676.218
6.371291	225	7945.799	7.787134	275	9711.532
6.399608	226	7981.114	7.815451	276	9746.847
6.427925	227	8016.428	7.843767	277	9782.161
6.456242	228	8051.743	7.872084	278	9817.476
6.484559	229	8087.058	7.900401	279	9852.791
6.512875	230	8122.372	7.928718	280	9888.105
6.541192	231	8157.687	7.957035	281	9923.420
6.569509	232	8193.002	7.985352	282	9958.735
6.597826	233	8228.316	8.013669	283	9994.049
6.626143	234	8263.631	8.041985	284	10029.36
6.654460	235	8298.946	8.070302	285	10064.68
6.682777	236	8334.260	8.098619	286	10099.99
6.711093	237	8369.575	8.126936	287	10135.31
6.739410	238	8404.890	8.155253	288	10170.62
6.767727	239	8440.204	8.183570	289	10205.94
6.796044	240	8475.519	8.211887	290	10241.25
6.824361	241	8510.834	8.240203	291	10276.57
6.852678	242	8546.148	8.268520	292	10311.88
6.880995	243	8581.463	8.296837	293	10347.20
6.909311	244	8616.778	8.325154	294	10382.51
6.937628	245	8652.092	8.353471	295	10417.83
6.965945	246	8687.407	8.381788	296	10453.14
6.994262	247	8722.722	8.410105	297	10488.45
7.022579	248	8758.036	8.438421	298	10523.77
7.050896	249	8793.351	8.466738	299	10559.08
7.079213	250	8828.666	8.495055	300	10594.40

$$km^3 \xrightarrow{\quad +9 \quad} \quad | \quad m^3 \quad | \quad \xleftarrow{\quad -3 \quad} \quad l$$
$$\xleftarrow{\quad -6 \quad} \quad cm^3$$

m^3 m^3/s	REF	ft^3 ft^3/s	m^3 m^3/s	REF	ft^3 ft^3/s
8.523372	301	10629.71	9.939214	351	12395.45
8.551689	302	10665.03	9.967531	352	12430.76
8.580006	303	10700.34	9.995848	353	12466.08
8.608322	304	10735.66	10.02416	354	12501.39
8.636639	305	10770.97	10.05248	355	12536.71
8.664956	306	10806.29	10.08080	356	12572.02
8.693273	307	10841.60	10.10912	357	12607.33
8.721590	308	10876.92	10.13743	358	12642.65
8.749907	309	10912.23	10.16575	359	12677.96
8.778224	310	10947.55	10.19407	360	12713.28
8.806540	311	10982.86	10.22238	361	12748.59
8.834857	312	11018.17	10.25070	362	12783.91
8.863174	313	11053.49	10.27902	363	12819.22
8.891491	314	11088.80	10.30733	364	12854.54
8.919808	315	11124.12	10.33565	365	12889.85
8.948125	316	11159.43	10.36397	366	12925.17
8.976442	317	11194.75	10.39228	367	12960.48
9.004758	318	11230.06	10.42060	368	12995.80
9.033075	319	11265.38	10.44892	369	13031.11
9.061392	320	11300.69	10.47723	370	13066.43
9.089709	321	11336.01	10.50555	371	13101.74
9.118026	322	11371.32	10.53387	372	13137.05
9.146343	323	11406.64	10.56219	373	13172.37
9.174659	324	11441.95	10.59050	374	13207.68
9.202976	325	11477.27	10.61882	375	13243.00
9.231293	326	11512.58	10.64714	376	13278.31
9.259610	327	11547.89	10.67545	377	13313.63
9.287927	328	11583.21	10.70377	378	13348.94
9.316244	329	11618.52	10.73209	379	13384.26
9.344561	330	11653.84	10.76040	380	13419.57
9.372877	331	11689.15	10.78872	381	13454.89
9.401194	332	11724.47	10.81704	382	13490.20
9.429511	333	11759.78	10.84535	383	13525.52
9.457828	334	11795.10	10.87367	384	13560.83
9.486145	335	11830.41	10.90199	385	13596.15
9.514462	336	11865.73	10.93030	386	13631.46
9.542778	337	11901.04	10.95862	387	13666.77
9.571095	338	11936.36	10.98694	388	13702.09
9.599412	339	11971.67	11.01525	389	13737.40
9.627729	340	12006.99	11.04357	390	13772.72
9.656046	341	12042.30	11.07189	391	13808.03
9.684363	342	12077.61	11.10021	392	13843.35
9.712680	343	12112.93	11.12852	393	13878.66
9.740996	344	12148.24	11.15684	394	13913.98
9.769313	345	12183.56	11.18516	395	13949.29
9.797630	346	12218.87	11.21347	396	13984.61
9.825947	347	12254.19	11.24179	397	14019.92
9.854264	348	12289.50	11.27011	398	14055.24
9.882581	349	12324.82	11.29842	399	14090.55
9.910897	350	12360.13	11.32674	400	14125.86

m^3		ft^3		m^3		ft^3
m^3/s	REF	ft^3/s		m^3/s	REF	ft^3/s
11.35506	401	14161.18		12.77090	451	15926.91
11.38337	402	14196.49		12.79922	452	15962.23
11.41169	403	14231.81		12.82753	453	15997.54
11.44001	404	14267.12		12.85585	454	16032.86
11.46832	405	14302.44		12.88417	455	16068.17
11.49664	406	14337.75		12.91248	456	16103.49
11.52496	407	14373.07		12.94080	457	16138.80
11.55327	408	14408.38		12.96912	458	16174.12
11.58159	409	14443.70		12.99743	459	16209.43
11.60991	410	14479.01		13.02575	460	16244.74
11.63823	411	14514.33		13.05407	461	16280.06
11.66654	412	14549.64		13.08238	462	16315.37
11.69486	413	14584.96		13.11070	463	16350.69
11.72318	414	14620.27		13.13902	464	16386.00
11.75149	415	14655.58		13.16734	465	16421.32
11.77981	416	14690.90		13.19565	466	16456.63
11.80813	417	14726.21		13.22397	467	16491.95
11.83644	418	14761.53		13.25229	468	16527.26
11.86476	419	14796.84		13.28060	469	16562.58
11.89308	420	14832.16		13.30892	470	16597.89
11.92139	421	14867.47		13.33724	471	16633.21
11.94971	422	14902.79		13.36555	472	16668.52
11.97803	423	14938.10		13.39387	473	16703.84
12.00634	424	14973.42		13.42219	474	16739.15
12.03466	425	15008.73		13.45050	475	16774.46
12.06298	426	15044.05		13.47882	476	16809.78
12.09130	427	15079.36		13.50714	477	16845.09
12.11961	428	15114.68		13.53545	478	16880.41
12.14793	429	15149.99		13.56377	479	16915.72
12.17625	430	15185.30		13.59209	480	16951.04
12.20456	431	15220.62		13.62040	481	16986.35
12.23288	432	15255.93		13.64872	482	17021.67
12.26120	433	15291.25		13.67704	483	17056.98
12.28951	434	15326.56		13.70536	484	17092.30
12.31783	435	15361.88		13.73367	485	17127.61
12.34615	436	15397.19		13.76199	486	17162.93
12.37446	437	15432.51		13.79031	487	17198.24
12.40278	438	15467.82		13.81862	488	17233.56
12.43110	439	15503.14		13.84694	489	17268.87
12.45941	440	15538.45		13.87526	490	17304.18
12.48773	441	15573.77		13.90357	491	17339.50
12.51605	442	15609.08		13.93189	492	17374.81
12.54436	443	15644.40		13.96021	493	17410.13
12.57268	444	15679.71		13.98852	494	17445.44
12.60100	445	15715.02		14.01684	495	17480.76
12.62932	446	15750.34		14.04516	496	17516.07
12.65763	447	15785.65		14.07347	497	17551.39
12.68595	448	15820.97		14.10179	498	17586.70
12.71427	449	15856.28		14.13011	499	17622.02
12.74258	450	15891.60		14.15843	500	17657.33

m³		ft³	m³		ft³
m³/s	REF	ft³/s	m³/s	REF	ft³/s
14.18674	501	17692.65	15.60258	551	19458.38
14.21506	502	17727.96	15.63090	552	19493.69
14.24338	503	17763.28	15.65922	553	19529.01
14.27169	504	17798.59	15.68753	554	19564.32
14.30001	505	17833.90	15.71585	555	19599.64
14.32833	506	17869.22	15.74417	556	19634.95
14.35664	507	17904.53	15.77249	557	19670.27
14.38496	508	17939.85	15.80080	558	19705.58
14.41328	509	17975.16	15.82912	559	19740.90
14.44159	510	18010.48	15.85744	560	19776.21
14.46991	511	18045.79	15.88575	561	19811.53
14.49823	512	18081.11	15.91407	562	19846.84
14.52654	513	18116.42	15.94239	563	19882.16
14.55486	514	18151.74	15.97070	564	19917.47
14.58318	515	18187.05	15.99902	565	19952.78
14.61149	516	18222.37	16.02734	566	19988.10
14.63981	517	18257.68	16.05565	567	20023.41
14.66813	518	18293.00	16.08397	568	20058.73
14.69645	519	18328.31	16.11229	569	20094.04
14.72476	520	18363.62	16.14060	570	20129.36
14.75308	521	18398.94	16.16892	571	20164.67
14.78140	522	18434.25	16.19724	572	20199.99
14.80971	523	18469.57	16.22556	573	20235.30
14.83803	524	18504.88	16.25387	574	20270.62
14.86635	525	18540.20	16.28219	575	20305.93
14.89466	526	18575.51	16.31051	576	20341.25
14.92298	527	18610.83	16.33882	577	20376.56
14.95130	528	18646.14	16.36714	578	20411.87
14.97961	529	18681.46	16.39546	579	20447.19
15.00793	530	18716.77	16.42377	580	20482.50
15.03625	531	18752.09	16.45209	581	20517.82
15.06456	532	18787.40	16.48041	582	20553.13
15.09288	533	18822.72	16.50872	583	20588.45
15.12120	534	18858.03	16.53704	584	20623.76
15.14951	535	18893.34	16.56536	585	20659.08
15.17783	536	18928.66	16.59367	586	20694.39
15.20615	537	18963.97	16.62199	587	20729.71
15.23447	538	18999.29	16.65031	588	20765.02
15.26278	539	19034.60	16.67862	589	20800.34
15.29110	540	19069.92	16.70694	590	20835.65
15.31942	541	19105.23	16.73526	591	20870.97
15.34773	542	19140.55	16.76358	592	20906.28
15.37605	543	19175.86	16.79189	593	20941.59
15.40437	544	19211.18	16.82021	594	20976.91
15.43268	545	19246.49	16.84853	595	21012.22
15.46100	546	19281.81	16.87684	596	21047.54
15.48932	547	19317.12	16.90516	597	21082.85
15.51763	548	19352.44	16.93348	598	21118.17
15.54595	549	19387.75	16.96179	599	21153.48
15.57427	550	19423.06	16.99011	600	21188.80

km³ ←	−9		m³	+3		l
				+6		cm³

m³ m³/s	REF	ft³ ft³/s	m³ m³/s	REF	ft³ ft³/s
17.01843	601	21224.11	18.43427	651	22989.85
17.04674	602	21259.43	18.46259	652	23025.16
17.07506	603	21294.74	18.49090	653	23060.47
17.10338	604	21330.06	18.51922	654	23095.79
17.13169	605	21365.37	18.54754	655	23131.10
17.16001	606	21400.69	18.57585	656	23166.42
17.18833	607	21436.00	18.60417	657	23201.73
17.21664	608	21471.31	18.63249	658	23237.05
17.24496	609	21506.63	18.66080	659	23272.36
17.27328	610	21541.94	18.68912	660	23307.68
17.30160	611	21577.26	18.71744	661	23342.99
17.32991	612	21612.57	18.74575	662	23378.31
17.35823	613	21647.89	18.77407	663	23413.62
17.38655	614	21683.20	18.80239	664	23448.94
17.41486	615	21718.52	18.83071	665	23484.25
17.44318	616	21753.83	18.85902	666	23519.57
17.47150	617	21789.15	18.88734	667	23554.88
17.49981	618	21824.46	18.91566	668	23590.19
17.52813	619	21859.78	18.94397	669	23625.51
17.55645	620	21895.09	18.97229	670	23660.82
17.58476	621	21930.41	19.00061	671	23696.14
17.61308	622	21965.72	19.02892	672	23731.45
17.64140	623	22001.03	19.05724	673	23766.77
17.66971	624	22036.35	19.08556	674	23802.08
17.69803	625	22071.66	19.11387	675	23837.40
17.72635	626	22106.98	19.14219	676	23872.71
17.75466	627	22142.29	19.17051	677	23908.03
17.78298	628	22177.61	19.19882	678	23943.34
17.81130	629	22212.92	19.22714	679	23978.66
17.83962	630	22248.24	19.25546	680	24013.97
17.86793	631	22283.55	19.28377	681	24049.29
17.89625	632	22318.87	19.31209	682	24084.60
17.92457	633	22354.18	19.34041	683	24119.91
17.95288	634	22389.50	19.36873	684	24155.23
17.98120	635	22424.81	19.39704	685	24190.54
18.00952	636	22460.13	19.42536	686	24225.86
18.03783	637	22495.44	19.45368	687	24261.17
18.06615	638	22530.75	19.48199	688	24296.49
18.09447	639	22566.07	19.51031	689	24331.80
18.12278	640	22601.38	19.53863	690	24367.12
18.15110	641	22636.70	19.56694	691	24402.43
18.17942	642	22672.01	19.59526	692	24437.75
18.20773	643	22707.33	19.62358	693	24473.06
18.23605	644	22742.64	19.65189	694	24508.38
18.26437	645	22777.96	19.68021	695	24543.69
18.29269	646	22813.27	19.70853	696	24579.01
18.32100	647	22848.59	19.73684	697	24614.32
18.34932	648	22883.90	19.76516	698	24649.63
18.37764	649	22919.22	19.79348	699	24684.95
18.40595	650	22954.53	19.82179	700	24720.26

m^3		ft^3		m^3		ft^3
m^3/s	REF	ft^3/s		m^3/s	REF	ft^3/s
19·85011	701	24755·58		21·26595	751	26521·31
19·87843	702	24790·89		21·29427	752	26556·63
19·90675	703	24826·21		21·32259	753	26591·94
19·93506	704	24861·52		21·35090	754	26627·26
19·96338	705	24896·84		21·37922	755	26662·57
19·99170	706	24932·15		21·40754	756	26697·88
20·02001	707	24967·47		21·43586	757	26733·20
20·04833	708	25002·78		21·46417	758	26768·51
20·07665	709	25038·10		21·49249	759	26803·83
20·10496	710	25073·41		21·52081	760	26839·14
20·13328	711	25108·73		21·54912	761	26874·46
20·16160	712	25144·04		21·57744	762	26909·77
20·18991	713	25179·35		21·60576	763	26945·09
20·21823	714	25214·67		21·63407	764	26980·40
20·24655	715	25249·98		21·66239	765	27015·72
20·27486	716	25285·30		21·69071	766	27051·03
20·30318	717	25320·61		21·71902	767	27086·35
20·33150	718	25355·93		21·74734	768	27121·66
20·35982	719	25391·24		21·77566	769	27156·98
20·38813	720	25426·56		21·80397	770	27192·29
20·41645	721	25461·87		21·83229	771	27227·60
20·44477	722	25497·19		21·86061	772	27262·92
20·47308	723	25532·50		21·88893	773	27298·23
20·50140	724	25567·82		21·91724	774	27333·55
20·52972	725	25603·13		21·94556	775	27368·86
20·55803	726	25638·44		21·97388	776	27404·18
20·58635	727	25673·76		22·00219	777	27439·49
20·61467	728	25709·07		22·03051	778	27474·81
20·64298	729	25744·39		22·05883	779	27510·12
20·67130	730	25779·70		22·08714	780	27545·44
20·69962	731	25815·02		22·11546	781	27580·75
20·72793	732	25850·33		22·14378	782	27616·07
20·75625	733	25885·65		22·17209	783	27651·38
20·78457	734	25920·96		22·20041	784	27686·70
20·81288	735	25956·28		22·22873	785	27722·01
20·84120	736	25991·59		22·25704	786	27757·32
20·86952	737	26026·91		22·28536	787	27792·64
20·89784	738	26062·22		22·31368	788	27827·95
20·92615	739	26097·54		22·34199	789	27863·27
20·95447	740	26132·85		22·37031	790	27898·58
20·98279	741	26168·16		22·39863	791	27933·90
21·01110	742	26203·48		22·42695	792	27969·21
21·03942	743	26238·79		22·45526	793	28004·53
21·06774	744	26274·11		22·48358	794	28039·84
21·09605	745	26309·42		22·51190	795	28075·16
21·12437	746	26344·74		22·54021	796	28110·47
21·15269	747	26380·05		22·56853	797	28145·79
21·18100	748	26415·37		22·59685	798	28181·10
21·20932	749	26450·68		22·62516	799	28216·42
21·23764	750	26486·00		22·65348	800	28251·73

m³			m³		
m³/s	REF	ft³/s	m³/s	REF	ft³/s
22.68180	801	28287.04	24.09764	851	30052.78
22.71011	802	28322.36	24.12596	852	30088.09
22.73843	803	28357.67	24.15427	853	30123.41
22.76675	804	28392.99	24.18259	854	30158.72
22.79506	805	28428.30	24.21091	855	30194.04
22.82338	806	28463.62	24.23922	856	30229.35
22.85170	807	28498.93	24.26754	857	30264.67
22.88001	808	28534.25	24.29586	858	30299.98
22.90833	809	28569.56	24.32417	859	30335.30
22.93665	810	28604.88	24.35249	860	30370.61
22.96497	811	28640.19	24.38081	861	30405.92
22.99328	812	28675.51	24.40912	862	30441.24
23.02160	813	28710.82	24.43744	863	30476.55
23.04992	814	28746.14	24.46576	864	30511.87
23.07823	815	28781.45	24.49408	865	30547.18
23.10655	816	28816.76	24.52239	866	30582.50
23.13487	817	28852.08	24.55071	867	30617.81
23.16318	818	28887.39	24.57903	868	30653.13
23.19150	819	28922.71	24.60734	869	30688.44
23.21982	820	28958.02	24.63566	870	30723.76
23.24813	821	28993.34	24.66398	871	30759.07
23.27645	822	29028.65	24.69229	872	30794.39
23.30477	823	29063.97	24.72061	873	30829.70
23.33308	824	29099.28	24.74893	874	30865.01
23.36140	825	29134.60	24.77724	875	30900.33
23.38972	826	29169.91	24.80556	876	30935.64
23.41804	827	29205.23	24.83388	877	30970.96
23.44635	828	29240.54	24.86219	878	31006.27
23.47467	829	29275.86	24.89051	879	31041.59
23.50299	830	29311.17	24.91883	880	31076.90
23.53130	831	29346.48	24.94714	881	31112.22
23.55962	832	29381.80	24.97546	882	31147.53
23.58794	833	29417.11	25.00378	883	31182.85
23.61625	834	29452.43	25.03210	884	31218.16
23.64457	835	29487.74	25.06041	885	31253.48
23.67289	836	29523.06	25.08873	886	31288.79
23.70120	837	29558.37	25.11705	887	31324.11
23.72952	838	29593.69	25.14536	888	31359.42
23.75784	839	29629.00	25.17368	889	31394.73
23.78615	840	29664.32	25.20200	890	31430.05
23.81447	841	29699.63	25.23031	891	31465.36
23.84279	842	29734.95	25.25863	892	31500.68
23.87110	843	29770.26	25.28695	893	31535.99
23.89942	844	29805.58	25.31526	894	31571.31
23.92774	845	29840.89	25.34358	895	31606.62
23.95606	846	29876.20	25.37190	896	31641.94
23.98437	847	29911.52	25.40021	897	31677.25
24.01269	848	29946.83	25.42853	898	31712.57
24.04101	849	29982.15	25.45685	899	31747.88
24.06932	850	30017.46	25.48517	900	31783.20

km³ ——————— +9 ——————→ m³ ←—————— −3 ——————— l
                                  ←—————— −6 ——————— cm³

m³ m³/s	REF	ft³ ft³/s	m³ m³/s	REF	ft³ ft³/s
25.51348	901	31818.51	26.92932	951	33584.24
25.54180	902	31853.83	26.95764	952	33619.56
25.57012	903	31889.14	26.98596	953	33654.87
25.59843	904	31924.45	27.01428	954	33690.19
25.62675	905	31959.77	27.04259	955	33725.50
25.65507	906	31995.08	27.07091	956	33760.82
25.68338	907	32030.40	27.09923	957	33796.13
25.71170	908	32065.71	27.12754	958	33831.45
25.74002	909	32101.03	27.15586	959	33866.76
25.76833	910	32136.34	27.18418	960	33902.08
25.79665	911	32171.66	27.21249	961	33937.39
25.82497	912	32206.97	27.24081	962	33972.71
25.85328	913	32242.29	27.26913	963	34008.02
25.88160	914	32277.60	27.29744	964	34043.33
25.90992	915	32312.92	27.32576	965	34078.65
25.93823	916	32348.23	27.35408	966	34113.96
25.96655	917	32383.55	27.38239	967	34149.28
25.99487	918	32418.86	27.41071	968	34184.59
26.02319	919	32454.17	27.43903	969	34219.91
26.05150	920	32489.49	27.46734	970	34255.22
26.07982	921	32524.80	27.49566	971	34290.54
26.10814	922	32560.12	27.52398	972	34325.85
26.13645	923	32595.43	27.55230	973	34361.17
26.16477	924	32630.75	27.58061	974	34396.48
26.19309	925	32666.06	27.60893	975	34431.80
26.22140	926	32701.38	27.63725	976	34467.11
26.24972	927	32736.69	27.66556	977	34502.43
26.27804	928	32772.01	27.69388	978	34537.74
26.30635	929	32807.32	27.72220	979	34573.05
26.33467	930	32842.64	27.75051	980	34608.37
26.36299	931	32877.95	27.77883	981	34643.68
26.39130	932	32913.27	27.80715	982	34679.00
26.41962	933	32948.58	27.83546	983	34714.31
26.44794	934	32983.89	27.86378	984	34749.63
26.47625	935	33019.21	27.89210	985	34784.94
26.50457	936	33054.52	27.92041	986	34820.26
26.53289	937	33089.84	27.94873	987	34855.57
26.56121	938	33125.15	27.97705	988	34890.89
26.58952	939	33160.47	28.00536	989	34926.20
26.61784	940	33195.78	28.03368	990	34961.52
26.64616	941	33231.10	28.06200	991	34996.83
26.67447	942	33266.41	28.09032	992	35032.15
26.70279	943	33301.73	28.11863	993	35067.46
26.73111	944	33337.04	28.14695	994	35102.77
26.75942	945	33372.36	28.17527	995	35138.09
26.78774	946	33407.67	28.20358	996	35173.40
26.81606	947	33442.99	28.23190	997	35208.72
26.84437	948	33478.30	28.26022	998	35244.03
26.87269	949	33513.61	28.28853	999	35279.35
26.90101	950	33548.93	28.31685	1000	35314.66

12

VOLUME

Cubic-yard — Cubic-metre
$(yd^3 — m^3)$
Cubic-yard — Cubic-kilometre
$(yd^3 — km^3)$
Cubic-yard — Litre
$(yd^3 — l)$

VOLUME FLOW RATE

Cubic-yard/Second — Cubic-metre/Second
$(yd^3/s — m^3/s)$
Cubic-yard/Second — Litre/Second
$(yd^3/s — l/s)$

Conversion Factors:
1 cubic-yard $(yd^3) = 0.7645549$ cubic-metre (m^3)
$(1\ m^3 = 1.307951\ yd^3)$

Volume Flow Rate Values

Volume flow rate is also represented by Table 12 as long as the time units are the same for both English and SI conversion quantities. Besides per second shown above, these consistent time units can also be in per minute (min), per hour (h), etc.

Volume Flow Rate Conversion Factor:
1 yd^3/s (or yd^3/min, etc.) = 0.7645549 m^3/s (or m^3/min, etc.)

In converting the quantities mentioned above to units containing km³ and l (litres), Diagram 12-A, which appears in the headings on alternate pages of this table, will assist the user. It is derived from: $1 \text{ m}^3 = (10^{-9}) \text{ km}^3 = (10^3) \text{ l}$ (litres); $1 \text{ l} = 1 \text{ dm}^3$.

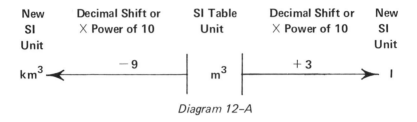

| New SI Unit | Decimal Shift or × Power of 10 | SI Table Unit | Decimal Shift or × Power of 10 | New SI Unit |

Diagram 12-A

Example (a): Convert 273 yd³ to m³, to l.
From table: $273 \text{ yd}^3 = 208.7235 \text{ m}^{3*} \approx 209 \text{ m}^3$
From Diagram 12-A: $208.7235 \text{ m}^3 = 208.7235 \times 10^3 \text{ l} = 208,723.5 \text{ l}$
$\approx 209,000 \text{ l}$
(m³ to litres by multiplying m³ value by 10^3 or by shifting decimal 3 places to the right.)

Diagram 12-B is also given at the top of alternate table pages:

$$\text{km}^3 \xrightarrow{\;\;+9\;\;} \quad \text{m}^3 \quad \xleftarrow{\;\;-3\;\;} \text{l}$$

Diagram 12-B

Diagram 12-B is the opposite of Diagram 12-A and is used to convert units containing km³ and l (litre) to those with m³. The equivalent with m³ is then applied to the REF column in the table to obtain yd³ quantities.

Example (b): Convert $8.59 \times 10^{-5} \text{ km}^3$ to yd³.
From Diagram 12-B: $8.59 \times 10^{-5} \text{ km}^3 = 8.59 \times 10^{-5} \times 10^9 \text{ m}^3$
$= 859 \times 10^2 \text{ m}^3$
From table: $859 \text{ m}^3 = 1123.530 \text{ yd}^{3*}$
Therefore: $859 \times 10^2 \text{ m}^3 = 1123.530 \times 10^2 \text{ yd}^3 = 112,353.0 \text{ yd}^3$
$\approx 112,000 \text{ yd}^3$.

---

*All conversion values obtained can be rounded to desired number of significant figures.

$$km^3 \longleftarrow \quad -9 \quad \Big| \quad m^3 \quad \Big| \quad +3 \quad \longrightarrow \quad I$$

m³ m³/s	REF	yd³ yd³/s	m³ m³/s	REF	yd³ yd³/s
0·7645549	1	1·307951	38·99230	51	66·70548
1·529110	2	2·615901	39·75686	52	68·01343
2·293665	3	3·923852	40·52141	53	69·32138
3·058220	4	5·231802	41·28596	54	70·62933
3·822775	5	6·539753	42·05052	55	71·93728
4·587329	6	7·847703	42·81507	56	73·24523
5·351884	7	9·155654	43·57963	57	74·55318
6·116439	8	10·46360	44·34418	58	75·86113
6·880994	9	11·77155	45·10874	59	77·16908
7·645549	10	13·07951	45·87329	60	78·47703
8·410104	11	14·38746	46·63785	61	79·78498
9·174659	12	15·69541	47·40240	62	81·09293
9·939214	13	17·00336	48·16696	63	82·40088
10·70377	14	18·31131	48·93151	64	83·70883
11·46832	15	19·61926	49·69607	65	85·01679
12·23288	16	20·92721	50·46062	66	86·32474
12·99743	17	22·23516	51·22518	67	87·63269
13·76199	18	23·54311	51·98973	68	88·94064
14·52654	19	24·85106	52·75429	69	90·24859
15·29110	20	26·15901	53·51884	70	91·55654
16·05565	21	27·46696	54·28340	71	92·86449
16·82021	22	28·77491	55·04795	72	94·17244
17·58476	23	30·08286	55·81251	73	95·48039
18·34932	24	31·39081	56·57706	74	96·78834
19·11387	25	32·69876	57·34162	75	98·09629
19·87843	26	34·00671	58·10617	76	99·40424
20·64298	27	35·31466	58·87073	77	100·7122
21·40754	28	36·62262	59·63528	78	102·0201
22·17209	29	37·93057	60·39984	79	103·3281
22·93665	30	39·23852	61·16439	80	104·6360
23·70120	31	40·54647	61·92895	81	105·9440
24·46576	32	41·85442	62·69350	82	107·2519
25·23031	33	43·16237	63·45806	83	108·5599
25·99487	34	44·47032	64·22261	84	109·8678
26·75942	35	45·77827	64·98717	85	111·1758
27·52398	36	47·08622	65·75172	86	112·4837
28·28853	37	48·39417	66·51628	87	113·7917
29·05309	38	49·70212	67·28083	88	115·0996
29·81764	39	51·01007	68·04539	89	116·4076
30·58220	40	52·31802	68·80994	90	117·7155
31·34675	41	53·62597	69·57450	91	119·0235
32·11131	42	54·93392	70·33905	92	120·3314
32·87586	43	56·24187	71·10361	93	121·6394
33·64042	44	57·54982	71·86816	94	122·9474
34·40497	45	58·85777	72·63272	95	124·2553
35·16953	46	60·16572	73·39727	96	125·5633
35·93408	47	61·47368	74·16183	97	126·8712
36·69864	48	62·78163	74·92638	98	128·1792
37·46319	49	64·08958	75·69094	99	129·4871
38·22775	50	65·39753	76·45549	100	130·7951

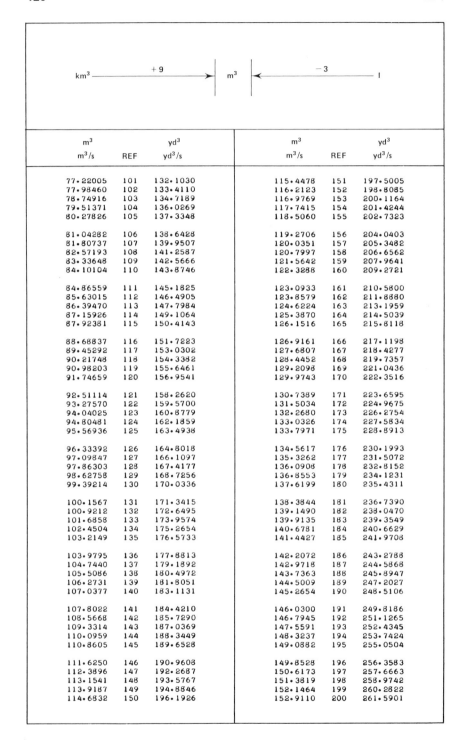

km³ ——————— + 9 ——————→ | m³ | ←—————— − 3 —————— l

m³ m³/s	REF	yd³ yd³/s	m³ m³/s	REF	yd³ yd³/s
77·22005	101	132·1030	115·4478	151	197·5005
77·98460	102	133·4110	116·2123	152	198·8085
78·74916	103	134·7189	116·9769	153	200·1164
79·51371	104	136·0269	117·7415	154	201·4244
80·27826	105	137·3348	118·5060	155	202·7323
81·04282	106	138·6428	119·2706	156	204·0403
81·80737	107	139·9507	120·0351	157	205·3482
82·57193	108	141·2587	120·7997	158	206·6562
83·33648	109	142·5666	121·5642	159	207·9641
84·10104	110	143·8746	122·3288	160	209·2721
84·86559	111	145·1825	123·0933	161	210·5800
85·63015	112	146·4905	123·8579	162	211·8880
86·39470	113	147·7984	124·6224	163	213·1959
87·15926	114	149·1064	125·3870	164	214·5039
87·92381	115	150·4143	126·1516	165	215·8118
88·68837	116	151·7223	126·9161	166	217·1198
89·45292	117	153·0302	127·6807	167	218·4277
90·21748	118	154·3382	128·4452	168	219·7357
90·98203	119	155·6461	129·2098	169	221·0436
91·74659	120	156·9541	129·9743	170	222·3516
92·51114	121	158·2620	130·7389	171	223·6595
93·27570	122	159·5700	131·5034	172	224·9675
94·04025	123	160·8779	132·2680	173	226·2754
94·80481	124	162·1859	133·0326	174	227·5834
95·56936	125	163·4938	133·7971	175	228·8913
96·33392	126	164·8018	134·5617	176	230·1993
97·09847	127	166·1097	135·3262	177	231·5072
97·86303	128	167·4177	136·0908	178	232·8152
98·62758	129	168·7256	136·8553	179	234·1231
99·39214	130	170·0336	137·6199	180	235·4311
100·1567	131	171·3415	138·3844	181	236·7390
100·9212	132	172·6495	139·1490	182	238·0470
101·6858	133	173·9574	139·9135	183	239·3549
102·4504	134	175·2654	140·6781	184	240·6629
103·2149	135	176·5733	141·4427	185	241·9708
103·9795	136	177·8813	142·2072	186	243·2788
104·7440	137	179·1892	142·9718	187	244·5868
105·5086	138	180·4972	143·7363	188	245·8947
106·2731	139	181·8051	144·5009	189	247·2027
107·0377	140	183·1131	145·2654	190	248·5106
107·8022	141	184·4210	146·0300	191	249·8186
108·5668	142	185·7290	146·7945	192	251·1265
109·3314	143	187·0369	147·5591	193	252·4345
110·0959	144	188·3449	148·3237	194	253·7424
110·8605	145	189·6528	149·0882	195	255·0504
111·6250	146	190·9608	149·8528	196	256·3583
112·3896	147	192·2687	150·6173	197	257·6663
113·1541	148	193·5767	151·3819	198	258·9742
113·9187	149	194·8846	152·1464	199	260·2822
114·6832	150	196·1926	152·9110	200	261·5901

m^3			m^3		
m^3/s	REF	yd^3/s	m^3/s	REF	yd^3/s
153.6755	201	262.8981	191.9033	251	328.2956
154.4401	202	264.2060	192.6678	252	329.6035
155.2046	203	265.5140	193.4324	253	330.9115
155.9692	204	266.8219	194.1969	254	332.2194
156.7338	205	268.1299	194.9615	255	333.5274
157.4983	206	269.4378	195.7261	256	334.8353
158.2629	207	270.7458	196.4906	257	336.1433
159.0274	208	272.0537	197.2552	258	337.4512
159.7920	209	273.3617	198.0197	259	338.7592
160.5565	210	274.6696	198.7843	260	340.0671
161.3211	211	275.9776	199.5488	261	341.3751
162.0856	212	277.2855	200.3134	262	342.6830
162.8502	213	278.5935	201.0779	263	343.9910
163.6147	214	279.9014	201.8425	264	345.2989
164.3793	215	281.2094	202.6070	265	346.6069
165.1439	216	282.5173	203.3716	266	347.9148
165.9084	217	283.8253	204.1362	267	349.2228
166.6730	218	285.1332	204.9007	268	350.5307
167.4375	219	286.4412	205.6653	269	351.8387
168.2021	220	287.7491	206.4298	270	353.1466
168.9666	221	289.0571	207.1944	271	354.4546
169.7312	222	290.3650	207.9589	272	355.7625
170.4957	223	291.6730	208.7235	273	357.0705
171.2603	224	292.9809	209.4880	274	358.3784
172.0249	225	294.2889	210.2526	275	359.6864
172.7894	226	295.5968	211.0172	276	360.9944
173.5540	227	296.9048	211.7817	277	362.3023
174.3185	228	298.2127	212.5463	278	363.6103
175.0831	229	299.5207	213.3108	279	364.9182
175.8476	230	300.8286	214.0754	280	366.2262
176.6122	231	302.1366	214.8399	281	367.5341
177.3767	232	303.4445	215.6045	282	368.8421
178.1413	233	304.7525	216.3690	283	370.1500
178.9058	234	306.0604	217.1336	284	371.4580
179.6704	235	307.3684	217.8981	285	372.7659
180.4350	236	308.6763	218.6627	286	374.0739
181.1995	237	309.9843	219.4273	287	375.3818
181.9641	238	311.2922	220.1918	288	376.6898
182.7286	239	312.6002	220.9564	289	377.9977
183.4932	240	313.9081	221.7209	290	379.3057
184.2577	241	315.2161	222.4855	291	380.6136
185.0223	242	316.5240	223.2500	292	381.9216
185.7868	243	317.8320	224.0146	293	383.2295
186.5514	244	319.1399	224.7791	294	384.5375
187.3160	245	320.4479	225.5437	295	385.8454
188.0805	246	321.7558	226.3083	296	387.1534
188.8451	247	323.0638	227.0728	297	388.4613
189.6096	248	324.3717	227.8374	298	389.7693
190.3742	249	325.6797	228.6019	299	391.0772
191.1387	250	326.9876	229.3665	300	392.3852

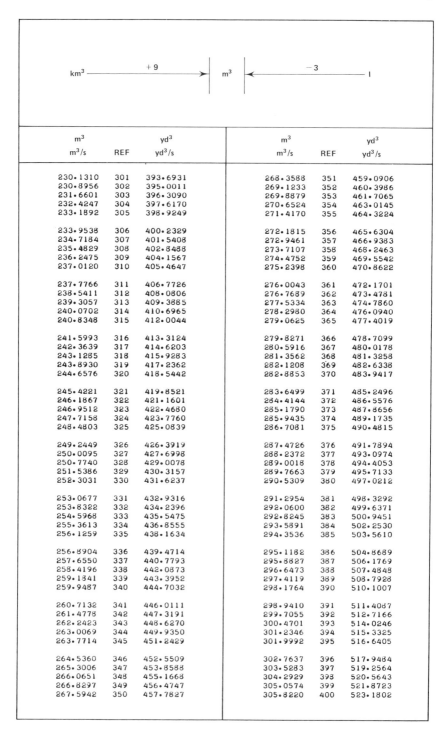

m³		yd³	m³		yd³
m³/s	REF	yd³/s	m³/s	REF	yd³/s
230.1310	301	393.6931	268.3588	351	459.0906
230.8956	302	395.0011	269.1233	352	460.3986
231.6601	303	396.3090	269.8879	353	461.7065
232.4247	304	397.6170	270.6524	354	463.0145
233.1892	305	398.9249	271.4170	355	464.3224
233.9538	306	400.2329	272.1815	356	465.6304
234.7184	307	401.5408	272.9461	357	466.9383
235.4829	308	402.8488	273.7107	358	468.2463
236.2475	309	404.1567	274.4752	359	469.5542
237.0120	310	405.4647	275.2398	360	470.8622
237.7766	311	406.7726	276.0043	361	472.1701
238.5411	312	408.0806	276.7689	362	473.4781
239.3057	313	409.3885	277.5334	363	474.7860
240.0702	314	410.6965	278.2980	364	476.0940
240.8348	315	412.0044	279.0625	365	477.4019
241.5993	316	413.3124	279.8271	366	478.7099
242.3639	317	414.6203	280.5916	367	480.0178
243.1285	318	415.9283	281.3562	368	481.3258
243.8930	319	417.2362	282.1208	369	482.6338
244.6576	320	418.5442	282.8853	370	483.9417
245.4221	321	419.8521	283.6499	371	485.2496
246.1867	322	421.1601	284.4144	372	486.5576
246.9512	323	422.4680	285.1790	373	487.8656
247.7158	324	423.7760	285.9435	374	489.1735
248.4803	325	425.0839	286.7081	375	490.4815
249.2449	326	426.3919	287.4726	376	491.7894
250.0095	327	427.6998	288.2372	377	493.0974
250.7740	328	429.0078	289.0018	378	494.4053
251.5386	329	430.3157	289.7663	379	495.7133
252.3031	330	431.6237	290.5309	380	497.0212
253.0677	331	432.9316	291.2954	381	498.3292
253.8322	332	434.2396	292.0600	382	499.6371
254.5968	333	435.5475	292.8245	383	500.9451
255.3613	334	436.8555	293.5891	384	502.2530
256.1259	335	438.1634	294.3536	385	503.5610
256.8904	336	439.4714	295.1182	386	504.8689
257.6550	337	440.7793	295.8827	387	506.1769
258.4196	338	442.0873	296.6473	388	507.4848
259.1841	339	443.3952	297.4119	389	508.7928
259.9487	340	444.7032	298.1764	390	510.1007
260.7132	341	446.0111	298.9410	391	511.4087
261.4778	342	447.3191	299.7055	392	512.7166
262.2423	343	448.6270	300.4701	393	514.0246
263.0069	344	449.9350	301.2346	394	515.3325
263.7714	345	451.2429	301.9992	395	516.6405
264.5360	346	452.5509	302.7637	396	517.9484
265.3006	347	453.8588	303.5283	397	519.2564
266.0651	348	455.1668	304.2929	398	520.5643
266.8297	349	456.4747	305.0574	399	521.8723
267.5942	350	457.7827	305.8220	400	523.1802

m³		m³	
m³/s	REF	yd³ m³/s	REF yd³/s

m³		m³	
m³/s	REF	yd³/s	

m³/s	REF	yd³/s	m³/s	REF	yd³/s
306·5865	401	524·4882	344·8143	451	589·8857
307·3511	402	525·7961	345·5788	452	591·1936
308·1156	403	527·1041	346·3434	453	592·5016
308·8802	404	528·4120	347·1079	454	593·8095
309·6447	405	529·7200	347·8725	455	595·1175
310·4093	406	531·0279	348·6370	456	596·4254
311·1738	407	532·3359	349·4016	457	597·7334
311·9384	408	533·6438	350·1661	458	599·0414
312·7030	409	534·9518	350·9307	459	600·3493
313·4675	410	536·2597	351·6953	460	601·6572
314·2321	411	537·5677	352·4598	461	602·9652
314·9966	412	538·8756	353·2244	462	604·2731
315·7612	413	540·1836	353·9889	463	605·5811
316·5257	414	541·4915	354·7535	464	606·8891
317·2903	415	542·7995	355·5180	465	608·1970
318·0548	416	544·1074	356·2826	466	609·5050
318·8194	417	545·4154	357·0471	467	610·8129
319·5840	418	546·7233	357·8117	468	612·1208
320·3485	419	548·0313	358·5762	469	613·4288
321·1131	420	549·3392	359·3408	470	614·7368
321·8776	421	550·6472	360·1054	471	616·0447
322·6422	422	551·9551	360·8699	472	617·3527
323·4067	423	553·2631	361·6345	473	618·6606
324·1713	424	554·5710	362·3990	474	619·9686
324·9358	425	555·8790	363·1636	475	621·2765
325·7004	426	557·1869	363·9281	476	622·5845
326·4649	427	558·4949	364·6927	477	623·8924
327·2295	428	559·8028	365·4572	478	625·2004
327·9941	429	561·1108	366·2218	479	626·5083
328·7586	430	562·4187	366·9864	480	627·8163
329·5232	431	563·7267	367·7509	481	629·1242
330·2877	432	565·0346	368·5155	482	630·4322
331·0523	433	566·3426	369·2800	483	631·7401
331·8168	434	567·6505	370·0446	484	633·0481
332·5814	435	568·9585	370·8091	485	634·3560
333·3459	436	570·2664	371·5737	486	635·6640
334·1105	437	571·5744	372·3382	487	636·9719
334·8750	438	572·8823	373·1028	488	638·2799
335·6396	439	574·1903	373·8673	489	639·5878
336·4042	440	575·4982	374·6319	490	640·8958
337·1687	441	576·8062	375·3965	491	642·2037
337·9333	442	578·1141	376·1610	492	643·5117
338·6978	443	579·4221	376·9256	493	644·8196
339·4624	444	580·7300	377·6901	494	646·1276
340·2269	445	582·0380	378·4547	495	647·4355
340·9915	446	583·3459	379·2192	496	648·7435
341·7560	447	584·6539	379·9838	497	650·0514
342·5206	448	585·9618	380·7483	498	651·3594
343·2852	449	587·2698	381·5129	499	652·6673
344·0497	450	588·5777	382·2775	500	653·9753

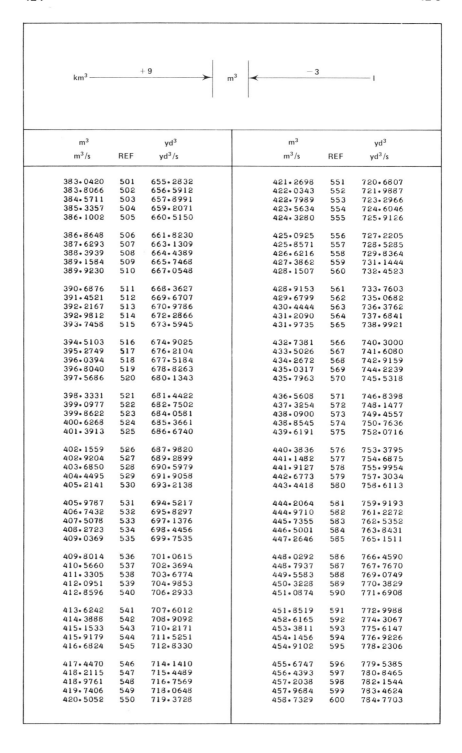

| km³ ——— +9 ———→ | m³ | ←——— −3 ——— l |

| m³ | | yd³ | | m³ | | yd³ |
m³/s	REF	yd³/s		m³/s	REF	yd³/s
383.0420	501	655.2832		421.2698	551	720.6807
383.8066	502	656.5912		422.0343	552	721.9887
384.5711	503	657.8991		422.7989	553	723.2966
385.3357	504	659.2071		423.5634	554	724.6046
386.1002	505	660.5150		424.3280	555	725.9126
386.8648	506	661.8230		425.0925	556	727.2205
387.6293	507	663.1309		425.8571	557	728.5285
388.3939	508	664.4389		426.6216	558	729.8364
389.1584	509	665.7468		427.3862	559	731.1444
389.9230	510	667.0548		428.1507	560	732.4523
390.6876	511	668.3627		428.9153	561	733.7603
391.4521	512	669.6707		429.6799	562	735.0682
392.2167	513	670.9786		430.4444	563	736.3762
392.9812	514	672.2866		431.2090	564	737.6841
393.7458	515	673.5945		431.9735	565	738.9921
394.5103	516	674.9025		432.7381	566	740.3000
395.2749	517	676.2104		433.5026	567	741.6080
396.0394	518	677.5184		434.2672	568	742.9159
396.8040	519	678.8263		435.0317	569	744.2239
397.5686	520	680.1343		435.7963	570	745.5318
398.3331	521	681.4422		436.5608	571	746.8398
399.0977	522	682.7502		437.3254	572	748.1477
399.8622	523	684.0581		438.0900	573	749.4557
400.6268	524	685.3661		438.8545	574	750.7636
401.3913	525	686.6740		439.6191	575	752.0716
402.1559	526	687.9820		440.3836	576	753.3795
402.9204	527	689.2899		441.1482	577	754.6875
403.6850	528	690.5979		441.9127	578	755.9954
404.4495	529	691.9058		442.6773	579	757.3034
405.2141	530	693.2138		443.4418	580	758.6113
405.9787	531	694.5217		444.2064	581	759.9193
406.7432	532	695.8297		444.9710	582	761.2272
407.5078	533	697.1376		445.7355	583	762.5352
408.2723	534	698.4456		446.5001	584	763.8431
409.0369	535	699.7535		447.2646	585	765.1511
409.8014	536	701.0615		448.0292	586	766.4590
410.5660	537	702.3694		448.7937	587	767.7670
411.3305	538	703.6774		449.5583	588	769.0749
412.0951	539	704.9853		450.3228	589	770.3829
412.8596	540	706.2933		451.0874	590	771.6908
413.6242	541	707.6012		451.8519	591	772.9988
414.3888	542	708.9092		452.6165	592	774.3067
415.1533	543	710.2171		453.3811	593	775.6147
415.9179	544	711.5251		454.1456	594	776.9226
416.6824	545	712.8330		454.9102	595	778.2306
417.4470	546	714.1410		455.6747	596	779.5385
418.2115	547	715.4489		456.4393	597	780.8465
418.9761	548	716.7569		457.2038	598	782.1544
419.7406	549	718.0648		457.9684	599	783.4624
420.5052	550	719.3728		458.7329	600	784.7703

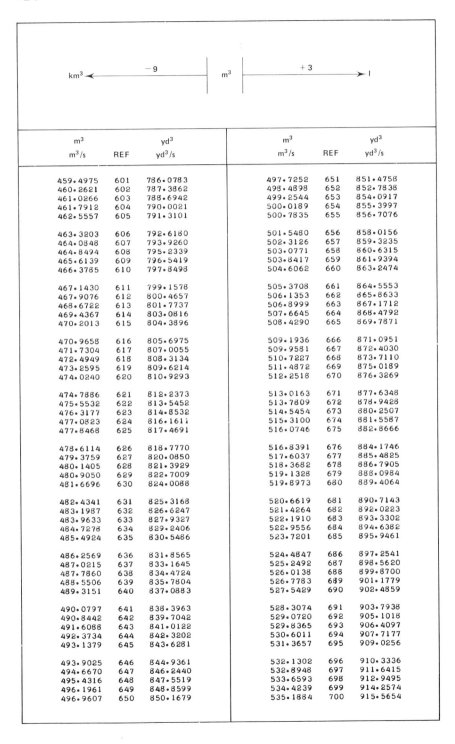

m³			m³		
m³/s	REF	yd³/s	m³/s	REF	yd³/s
459•4975	601	786•0783	497•7252	651	851•4758
460•2621	602	787•3862	498•4898	652	852•7838
461•0266	603	788•6942	499•2544	653	854•0917
461•7912	604	790•0021	500•0189	654	855•3997
462•5557	605	791•3101	500•7835	655	856•7076
463•3203	606	792•6180	501•5480	656	858•0156
464•0848	607	793•9260	502•3126	657	859•3235
464•8494	608	795•2339	503•0771	658	860•6315
465•6139	609	796•5419	503•8417	659	861•9394
466•3785	610	797•8498	504•6062	660	863•2474
467•1430	611	799•1578	505•3708	661	864•5553
467•9076	612	800•4657	506•1353	662	865•8633
468•6722	613	801•7737	506•8999	663	867•1712
469•4367	614	803•0816	507•6645	664	868•4792
470•2013	615	804•3896	508•4290	665	869•7871
470•9658	616	805•6975	509•1936	666	871•0951
471•7304	617	807•0055	509•9581	667	872•4030
472•4949	618	808•3134	510•7227	668	873•7110
473•2595	619	809•6214	511•4872	669	875•0189
474•0240	620	810•9293	512•2518	670	876•3269
474•7886	621	812•2373	513•0163	671	877•6348
475•5532	622	813•5452	513•7809	672	878•9428
476•3177	623	814•8532	514•5454	673	880•2507
477•0823	624	816•1611	515•3100	674	881•5587
477•8468	625	817•4691	516•0746	675	882•8666
478•6114	626	818•7770	516•8391	676	884•1746
479•3759	627	820•0850	517•6037	677	885•4825
480•1405	628	821•3929	518•3682	678	886•7905
480•9050	629	822•7009	519•1328	679	888•0984
481•6696	630	824•0088	519•8973	680	889•4064
482•4341	631	825•3168	520•6619	681	890•7143
483•1987	632	826•6247	521•4264	682	892•0223
483•9633	633	827•9327	522•1910	683	893•3302
484•7278	634	829•2406	522•9556	684	894•6382
485•4924	635	830•5486	523•7201	685	895•9461
486•2569	636	831•8565	524•4847	686	897•2541
487•0215	637	833•1645	525•2492	687	898•5620
487•7860	638	834•4724	526•0138	688	899•8700
488•5506	639	835•7804	526•7783	689	901•1779
489•3151	640	837•0883	527•5429	690	902•4859
490•0797	641	838•3963	528•3074	691	903•7938
490•8442	642	839•7042	529•0720	692	905•1018
491•6088	643	841•0122	529•8365	693	906•4097
492•3734	644	842•3202	530•6011	694	907•7177
493•1379	645	843•6281	531•3657	695	909•0256
493•9025	646	844•9361	532•1302	696	910•3336
494•6670	647	846•2440	532•8948	697	911•6415
495•4316	648	847•5519	533•6593	698	912•9495
496•1961	649	848•8599	534•4239	699	914•2574
496•9607	650	850•1679	535•1884	700	915•5654

$$km^3 \xrightarrow{+9} m^3 \xleftarrow{-3} l$$

m³ m³/s	REF	yd³ yd³/s	m³ m³/s	REF	yd³ yd³/s
535.9530	701	916.8733	574.1807	751	982.2709
536.7175	702	918.1813	574.9453	752	983.5788
537.4821	703	919.4892	575.7098	753	984.8868
538.2467	704	920.7972	576.4744	754	986.1947
539.0112	705	922.1051	577.2390	755	987.5027
539.7758	706	923.4131	578.0035	756	988.8106
540.5403	707	924.7210	578.7681	757	990.1186
541.3049	708	926.0290	579.5326	758	991.4265
542.0694	709	927.3369	580.2972	759	992.7345
542.8340	710	928.6449	581.0617	760	994.0424
543.5985	711	929.9528	581.8263	761	995.3504
544.3631	712	931.2608	582.5908	762	996.6583
545.1276	713	932.5687	583.3554	763	997.9663
545.8922	714	933.8767	584.1199	764	999.2742
546.6568	715	935.1846	584.8845	765	1000.582
547.4213	716	936.4926	585.6491	766	1001.890
548.1859	717	937.8005	586.4136	767	1003.198
548.9504	718	939.1085	587.1782	768	1004.506
549.7150	719	940.4164	587.9427	769	1005.814
550.4795	720	941.7244	588.7073	770	1007.122
551.2441	721	943.0323	589.4718	771	1008.430
552.0086	722	944.3403	590.2364	772	1009.738
552.7732	723	945.6482	591.0009	773	1011.046
553.5378	724	946.9562	591.7655	774	1012.354
554.3023	725	948.2641	592.5301	775	1013.662
555.0669	726	949.5721	593.2946	776	1014.970
555.8314	727	950.8800	594.0592	777	1016.278
556.5960	728	952.1880	594.8237	778	1017.586
557.3605	729	953.4959	595.5883	779	1018.893
558.1251	730	954.8039	596.3528	780	1020.201
558.8896	731	956.1118	597.1174	781	1021.509
559.6542	732	957.4198	597.8819	782	1022.817
560.4187	733	958.7277	598.6465	783	1024.125
561.1833	734	960.0357	599.4110	784	1025.433
561.9479	735	961.3437	600.1756	785	1026.741
562.7124	736	962.6516	600.9402	786	1028.049
563.4770	737	963.9595	601.7047	787	1029.357
564.2415	738	965.2675	602.4693	788	1030.665
565.0061	739	966.5754	603.2338	789	1031.973
565.7706	740	967.8834	603.9984	790	1033.281
566.5352	741	969.1914	604.7629	791	1034.589
567.2997	742	970.4993	605.5275	792	1035.897
568.0643	743	971.8073	606.2920	793	1037.205
568.8288	744	973.1152	607.0566	794	1038.513
569.5934	745	974.4232	607.8211	795	1039.821
570.3580	746	975.7311	608.5857	796	1041.129
571.1225	747	977.0391	609.3503	797	1042.437
571.8871	748	978.3470	610.1148	798	1043.745
572.6516	749	979.6550	610.8794	799	1045.052
573.4162	750	980.9629	611.6439	800	1046.360

km³ ← — 9 — | m³ | + 3 → l

m³			m³		
m³/s	REF	yd³/s	m³/s	REF	yd³/s
612.4085	801	1047.668	650.6362	851	1113.066
613.1730	802	1048.976	651.4008	852	1114.374
613.9376	803	1050.284	652.1653	853	1115.682
614.7021	804	1051.592	652.9299	854	1116.990
615.4667	805	1052.900	653.6944	855	1118.298
616.2313	806	1054.208	654.4590	856	1119.606
616.9958	807	1055.516	655.2235	857	1120.914
617.7604	808	1056.824	655.9881	858	1122.222
618.5249	809	1058.132	656.7527	859	1123.530
619.2895	810	1059.440	657.5172	860	1124.837
620.0540	811	1060.748	658.2818	861	1126.145
620.8186	812	1062.056	659.0463	862	1127.453
621.5831	813	1063.364	659.8109	863	1128.761
622.3477	814	1064.672	660.5754	864	1130.069
623.1122	815	1065.980	661.3400	865	1131.377
623.8768	816	1067.288	662.1045	866	1132.685
624.6414	817	1068.596	662.8691	867	1133.993
625.4059	818	1069.904	663.6337	868	1135.301
626.1705	819	1071.211	664.3982	869	1136.609
626.9350	820	1072.519	665.1628	870	1137.917
627.6996	821	1073.827	665.9273	871	1139.225
628.4641	822	1075.135	666.6919	872	1140.533
629.2287	823	1076.443	667.4564	873	1141.841
629.9932	824	1077.751	668.2210	874	1143.149
630.7578	825	1079.059	668.9855	875	1144.457
631.5223	826	1080.367	669.7501	876	1145.765
632.2869	827	1081.675	670.5146	877	1147.073
633.0515	828	1082.983	671.2792	878	1148.381
633.8160	829	1084.291	672.0438	879	1149.689
634.5806	830	1085.599	672.8083	880	1150.996
635.3451	831	1086.907	673.5729	881	1152.304
636.1097	832	1088.215	674.3374	882	1153.612
636.8742	833	1089.523	675.1020	883	1154.920
637.6388	834	1090.831	675.8665	884	1156.228
638.4033	835	1092.139	676.6311	885	1157.536
639.1679	836	1093.447	677.3956	886	1158.844
639.9325	837	1094.755	678.1602	887	1160.152
640.6970	838	1096.063	678.9248	888	1161.460
641.4616	839	1097.370	679.6893	889	1162.768
642.2261	840	1098.678	680.4539	890	1164.076
642.9907	841	1099.986	681.2184	891	1165.384
643.7552	842	1101.294	681.9830	892	1166.692
644.5198	843	1102.602	682.7475	893	1168.000
645.2843	844	1103.910	683.5121	894	1169.308
646.0489	845	1105.218	684.2766	895	1170.616
646.8134	846	1106.526	685.0412	896	1171.924
647.5780	847	1107.834	685.8057	897	1173.232
648.3426	848	1109.142	686.5703	898	1174.540
649.1071	849	1110.450	687.3349	899	1175.848
649.8717	850	1111.758	688.0994	900	1177.155

km³ ———————— + 9 ————————→ m³ ←———————— − 3 ———————— l

m³		yd³	m³		yd³
m³/s	REF	yd³/s	m³/s	REF	yd³/s
688.8640	901	1178.463	727.0917	951	1243.861
689.6285	902	1179.771	727.8563	952	1245.169
690.3931	903	1181.079	728.6208	953	1246.477
691.1576	904	1182.387	729.3854	954	1247.785
691.9222	905	1183.695	730.1499	955	1249.093
692.6867	906	1185.003	730.9145	956	1250.401
693.4513	907	1186.311	731.6790	957	1251.709
694.2159	908	1187.619	732.4436	958	1253.017
694.9804	909	1188.927	733.2082	959	1254.325
695.7450	910	1190.235	733.9727	960	1255.633
696.5095	911	1191.543	734.7373	961	1256.940
697.2741	912	1192.851	735.5018	962	1258.248
698.0386	913	1194.159	736.2664	963	1259.556
698.8032	914	1195.467	737.0309	964	1260.864
699.5677	915	1196.775	737.7955	965	1262.172
700.3323	916	1198.083	738.5600	966	1263.480
701.0968	917	1199.391	739.3246	967	1264.788
701.8614	918	1200.699	740.0891	968	1266.096
702.6260	919	1202.007	740.8537	969	1267.404
703.3905	920	1203.314	741.6183	970	1268.712
704.1551	921	1204.622	742.3828	971	1270.020
704.9196	922	1205.930	743.1474	972	1271.328
705.6842	923	1207.238	743.9119	973	1272.636
706.4487	924	1208.546	744.6765	974	1273.944
707.2133	925	1209.854	745.4410	975	1275.252
707.9778	926	1211.162	746.2056	976	1276.560
708.7424	927	1212.470	746.9701	977	1277.868
709.5070	928	1213.778	747.7347	978	1279.176
710.2715	929	1215.086	748.4993	979	1280.484
711.0361	930	1216.394	749.2638	980	1281.792
711.8006	931	1217.702	750.0284	981	1283.099
712.5652	932	1219.010	750.7929	982	1284.407
713.3297	933	1220.318	751.5575	983	1285.715
714.0943	934	1221.626	752.3220	984	1287.023
714.8588	935	1222.934	753.0866	985	1288.331
715.6234	936	1224.242	753.8511	986	1289.639
716.3879	937	1225.550	754.6157	987	1290.947
717.1525	938	1226.858	755.3802	988	1292.255
717.9171	939	1228.166	756.1448	989	1293.563
718.6816	940	1229.474	756.9094	990	1294.871
719.4462	941	1230.781	757.6739	991	1296.179
720.2107	942	1232.089	758.4385	992	1297.487
720.9753	943	1233.397	759.2030	993	1298.795
721.7398	944	1234.705	759.9676	994	1300.103
722.5044	945	1236.013	760.7321	995	1301.411
723.2689	946	1237.321	761.4967	996	1302.719
724.0335	947	1238.629	762.2612	997	1304.027
724.7980	948	1239.937	763.0258	998	1305.335
725.5626	949	1241.245	763.7903	999	1306.643
726.3272	950	1242.553	764.5549	1000	1307.951

13

VOLUME

Gallon (U.S. liquid) — Cubic-metre
$$(gal - m^3)$$
Gallon (U.S. liquid) — Cubic-centimetre
$$(gal - cm^3)$$
Gallon (U.S. liquid) — Litre
$$(gal - l)$$

VOLUME FLOW RATE

Gallon/Second — Cubic-metre/Second
$$(gal/s - m^3/s)$$
Gallon/Second — Cubic-centimetre/Second
$$(gal/s - cm^3/s)$$
Gallon/Second — Litre/Second
$$(gal/s - l/s)$$

Conversion Factors:

1 U.S. Liquid Gallon (gal) = 0.003785412 cubic metre (m^3)
(1 m^3 = 264.1720 gal)

Volume Flow Rate Values

Volume flow rate, in this case U.S. liquid, is represented by Table 13 as long as the time units are the same for both English and SI conversion quantities. Besides per second shown above, these consistent time units can also be in per minute (min), per hour (h), per day (d), etc.

Volume Flow Rate Conversion Factor:

1 gal/s (or gal/min, etc.) = 0.003785412 m^3/s (or m^3/min, etc.)

In converting m³ quantities to units containing cm³ and 1 (litres), Diagram 13-A, which appears in the headings on alternate pages of this table, will assist the user. It is derived from: $1 \text{ m}^3 = (10^6) \text{ cm}^3 = (10^3)1 \text{ (litres)}; 1\,1 = 1\text{dm}^3$.

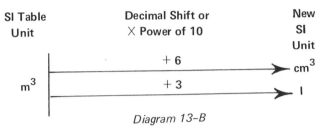

SI Table Unit	Decimal Shift or × Power of 10	New SI Unit
	+ 6	cm³
m³	+ 3	1

Diagram 13–B

Example (a): Convert 950 gal/min to m³/min, cm³/min, 1/min.
  From table: 950 gal/min = 3.596141 m³/min* ≈ 3.60 m³/min
  From Diagram 13-A: 1) 3.596141 m³/min = 3.596141 × 10⁶ cm³/min
                          = 3,596,141 cm³/min ≈ 3.60 × 10⁶ cm³/min

          2) 3.596141 m³/min = 3.596141 × 10³ 1/min
              = 3,596.141 1/min ≈ 3,600 1/min

Diagram 13-B is also given at the top of alternate table pages:

	− 6	cm³
m³	− 3	1

Diagram 13–B

It is the opposite of Diagram 13-A and is used to convert units containing cm³ and 1 (litre) to those with m³. The equivalent with m³ is then applied to the REF column in the table to obtain U.S. gallon quantities.

Example (b): Convert 739.5 1 to gal.
  From Diagram 13-B: 739.5 1 = 739.5 × 10⁻³ m³
  Since: 739.5 m³ = (730 + 9.5) m³ :
  From table: 730 m³ = 192,845.6 gal*
          95 m³ = 25,096.34 gal*

By shifting decimals where necessary and adding:
    730.0 m³ = 192,845.6   gal
      9.5 m³ =   2,509.634 gal
  739.5 m³ = 195,355.23   gal
  Therefore: 739.5 × 10⁻³ m³ = 195,355.2 × 10⁻³ gal = 195.3552 gal
          ≈ 195.4 gal.

*All conversion values obtained can be rounded to desired number of significant figures.

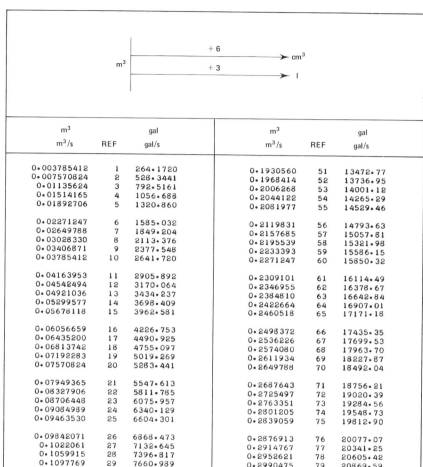

m³ / m³/s	REF	gal / gal/s		m³ / m³/s	REF	gal / gal/s
0.003785412	1	264.1720		0.1930560	51	13472.77
0.007570824	2	528.3441		0.1968414	52	13736.95
0.01135624	3	792.5161		0.2006268	53	14001.12
0.01514165	4	1056.688		0.2044122	54	14265.29
0.01892706	5	1320.860		0.2081977	55	14529.46
0.02271247	6	1585.032		0.2119831	56	14793.63
0.02649788	7	1849.204		0.2157685	57	15057.81
0.03028330	8	2113.376		0.2195539	58	15321.98
0.03406871	9	2377.548		0.2233393	59	15586.15
0.03785412	10	2641.720		0.2271247	60	15850.32
0.04163953	11	2905.892		0.2309101	61	16114.49
0.04542494	12	3170.064		0.2346955	62	16378.67
0.04921036	13	3434.237		0.2384810	63	16642.84
0.05299577	14	3698.409		0.2422664	64	16907.01
0.05678118	15	3962.581		0.2460518	65	17171.18
0.06056659	16	4226.753		0.2498372	66	17435.35
0.06435200	17	4490.925		0.2536226	67	17699.53
0.06813742	18	4755.097		0.2574080	68	17963.70
0.07192283	19	5019.269		0.2611934	69	18227.87
0.07570824	20	5283.441		0.2649788	70	18492.04
0.07949365	21	5547.613		0.2687643	71	18756.21
0.08327906	22	5811.785		0.2725497	72	19020.39
0.08706448	23	6075.957		0.2763351	73	19284.56
0.09084989	24	6340.129		0.2801205	74	19548.73
0.09463530	25	6604.301		0.2839059	75	19812.90
0.09842071	26	6868.473		0.2876913	76	20077.07
0.1022061	27	7132.645		0.2914767	77	20341.25
0.1059915	28	7396.817		0.2952621	78	20605.42
0.1097769	29	7660.989		0.2990475	79	20869.59
0.1135624	30	7925.161		0.3028330	80	21133.76
0.1173478	31	8189.333		0.3066184	81	21397.94
0.1211332	32	8453.505		0.3104038	82	21662.11
0.1249186	33	8717.677		0.3141892	83	21926.28
0.1287040	34	8981.849		0.3179746	84	22190.45
0.1324894	35	9246.021		0.3217600	85	22454.62
0.1362748	36	9510.193		0.3255454	86	22718.80
0.1400602	37	9774.365		0.3293308	87	22982.97
0.1438457	38	10038.54		0.3331163	88	23247.14
0.1476311	39	10302.71		0.3369017	89	23511.31
0.1514165	40	10566.88		0.3406871	90	23775.48
0.1552019	41	10831.05		0.3444725	91	24039.66
0.1589873	42	11095.23		0.3482579	92	24303.83
0.1627727	43	11359.40		0.3520433	93	24568.00
0.1665581	44	11623.57		0.3558287	94	24832.17
0.1703435	45	11887.74		0.3596141	95	25096.34
0.1741290	46	12151.91		0.3633996	96	25360.52
0.1779144	47	12416.09		0.3671850	97	25624.69
0.1816998	48	12680.26		0.3709704	98	25888.86
0.1854852	49	12944.43		0.3747558	99	26153.03
0.1892706	50	13208.60		0.3785412	100	26417.20

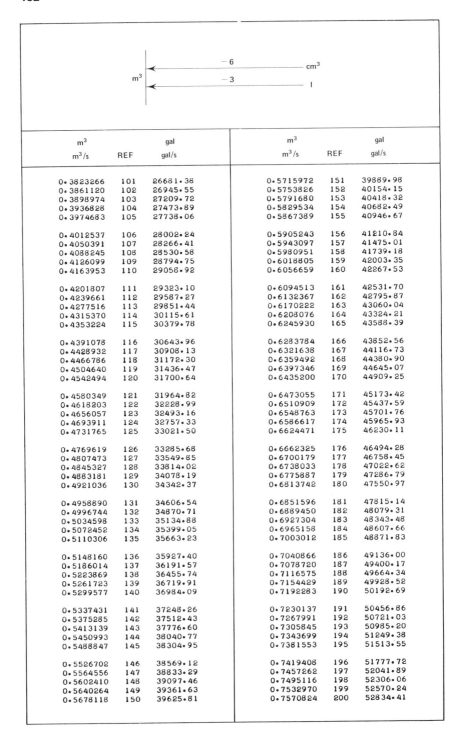

m³ m³/s	REF	gal gal/s	m³ m³/s	REF	gal gal/s
0.3823266	101	26681.38	0.5715972	151	39889.98
0.3861120	102	26945.55	0.5753826	152	40154.15
0.3898974	103	27209.72	0.5791680	153	40418.32
0.3936828	104	27473.89	0.5829534	154	40682.49
0.3974683	105	27738.06	0.5867389	155	40946.67
0.4012537	106	28002.24	0.5905243	156	41210.84
0.4050391	107	28266.41	0.5943097	157	41475.01
0.4088245	108	28530.58	0.5980951	158	41739.18
0.4126099	109	28794.75	0.6018805	159	42003.35
0.4163953	110	29058.92	0.6056659	160	42267.53
0.4201807	111	29323.10	0.6094513	161	42531.70
0.4239661	112	29587.27	0.6132367	162	42795.87
0.4277516	113	29851.44	0.6170222	163	43060.04
0.4315370	114	30115.61	0.6208076	164	43324.21
0.4353224	115	30379.78	0.6245930	165	43588.39
0.4391078	116	30643.96	0.6283784	166	43852.56
0.4428932	117	30908.13	0.6321638	167	44116.73
0.4466786	118	31172.30	0.6359492	168	44380.90
0.4504640	119	31436.47	0.6397346	169	44645.07
0.4542494	120	31700.64	0.6435200	170	44909.25
0.4580349	121	31964.82	0.6473055	171	45173.42
0.4618203	122	32228.99	0.6510909	172	45437.59
0.4656057	123	32493.16	0.6548763	173	45701.76
0.4693911	124	32757.33	0.6586617	174	45965.93
0.4731765	125	33021.50	0.6624471	175	46230.11
0.4769619	126	33285.68	0.6662325	176	46494.28
0.4807473	127	33549.85	0.6700179	177	46758.45
0.4845327	128	33814.02	0.6738033	178	47022.62
0.4883181	129	34078.19	0.6775887	179	47286.79
0.4921036	130	34342.37	0.6813742	180	47550.97
0.4958890	131	34606.54	0.6851596	181	47815.14
0.4996744	132	34870.71	0.6889450	182	48079.31
0.5034598	133	35134.88	0.6927304	183	48343.48
0.5072452	134	35399.05	0.6965158	184	48607.66
0.5110306	135	35663.23	0.7003012	185	48871.83
0.5148160	136	35927.40	0.7040866	186	49136.00
0.5186014	137	36191.57	0.7078720	187	49400.17
0.5223869	138	36455.74	0.7116575	188	49664.34
0.5261723	139	36719.91	0.7154429	189	49928.52
0.5299577	140	36984.09	0.7192283	190	50192.69
0.5337431	141	37248.26	0.7230137	191	50456.86
0.5375285	142	37512.43	0.7267991	192	50721.03
0.5413139	143	37776.60	0.7305845	193	50985.20
0.5450993	144	38040.77	0.7343699	194	51249.38
0.5488847	145	38304.95	0.7381553	195	51513.55
0.5526702	146	38569.12	0.7419408	196	51777.72
0.5564556	147	38833.29	0.7457262	197	52041.89
0.5602410	148	39097.46	0.7495116	198	52306.06
0.5640264	149	39361.63	0.7532970	199	52570.24
0.5678118	150	39625.81	0.7570824	200	52834.41

m³ m³/s	REF	gal gal/s	m³ m³/s	REF	gal gal/s
0.7608678	201	53098.58	0.9501384	251	66307.18
0.7646532	202	53362.75	0.9539238	252	66571.35
0.7684386	203	53626.92	0.9577092	253	66835.53
0.7722240	204	53891.10	0.9614946	254	67099.70
0.7760095	205	54155.27	0.9652801	255	67363.87
0.7797949	206	54419.44	0.9690655	256	67628.04
0.7835803	207	54683.61	0.9728509	257	67892.21
0.7873657	208	54947.78	0.9766363	258	68156.39
0.7911511	209	55211.96	0.9804217	259	68420.56
0.7949365	210	55476.13	0.9842071	260	68684.73
0.7987219	211	55740.30	0.9879925	261	68948.90
0.8025073	212	56004.47	0.9917779	262	69213.07
0.8062928	213	56268.64	0.9955634	263	69477.25
0.8100782	214	56532.82	0.9993488	264	69741.42
0.8138636	215	56796.99	1.003134	265	70005.59
0.8176490	216	57061.16	1.006920	266	70269.76
0.8214344	217	57325.33	1.010705	267	70533.93
0.8252198	218	57589.50	1.014490	268	70798.11
0.8290052	219	57853.68	1.018276	269	71062.28
0.8327906	220	58117.85	1.022061	270	71326.45
0.8365761	221	58382.02	1.025847	271	71590.62
0.8403615	222	58646.19	1.029632	272	71854.79
0.8441469	223	58910.36	1.033417	273	72118.97
0.8479323	224	59174.54	1.037203	274	72383.14
0.8517177	225	59438.71	1.040988	275	72647.31
0.8555031	226	59702.88	1.044774	276	72911.48
0.8592885	227	59967.05	1.048559	277	73175.66
0.8630739	228	60231.23	1.052345	278	73439.83
0.8668593	229	60495.40	1.056130	279	73704.00
0.8706448	230	60759.57	1.059915	280	73968.17
0.8744302	231	61023.74	1.063701	281	74232.34
0.8782156	232	61287.91	1.067486	282	74496.51
0.8820010	233	61552.08	1.071272	283	74760.69
0.8857864	234	61816.26	1.075057	284	75024.86
0.8895718	235	62080.43	1.078842	285	75289.03
0.8933572	236	62344.60	1.082628	286	75553.20
0.8971426	237	62608.77	1.086413	287	75817.37
0.9009281	238	62872.95	1.090199	288	76081.55
0.9047135	239	63137.12	1.093984	289	76345.72
0.9084989	240	63401.29	1.097769	290	76609.89
0.9122843	241	63665.46	1.101555	291	76874.06
0.9160697	242	63929.63	1.105340	292	77138.24
0.9198551	243	64193.81	1.109126	293	77402.41
0.9236405	244	64457.98	1.112911	294	77666.58
0.9274259	245	64722.15	1.116697	295	77930.75
0.9312114	246	64986.32	1.120482	296	78194.92
0.9349968	247	65250.49	1.124267	297	78459.10
0.9387822	248	65514.67	1.128053	298	78723.27
0.9425676	249	65778.84	1.131838	299	78987.44
0.9463530	250	66043.01	1.135624	300	79251.61

m^3 m^3/s	REF	gal gal/s	m^3 m^3/s	REF	gal gal/s
1.139409	301	79515.78	1.328680	351	92724.39
1.143194	302	79779.96	1.332465	352	92988.56
1.146980	303	80044.13	1.336250	353	93252.73
1.150765	304	80308.30	1.340036	354	93516.90
1.154551	305	80572.47	1.343821	355	93781.07
1.158336	306	80836.64	1.347607	356	94045.25
1.162121	307	81100.82	1.351392	357	94309.42
1.165907	308	81364.99	1.355177	358	94573.59
1.169692	309	81629.16	1.358963	359	94837.76
1.173478	310	81893.33	1.362748	360	95101.93
1.177263	311	82157.50	1.366534	361	95366.11
1.181049	312	82421.68	1.370319	362	95630.28
1.184834	313	82685.85	1.374105	363	95894.45
1.188619	314	82950.02	1.377890	364	96158.62
1.192405	315	83214.19	1.381675	365	96422.79
1.196190	316	83478.36	1.385461	366	96686.97
1.199976	317	83742.54	1.389246	367	96951.14
1.203761	318	84006.71	1.393032	368	97215.31
1.207546	319	84270.88	1.396817	369	97479.48
1.211332	320	84535.05	1.400602	370	97743.65
1.215117	321	84799.22	1.404388	371	98007.83
1.218903	322	85063.40	1.408173	372	98272.00
1.222688	323	85327.57	1.411959	373	98536.17
1.226473	324	85591.74	1.415744	374	98800.34
1.230259	325	85855.91	1.419529	375	99064.51
1.234044	326	86120.08	1.423315	376	99328.69
1.237830	327	86384.26	1.427100	377	99592.86
1.241615	328	86648.43	1.430886	378	99857.03
1.245401	329	86912.60	1.434671	379	100121.2
1.249186	330	87176.77	1.438457	380	100385.4
1.252971	331	87440.95	1.442242	381	100649.5
1.256757	332	87705.12	1.446027	382	100913.7
1.260542	333	87969.29	1.449813	383	101177.9
1.264328	334	88233.46	1.453598	384	101442.1
1.268113	335	88497.63	1.457384	385	101706.2
1.271898	336	88761.80	1.461169	386	101970.4
1.275684	337	89025.98	1.464954	387	102234.6
1.279469	338	89290.15	1.468740	388	102498.8
1.283255	339	89554.32	1.472525	389	102762.9
1.287040	340	89818.49	1.476311	390	103027.1
1.290825	341	90082.67	1.480096	391	103291.3
1.294611	342	90346.84	1.483882	392	103555.4
1.298396	343	90611.01	1.487667	393	103819.6
1.302182	344	90875.18	1.491452	394	104083.8
1.305967	345	91139.35	1.495238	395	104348.0
1.309753	346	91403.53	1.499023	396	104612.1
1.313538	347	91667.70	1.502809	397	104876.3
1.317323	348	91931.87	1.506594	398	105140.5
1.321109	349	92196.04	1.510379	399	105404.6
1.324894	350	92460.21	1.514165	400	105668.8

m³		gal	m³		gal
m³/s	REF	gal/s	m³/s	REF	gal/s
1.517950	401	105933.0	1.707221	451	119141.6
1.521736	402	106197.2	1.711006	452	119405.8
1.525521	403	106461.3	1.714792	453	119669.9
1.529306	404	106725.5	1.718577	454	119934.1
1.533092	405	106989.7	1.722362	455	120198.3
1.536877	406	107253.8	1.726148	456	120462.5
1.540663	407	107518.0	1.729933	457	120726.6
1.544448	408	107782.2	1.733719	458	120990.8
1.548234	409	108046.4	1.737504	459	121255.0
1.552019	410	108310.5	1.741290	460	121519.1
1.555804	411	108574.7	1.745075	461	121783.3
1.559590	412	108838.9	1.748860	462	122047.5
1.563375	413	109103.1	1.752646	463	122311.7
1.567161	414	109367.2	1.756431	464	122575.8
1.570946	415	109631.4	1.760217	465	122840.0
1.574731	416	109895.6	1.764002	466	123104.2
1.578517	417	110159.7	1.767787	467	123368.3
1.582302	418	110423.9	1.771573	468	123632.5
1.586088	419	110688.1	1.775358	469	123896.7
1.589873	420	110952.3	1.779144	470	124160.9
1.593658	421	111216.4	1.782929	471	124425.0
1.597444	422	111480.6	1.786714	472	124689.2
1.601229	423	111744.8	1.790500	473	124953.4
1.605015	424	112008.9	1.794285	474	125217.5
1.608800	425	112273.1	1.798071	475	125481.7
1.612586	426	112537.3	1.801856	476	125745.9
1.616371	427	112801.5	1.805642	477	126010.1
1.620156	428	113065.6	1.809427	478	126274.2
1.623942	429	113329.8	1.813212	479	126538.4
1.627727	430	113594.0	1.816998	480	126802.6
1.631513	431	113858.1	1.820783	481	127066.8
1.635298	432	114122.3	1.824569	482	127330.9
1.639083	433	114386.5	1.828354	483	127595.1
1.642869	434	114650.7	1.832139	484	127859.3
1.646654	435	114914.8	1.835925	485	128123.4
1.650440	436	115179.0	1.839710	486	128387.6
1.654225	437	115443.2	1.843496	487	128651.8
1.658010	438	115707.4	1.847281	488	128916.0
1.661796	439	115971.5	1.851066	489	129180.1
1.665581	440	116235.7	1.854852	490	129444.3
1.669367	441	116499.9	1.858637	491	129708.5
1.673152	442	116764.0	1.862423	492	129972.6
1.676938	443	117028.2	1.866208	493	130236.8
1.680723	444	117292.4	1.869994	494	130501.0
1.684508	445	117556.6	1.873779	495	130765.2
1.688294	446	117820.7	1.877564	496	131029.3
1.692079	447	118084.9	1.881350	497	131293.5
1.695865	448	118349.1	1.885135	498	131557.7
1.699650	449	118613.2	1.888921	499	131821.8
1.703435	450	118877.4	1.892706	500	132086.0

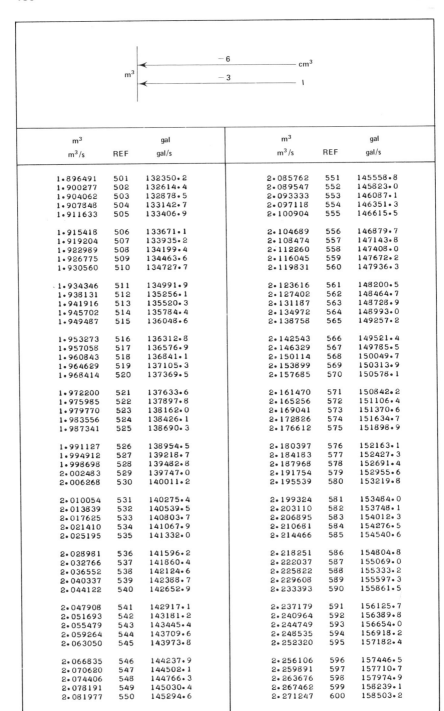

m³		gal	m³		gal
m³/s	REF	gal/s	m³/s	REF	gal/s
1.896491	501	132350.2	2.085762	551	145558.8
1.900277	502	132614.4	2.089547	552	145823.0
1.904062	503	132878.5	2.093333	553	146087.1
1.907848	504	133142.7	2.097118	554	146351.3
1.911633	505	133406.9	2.100904	555	146615.5
1.915418	506	133671.1	2.104689	556	146879.7
1.919204	507	133935.2	2.108474	557	147143.8
1.922989	508	134199.4	2.112260	558	147408.0
1.926775	509	134463.6	2.116045	559	147672.2
1.930560	510	134727.7	2.119831	560	147936.3
1.934346	511	134991.9	2.123616	561	148200.5
1.938131	512	135256.1	2.127402	562	148464.7
1.941916	513	135520.3	2.131187	563	148728.9
1.945702	514	135784.4	2.134972	564	148993.0
1.949487	515	136048.6	2.138758	565	149257.2
1.953273	516	136312.8	2.142543	566	149521.4
1.957058	517	136576.9	2.146329	567	149785.5
1.960843	518	136841.1	2.150114	568	150049.7
1.964629	519	137105.3	2.153899	569	150313.9
1.968414	520	137369.5	2.157685	570	150578.1
1.972200	521	137633.6	2.161470	571	150842.2
1.975985	522	137897.8	2.165256	572	151106.4
1.979770	523	138162.0	2.169041	573	151370.6
1.983556	524	138426.1	2.172826	574	151634.7
1.987341	525	138690.3	2.176612	575	151898.9
1.991127	526	138954.5	2.180397	576	152163.1
1.994912	527	139218.7	2.184183	577	152427.3
1.998698	528	139482.8	2.187968	578	152691.4
2.002483	529	139747.0	2.191754	579	152955.6
2.006268	530	140011.2	2.195539	580	153219.8
2.010054	531	140275.4	2.199324	581	153484.0
2.013839	532	140539.5	2.203110	582	153748.1
2.017625	533	140803.7	2.206895	583	154012.3
2.021410	534	141067.9	2.210681	584	154276.5
2.025195	535	141332.0	2.214466	585	154540.6
2.028981	536	141596.2	2.218251	586	154804.8
2.032766	537	141860.4	2.222037	587	155069.0
2.036552	538	142124.6	2.225822	588	155333.2
2.040337	539	142388.7	2.229608	589	155597.3
2.044122	540	142652.9	2.233393	590	155861.5
2.047908	541	142917.1	2.237179	591	156125.7
2.051693	542	143181.2	2.240964	592	156389.8
2.055479	543	143445.4	2.244749	593	156654.0
2.059264	544	143709.6	2.248535	594	156918.2
2.063050	545	143973.8	2.252320	595	157182.4
2.066835	546	144237.9	2.256106	596	157446.5
2.070620	547	144502.1	2.259891	597	157710.7
2.074406	548	144766.3	2.263676	598	157974.9
2.078191	549	145030.4	2.267462	599	158239.1
2.081977	550	145294.6	2.271247	600	158503.2

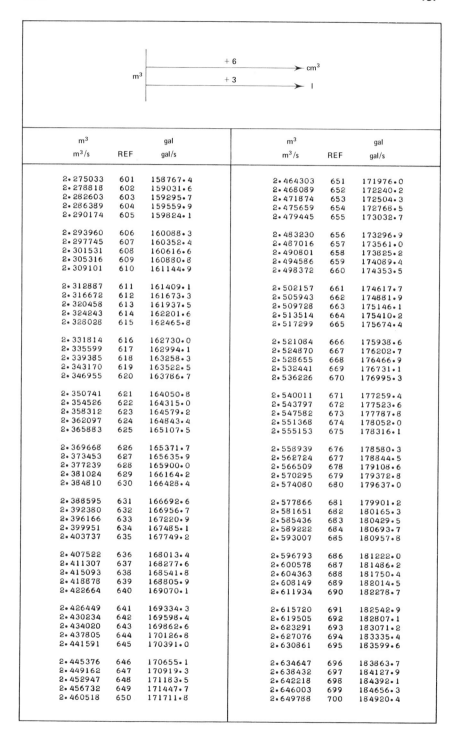

m³		gal	m³		gal
m³/s	REF	gal/s	m³/s	REF	gal/s
2.275033	601	158767.4	2.464303	651	171976.0
2.278818	602	159031.6	2.468089	652	172240.2
2.282603	603	159295.7	2.471874	653	172504.3
2.286389	604	159559.9	2.475659	654	172768.5
2.290174	605	159824.1	2.479445	655	173032.7
2.293960	606	160088.3	2.483230	656	173296.9
2.297745	607	160352.4	2.487016	657	173561.0
2.301531	608	160616.6	2.490801	658	173825.2
2.305316	609	160880.8	2.494586	659	174089.4
2.309101	610	161144.9	2.498372	660	174353.5
2.312887	611	161409.1	2.502157	661	174617.7
2.316672	612	161673.3	2.505943	662	174881.9
2.320458	613	161937.5	2.509728	663	175146.1
2.324243	614	162201.6	2.513514	664	175410.2
2.328028	615	162465.8	2.517299	665	175674.4
2.331814	616	162730.0	2.521084	666	175938.6
2.335599	617	162994.1	2.524870	667	176202.7
2.339385	618	163258.3	2.528655	668	176466.9
2.343170	619	163522.5	2.532441	669	176731.1
2.346955	620	163786.7	2.536226	670	176995.3
2.350741	621	164050.8	2.540011	671	177259.4
2.354526	622	164315.0	2.543797	672	177523.6
2.358312	623	164579.2	2.547582	673	177787.8
2.362097	624	164843.4	2.551368	674	178052.0
2.365883	625	165107.5	2.555153	675	178316.1
2.369668	626	165371.7	2.558939	676	178580.3
2.373453	627	165635.9	2.562724	677	178844.5
2.377239	628	165900.0	2.566509	678	179108.6
2.381024	629	166164.2	2.570295	679	179372.8
2.384810	630	166428.4	2.574080	680	179637.0
2.388595	631	166692.6	2.577866	681	179901.2
2.392380	632	166956.7	2.581651	682	180165.3
2.396166	633	167220.9	2.585436	683	180429.5
2.399951	634	167485.1	2.589222	684	180693.7
2.403737	635	167749.2	2.593007	685	180957.8
2.407522	636	168013.4	2.596793	686	181222.0
2.411307	637	168277.6	2.600578	687	181486.2
2.415093	638	168541.8	2.604363	688	181750.4
2.418878	639	168805.9	2.608149	689	182014.5
2.422664	640	169070.1	2.611934	690	182278.7
2.426449	641	169334.3	2.615720	691	182542.9
2.430234	642	169598.4	2.619505	692	182807.1
2.434020	643	169862.6	2.623291	693	183071.2
2.437805	644	170126.8	2.627076	694	183335.4
2.441591	645	170391.0	2.630861	695	183599.6
2.445376	646	170655.1	2.634647	696	183863.7
2.449162	647	170919.3	2.638432	697	184127.9
2.452947	648	171183.5	2.642218	698	184392.1
2.456732	649	171447.7	2.646003	699	184656.3
2.460518	650	171711.8	2.649788	700	184920.4

$$m^3 \xleftarrow{\quad -6 \quad} cm^3$$
$$m^3 \xleftarrow{\quad -3 \quad} l$$

m³ / m³/s	REF	gal / gal/s	m³ / m³/s	REF	gal / gal/s
2.653574	701	185184.6	2.842844	751	198393.2
2.657359	702	185448.8	2.846630	752	198657.4
2.661145	703	185712.9	2.850415	753	198921.5
2.664930	704	185977.1	2.854201	754	199185.7
2.668715	705	186241.3	2.857986	755	199449.9
2.672501	706	186505.5	2.861771	756	199714.1
2.676286	707	186769.6	2.865557	757	199978.2
2.680072	708	187033.8	2.869342	758	200242.4
2.683857	709	187298.0	2.873128	759	200506.6
2.687643	710	187562.1	2.876913	760	200770.7
2.691428	711	187826.3	2.880699	761	201034.9
2.695213	712	188090.5	2.884484	762	201299.1
2.698999	713	188354.7	2.888269	763	201563.3
2.702784	714	188618.8	2.892055	764	201827.4
2.706570	715	188883.0	2.895840	765	202091.6
2.710355	716	189147.2	2.899626	766	202355.8
2.714140	717	189411.4	2.903411	767	202620.0
2.717926	718	189675.5	2.907196	768	202884.1
2.721711	719	189939.7	2.910982	769	203148.3
2.725497	720	190203.9	2.914767	770	203412.5
2.729282	721	190468.0	2.918553	771	203676.6
2.733067	722	190732.2	2.922338	772	203940.8
2.736853	723	190996.4	2.926123	773	204205.0
2.740638	724	191260.6	2.929909	774	204469.2
2.744424	725	191524.7	2.933694	775	204733.3
2.748209	726	191788.9	2.937480	776	204997.5
2.751995	727	192053.1	2.941265	777	205261.7
2.755780	728	192317.2	2.945051	778	205525.8
2.759565	729	192581.4	2.948836	779	205790.0
2.763351	730	192845.6	2.952621	780	206054.2
2.767136	731	193109.8	2.956407	781	206318.4
2.770922	732	193373.9	2.960192	782	206582.5
2.774707	733	193638.1	2.963978	783	206846.7
2.778492	734	193902.3	2.967763	784	207110.9
2.782278	735	194166.4	2.971548	785	207375.1
2.786063	736	194430.6	2.975334	786	207639.2
2.789849	737	194694.8	2.979119	787	207903.4
2.793634	738	194959.0	2.982905	788	208167.6
2.797419	739	195223.1	2.986690	789	208431.7
2.801205	740	195487.3	2.990475	790	208695.9
2.804990	741	195751.5	2.994261	791	208960.1
2.808776	742	196015.7	2.998046	792	209224.3
2.812561	743	196279.8	3.001832	793	209488.4
2.816347	744	196544.0	3.005617	794	209752.6
2.820132	745	196808.2	3.009403	795	210016.8
2.823917	746	197072.3	3.013188	796	210280.9
2.827703	747	197336.5	3.016973	797	210545.1
2.831488	748	197600.7	3.020759	798	210809.3
2.835274	749	197864.9	3.024544	799	211073.5
2.839059	750	198129.0	3.028330	800	211337.6

m^3 m^3/s	REF	gal gal/s	m^3 m^3/s	REF	gal gal/s
3.032115	801	211601.8	3.221386	851	224810.4
3.035900	802	211866.0	3.225171	852	225074.6
3.039686	803	212130.1	3.228956	853	225338.7
3.043471	804	212394.3	3.232742	854	225602.9
3.047257	805	212658.5	3.236527	855	225867.1
3.051042	806	212922.7	3.240313	856	226131.3
3.054827	807	213186.8	3.244098	857	226395.4
3.058613	808	213451.0	3.247883	858	226659.6
3.062398	809	213715.2	3.251669	859	226923.8
3.066184	810	213979.4	3.255454	860	227188.0
3.069969	811	214243.5	3.259240	861	227452.1
3.073755	812	214507.7	3.263025	862	227716.3
3.077540	813	214771.9	3.266811	863	227980.5
3.081325	814	215036.0	3.270596	864	228244.6
3.085111	815	215300.2	3.274381	865	228508.8
3.088896	816	215564.4	3.278167	866	228773.0
3.092682	817	215828.6	3.281952	867	229037.2
3.096467	818	216092.7	3.285738	868	229301.3
3.100252	819	216356.9	3.289523	869	229565.5
3.104038	820	216621.1	3.293308	870	229829.7
3.107823	821	216885.2	3.297094	871	230093.8
3.111609	822	217149.4	3.300879	872	230358.0
3.115394	823	217413.6	3.304665	873	230622.2
3.119179	824	217677.8	3.308450	874	230886.4
3.122965	825	217941.9	3.312236	875	231150.5
3.126750	826	218206.1	3.316021	876	231414.7
3.130536	827	218470.3	3.319806	877	231678.9
3.134321	828	218734.4	3.323592	878	231943.1
3.138107	829	218998.6	3.327377	879	232207.2
3.141892	830	219262.8	3.331163	880	232471.4
3.145677	831	219527.0	3.334948	881	232735.6
3.149463	832	219791.1	3.338733	882	232999.7
3.153248	833	220055.3	3.342519	883	233263.9
3.157034	834	220319.5	3.346304	884	233528.1
3.160819	835	220583.7	3.350090	885	233792.3
3.164604	836	220847.8	3.353875	886	234056.4
3.168390	837	221112.0	3.357660	887	234320.6
3.172175	838	221376.2	3.361446	888	234584.8
3.175961	839	221640.3	3.365231	889	234848.9
3.179746	840	221904.5	3.369017	890	235113.1
3.183531	841	222168.7	3.372802	891	235377.3
3.187317	842	222432.9	3.376588	892	235641.5
3.191102	843	222697.0	3.380373	893	235905.6
3.194888	844	222961.2	3.384158	894	236169.8
3.198673	845	223225.4	3.387944	895	236434.0
3.202459	846	223489.5	3.391729	896	236698.1
3.206244	847	223753.7	3.395515	897	236962.3
3.210029	848	224017.9	3.399300	898	237226.5
3.213815	849	224282.1	3.403085	899	237490.7
3.217600	850	224546.2	3.406871	900	237754.8

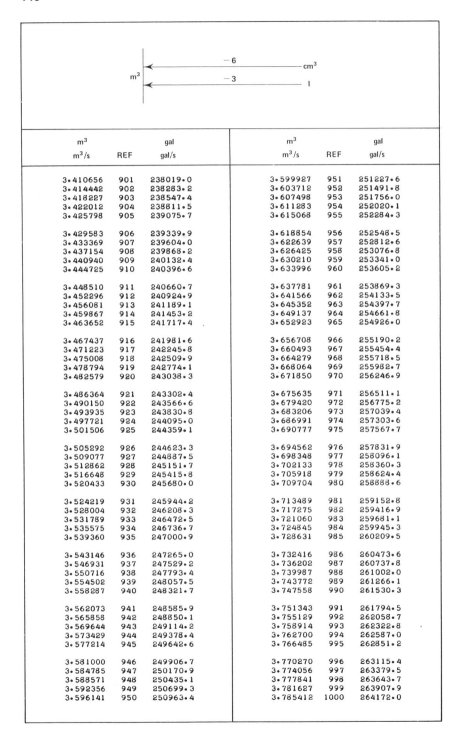

m³		gal	m³		gal
m³/s	REF	gal/s	m³/s	REF	gal/s
3.410656	901	238019.0	3.599927	951	251227.6
3.414442	902	238283.2	3.603712	952	251491.8
3.418227	903	238547.4	3.607498	953	251756.0
3.422012	904	238811.5	3.611283	954	252020.1
3.425798	905	239075.7	3.615068	955	252284.3
3.429583	906	239339.9	3.618854	956	252548.5
3.433369	907	239604.0	3.622639	957	252812.6
3.437154	908	239868.2	3.626425	958	253076.8
3.440940	909	240132.4	3.630210	959	253341.0
3.444725	910	240396.6	3.633996	960	253605.2
3.448510	911	240660.7	3.637781	961	253869.3
3.452296	912	240924.9	3.641566	962	254133.5
3.456081	913	241189.1	3.645352	963	254397.7
3.459867	914	241453.2	3.649137	964	254661.8
3.463652	915	241717.4	3.652923	965	254926.0
3.467437	916	241981.6	3.656708	966	255190.2
3.471223	917	242245.8	3.660493	967	255454.4
3.475008	918	242509.9	3.664279	968	255718.5
3.478794	919	242774.1	3.668064	969	255982.7
3.482579	920	243038.3	3.671850	970	256246.9
3.486364	921	243302.4	3.675635	971	256511.1
3.490150	922	243566.6	3.679420	972	256775.2
3.493935	923	243830.8	3.683206	973	257039.4
3.497721	924	244095.0	3.686991	974	257303.6
3.501506	925	244359.1	3.690777	975	257567.7
3.505292	926	244623.3	3.694562	976	257831.9
3.509077	927	244887.5	3.698348	977	258096.1
3.512862	928	245151.7	3.702133	978	258360.3
3.516648	929	245415.8	3.705918	979	258624.4
3.520433	930	245680.0	3.709704	980	258888.6
3.524219	931	245944.2	3.713489	981	259152.8
3.528004	932	246208.3	3.717275	982	259416.9
3.531789	933	246472.5	3.721060	983	259681.1
3.535575	934	246736.7	3.724845	984	259945.3
3.539360	935	247000.9	3.728631	985	260209.5
3.543146	936	247265.0	3.732416	986	260473.6
3.546931	937	247529.2	3.736202	987	260737.8
3.550716	938	247793.4	3.739987	988	261002.0
3.554502	939	248057.5	3.743772	989	261266.1
3.558287	940	248321.7	3.747558	990	261530.3
3.562073	941	248585.9	3.751343	991	261794.5
3.565858	942	248850.1	3.755129	992	262058.7
3.569644	943	249114.2	3.758914	993	262322.8
3.573429	944	249378.4	3.762700	994	262587.0
3.577214	945	249642.6	3.766485	995	262851.2
3.581000	946	249906.7	3.770270	996	263115.4
3.584785	947	250170.9	3.774056	997	263379.5
3.588571	948	250435.1	3.777841	998	263643.7
3.592356	949	250699.3	3.781627	999	263907.9
3.596141	950	250963.4	3.785412	1000	264172.0

VOLUME

Gallon (British) — Cubic-metre
(U.K. gal — m^3)

Gallon (British) — Cubic-centimetre
(U.K. gal — cm^3)

Gallon (British) — Litre
(U.K. gal — l)

VOLUME FLOW RATE

Gallon/Second — Cubic-metre/Second
(U.K. gal/s — m^3/s)

Gallon/Second — Cubic-centimetre/Second
(U.K. gal/s — cm^3/s)

Gallon/Second — Litre/Second
(U.K. gal/s — l/s)

Conversion Factors:
1 British gallon (U.K. gal) = 0.004546092 cubic metre (m^3)
(1 m^3 = 219.9692 U.K. gal)

Volume Flow Rate Values

Volume flow rate, in this case British liquid, is represented by Table 14 as long as the time units are the same for both English and SI conversion quantities. Besides per second shown above, these consistent time units can also be in per minute (min), per hour (h), per day (d), etc.

Volume Flow Rate Conversion Factor:

1 U.K. gal/sec (or U.K. gal/min, etc.) = 0.004546092 m^2/s (or m^3/min, etc.)

In converting m^3 quantities to units containing cm^3 and 1 (litres), Diagram 14-A, which appears in the headings on alternate pages of this table, will assist the user. It is derived from: $1 \, m^3 = (10^6) \, cm^2 = (10^3) \, l \, (litre); 1 \, l = 1 \, dm^3$.

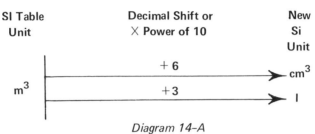

SI Table	Decimal Shift or	New
Unit	$\times$ Power of 10	Si
		Unit

Diagram 14–A

Example (a): Convert 775 U.K. gal/h to m^3/h, cm^3/h, l/h.

From table: $775 \text{ U.K. gal/h} = 3.523221 \, m^3/h^* \approx 3.52 \, m^3/h$

From Diagram 14-A: 1) $3.523221 \, m^3/h = 3.523221 \times 10^6 \, cm^3/h$
$$= 3,523,221 \, cm^3/h \approx 3.52 \times 10^6 \, cm^3/h$$
(m^3 to cm^3 by multiplying m^3 value by 10^6 or shifting decimal 6 places to the right.)

2) $3.523221 \, m^3/h = 3.523221 \times 10^3 \, l/h$
$$= 3,523.221 \, l/h \approx 3,520 \, l/h$$
(m^3/h to l/h by multiplying m^3/h value by 10^3 or shifting decimal 3 places to the right.)

Diagram 14-B is also given at the top of alternate table pages:

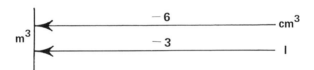

Diagram   14-B

It is the opposite of Diagram 14-A and is used to convert units containing cm^3 and 1 (litre) to those with m^3. The equivalent with m^3 is then applied to the REF column in the table to obtain British gallon quantities.

Example (b): Convert 14.3 l to U.K. gal.

From Diagram 14-B: $14.3 \, l = 14.3 \times 10^{-3} \, m^3 = 143 \times 10^{-4} \, m^3$
From table: $143 \, m^3 = 31,455.59 \text{ U.K. gal}^*$
Therefore: $143 \times 10^{-4} \, m^3 = 31,455.59 \times 10^{-4} = 3.145559 \text{ U.K. gal}$
$\approx 3.15 \text{ U.K. gal.}$

---

*All conversion values obtained can be rounded to desired number of significant figures.

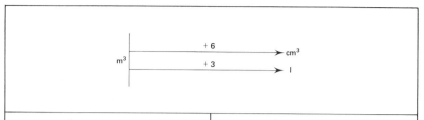

m³ m³/s	REF	U.K. gal U.K. gal/s	m³ m³/s	REF	U.K. gal U.K. gal/s
0.004546092	1	219.9692	0.2318507	51	11218.43
0.009092184	2	439.9383	0.2363968	52	11438.40
0.01363828	3	659.9075	0.2409429	53	11658.36
0.01818437	4	879.8766	0.2454890	54	11878.33
0.02273046	5	1099.846	0.2500351	55	12098.30
0.02727655	6	1319.815	0.2545812	56	12318.27
0.03182264	7	1539.784	0.2591272	57	12538.24
0.03636874	8	1759.753	0.2636733	58	12758.21
0.04091483	9	1979.722	0.2682194	59	12978.18
0.04546092	10	2199.691	0.2727655	60	13198.15
0.05000701	11	2419.661	0.2773116	61	13418.12
0.05455310	12	2639.630	0.2818577	62	13638.09
0.05909920	13	2859.599	0.2864038	63	13858.06
0.06364529	14	3079.568	0.2909499	64	14078.03
0.06819138	15	3299.537	0.2954960	65	14297.99
0.07273747	16	3519.506	0.3000421	66	14517.96
0.07728356	17	3739.476	0.3045882	67	14737.93
0.08182966	18	3959.445	0.3091343	68	14957.90
0.08637575	19	4179.414	0.3136803	69	15177.87
0.09092184	20	4399.383	0.3182264	70	15397.84
0.09546793	21	4619.352	0.3227725	71	15617.81
0.1000140	22	4839.321	0.3273186	72	15837.78
0.1045601	23	5059.290	0.3318647	73	16057.75
0.1091062	24	5279.260	0.3364108	74	16277.72
0.1136523	25	5499.229	0.3409569	75	16497.69
0.1181984	26	5719.198	0.3455030	76	16717.66
0.1227445	27	5939.167	0.3500491	77	16937.62
0.1272906	28	6159.136	0.3545952	78	17157.59
0.1318367	29	6379.105	0.3591413	79	17377.56
0.1363828	30	6599.075	0.3636874	80	17597.53
0.1409289	31	6819.044	0.3682335	81	17817.50
0.1454749	32	7039.013	0.3727795	82	18037.47
0.1500210	33	7258.982	0.3773256	83	18257.44
0.1545671	34	7478.951	0.3818717	84	18477.41
0.1591132	35	7698.920	0.3864178	85	18697.38
0.1636593	36	7918.889	0.3909639	86	18917.35
0.1682054	37	8138.859	0.3955100	87	19137.32
0.1727515	38	8358.828	0.4000561	88	19357.29
0.1772976	39	8578.797	0.4046022	89	19577.25
0.1818437	40	8798.766	0.4091483	90	19797.22
0.1863898	41	9018.735	0.4136944	91	20017.19
0.1909359	42	9238.704	0.4182405	92	20237.16
0.1954820	43	9458.673	0.4227866	93	20457.13
0.2000280	44	9678.643	0.4273326	94	20677.10
0.2045741	45	9898.612	0.4318787	95	20897.07
0.2091202	46	10118.58	0.4364248	96	21117.04
0.2136663	47	10338.55	0.4409709	97	21337.01
0.2182124	48	10558.52	0.4455170	98	21556.98
0.2227585	49	10778.49	0.4500631	99	21776.95
0.2273046	50	10998.46	0.4546092	100	21996.92

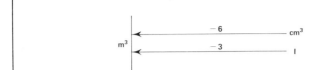

m³		U.K. gal	m³		U.K. gal
m³/s	REF	U.K. gal/s	m³/s	REF	U.K. gal/s
0.4591553	101	22216.88	0.6864599	151	33215.34
0.4637014	102	22436.85	0.6910060	152	33435.31
0.4682475	103	22656.82	0.6955521	153	33655.28
0.4727936	104	22876.79	0.7000982	154	33875.25
0.4773397	105	23096.76	0.7046443	155	34095.22
0.4818858	106	23316.73	0.7091904	156	34315.19
0.4864318	107	23536.70	0.7137364	157	34535.16
0.4909779	108	23756.67	0.7182825	158	34755.13
0.4955240	109	23976.64	0.7228286	159	34975.09
0.5000701	110	24196.61	0.7273747	160	35195.06
0.5046162	111	24416.58	0.7319208	161	35415.03
0.5091623	112	24636.54	0.7364669	162	35635.00
0.5137084	113	24856.51	0.7410130	163	35854.97
0.5182545	114	25076.48	0.7455591	164	36074.94
0.5228006	115	25296.45	0.7501052	165	36294.91
0.5273467	116	25516.42	0.7546513	166	36514.88
0.5318928	117	25736.39	0.7591974	167	36734.85
0.5364389	118	25956.36	0.7637435	168	36954.82
0.5409850	119	26176.33	0.7682896	169	37174.79
0.5455310	120	26396.30	0.7728356	170	37394.76
0.5500771	121	26616.27	0.7773817	171	37614.72
0.5546232	122	26836.24	0.7819278	172	37834.69
0.5591693	123	27056.21	0.7864739	173	38054.66
0.5637154	124	27276.17	0.7910200	174	38274.63
0.5682615	125	27496.14	0.7955661	175	38494.60
0.5728076	126	27716.11	0.8001122	176	38714.57
0.5773537	127	27936.08	0.8046583	177	38934.54
0.5818998	128	28156.05	0.8092044	178	39154.51
0.5864459	129	28376.02	0.8137505	179	39374.48
0.5909920	130	28595.99	0.8182966	180	39594.45
0.5955380	131	28815.96	0.8228427	181	39814.42
0.6000841	132	29035.93	0.8273887	182	40034.39
0.6046302	133	29255.90	0.8319348	183	40254.35
0.6091763	134	29475.87	0.8364809	184	40474.32
0.6137224	135	29695.84	0.8410270	185	40694.29
0.6182685	136	29915.80	0.8455731	186	40914.26
0.6228146	137	30135.77	0.8501192	187	41134.23
0.6273607	138	30355.74	0.8546653	188	41354.20
0.6319068	139	30575.71	0.8592114	189	41574.17
0.6364529	140	30795.68	0.8637575	190	41794.14
0.6409990	141	31015.65	0.8683036	191	42014.11
0.6455451	142	31235.62	0.8728497	192	42234.08
0.6500912	143	31455.59	0.8773958	193	42454.05
0.6546372	144	31675.56	0.8819418	194	42674.02
0.6591833	145	31895.53	0.8864879	195	42893.98
0.6637294	146	32115.50	0.8910340	196	43113.95
0.6682755	147	32335.47	0.8955801	197	43333.92
0.6728216	148	32555.43	0.9001262	198	43553.89
0.6773677	149	32775.40	0.9046723	199	43773.86
0.6819138	150	32995.37	0.9092184	200	43993.83

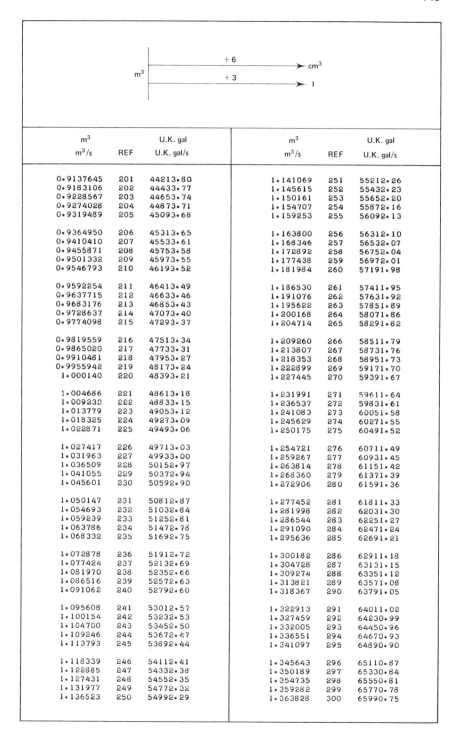

m^3		U.K. gal	m^3		U.K. gal
m^3/s	REF	U.K. gal/s	m^3/s	REF	U.K. gal/s
0.9137645	201	44213.80	1.141069	251	55212.26
0.9183106	202	44433.77	1.145615	252	55432.23
0.9228567	203	44653.74	1.150161	253	55652.20
0.9274028	204	44873.71	1.154707	254	55872.16
0.9319489	205	45093.68	1.159253	255	56092.13
0.9364950	206	45313.65	1.163800	256	56312.10
0.9410410	207	45533.61	1.168346	257	56532.07
0.9455871	208	45753.58	1.172892	258	56752.04
0.9501332	209	45973.55	1.177438	259	56972.01
0.9546793	210	46193.52	1.181984	260	57191.98
0.9592254	211	46413.49	1.186530	261	57411.95
0.9637715	212	46633.46	1.191076	262	57631.92
0.9683176	213	46853.43	1.195622	263	57851.89
0.9728637	214	47073.40	1.200168	264	58071.86
0.9774098	215	47293.37	1.204714	265	58291.82
0.9819559	216	47513.34	1.209260	266	58511.79
0.9865020	217	47733.31	1.213807	267	58731.76
0.9910481	218	47953.27	1.218353	268	58951.73
0.9955942	219	48173.24	1.222899	269	59171.70
1.000140	220	48393.21	1.227445	270	59391.67
1.004686	221	48613.18	1.231991	271	59611.64
1.009232	222	48833.15	1.236537	272	59831.61
1.013779	223	49053.12	1.241083	273	60051.58
1.018325	224	49273.09	1.245629	274	60271.55
1.022871	225	49493.06	1.250175	275	60491.52
1.027417	226	49713.03	1.254721	276	60711.49
1.031963	227	49933.00	1.259267	277	60931.45
1.036509	228	50152.97	1.263814	278	61151.42
1.041055	229	50372.94	1.268360	279	61371.39
1.045601	230	50592.90	1.272906	280	61591.36
1.050147	231	50812.87	1.277452	281	61811.33
1.054693	232	51032.84	1.281998	282	62031.30
1.059239	233	51252.81	1.286544	283	62251.27
1.063786	234	51472.78	1.291090	284	62471.24
1.068332	235	51692.75	1.295636	285	62691.21
1.072878	236	51912.72	1.300182	286	62911.18
1.077424	237	52132.69	1.304728	287	63131.15
1.081970	238	52352.66	1.309274	288	63351.12
1.086516	239	52572.63	1.313821	289	63571.08
1.091062	240	52792.60	1.318367	290	63791.05
1.095608	241	53012.57	1.322913	291	64011.02
1.100154	242	53232.53	1.327459	292	64230.99
1.104700	243	53452.50	1.332005	293	64450.96
1.109246	244	53672.47	1.336551	294	64670.93
1.113793	245	53892.44	1.341097	295	64890.90
1.118339	246	54112.41	1.345643	296	65110.87
1.122885	247	54332.38	1.350189	297	65330.84
1.127431	248	54552.35	1.354735	298	65550.81
1.131977	249	54772.32	1.359282	299	65770.78
1.136523	250	54992.29	1.363828	300	65990.75

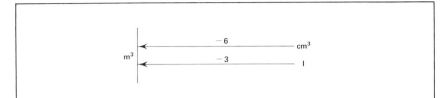

| m^3 | | U.K. gal | m^3 | | U.K. gal |
m^3/s	REF	U.K. gal/s	m^3/s	REF	U.K. gal/s
1·368374	301	66210·71	1·595678	351	77209·17
1·372920	302	66430·68	1·600224	352	77429·14
1·377466	303	66650·65	1·604770	353	77649·11
1·382012	304	66870·62	1·609317	354	77869·08
1·386558	305	67090·59	1·613863	355	78089·05
1·391104	306	67310·56	1·618409	356	78309·02
1·395650	307	67530·53	1·622955	357	78528·99
1·400196	308	67750·50	1·627501	358	78748·96
1·404742	309	67970·47	1·632047	359	78968·92
1·409289	310	68190·44	1·636593	360	79188·89
1·413835	311	68410·41	1·641139	361	79408·86
1·418381	312	68630·37	1·645685	362	79628·83
1·422927	313	68850·34	1·650231	363	79848·80
1·427473	314	69070·31	1·654777	364	80068·77
1·432019	315	69290·28	1·659324	365	80288·74
1·436565	316	69510·25	1·663870	366	80508·71
1·441111	317	69730·22	1·668416	367	80728·68
1·445657	318	69950·19	1·672962	368	80948·65
1·450203	319	70170·16	1·677508	369	81168·62
1·454749	320	70390·13	1·682054	370	81388·59
1·459296	321	70610·10	1·686600	371	81608·55
1·463842	322	70830·07	1·691146	372	81828·52
1·468388	323	71050·04	1·695692	373	82048·49
1·472934	324	71270·00	1·700238	374	82268·46
1·477480	325	71489·97	1·704784	375	82488·43
1·482026	326	71709·94	1·709331	376	82708·40
1·486572	327	71929·91	1·713877	377	82928·37
1·491118	328	72149·88	1·718423	378	83148·34
1·495664	329	72369·85	1·722969	379	83368·31
1·500210	330	72589·82	1·727515	380	83588·28
1·504756	331	72809·79	1·732061	381	83808·25
1·509303	332	73029·76	1·736607	382	84028·22
1·513849	333	73249·73	1·741153	383	84248·18
1·518395	334	73469·70	1·745699	384	84468·15
1·522941	335	73689·67	1·750245	385	84688·12
1·527487	336	73909·63	1·754792	386	84908·09
1·532033	337	74129·60	1·759338	387	85128·06
1·536579	338	74349·57	1·763884	388	85348·03
1·541125	339	74569·54	1·768430	389	85568·00
1·545671	340	74789·51	1·772976	390	85787·97
1·550217	341	75009·48	1·777522	391	86007·94
1·554763	342	75229·45	1·782068	392	86227·91
1·559310	343	75449·42	1·786614	393	86447·88
1·563856	344	75669·39	1·791160	394	86667·85
1·568402	345	75889·36	1·795706	395	86887·81
1·572948	346	76109·33	1·800252	396	87107·78
1·577494	347	76329·29	1·804799	397	87327·75
1·582040	348	76549·26	1·809345	398	87547·72
1·586586	349	76769·23	1·813891	399	87767·69
1·591132	350	76989·20	1·818437	400	87987·66

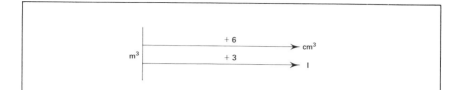

m³ m³/s	REF	U.K. gal U.K. gal/s	m³ m³/s	REF	U.K. gal U.K. gal/s
1.822983	401	88207.63	2.050287	451	99206.09
1.827529	402	88427.60	2.054834	452	99426.06
1.832075	403	88647.57	2.059380	453	99646.03
1.836621	404	88867.54	2.063926	454	99865.99
1.841167	405	89087.51	2.068472	455	100086.0
1.845713	406	89307.48	2.073018	456	100305.9
1.850259	407	89527.44	2.077564	457	100525.9
1.854806	408	89747.41	2.082110	458	100745.9
1.859352	409	89967.38	2.086656	459	100965.8
1.863898	410	90187.35	2.091202	460	101185.8
1.868444	411	90407.32	2.095748	461	101405.8
1.872990	412	90627.29	2.100295	462	101625.7
1.877536	413	90847.26	2.104841	463	101845.7
1.882082	414	91067.23	2.109387	464	102065.7
1.886628	415	91287.20	2.113933	465	102285.7
1.891174	416	91507.17	2.118479	466	102505.6
1.895720	417	91727.14	2.123025	467	102725.6
1.900266	418	91947.10	2.127571	468	102945.6
1.904813	419	92167.07	2.132117	469	103165.5
1.909359	420	92387.04	2.136663	470	103385.5
1.913905	421	92607.01	2.141209	471	103605.5
1.918451	422	92826.98	2.145755	472	103825.4
1.922997	423	93046.95	2.150302	473	104045.4
1.927543	424	93266.92	2.154848	474	104265.4
1.932089	425	93486.89	2.159394	475	104485.3
1.936635	426	93706.86	2.163940	476	104705.3
1.941181	427	93926.83	2.168486	477	104925.3
1.945727	428	94146.80	2.173032	478	105145.3
1.950273	429	94366.77	2.177578	479	105365.2
1.954820	430	94586.73	2.182124	480	105585.2
1.959366	431	94806.70	2.186670	481	105805.2
1.963912	432	95026.67	2.191216	482	106025.1
1.968458	433	95246.64	2.195762	483	106245.1
1.973004	434	95466.61	2.200309	484	106465.1
1.977550	435	95686.58	2.204855	485	106685.0
1.982096	436	95906.55	2.209401	486	106905.0
1.986642	437	96126.52	2.213947	487	107125.0
1.991188	438	96346.49	2.218493	488	107344.9
1.995734	439	96566.46	2.223039	489	107564.9
2.000280	440	96786.43	2.227585	490	107784.9
2.004827	441	97006.40	2.232131	491	108004.9
2.009373	442	97226.36	2.236677	492	108224.8
2.013919	443	97446.33	2.241223	493	108444.8
2.018465	444	97666.30	2.245769	494	108664.8
2.023011	445	97886.27	2.250316	495	108884.7
2.027557	446	98106.24	2.254862	496	109104.7
2.032103	447	98326.21	2.259408	497	109324.7
2.036649	448	98546.18	2.263954	498	109544.6
2.041195	449	98766.15	2.268500	499	109764.6
2.045741	450	98986.12	2.273046	500	109984.6

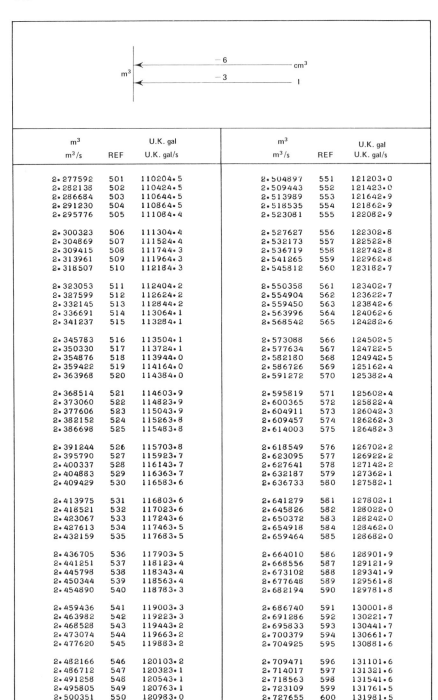

m³ m³/s	REF	U.K. gal U.K. gal/s	m³ m³/s	REF	U.K. gal U.K. gal/s
2.277592	501	110204.5	2.504897	551	121203.0
2.282138	502	110424.5	2.509443	552	121423.0
2.286684	503	110644.5	2.513989	553	121642.9
2.291230	504	110864.5	2.518535	554	121862.9
2.295776	505	111084.4	2.523081	555	122082.9
2.300323	506	111304.4	2.527627	556	122302.8
2.304869	507	111524.4	2.532173	557	122522.8
2.309415	508	111744.3	2.536719	558	122742.8
2.313961	509	111964.3	2.541265	559	122962.8
2.318507	510	112184.3	2.545812	560	123182.7
2.323053	511	112404.2	2.550358	561	123402.7
2.327599	512	112624.2	2.554904	562	123622.7
2.332145	513	112844.2	2.559450	563	123842.6
2.336691	514	113064.1	2.563996	564	124062.6
2.341237	515	113284.1	2.568542	565	124282.6
2.345783	516	113504.1	2.573088	566	124502.5
2.350330	517	113724.1	2.577634	567	124722.5
2.354876	518	113944.0	2.582180	568	124942.5
2.359422	519	114164.0	2.586726	569	125162.4
2.363968	520	114384.0	2.591272	570	125382.4
2.368514	521	114603.9	2.595819	571	125602.4
2.373060	522	114823.9	2.600365	572	125822.4
2.377606	523	115043.9	2.604911	573	126042.3
2.382152	524	115263.8	2.609457	574	126262.3
2.386698	525	115483.8	2.614003	575	126482.3
2.391244	526	115703.8	2.618549	576	126702.2
2.395790	527	115923.7	2.623095	577	126922.2
2.400337	528	116143.7	2.627641	578	127142.2
2.404883	529	116363.7	2.632187	579	127362.1
2.409429	530	116583.6	2.636733	580	127582.1
2.413975	531	116803.6	2.641279	581	127802.1
2.418521	532	117023.6	2.645826	582	128022.0
2.423067	533	117243.6	2.650372	583	128242.0
2.427613	534	117463.5	2.654918	584	128462.0
2.432159	535	117683.5	2.659464	585	128682.0
2.436705	536	117903.5	2.664010	586	128901.9
2.441251	537	118123.4	2.668556	587	129121.9
2.445798	538	118343.4	2.673102	588	129341.9
2.450344	539	118563.4	2.677648	589	129561.8
2.454890	540	118783.3	2.682194	590	129781.8
2.459436	541	119003.3	2.686740	591	130001.8
2.463982	542	119223.3	2.691286	592	130221.7
2.468528	543	119443.2	2.695833	593	130441.7
2.473074	544	119663.2	2.700379	594	130661.7
2.477620	545	119883.2	2.704925	595	130881.6
2.482166	546	120103.2	2.709471	596	131101.6
2.486712	547	120323.1	2.714017	597	131321.6
2.491258	548	120543.1	2.718563	598	131541.6
2.495805	549	120763.1	2.723109	599	131761.5
2.500351	550	120983.0	2.727655	600	131981.5

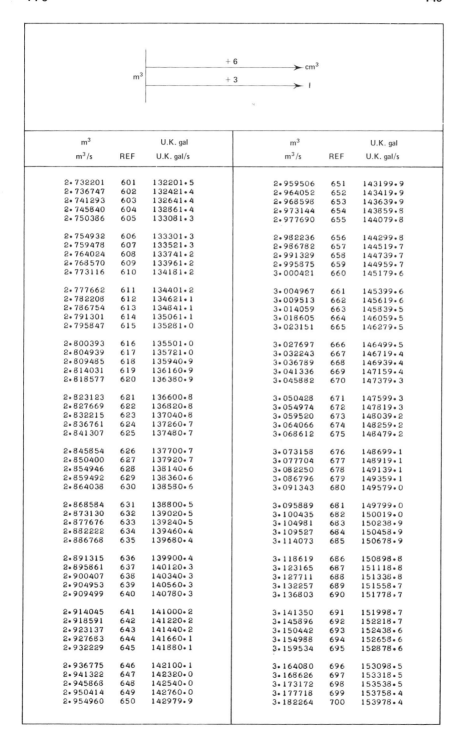

m³		U.K. gal	m³		U.K. gal
m³/s	REF	U.K. gal/s	m³/s	REF	U.K. gal/s
2.732201	601	132201.5	2.959506	651	143199.9
2.736747	602	132421.4	2.964052	652	143419.9
2.741293	603	132641.4	2.968598	653	143639.9
2.745840	604	132861.4	2.973144	654	143859.8
2.750386	605	133081.3	2.977690	655	144079.8
2.754932	606	133301.3	2.982236	656	144299.8
2.759478	607	133521.3	2.986782	657	144519.7
2.764024	608	133741.2	2.991329	658	144739.7
2.768570	609	133961.2	2.995875	659	144959.7
2.773116	610	134181.2	3.000421	660	145179.6
2.777662	611	134401.2	3.004967	661	145399.6
2.782208	612	134621.1	3.009513	662	145619.6
2.786754	613	134841.1	3.014059	663	145839.5
2.791301	614	135061.1	3.018605	664	146059.5
2.795847	615	135281.0	3.023151	665	146279.5
2.800393	616	135501.0	3.027697	666	146499.5
2.804939	617	135721.0	3.032243	667	146719.4
2.809485	618	135940.9	3.036789	668	146939.4
2.814031	619	136160.9	3.041336	669	147159.4
2.818577	620	136380.9	3.045882	670	147379.3
2.823123	621	136600.8	3.050428	671	147599.3
2.827669	622	136820.8	3.054974	672	147819.3
2.832215	623	137040.8	3.059520	673	148039.2
2.836761	624	137260.7	3.064066	674	148259.2
2.841307	625	137480.7	3.068612	675	148479.2
2.845854	626	137700.7	3.073158	676	148699.1
2.850400	627	137920.7	3.077704	677	148919.1
2.854946	628	138140.6	3.082250	678	149139.1
2.859492	629	138360.6	3.086796	679	149359.1
2.864038	630	138580.6	3.091343	680	149579.0
2.868584	631	138800.5	3.095889	681	149799.0
2.873130	632	139020.5	3.100435	682	150019.0
2.877676	633	139240.5	3.104981	683	150238.9
2.882222	634	139460.4	3.109527	684	150458.9
2.886768	635	139680.4	3.114073	685	150678.9
2.891315	636	139900.4	3.118619	686	150898.8
2.895861	637	140120.3	3.123165	687	151118.8
2.900407	638	140340.3	3.127711	688	151338.8
2.904953	639	140560.3	3.132257	689	151558.7
2.909499	640	140780.3	3.136803	690	151778.7
2.914045	641	141000.2	3.141350	691	151998.7
2.918591	642	141220.2	3.145896	692	152218.7
2.923137	643	141440.2	3.150442	693	152438.6
2.927683	644	141660.1	3.154988	694	152658.6
2.932229	645	141880.1	3.159534	695	152878.6
2.936775	646	142100.1	3.164080	696	153098.5
2.941322	647	142320.0	3.168626	697	153318.5
2.945868	648	142540.0	3.173172	698	153538.5
2.950414	649	142760.0	3.177718	699	153758.4
2.954960	650	142979.9	3.182264	700	153978.4

m³ m³/s	REF	U.K. gal U.K. gal/s	m³ m³/s	REF	U.K. gal U.K. gal/s
3·186810	701	154198·4	3·414115	751	165196·8
3·191357	702	154418·3	3·418661	752	165416·8
3·195903	703	154638·3	3·423207	753	165636·8
3·200449	704	154858·3	3·427753	754	165856·7
3·204995	705	155078·3	3·432299	755	166076·7
3·209541	706	155298·2	3·436846	756	166296·7
3·214087	707	155518·2	3·441392	757	166516·6
3·218633	708	155738·2	3·445938	758	166736·6
3·223179	709	155958·1	3·450484	759	166956·6
3·227725	710	156178·1	3·455030	760	167176·6
3·232271	711	156398·1	3·459576	761	167396·5
3·236818	712	156618·0	3·464122	762	167616·5
3·241364	713	156838·0	3·468668	763	167836·5
3·245910	714	157058·0	3·473214	764	168056·4
3·250456	715	157277·9	3·477760	765	168276·4
3·255002	716	157497·9	3·482306	766	168496·4
3·259548	717	157717·9	3·486853	767	168716·3
3·264094	718	157937·8	3·491399	768	168936·3
3·268640	719	158157·8	3·495945	769	169156·3
3·273186	720	158377·8	3·500491	770	169376·2
3·277732	721	158597·8	3·505037	771	169596·2
3·282278	722	158817·7	3·509583	772	169816·2
3·286825	723	159037·7	3·514129	773	170036·2
3·291371	724	159257·7	3·518675	774	170256·1
3·295917	725	159477·6	3·523221	775	170476·1
3·300463	726	159697·6	3·527767	776	170696·1
3·305009	727	159917·6	3·532313	777	170916·0
3·309555	728	160137·5	3·536860	778	171136·0
3·314101	729	160357·5	3·541406	779	171356·0
3·318647	730	160577·5	3·545952	780	171575·9
3·323193	731	160797·4	3·550498	781	171795·9
3·327739	732	161017·4	3·555044	782	172015·9
3·332285	733	161237·4	3·559590	783	172235·8
3·336832	734	161457·4	3·564136	784	172455·8
3·341378	735	161677·3	3·568682	785	172675·8
3·345924	736	161897·3	3·573228	786	172895·8
3·350470	737	162117·3	3·577774	787	173115·7
3·355016	738	162337·2	3·582321	788	173335·7
3·359562	739	162557·2	3·586867	789	173555·7
3·364108	740	162777·2	3·591413	790	173775·6
3·368654	741	162997·1	3·595959	791	173995·6
3·373200	742	163217·1	3·600505	792	174215·6
3·377746	743	163437·1	3·605051	793	174435·5
3·382292	744	163657·0	3·609597	794	174655·5
3·386839	745	163877·0	3·614143	795	174875·5
3·391385	746	164097·0	3·618689	796	175095·4
3·395931	747	164317·0	3·623235	797	175315·4
3·400477	748	164536·9	3·627781	798	175535·4
3·405023	749	164756·9	3·632328	799	175755·4
3·409569	750	164976·9	3·636874	800	175975·3

m³		U.K. gal		m³		U.K. gal
m³/s	REF	U.K. gal/s		m³/s	REF	U.K. gal/s
3.641420	801	176195.3		3.868724	851	187193.7
3.645966	802	176415.3		3.873270	852	187413.7
3.650512	803	176635.2		3.877816	853	187633.7
3.655058	804	176855.2		3.882363	854	187853.7
3.659604	805	177075.2		3.886909	855	188073.6
3.664150	806	177295.1		3.891455	856	188293.6
3.668696	807	177515.1		3.896001	857	188513.6
3.673242	808	177735.1		3.900547	858	188733.5
3.677788	809	177955.0		3.905093	859	188953.5
3.682335	810	178175.0		3.909639	860	189173.5
3.686881	811	178395.0		3.914185	861	189393.4
3.691427	812	178615.0		3.918731	862	189613.4
3.695973	813	178834.9		3.923277	863	189833.4
3.700519	814	179054.9		3.927823	864	190053.3
3.705065	815	179274.9		3.932370	865	190273.3
3.709611	816	179494.8		3.936916	866	190493.3
3.714157	817	179714.8		3.941462	867	190713.3
3.718703	818	179934.8		3.946008	868	190933.2
3.723249	819	180154.7		3.950554	869	191153.2
3.727795	820	180374.7		3.955100	870	191373.2
3.732342	821	180594.7		3.959646	871	191593.1
3.736888	822	180814.6		3.964192	872	191813.1
3.741434	823	181034.6		3.968738	873	192033.1
3.745980	824	181254.6		3.973284	874	192253.0
3.750526	825	181474.5		3.977830	875	192473.0
3.755072	826	181694.5		3.982377	876	192693.0
3.759618	827	181914.5		3.986923	877	192912.9
3.764164	828	182134.5		3.991469	878	193132.9
3.768710	829	182354.4		3.996015	879	193352.9
3.773256	830	182574.4		4.000561	880	193572.9
3.777802	831	182794.4		4.005107	881	193792.8
3.782349	832	183014.3		4.009653	882	194012.8
3.786895	833	183234.3		4.014199	883	194232.8
3.791441	834	183454.3		4.018745	884	194452.7
3.795987	835	183674.2		4.023291	885	194672.7
3.800533	836	183894.2		4.027838	886	194892.7
3.805079	837	184114.2		4.032384	887	195112.6
3.809625	838	184334.1		4.036930	888	195332.6
3.814171	839	184554.1		4.041476	889	195552.6
3.818717	840	184774.1		4.046022	890	195772.5
3.823263	841	184994.1		4.050568	891	195992.5
3.827809	842	185214.0		4.055114	892	196212.5
3.832356	843	185434.0		4.059660	893	196432.5
3.836902	844	185654.0		4.064206	894	196652.4
3.841448	845	185873.9		4.068752	895	196872.4
3.845994	846	186093.9		4.073298	896	197092.4
3.850540	847	186313.9		4.077845	897	197312.3
3.855086	848	186533.8		4.082391	898	197532.3
3.859632	849	186753.8		4.086937	899	197752.3
3.864178	850	186973.8		4.091483	900	197972.2

m^3		U.K. gal	m^3		U.K. gal
m^3/s	REF	U.K. gal/s	m^3/s	REF	U.K. gal/s
4·096029	901	198192·2	4·323334	951	209190·7
4·100575	902	198412·2	4·327880	952	209410·6
4·105121	903	198632·1	4·332426	953	209630·6
4·109667	904	198852·1	4·336972	954	209850·6
4·114213	905	199072·1	4·341518	955	210070·5
4·118759	906	199292·1	4·346064	956	210290·5
4·123305	907	199512·0	4·350610	957	210510·5
4·127852	908	199732·0	4·355156	958	210730·4
4·132398	909	199952·0	4·359702	959	210950·4
4·136944	910	200171·9	4·364248	960	211170·4
4·141490	911	200391·9	4·368794	961	211390·4
4·146036	912	200611·9	4·373340	962	211610·3
4·150582	913	200831·8	4·377887	963	211830·3
4·155128	914	201051·8	4·382433	964	212050·3
4·159674	915	201271·8	4·386979	965	212270·2
4·164220	916	201491·7	4·391525	966	212490·2
4·168766	917	201711·7	4·396071	967	212710·2
4·173312	918	201931·7	4·400617	968	212930·1
4·177859	919	202151·6	4·405163	969	213150·1
4·182405	920	202371·6	4·409709	970	213370·1
4·186951	921	202591·6	4·414255	971	213590·0
4·191497	922	202811·6	4·418801	972	213810·0
4·196043	923	203031·5	4·423348	973	214030·0
4·200589	924	203251·5	4·427894	974	214250·0
4·205135	925	203471·5	4·432440	975	214469·9
4·209681	926	203691·4	4·436986	976	214689·9
4·214227	927	203911·4	4·441532	977	214909·9
4·218773	928	204131·4	4·446078	978	215129·8
4·223319	929	204351·3	4·450624	979	215349·8
4·227866	930	204571·3	4·455170	980	215569·8
4·232412	931	204791·3	4·459716	981	215789·7
4·236958	932	205011·2	4·464262	982	216009·7
4·241504	933	205231·2	4·468808	983	216229·7
4·246050	934	205451·2	4·473355	984	216449·6
4·250596	935	205671·2	4·477901	985	216669·6
4·255142	936	205891·1	4·482447	986	216889·6
4·259688	937	206111·1	4·486993	987	217109·6
4·264234	938	206331·1	4·491539	988	217329·5
4·268780	939	206551·0	4·496085	989	217549·5
4·273327	940	206771·0	4·500631	990	217769·5
4·277873	941	206991·0	4·505177	991	217989·4
4·282419	942	207210·9	4·509723	992	218209·4
4·286965	943	207430·9	4·514269	993	218429·4
4·291511	944	207650·9	4·518815	994	218649·3
4·296057	945	207870·8	4·523362	995	218869·3
4·300603	946	208090·8	4·527908	996	219089·3
4·305149	947	208310·8	4·532454	997	219309·2
4·309695	948	208530·8	4·537000	998	219529·2
4·314241	949	208750·7	4·541546	999	219749·2
4·318787	950	208970·7	4·546092	1000	219969·2

15

VELOCITY

Foot/Minute — Metre/Second
  (ft/min — m/s)
Foot/Minute — Centimetre/Second
  (ft/min — cm/s)
Foot/Minute — Millimetre/Second
  (ft/min — mm/s)

Conversion Factors:
  1 foot/minute (ft/min) = 0.005080 metre/second (m/s)
  (1 m/s = 196.8504 ft/min)

In converting m/s quantities to cm/s and mm/s, Diagram 15-A, which appears in the headings on alternate pages of this table, will assist the user. It is derived from: $1 \text{ m/s} = (10^2) \text{ cm/s} = (10^3) \text{ mm/s}$.

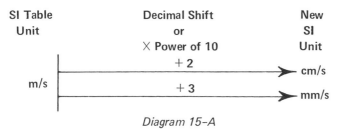

Diagram 15-A

Example (a): Convert 9.49 ft/min to m/s; cm/s; mm/s.
  Since: 9.49 ft/min = 949×10^{-2} ft/min:
  From table: 949 ft/min = 4.82092 m/s*
  Therefore: 949×10^{-2} ft/min = 0.0482092 m/s ≈ 0.0482 m/s.

---

*All conversion values obtained can be rounded to desired number of significant figures.

From Diagram 15-A: 1) 0.0482092 m/s = 0.0482092 × 10² cm/s
= 4.82092 cm/s ≈ 4.82 cm/s
(m/s to cm/s by multiplying m/s value by 10² or by
shifting its decimal 2 places to the right.)

2) 0.0482092 m/s = 0.0482092 × 10³ mm/s
= 48.2092 mm/s ≈ 48.2 mm/s
(m/s to mm/s by multiplying m/s value by 10³ or
by shifting its decimal 3 places to the right.)

Diagram 15-B is given at the top of alternate table pages:

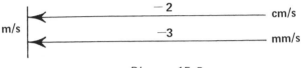

Diagram 15-B

It is the opposite of Diagram 15-A and is used to convert cm/s and mm/s
values to m/s. The m/s equivalent is then applied to the REF column in the
table to obtain ft/min quantities.

Example (*b*): Convert 26.5 cm/s to ft/min.
From Diagram 15-B: 26.5 cm/s = 26.5 × 10⁻² m/s = 265 × 10⁻³ m/s
From table: 265 m/s = 52, 165.35 ft/min*
Therefore: 265 × 10⁻³ m/s = 52,165.35 × 10⁻³ ft/min = 52.16535 ft/min
≈ 52.2 ft/min.

---

*All conversion values obtained can be rounded to desired number of significant figures.

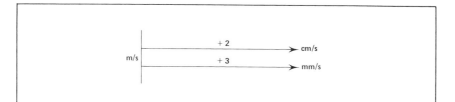

m/s	REF	ft/min		m/s	REF	ft/min
0.005080000	1	196.8504		0.2590800	51	10039.37
0.01016000	2	393.7008		0.2641600	52	10236.22
0.01524000	3	590.5512		0.2692400	53	10433.07
0.02032000	4	787.4016		0.2743200	54	10629.92
0.02540000	5	984.2520		0.2794000	55	10826.77
0.03048000	6	1181.102		0.2844800	56	11023.62
0.03556000	7	1377.953		0.2895600	57	11220.47
0.04064000	8	1574.803		0.2946400	58	11417.32
0.04572000	9	1771.654		0.2997200	59	11614.17
0.05080000	10	1968.504		0.3048000	60	11811.02
0.05588000	11	2165.354		0.3098800	61	12007.87
0.06096000	12	2362.205		0.3149600	62	12204.72
0.06604000	13	2559.055		0.3200400	63	12401.57
0.07112000	14	2755.906		0.3251200	64	12598.43
0.07620000	15	2952.756		0.3302000	65	12795.28
0.08128000	16	3149.606		0.3352800	66	12992.13
0.08636000	17	3346.457		0.3403600	67	13188.98
0.09144000	18	3543.307		0.3454400	68	13385.83
0.09652000	19	3740.158		0.3505200	69	13582.68
0.1016000	20	3937.008		0.3556000	70	13779.53
0.1066800	21	4133.858		0.3606800	71	13976.38
0.1117600	22	4330.709		0.3657600	72	14173.23
0.1168400	23	4527.559		0.3708400	73	14370.08
0.1219200	24	4724.409		0.3759200	74	14566.93
0.1270000	25	4921.260		0.3810000	75	14763.78
0.1320800	26	5118.110		0.3860800	76	14960.63
0.1371600	27	5314.961		0.3911600	77	15157.48
0.1422400	28	5511.811		0.3962400	78	15354.33
0.1473200	29	5708.661		0.4013200	79	15551.18
0.1524000	30	5905.512		0.4064000	80	15748.03
0.1574800	31	6102.362		0.4114800	81	15944.88
0.1625600	32	6299.213		0.4165600	82	16141.73
0.1676400	33	6496.063		0.4216400	83	16338.58
0.1727200	34	6692.913		0.4267200	84	16535.43
0.1778000	35	6889.764		0.4318000	85	16732.28
0.1828800	36	7086.614		0.4368800	86	16929.13
0.1879600	37	7283.465		0.4419600	87	17125.98
0.1930400	38	7480.315		0.4470400	88	17322.83
0.1981200	39	7677.165		0.4521200	89	17519.69
0.2032000	40	7874.016		0.4572000	90	17716.54
0.2082800	41	8070.866		0.4622800	91	17913.39
0.2133600	42	8267.717		0.4673600	92	18110.24
0.2184400	43	8464.567		0.4724400	93	18307.09
0.2235200	44	8661.417		0.4775200	94	18503.94
0.2286000	45	8858.268		0.4826000	95	18700.79
0.2336800	46	9055.118		0.4876800	96	18897.64
0.2387600	47	9251.969		0.4927600	97	19094.49
0.2438400	48	9448.819		0.4978400	98	19291.34
0.2489200	49	9645.669		0.5029200	99	19488.19
0.2540000	50	9842.520		0.5080000	100	19685.04

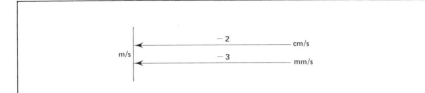

m/s	REF	ft/min		m/s	REF	ft/min
0.5130800	101	19881.89		0.7670800	151	29724.41
0.5181600	102	20078.74		0.7721600	152	29921.26
0.5232400	103	20275.59		0.7772400	153	30118.11
0.5283200	104	20472.44		0.7823200	154	30314.96
0.5334000	105	20669.29		0.7874000	155	30511.81
0.5384800	106	20866.14		0.7924800	156	30708.66
0.5435600	107	21062.99		0.7975600	157	30905.51
0.5486400	108	21259.84		0.8026400	158	31102.36
0.5537200	109	21456.69		0.8077200	159	31299.21
0.5588000	110	21653.54		0.8128000	160	31496.06
0.5638800	111	21850.39		0.8178800	161	31692.91
0.5689600	112	22047.24		0.8229600	162	31889.76
0.5740400	113	22244.09		0.8280400	163	32086.61
0.5791200	114	22440.95		0.8331200	164	32283.46
0.5842000	115	22637.80		0.8382000	165	32480.32
0.5892800	116	22834.65		0.8432800	166	32677.17
0.5943600	117	23031.50		0.8483600	167	32874.02
0.5994400	118	23228.35		0.8534400	168	33070.87
0.6045200	119	23425.20		0.8585200	169	33267.72
0.6096000	120	23622.05		0.8636000	170	33464.57
0.6146800	121	23818.90		0.8686800	171	33661.42
0.6197600	122	24015.75		0.8737600	172	33858.27
0.6248400	123	24212.60		0.8788400	173	34055.12
0.6299200	124	24409.45		0.8839200	174	34251.97
0.6350000	125	24606.30		0.8890000	175	34448.82
0.6400800	126	24803.15		0.8940800	176	34645.67
0.6451600	127	25000.00		0.8991600	177	34842.52
0.6502400	128	25196.85		0.9042400	178	35039.37
0.6553200	129	25393.70		0.9093200	179	35236.22
0.6604000	130	25590.55		0.9144000	180	35433.07
0.6654800	131	25787.40		0.9194800	181	35629.92
0.6705600	132	25984.25		0.9245600	182	35826.77
0.6756400	133	26181.10		0.9296400	183	36023.62
0.6807200	134	26377.95		0.9347200	184	36220.47
0.6858000	135	26574.80		0.9398000	185	36417.32
0.6908800	136	26771.65		0.9448800	186	36614.17
0.6959600	137	26968.50		0.9499600	187	36811.02
0.7010400	138	27165.35		0.9550400	188	37007.87
0.7061200	139	27362.20		0.9601200	189	37204.72
0.7112000	140	27559.06		0.9652000	190	37401.58
0.7162800	141	27755.91		0.9702800	191	37598.43
0.7213600	142	27952.76		0.9753600	192	37795.28
0.7264400	143	28149.61		0.9804400	193	37992.13
0.7315200	144	28346.46		0.9855200	194	38188.98
0.7366000	145	28543.31		0.9906000	195	38385.83
0.7416800	146	28740.16		0.9956800	196	38582.68
0.7467600	147	28937.01		1.000760	197	38779.53
0.7518400	148	29133.86		1.005840	198	38976.38
0.7569200	149	29330.71		1.010920	199	39173.23
0.7620000	150	29527.56		1.016000	200	39370.08

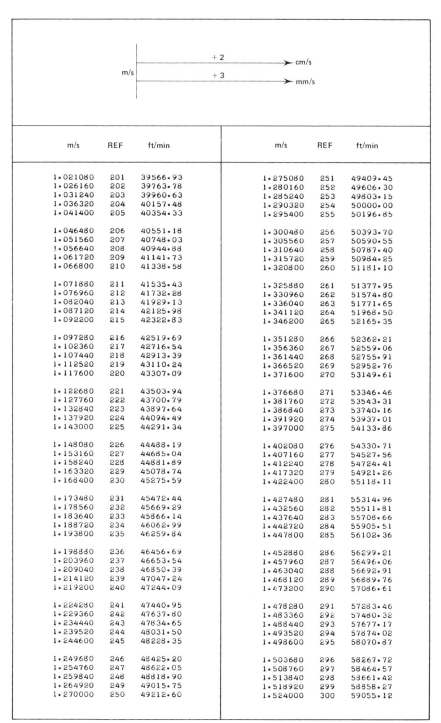

m/s	REF	ft/min		m/s	REF	ft/min
1.021080	201	39566.93		1.275080	251	49409.45
1.026160	202	39763.78		1.280160	252	49606.30
1.031240	203	39960.63		1.285240	253	49803.15
1.036320	204	40157.48		1.290320	254	50000.00
1.041400	205	40354.33		1.295400	255	50196.85
1.046480	206	40551.18		1.300480	256	50393.70
1.051560	207	40748.03		1.305560	257	50590.55
1.056640	208	40944.88		1.310640	258	50787.40
1.061720	209	41141.73		1.315720	259	50984.25
1.066800	210	41338.58		1.320800	260	51181.10
1.071880	211	41535.43		1.325880	261	51377.95
1.076960	212	41732.28		1.330960	262	51574.80
1.082040	213	41929.13		1.336040	263	51771.65
1.087120	214	42125.98		1.341120	264	51968.50
1.092200	215	42322.83		1.346200	265	52165.35
1.097280	216	42519.69		1.351280	266	52362.21
1.102360	217	42716.54		1.356360	267	52559.06
1.107440	218	42913.39		1.361440	268	52755.91
1.112520	219	43110.24		1.366520	269	52952.76
1.117600	220	43307.09		1.371600	270	53149.61
1.122680	221	43503.94		1.376680	271	53346.46
1.127760	222	43700.79		1.381760	272	53543.31
1.132840	223	43897.64		1.386840	273	53740.16
1.137920	224	44094.49		1.391920	274	53937.01
1.143000	225	44291.34		1.397000	275	54133.86
1.148080	226	44488.19		1.402080	276	54330.71
1.153160	227	44685.04		1.407160	277	54527.56
1.158240	228	44881.89		1.412240	278	54724.41
1.163320	229	45078.74		1.417320	279	54921.26
1.168400	230	45275.59		1.422400	280	55118.11
1.173480	231	45472.44		1.427480	281	55314.96
1.178560	232	45669.29		1.432560	282	55511.81
1.183640	233	45866.14		1.437640	283	55708.66
1.188720	234	46062.99		1.442720	284	55905.51
1.193800	235	46259.84		1.447800	285	56102.36
1.198880	236	46456.69		1.452880	286	56299.21
1.203960	237	46653.54		1.457960	287	56496.06
1.209040	238	46850.39		1.463040	288	56692.91
1.214120	239	47047.24		1.468120	289	56889.76
1.219200	240	47244.09		1.473200	290	57086.61
1.224280	241	47440.95		1.478280	291	57283.46
1.229360	242	47637.80		1.483360	292	57480.32
1.234440	243	47834.65		1.488440	293	57677.17
1.239520	244	48031.50		1.493520	294	57874.02
1.244600	245	48228.35		1.498600	295	58070.87
1.249680	246	48425.20		1.503680	296	58267.72
1.254760	247	48622.05		1.508760	297	58464.57
1.259840	248	48818.90		1.513840	298	58661.42
1.264920	249	49015.75		1.518920	299	58858.27
1.270000	250	49212.60		1.524000	300	59055.12

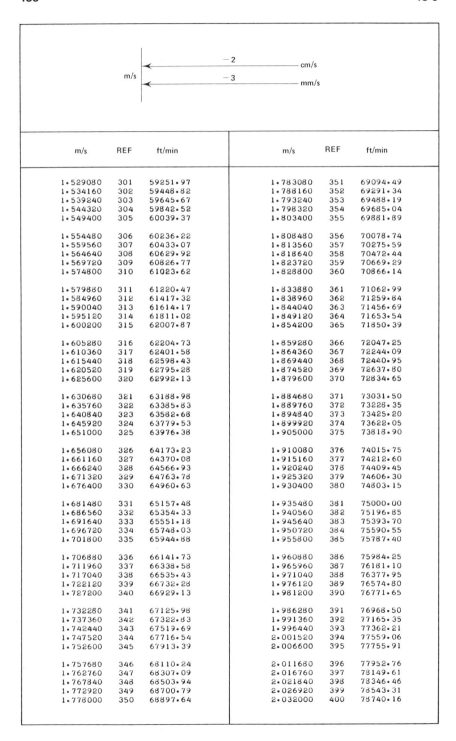

m/s	REF	ft/min	m/s	REF	ft/min
1.529080	301	59251.97	1.783080	351	69094.49
1.534160	302	59448.82	1.788160	352	69291.34
1.539240	303	59645.67	1.793240	353	69488.19
1.544320	304	59842.52	1.798320	354	69685.04
1.549400	305	60039.37	1.803400	355	69881.89
1.554480	306	60236.22	1.808480	356	70078.74
1.559560	307	60433.07	1.813560	357	70275.59
1.564640	308	60629.92	1.818640	358	70472.44
1.569720	309	60826.77	1.823720	359	70669.29
1.574800	310	61023.62	1.828800	360	70866.14
1.579880	311	61220.47	1.833880	361	71062.99
1.584960	312	61417.32	1.838960	362	71259.84
1.590040	313	61614.17	1.844040	363	71456.69
1.595120	314	61811.02	1.849120	364	71653.54
1.600200	315	62007.87	1.854200	365	71850.39
1.605280	316	62204.73	1.859280	366	72047.25
1.610360	317	62401.58	1.864360	367	72244.09
1.615440	318	62598.43	1.869440	368	72440.95
1.620520	319	62795.28	1.874520	369	72637.80
1.625600	320	62992.13	1.879600	370	72834.65
1.630680	321	63188.98	1.884680	371	73031.50
1.635760	322	63385.83	1.889760	372	73228.35
1.640840	323	63582.68	1.894840	373	73425.20
1.645920	324	63779.53	1.899920	374	73622.05
1.651000	325	63976.38	1.905000	375	73818.90
1.656080	326	64173.23	1.910080	376	74015.75
1.661160	327	64370.08	1.915160	377	74212.60
1.666240	328	64566.93	1.920240	378	74409.45
1.671320	329	64763.78	1.925320	379	74606.30
1.676400	330	64960.63	1.930400	380	74803.15
1.681480	331	65157.48	1.935480	381	75000.00
1.686560	332	65354.33	1.940560	382	75196.85
1.691640	333	65551.18	1.945640	383	75393.70
1.696720	334	65748.03	1.950720	384	75590.55
1.701800	335	65944.88	1.955800	385	75787.40
1.706880	336	66141.73	1.960880	386	75984.25
1.711960	337	66338.58	1.965960	387	76181.10
1.717040	338	66535.43	1.971040	388	76377.95
1.722120	339	66732.28	1.976120	389	76574.80
1.727200	340	66929.13	1.981200	390	76771.65
1.732280	341	67125.98	1.986280	391	76968.50
1.737360	342	67322.83	1.991360	392	77165.35
1.742440	343	67519.69	1.996440	393	77362.21
1.747520	344	67716.54	2.001520	394	77559.06
1.752600	345	67913.39	2.006600	395	77755.91
1.757680	346	68110.24	2.011680	396	77952.76
1.762760	347	68307.09	2.016760	397	78149.61
1.767840	348	68503.94	2.021840	398	78346.46
1.772920	349	68700.79	2.026920	399	78543.31
1.778000	350	68897.64	2.032000	400	78740.16

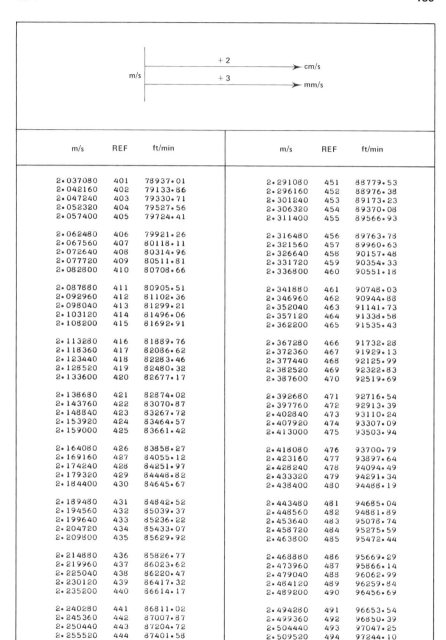

m/s	REF	ft/min		m/s	REF	ft/min
2.037080	401	78937.01		2.291080	451	88779.53
2.042160	402	79133.86		2.296160	452	88976.38
2.047240	403	79330.71		2.301240	453	89173.23
2.052320	404	79527.56		2.306320	454	89370.08
2.057400	405	79724.41		2.311400	455	89566.93
2.062480	406	79921.26		2.316480	456	89763.78
2.067560	407	80118.11		2.321560	457	89960.63
2.072640	408	80314.96		2.326640	458	90157.48
2.077720	409	80511.81		2.331720	459	90354.33
2.082800	410	80708.66		2.336800	460	90551.18
2.087880	411	80905.51		2.341880	461	90748.03
2.092960	412	81102.36		2.346960	462	90944.88
2.098040	413	81299.21		2.352040	463	91141.73
2.103120	414	81496.06		2.357120	464	91338.58
2.108200	415	81692.91		2.362200	465	91535.43
2.113280	416	81889.76		2.367280	466	91732.28
2.118360	417	82086.62		2.372360	467	91929.13
2.123440	418	82283.46		2.377440	468	92125.99
2.128520	419	82480.32		2.382520	469	92322.83
2.133600	420	82677.17		2.387600	470	92519.69
2.138680	421	82874.02		2.392680	471	92716.54
2.143760	422	83070.87		2.397760	472	92913.39
2.148840	423	83267.72		2.402840	473	93110.24
2.153920	424	83464.57		2.407920	474	93307.09
2.159000	425	83661.42		2.413000	475	93503.94
2.164080	426	83858.27		2.418080	476	93700.79
2.169160	427	84055.12		2.423160	477	93897.64
2.174240	428	84251.97		2.428240	478	94094.49
2.179320	429	84448.82		2.433320	479	94291.34
2.184400	430	84645.67		2.438400	480	94488.19
2.189480	431	84842.52		2.443480	481	94685.04
2.194560	432	85039.37		2.448560	482	94881.89
2.199640	433	85236.22		2.453640	483	95078.74
2.204720	434	85433.07		2.458720	484	95275.59
2.209800	435	85629.92		2.463800	485	95472.44
2.214880	436	85826.77		2.468880	486	95669.29
2.219960	437	86023.62		2.473960	487	95866.14
2.225040	438	86220.47		2.479040	488	96062.99
2.230120	439	86417.32		2.484120	489	96259.84
2.235200	440	86614.17		2.489200	490	96456.69
2.240280	441	86811.02		2.494280	491	96653.54
2.245360	442	87007.87		2.499360	492	96850.39
2.250440	443	87204.72		2.504440	493	97047.25
2.255520	444	87401.58		2.509520	494	97244.10
2.260600	445	87598.43		2.514600	495	97440.95
2.265680	446	87795.28		2.519680	496	97637.80
2.270760	447	87992.13		2.524760	497	97834.65
2.275840	448	88188.98		2.529840	498	98031.50
2.280920	449	88385.83		2.534920	499	98228.35
2.286000	450	88582.68		2.540000	500	98425.20

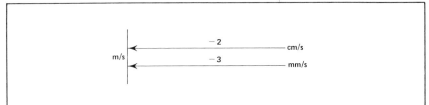

m/s	REF	ft/min		m/s	REF	ft/min
2.545080	501	98622.05		2.799080	551	108464.6
2.550160	502	98818.90		2.804160	552	108661.4
2.555240	503	99015.75		2.809240	553	108858.3
2.560320	504	99212.60		2.814320	554	109055.1
2.565400	505	99409.45		2.819400	555	109252.0
2.570480	506	99606.30		2.824480	556	109448.8
2.575560	507	99803.15		2.829560	557	109645.7
2.580640	508	00000.00		2.834640	558	109842.5
2.585720	509	100196.9		2.839720	559	110039.4
2.590800	510	100393.7		2.844800	560	110236.2
2.595880	511	100590.6		2.849880	561	110433.1
2.600960	512	100787.4		2.854960	562	110629.9
2.606040	513	100984.3		2.860040	563	110826.8
2.611120	514	101181.1		2.865120	564	111023.6
2.616200	515	101378.0		2.870200	565	111220.5
2.621280	516	101574.8		2.875280	566	111417.3
2.626360	517	101771.7		2.880360	567	111614.2
2.631440	518	101968.5		2.885440	568	111811.0
2.636520	519	102165.4		2.890520	569	112007.9
2.641600	520	102362.2		2.895600	570	112204.7
2.646680	521	102559.1		2.900680	571	112401.6
2.651760	522	102755.9		2.905760	572	112598.4
2.656840	523	102952.8		2.910840	573	112795.3
2.661920	524	103149.6		2.915920	574	112992.1
2.667000	525	103346.5		2.921000	575	113189.0
2.672080	526	103543.3		2.926080	576	113385.8
2.677160	527	103740.2		2.931160	577	113582.7
2.682240	528	103937.0		2.936240	578	113779.5
2.687320	529	104133.9		2.941320	579	113976.4
2.692400	530	104330.7		2.946400	580	114173.2
2.697480	531	104527.6		2.951480	581	114370.1
2.702560	532	104724.4		2.956560	582	114566.9
2.707640	533	104921.3		2.961640	583	114763.8
2.712720	534	105118.1		2.966720	584	114960.6
2.717800	535	105315.0		2.971800	585	115157.5
2.722880	536	105511.8		2.976880	586	115354.3
2.727960	537	105708.7		2.981960	587	115551.2
2.733040	538	105905.5		2.987040	588	115748.0
2.738120	539	106102.4		2.992120	589	115944.9
2.743200	540	106299.2		2.997200	590	116141.7
2.748280	541	106496.1		3.002280	591	116338.6
2.753360	542	106692.9		3.007360	592	116535.4
2.758440	543	106889.8		3.012440	593	116732.3
2.763520	544	107086.6		3.017520	594	116929.1
2.768600	545	107283.5		3.022600	595	117126.0
2.773680	546	107480.3		3.027680	596	117322.8
2.778760	547	107677.2		3.032760	597	117519.7
2.783840	548	107874.0		3.037840	598	117716.5
2.788920	549	108070.9		3.042920	599	117913.4
2.794000	550	108267.7		3.048000	600	118110.2

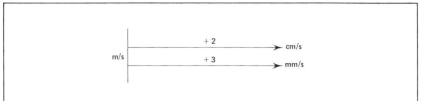

m/s	REF	ft/min		m/s	REF	ft/min
3.053080	601	118307.1		3.307080	651	121149.6
3.058160	602	118503.9		3.312160	652	128346.5
3.063240	603	118700.8		3.317240	653	128543.3
3.068320	604	118897.6		3.322320	654	128740.2
3.073400	605	119094.5		3.327400	655	128937.0
3.078480	606	119291.3		3.332480	656	129133.9
3.083560	607	119488.2		3.337560	657	129330.7
3.088640	608	119685.0		3.342640	658	129527.6
3.093720	609	119881.9		3.347720	659	129724.4
3.098800	610	120078.7		3.352800	660	129921.3
3.103880	611	120275.6		3.357880	661	130118.1
3.108960	612	120472.4		3.362960	662	130315.0
3.114040	613	120669.3		3.368040	663	130511.8
3.119120	614	120866.1		3.373120	664	130708.7
3.124200	615	121063.0		3.378200	665	130905.5
3.129280	616	121259.8		3.383280	666	131102.4
3.134360	617	121456.7		3.388360	667	131299.2
3.139440	618	121653.5		3.393440	668	131496.1
3.144520	619	121850.4		3.398520	669	131692.9
3.149600	620	122047.2		3.403600	670	131889.8
3.154680	621	122244.1		3.408680	671	132086.6
3.159760	622	122440.9		3.413760	672	132283.5
3.164840	623	122637.8		3.418840	673	132480.3
3.169920	624	122834.6		3.423920	674	132677.2
3.175000	625	123031.5		3.429000	675	132874.0
3.180080	626	123228.3		3.434080	676	133070.9
3.185160	627	123425.2		3.439160	677	133267.7
3.190240	628	123622.0		3.444240	678	133464.6
3.195320	629	123818.9		3.449320	679	133661.4
3.200400	630	124015.7		3.454400	680	133858.3
3.205480	631	124212.6		3.459480	681	134055.1
3.210560	632	124409.5		3.464560	682	134252.0
3.215640	633	124606.3		3.469640	683	134448.8
3.220720	634	124803.2		3.474720	684	134645.7
3.225800	635	125000.0		3.479800	685	134842.5
3.230880	636	125196.9		3.484880	686	135039.4
3.235960	637	125393.7		3.489960	687	135236.2
3.241040	638	125590.6		3.495040	688	135433.1
3.246120	639	125787.4		3.500120	689	135629.9
3.251200	640	125984.3		3.505200	690	135826.8
3.256280	641	126181.1		3.510280	691	136023.6
3.261360	642	126378.0		3.515360	692	136220.5
3.266440	643	126574.8		3.520440	693	136417.3
3.271520	644	126771.7		3.525520	694	136614.2
3.276600	645	126968.5		3.530600	695	136811.0
3.281680	646	127165.4		3.535680	696	137007.9
3.286760	647	127362.2		3.540760	697	137204.7
3.291840	648	127559.1		3.545840	698	137401.6
3.296920	649	127755.9		3.550920	699	137598.4
3.302000	650	127952.8		3.556000	700	137795.3

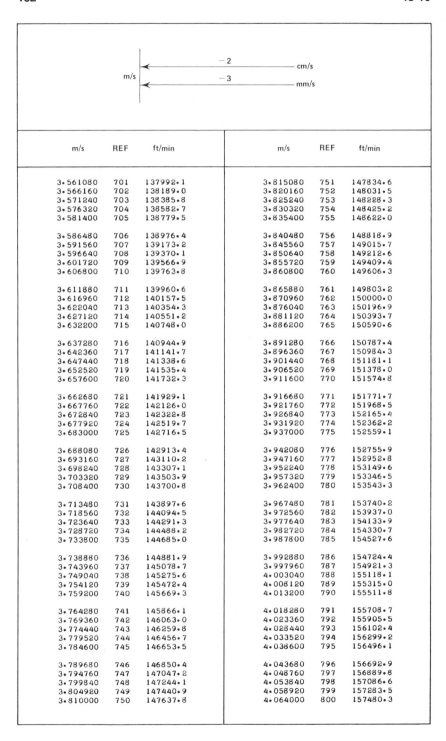

m/s	REF	ft/min	m/s	REF	ft/min
3.561080	701	137992.1	3.815080	751	147834.6
3.566160	702	138189.0	3.820160	752	148031.5
3.571240	703	138385.8	3.825240	753	148228.3
3.576320	704	138582.7	3.830320	754	148425.2
3.581400	705	138779.5	3.835400	755	148622.0
3.586480	706	138976.4	3.840480	756	148818.9
3.591560	707	139173.2	3.845560	757	149015.7
3.596640	708	139370.1	3.850640	758	149212.6
3.601720	709	139566.9	3.855720	759	149409.4
3.606800	710	139763.8	3.860800	760	149606.3
3.611880	711	139960.6	3.865880	761	149803.2
3.616960	712	140157.5	3.870960	762	150000.0
3.622040	713	140354.3	3.876040	763	150196.9
3.627120	714	140551.2	3.881120	764	150393.7
3.632200	715	140748.0	3.886200	765	150590.6
3.637280	716	140944.9	3.891280	766	150787.4
3.642360	717	141141.7	3.896360	767	150984.3
3.647440	718	141338.6	3.901440	768	151181.1
3.652520	719	141535.4	3.906520	769	151378.0
3.657600	720	141732.3	3.911600	770	151574.8
3.662680	721	141929.1	3.916680	771	151771.7
3.667760	722	142126.0	3.921760	772	151968.5
3.672840	723	142322.8	3.926840	773	152165.4
3.677920	724	142519.7	3.931920	774	152362.2
3.683000	725	142716.5	3.937000	775	152559.1
3.688080	726	142913.4	3.942080	776	152755.9
3.693160	727	143110.2	3.947160	777	152952.8
3.698240	728	143307.1	3.952240	778	153149.6
3.703320	729	143503.9	3.957320	779	153346.5
3.708400	730	143700.8	3.962400	780	153543.3
3.713480	731	143897.6	3.967480	781	153740.2
3.718560	732	144094.5	3.972560	782	153937.0
3.723640	733	144291.3	3.977640	783	154133.9
3.728720	734	144488.2	3.982720	784	154330.7
3.733800	735	144685.0	3.987800	785	154527.6
3.738880	736	144881.9	3.992880	786	154724.4
3.743960	737	145078.7	3.997960	787	154921.3
3.749040	738	145275.6	4.003040	788	155118.1
3.754120	739	145472.4	4.008120	789	155315.0
3.759200	740	145669.3	4.013200	790	155511.8
3.764280	741	145866.1	4.018280	791	155708.7
3.769360	742	146063.0	4.023360	792	155905.5
3.774440	743	146259.8	4.028440	793	156102.4
3.779520	744	146456.7	4.033520	794	156299.2
3.784600	745	146653.5	4.038600	795	156496.1
3.789680	746	146850.4	4.043680	796	156692.9
3.794760	747	147047.2	4.048760	797	156889.8
3.799840	748	147244.1	4.053840	798	157086.6
3.804920	749	147440.9	4.058920	799	157283.5
3.810000	750	147637.8	4.064000	800	157480.3

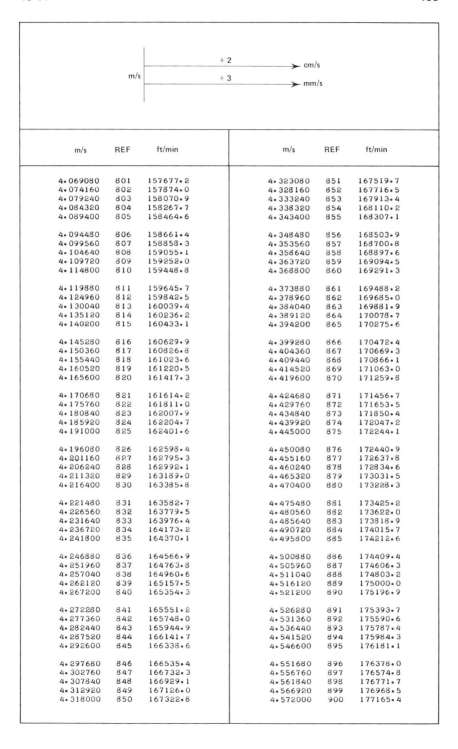

m/s	REF	ft/min	m/s	REF	ft/min
4.069080	801	157677.2	4.323080	851	167519.7
4.074160	802	157874.0	4.328160	852	167716.5
4.079240	803	158070.9	4.333240	853	167913.4
4.084320	804	158267.7	4.338320	854	168110.2
4.089400	805	158464.6	4.343400	855	168307.1
4.094480	806	158661.4	4.348480	856	168503.9
4.099560	807	158858.3	4.353560	857	168700.8
4.104640	808	159055.1	4.358640	858	168897.6
4.109720	809	159252.0	4.363720	859	169094.5
4.114800	810	159448.8	4.368800	860	169291.3
4.119880	811	159645.7	4.373880	861	169488.2
4.124960	812	159842.5	4.378960	862	169685.0
4.130040	813	160039.4	4.384040	863	169881.9
4.135120	814	160236.2	4.389120	864	170078.7
4.140200	815	160433.1	4.394200	865	170275.6
4.145280	816	160629.9	4.399280	866	170472.4
4.150360	817	160826.8	4.404360	867	170669.3
4.155440	818	161023.6	4.409440	868	170866.1
4.160520	819	161220.5	4.414520	869	171063.0
4.165600	820	161417.3	4.419600	870	171259.8
4.170680	821	161614.2	4.424680	871	171456.7
4.175760	822	161811.0	4.429760	872	171653.5
4.180840	823	162007.9	4.434840	873	171850.4
4.185920	824	162204.7	4.439920	874	172047.2
4.191000	825	162401.6	4.445000	875	172244.1
4.196080	826	162598.4	4.450080	876	172440.9
4.201160	827	162795.3	4.455160	877	172637.8
4.206240	828	162992.1	4.460240	878	172834.6
4.211320	829	163189.0	4.465320	879	173031.5
4.216400	830	163385.8	4.470400	880	173228.3
4.221480	831	163582.7	4.475480	881	173425.2
4.226560	832	163779.5	4.480560	882	173622.0
4.231640	833	163976.4	4.485640	883	173818.9
4.236720	834	164173.2	4.490720	884	174015.7
4.241800	835	164370.1	4.495800	885	174212.6
4.246880	836	164566.9	4.500880	886	174409.4
4.251960	837	164763.8	4.505960	887	174606.3
4.257040	838	164960.6	4.511040	888	174803.2
4.262120	839	165157.5	4.516120	889	175000.0
4.267200	840	165354.3	4.521200	890	175196.9
4.272280	841	165551.2	4.526280	891	175393.7
4.277360	842	165748.0	4.531360	892	175590.6
4.282440	843	165944.9	4.536440	893	175787.4
4.287520	844	166141.7	4.541520	894	175984.3
4.292600	845	166338.6	4.546600	895	176181.1
4.297680	846	166535.4	4.551680	896	176378.0
4.302760	847	166732.3	4.556760	897	176574.8
4.307840	848	166929.1	4.561840	898	176771.7
4.312920	849	167126.0	4.566920	899	176968.5
4.318000	850	167322.8	4.572000	900	177165.4

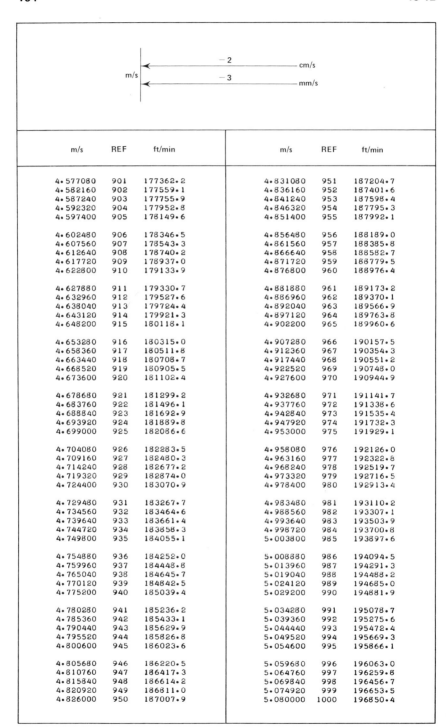

m/s	REF	ft/min	m/s	REF	ft/min
4.577080	901	177362.2	4.831080	951	187204.7
4.582160	902	177559.1	4.836160	952	187401.6
4.587240	903	177755.9	4.841240	953	187598.4
4.592320	904	177952.8	4.846320	954	187795.3
4.597400	905	178149.6	4.851400	955	187992.1
4.602480	906	178346.5	4.856480	956	188189.0
4.607560	907	178543.3	4.861560	957	188385.8
4.612640	908	178740.2	4.866640	958	188582.7
4.617720	909	178937.0	4.871720	959	188779.5
4.622800	910	179133.9	4.876800	960	188976.4
4.627880	911	179330.7	4.881880	961	189173.2
4.632960	912	179527.6	4.886960	962	189370.1
4.638040	913	179724.4	4.892040	963	189566.9
4.643120	914	179921.3	4.897120	964	189763.8
4.648200	915	180118.1	4.902200	965	189960.6
4.653280	916	180315.0	4.907280	966	190157.5
4.658360	917	180511.8	4.912360	967	190354.3
4.663440	918	180708.7	4.917440	968	190551.2
4.668520	919	180905.5	4.922520	969	190748.0
4.673600	920	181102.4	4.927600	970	190944.9
4.678680	921	181299.2	4.932680	971	191141.7
4.683760	922	181496.1	4.937760	972	191338.6
4.688840	923	181692.9	4.942840	973	191535.4
4.693920	924	181889.8	4.947920	974	191732.3
4.699000	925	182086.6	4.953000	975	191929.1
4.704080	926	182283.5	4.958080	976	192126.0
4.709160	927	182480.3	4.963160	977	192322.8
4.714240	928	182677.2	4.968240	978	192519.7
4.719320	929	182874.0	4.973320	979	192716.5
4.724400	930	183070.9	4.978400	980	192913.4
4.729480	931	183267.7	4.983480	981	193110.2
4.734560	932	183464.6	4.988560	982	193307.1
4.739640	933	183661.4	4.993640	983	193503.9
4.744720	934	183858.3	4.998720	984	193700.8
4.749800	935	184055.1	5.003800	985	193897.6
4.754880	936	184252.0	5.008880	986	194094.5
4.759960	937	184448.8	5.013960	987	194291.3
4.765040	938	184645.7	5.019040	988	194488.2
4.770120	939	184842.5	5.024120	989	194685.0
4.775200	940	185039.4	5.029200	990	194881.9
4.780280	941	185236.2	5.034280	991	195078.7
4.785360	942	185433.1	5.039360	992	195275.6
4.790440	943	185629.9	5.044440	993	195472.4
4.795520	944	185826.8	5.049520	994	195669.3
4.800600	945	186023.6	5.054600	995	195866.1
4.805680	946	186220.5	5.059680	996	196063.0
4.810760	947	186417.3	5.064760	997	196259.8
4.815840	948	186614.2	5.069840	998	196456.7
4.820920	949	186811.0	5.074920	999	196653.5
4.826000	950	187007.9	5.080000	1000	196850.4

16

VOLUME FLOW RATE

Cubic-foot/Minute — Cubic-metre/Second
  $(\text{ft}^3/\text{min} - \text{m}^3/\text{s})$

Cubic-foot/Minute — Cubic-centimetre/Second
  $(\text{ft}^3/\text{min} - \text{cm}^3/\text{s})$

Cubic-foot/Minute — Cubic-millimetre/Second
  $(\text{ft}^3/\text{min} - \text{mm}^3/\text{s})$

Cubic-foot/Minute — Litre/Second
  $(\text{ft}^3/\text{min} - \text{l/s})$

Conversion Factors:
  1 cubic-foot/minute $(\text{ft}^3/\text{min}) = 0.0004719474$ cubic-metre/second (m^3/s)
  $(1\ \text{m}^3/\text{s} = 2118.880\ \text{ft}^3/\text{min})$

In converting m^3/s quantities to cm^3/s, mm^3/s, and l/s, Diagram 16-A, which appears in the headings on alternate pages of this table, will assist the user. It is derived from: $1\ \text{m}^3/\text{s} = (10^6)\ \text{cm}^3/\text{s} = (10^9)\ \text{mm}^3/\text{s} = (10^3)\ \text{l/s}$.

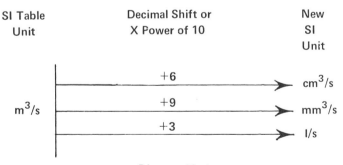

SI Table Unit	Decimal Shift or X Power of 10	New SI Unit
	+6	cm^3/s
m^3/s	+9	mm^3/s
	+3	l/s

Diagram 16–A

Example (*a*): Convert 31.3 ft^3/min to m^3/s, cm^3/s, l/s

  Since: 31.3 ft/min = 313 × 10^{-1} ft/min:

  From table: 313 ft^3/min = 0.1477195 m^3/s*

  Therefore: 313 × 10^{-1} ft^3/min = 0.01477195 m^3/s ≈ 0.0148 m^3/s.

  From Diagram 16-A: 1) 0.01477195 m^3/s = 0.01477195 × 10^6 cm^3/s

                        = 14,771.95 cm^3/s ≈ 14,800 cm^3/s

                        (m^3/s to cm^3/s by multiplying m^3/s value by 10^6 or

                        by shifting its decimal 6 places to the right.)

                   2) 0.01477195 m^3/s = 0.01477195 × 10^3 l/s

                        = 14.77195 l/s ≈ 14.8 l/s

                        (m^3/s to l/s by multiplying m^3/s value by 10^3 or by

                        shifting its decimal 3 places to the right.)

  Diagram 16-B is also given at the top of alternate table pages:

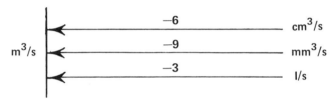

Diagram 16–B

It is the opposite of Diagram 16-A and is used to convert cm^3/s, mm^3/s, l/s values to m^3/s. The m^3/s equivalent is then applied to the REF column in the table to obtain ft^3/s quantities.

Example (*b*): Convert 1500 mm^3/s to ft^3/min.

  From Diagram 16-B: 1500 mm^3/s = 1500 × 10^{-9} m^3/s = 150 × 10^{-8} m^3/s

  From table: 150 m^3/s = 317,832 ft^3/min*

  Therefore: 150 × 10^{-8} m^3/s = 317,832 × 10^{-8} ft^3/min

             = 0.00317832 ft^3/min* ≈ 3.18 × 10^{-3} ft^3/min.

---

*All conversion values obtained can be rounded to desired number of significant figures.

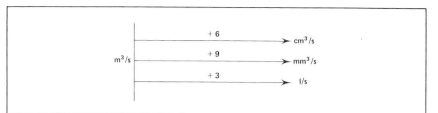

m³/s	REF	ft³/min		m³/s	REF	ft³/min
0.0004719474	1	2118.880		0.02406932	51	108062.9
0.0009438948	2	4237.760		0.02454126	52	110181.8
0.001415842	3	6356.641		0.02501321	53	112300.7
0.001887790	4	8475.521		0.02548516	54	114419.5
0.002359737	5	10594.40		0.02595711	55	116538.4
0.002831684	6	12713.28		0.02642905	56	118657.3
0.003303632	7	14832.16		0.02690100	57	120776.2
0.003775579	8	16951.04		0.02737295	58	122895.1
0.004247527	9	19069.92		0.02784490	59	125013.9
0.004719474	10	21188.80		0.02831684	60	127132.8
0.005191421	11	23307.68		0.02878879	61	129251.7
0.005663369	12	25426.56		0.02926074	62	131370.6
0.006135316	13	27545.44		0.02973269	63	133489.5
0.006607264	14	29664.32		0.03020463	64	135608.3
0.007079211	15	31783.20		0.03067658	65	137727.2
0.007551158	16	33902.08		0.03114853	66	139846.1
0.008023106	17	36020.96		0.03162048	67	141965.0
0.008495053	18	38139.84		0.03209242	68	144083.9
0.008967001	19	40258.72		0.03256437	69	146202.7
0.009438948	20	42377.60		0.03303632	70	148321.6
0.009910895	21	44496.48		0.03350827	71	150440.5
0.01038284	22	46615.36		0.03398021	72	152559.4
0.01085479	23	48734.25		0.03445216	73	154678.3
0.01132674	24	50853.13		0.03492411	74	156797.1
0.01179868	25	52972.01		0.03539605	75	158916.0
0.01227063	26	55090.89		0.03586800	76	161034.9
0.01274258	27	57209.77		0.03633995	77	163153.8
0.01321453	28	59328.65		0.03681190	78	165272.7
0.01368647	29	61447.53		0.03728384	79	167391.5
0.01415842	30	63566.41		0.03775579	80	169510.4
0.01463037	31	65685.29		0.03822774	81	171629.3
0.01510232	32	67804.17		0.03869969	82	173748.2
0.01557426	33	69923.05		0.03917163	83	175867.1
0.01604621	34	72041.93		0.03964358	84	177985.9
0.01651816	35	74160.81		0.04011553	85	180104.8
0.01699011	36	76279.69		0.04058748	86	182223.7
0.01746205	37	78398.57		0.04105942	87	184342.6
0.01793400	38	80517.45		0.04153137	88	186461.5
0.01840595	39	82636.33		0.04200332	89	188580.3
0.01887790	40	84755.21		0.04247527	90	190699.2
0.01934984	41	86874.09		0.04294721	91	192818.1
0.01982179	42	88992.97		0.04341916	92	194937.0
0.02029374	43	91111.85		0.04389111	93	197055.9
0.02076569	44	93230.73		0.04436306	94	199174.7
0.02123763	45	95349.61		0.04483500	95	201293.6
0.02170958	46	97468.49		0.04530695	96	203412.5
0.02218153	47	99587.37		0.04577890	97	205531.4
0.02265348	48	101706.3		0.04625084	98	207650.3
0.02312542	49	103825.1		0.04672279	99	209769.1
0.02359737	50	105944.0		0.04719474	100	211888.0

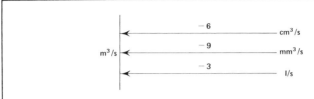

m³/s	REF	ft³/min		m³/s	REF	ft³/min
0.04766669	101	214006.9		0.07126406	151	319950.9
0.04813863	102	216125.8		0.07173600	152	322069.8
0.04861058	103	218244.7		0.07220795	153	324188.7
0.04908253	104	220363.5		0.07267990	154	326307.6
0.04955448	105	222482.4		0.07315185	155	328426.4
0.05002642	106	224601.3		0.07362379	156	330545.3
0.05049837	107	226720.2		0.07409574	157	332664.2
0.05097032	108	228839.1		0.07456769	158	334783.1
0.05144227	109	230957.9		0.07503964	159	336902.0
0.05191421	110	233076.8		0.07551158	160	339020.8
0.05238616	111	235195.7		0.07598353	161	341139.7
0.05285811	112	237314.6		0.07645548	162	343258.6
0.05333006	113	239433.5		0.07692743	163	345377.5
0.05380200	114	241552.3		0.07739937	164	347496.4
0.05427395	115	243671.2		0.07787132	165	349615.2
0.05474590	116	245790.1		0.07834327	166	351734.1
0.05521785	117	247909.0		0.07881522	167	353853.0
0.05568979	118	250027.9		0.07928716	168	355971.9
0.05616174	119	252146.7		0.07975911	169	358090.6
0.05663369	120	254265.6		0.08023106	170	360209.6
0.05710564	121	256384.5		0.08070301	171	362328.5
0.05757758	122	258503.4		0.08117495	172	364447.4
0.05804953	123	260622.3		0.08164690	173	366566.3
0.05852148	124	262741.1		0.08211885	174	368685.2
0.05899342	125	264860.0		0.08259080	175	370804.0
0.05946537	126	266978.9		0.08306274	176	372922.9
0.05993732	127	269097.8		0.08353469	177	375041.8
0.06040927	128	271216.7		0.08400664	178	377160.7
0.06088121	129	273335.5		0.08447858	179	379279.6
0.06135316	130	275454.4		0.08495053	180	381398.4
0.06182511	131	277573.3		0.08542248	181	383517.3
0.06229706	132	279692.2		0.08589443	182	385636.2
0.06276900	133	281811.1		0.08636637	183	387755.1
0.06324095	134	283929.9		0.08683832	184	389874.0
0.06371290	135	286048.8		0.08731027	185	391992.8
0.06418485	136	288167.7		0.08778222	186	394111.7
0.06465679	137	290286.6		0.08825416	187	396230.6
0.06512874	138	292405.5		0.08872611	188	398349.5
0.06560069	139	294524.4		0.08919806	189	400468.4
0.06607264	140	296643.2		0.08967001	190	402587.2
0.06654458	141	298762.1		0.09014195	191	404706.1
0.06701653	142	300881.0		0.09061390	192	406825.0
0.06748848	143	302999.9		0.09108585	193	408943.9
0.06796043	144	305118.7		0.09155780	194	411062.8
0.06843237	145	307237.6		0.09202974	195	413181.6
0.06890432	146	309356.5		0.09250169	196	415300.5
0.06937627	147	311475.4		0.09297364	197	417419.4
0.06984822	148	313594.3		0.09344559	198	419538.3
0.07032016	149	315713.2		0.09391753	199	421657.2
0.07079211	150	317832.0		0.09438948	200	423776.0

m³/s			
	+6		cm³/s
	+9		mm³/s
	+3		l/s

m³/s	REF	ft³/min		m³/s	REF	ft³/min
0·09486143	201	425894·9		0·1184588	251	531838·9
0·09533337	202	428013·8		0·1189307	252	533957·8
0·09580532	203	430132·7		0·1194027	253	536076·7
0·09627727	204	432251·6		0·1198746	254	538195·6
0·09674922	205	434370·4		0·1203466	255	540314·5
0·09722116	206	436489·3		0·1208185	256	542433·3
0·09769311	207	438608·2		0·1212905	257	544552·2
0·09816506	208	440727·1		0·1217624	258	546671·1
0·09863701	209	442846·0		0·1222344	259	548790·0
0·09910895	210	444964·8		0·1227063	260	550908·9
0·09958090	211	447083·7		0·1231783	261	553027·7
0·1000528	212	449202·6		0·1236502	262	555146·6
0·1005248	213	451321·5		0·1241222	263	557265·5
0·1009967	214	453440·4		0·1245941	264	559384·4
0·1014687	215	455559·2		0·1250661	265	561503·3
0·1019406	216	457678·1		0·1255380	266	563622·1
0·1024126	217	459797·0		0·1260100	267	565741·0
0·1028845	218	461915·9		0·1264819	268	567859·9
0·1033565	219	464034·8		0·1269538	269	569978·8
0·1038284	220	466153·6		0·1274258	270	572097·7
0·1043004	221	468272·5		0·1278977	271	574216·5
0·1047723	222	470391·4		0·1283697	272	576335·4
0·1052443	223	472510·3		0·1288416	273	578454·3
0·1057162	224	474629·2		0·1293136	274	580573·2
0·1061882	225	476748·1		0·1297855	275	582692·1
0·1066601	226	478866·9		0·1302575	276	584810·9
0·1071321	227	480985·8		0·1307294	277	586929·8
0·1076040	228	483104·7		0·1312014	278	589048·7
0·1080760	229	485223·6		0·1316733	279	591167·6
0·1085479	230	487342·4		0·1321453	280	593286·5
0·1090198	231	489461·3		0·1326172	281	595405·3
0·1094918	232	491580·2		0·1330892	282	597524·2
0·1099637	233	493699·1		0·1335611	283	599643·1
0·1104357	234	495818·0		0·1340331	284	601762·0
0·1109076	235	497936·9		0·1345050	285	603880·9
0·1113796	236	500055·7		0·1349770	286	605999·7
0·1118515	237	502174·6		0·1354489	287	608118·6
0·1123235	238	504293·5		0·1359209	288	610237·5
0·1127954	239	506412·4		0·1363928	289	612356·4
0·1132674	240	508531·3		0·1368647	290	614475·3
0·1137393	241	510650·1		0·1373367	291	616594·1
0·1142113	242	512769·0		0·1378086	292	618713·0
0·1146832	243	514887·9		0·1382806	293	620831·9
0·1151552	244	517006·8		0·1387525	294	622950·8
0·1156271	245	519125·7		0·1392245	295	625069·7
0·1160991	246	521244·5		0·1396964	296	627188·5
0·1165710	247	523363·4		0·1401684	297	629307·4
0·1170430	248	525482·3		0·1406403	298	631426·3
0·1175149	249	527601·2		0·1411123	299	633545·2
0·1179868	250	529720·1		0·1415842	300	635664·1

```
                              ← ——————— −6 ——————— cm³/s
                   m³/s   ← ——————— −9 ——————— mm³/s
                              ← ——————— −3 ——————— l/s
```

m³/s	REF	ft³/min	m³/s	REF	ft³/min
0.1420562	301	637782.9	0.1656535	351	743727.0
0.1425281	302	639901.8	0.1661255	352	745845.8
0.1430001	303	642020.7	0.1665974	353	747964.7
0.1434720	304	644139.6	0.1670694	354	750083.6
0.1439440	305	646258.5	0.1675413	355	752202.5
0.1444159	306	648377.3	0.1680133	356	754321.4
0.1448879	307	650496.2	0.1684852	357	756440.2
0.1453598	308	652615.1	0.1689572	358	758559.1
0.1458317	309	654734.0	0.1694291	359	760678.0
0.1463037	310	656852.9	0.1699011	360	762796.9
0.1467756	311	658971.7	0.1703730	361	764915.8
0.1472476	312	661090.6	0.1708450	362	767034.6
0.1477195	313	663209.5	0.1713169	363	769153.5
0.1481915	314	665328.4	0.1717889	364	771272.4
0.1486634	315	667447.3	0.1722608	365	773391.3
0.1491354	316	669566.1	0.1727327	366	775510.2
0.1496073	317	671685.0	0.1732047	367	777629.0
0.1500793	318	673803.9	0.1736766	368	779747.9
0.1505512	319	675922.8	0.1741486	369	781866.8
0.1510232	320	678041.7	0.1746205	370	783985.7
0.1514951	321	680160.5	0.1750925	371	786104.6
0.1519671	322	682279.4	0.1755644	372	788223.4
0.1524390	323	684398.3	0.1760364	373	790342.3
0.1529110	324	686517.2	0.1765083	374	792461.2
0.1533829	325	688636.1	0.1769803	375	794580.1
0.1538549	326	690755.0	0.1774522	376	796699.0
0.1543268	327	692873.8	0.1779242	377	798817.8
0.1547987	328	694992.7	0.1783961	378	800936.7
0.1552707	329	697111.6	0.1788681	379	803055.6
0.1557426	330	699230.5	0.1793400	380	805174.5
0.1562146	331	701349.4	0.1798120	381	807293.4
0.1566865	332	703468.2	0.1802839	382	809412.2
0.1571585	333	705587.1	0.1807559	383	811531.1
0.1576304	334	707706.0	0.1812278	384	813650.0
0.1581024	335	709824.9	0.1816997	385	815768.9
0.1585743	336	711943.7	0.1821717	386	817887.8
0.1590463	337	714062.6	0.1826464	387	820006.6
0.1595182	338	716181.5	0.1831156	388	822125.5
0.1599902	339	718300.4	0.1835875	389	824244.4
0.1604621	340	720419.3	0.1840595	390	826363.3
0.1609341	341	722538.2	0.1845314	391	828482.2
0.1614060	342	724657.0	0.1850034	392	830601.0
0.1618780	343	726775.9	0.1854753	393	832719.9
0.1623499	344	728894.8	0.1859473	394	834838.8
0.1628219	345	731013.7	0.1864192	395	836957.7
0.1632938	346	733132.6	0.1868912	396	839076.6
0.1637657	347	735251.4	0.1873631	397	841195.4
0.1642377	348	737370.3	0.1878351	398	843314.3
0.1647096	349	739489.2	0.1883070	399	845433.2
0.1651816	350	741608.1	0.1887790	400	847552.1

m³/s	REF	ft³/min		m³/s	REF	ft³/min
0·1892509	401	849671·0		0·2128483	451	955615·0
0·1897229	402	851789·9		0·2133202	452	957733·9
0·1901948	403	853908·7		0·2137922	453	959852·7
0·1906667	404	856027·6		0·2142641	454	961971·6
0·1911387	405	858146·5		0·2147361	455	964090·5
0·1916106	406	860265·4		0·2152080	456	966209·4
0·1920826	407	862384·2		0·2156800	457	968328·3
0·1925545	408	864503·1		0·2161519	458	970447·1
0·1930265	409	866622·0		0·2166239	459	972566·0
0·1934984	410	868740·9		0·2170958	460	974684·9
0·1939704	411	870859·8		0·2175678	461	976803·8
0·1944423	412	872978·6		0·2180397	462	978922·7
0·1949143	413	875097·5		0·2185116	463	981041·5
0·1953862	414	877216·4		0·2189836	464	983160·4
0·1958582	415	879335·3		0·2194555	465	985279·3
0·1963301	416	881454·2		0·2199275	466	987398·2
0·1968021	417	883573·1		0·2203994	467	989517·1
0·1972740	418	885691·9		0·2208714	468	991635·9
0·1977460	419	887810·8		0·2213433	469	993754·8
0·1982179	420	889929·7		0·2218153	470	995873·7
0·1986899	421	892048·6		0·2222872	471	997992·6
0·1991618	422	894167·5		0·2227592	472	1000111·
0·1996337	423	896286·3		0·2232311	473	1002230·
0·2001057	424	898405·2		0·2237031	474	1004349·
0·2005776	425	900524·1		0·2241750	475	1006468·
0·2010496	426	902643·0		0·2246470	476	1008587·
0·2015215	427	904761·9		0·2251189	477	1010706·
0·2019935	428	906880·7		0·2255909	478	1012825·
0·2024654	429	908999·6		0·2260628	479	1014944·
0·2029374	430	911118·5		0·2265348	480	1017063·
0·2034093	431	913237·4		0·2270067	481	1019181·
0·2038813	432	915356·3		0·2274786	482	1021300·
0·2043532	433	917475·1		0·2279506	483	1023419·
0·2048252	434	919594·0		0·2284225	484	1025538·
0·2052971	435	921712·9		0·2288945	485	1027657·
0·2057691	436	923831·8		0·2293664	486	1029776·
0·2062410	437	925950·7		0·2298384	487	1031895·
0·2067130	438	928069·5		0·2303103	488	1034014·
0·2071849	439	930188·4		0·2307823	489	1036132·
0·2076569	440	932307·3		0·2312542	490	1038251·
0·2081288	441	934426·2		0·2317262	491	1040370·
0·2086008	442	936545·1		0·2321981	492	1042489·
0·2090727	443	938663·9		0·2326701	493	1044608·
0·2095446	444	940782·8		0·2331420	494	1046727·
0·2100166	445	942901·7		0·2336140	495	1048846·
0·2104885	446	945020·6		0·2340859	496	1050965·
0·2109605	447	947139·5		0·2345579	497	1053083·
0·2114324	448	949258·3		0·2350298	498	1055202·
0·2119044	449	951377·2		0·2355018	499	1057321·
0·2123763	450	953496·1		0·2359737	500	1059440·

m³/s	REF	ft³/min		m³/s	REF	ft³/min
0.2364456	501	1061559.		0.2600430	551	1167503.
0.2369176	502	1063678.		0.2605150	552	1169622.
0.2373895	503	1065797.		0.2609869	553	1171741.
0.2378615	504	1067916.		0.2614589	554	1173860.
0.2383334	505	1070035.		0.2619308	555	1175979.
0.2388054	506	1072153.		0.2624028	556	1178097.
0.2392773	507	1074272.		0.2628747	557	1180216.
0.2397493	508	1076391.		0.2633466	558	1182335.
0.2402212	509	1078510.		0.2638186	559	1184454.
0.2406932	510	1080629.		0.2642905	560	1186573.
0.2411651	511	1082748.		0.2647625	561	1188692.
0.2416371	512	1084867.		0.2652344	562	1190811.
0.2421090	513	1086986.		0.2657064	563	1192930.
0.2425810	514	1089104.		0.2661783	564	1195048.
0.2430529	515	1091223.		0.2666503	565	1197167.
0.2435249	516	1093342.		0.2671222	566	1199286.
0.2439968	517	1095461.		0.2675942	567	1201405.
0.2444688	518	1097580.		0.2680661	568	1203524.
0.2449407	519	1099699.		0.2685381	569	1205643.
0.2454126	520	1101818.		0.2690100	570	1207762.
0.2458846	521	1103937.		0.2694820	571	1209881.
0.2463565	522	1106055.		0.2699539	572	1211999.
0.2468285	523	1108174.		0.2704259	573	1214118.
0.2473004	524	1110293.		0.2708978	574	1216237.
0.2477724	525	1112412.		0.2713698	575	1218356.
0.2482443	526	1114531.		0.2718417	576	1220475.
0.2487163	527	1116650.		0.2723136	577	1222594.
0.2491882	528	1118769.		0.2727856	578	1224713.
0.2496602	529	1120888.		0.2732575	579	1226832.
0.2501321	530	1123007.		0.2737295	580	1228951.
0.2506041	531	1125125.		0.2742014	581	1231069.
0.2510760	532	1127244.		0.2746734	582	1233188.
0.2515480	533	1129363.		0.2751453	583	1235307.
0.2520199	534	1131482.		0.2756173	584	1237426.
0.2524919	535	1133601.		0.2760892	585	1239545.
0.2529638	536	1135720.		0.2765612	586	1241664.
0.2534358	537	1137839.		0.2770331	587	1243783.
0.2539077	538	1139958.		0.2775051	588	1245902.
0.2543796	539	1142076.		0.2779770	589	1248020.
0.2548516	540	1144195.		0.2784490	590	1250139.
0.2553235	541	1146314.		0.2789209	591	1252258.
0.2557955	542	1148433.		0.2793929	592	1254377.
0.2562674	543	1150552.		0.2798648	593	1256496.
0.2567394	544	1152671.		0.2803368	594	1258615.
0.2572113	545	1154790.		0.2808087	595	1260734.
0.2576833	546	1156909.		0.2812806	596	1262853.
0.2581552	547	1159027.		0.2817526	597	1264971.
0.2586272	548	1161146.		0.2822245	598	1267090.
0.2590991	549	1163265.		0.2826965	599	1269209.
0.2595711	550	1165384.		0.2831684	600	1271328.

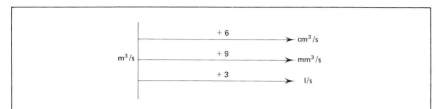

m³/s	REF	ft³/min		m³/s	REF	ft³/min
0.2836404	601	1273447.		0.3072378	651	1379391.
0.2841123	602	1275566.		0.3077097	652	1381510.
0.2845843	603	1277685.		0.3081817	653	1383629.
0.2850562	604	1279804.		0.3086536	654	1385748.
0.2855282	605	1281923.		0.3091255	655	1387867.
0.2860001	606	1284041.		0.3095975	656	1389985.
0.2864721	607	1286160.		0.3100694	657	1392104.
0.2869440	608	1288279.		0.3105414	658	1394223.
0.2874160	609	1290398.		0.3110133	659	1396342.
0.2878879	610	1292517.		0.3114853	660	1398461.
0.2883599	611	1294636.		0.3119572	661	1400580.
0.2888318	612	1296755.		0.3124292	662	1402699.
0.2893038	613	1298874.		0.3129011	663	1404818.
0.2897757	614	1300992.		0.3133731	664	1406936.
0.2902476	615	1303111.		0.3138450	665	1409055.
0.2907196	616	1305230.		0.3143170	666	1411174.
0.2911915	617	1307349.		0.3147889	667	1413293.
0.2916635	618	1309468.		0.3152609	668	1415412.
0.2921354	619	1311587.		0.3157328	669	1417531.
0.2926074	620	1313706.		0.3162048	670	1419650.
0.2930793	621	1315825.		0.3166767	671	1421769.
0.2935513	622	1317944.		0.3171487	672	1423888.
0.2940232	623	1320062.		0.3176206	673	1426006.
0.2944952	624	1322181.		0.3180925	674	1428125.
0.2949671	625	1324300.		0.3185645	675	1430244.
0.2954391	626	1326419.		0.3190364	676	1432363.
0.2959110	627	1328538.		0.3195084	677	1434482.
0.2963830	628	1330657.		0.3199803	678	1436601.
0.2968549	629	1332776.		0.3204523	679	1438720.
0.2973269	630	1334895.		0.3209242	680	1440839.
0.2977988	631	1337013.		0.3213962	681	1442957.
0.2982708	632	1339132.		0.3218681	682	1445076.
0.2987427	633	1341251.		0.3223401	683	1447195.
0.2992146	634	1343370.		0.3228120	684	1449314.
0.2996866	635	1345489.		0.3232840	685	1451433.
0.3001585	636	1347608.		0.3237559	686	1453552.
0.3006305	637	1349727.		0.3242279	687	1455671.
0.3011024	638	1351846.		0.3246998	688	1457790.
0.3015744	639	1353964.		0.3251718	689	1459908.
0.3020463	640	1356083.		0.3256437	690	1462027.
0.3025183	641	1358202.		0.3261157	691	1464146.
0.3029902	642	1360321.		0.3265876	692	1466265.
0.3034622	643	1362440.		0.3270595	693	1468384.
0.3039341	644	1364559.		0.3275315	694	1470503.
0.3044061	645	1366678.		0.3280034	695	1472622.
0.3048780	646	1368797.		0.3284754	696	1474741.
0.3053500	647	1370916.		0.3289473	697	1476860.
0.3058219	648	1373034.		0.3294193	698	1478978.
0.3062939	649	1375153.		0.3298912	699	1481097.
0.3067658	650	1377272.		0.3303632	700	1483216.

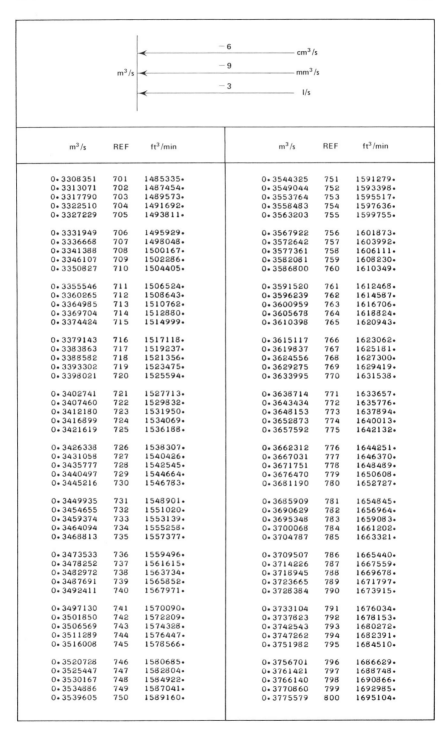

m³/s	REF	ft³/min	m³/s	REF	ft³/min
0.3308351	701	1485335.	0.3544325	751	1591279.
0.3313071	702	1487454.	0.3549044	752	1593398.
0.3317790	703	1489573.	0.3553764	753	1595517.
0.3322510	704	1491692.	0.3558483	754	1597636.
0.3327229	705	1493811.	0.3563203	755	1599755.
0.3331949	706	1495929.	0.3567922	756	1601873.
0.3336668	707	1498048.	0.3572642	757	1603992.
0.3341388	708	1500167.	0.3577361	758	1606111.
0.3346107	709	1502286.	0.3582081	759	1608230.
0.3350827	710	1504405.	0.3586800	760	1610349.
0.3355546	711	1506524.	0.3591520	761	1612468.
0.3360265	712	1508643.	0.3596239	762	1614587.
0.3364985	713	1510762.	0.3600959	763	1616706.
0.3369704	714	1512880.	0.3605678	764	1618824.
0.3374424	715	1514999.	0.3610398	765	1620943.
0.3379143	716	1517118.	0.3615117	766	1623062.
0.3383863	717	1519237.	0.3619837	767	1625181.
0.3388582	718	1521356.	0.3624556	768	1627300.
0.3393302	719	1523475.	0.3629275	769	1629419.
0.3398021	720	1525594.	0.3633995	770	1631538.
0.3402741	721	1527713.	0.3638714	771	1633657.
0.3407460	722	1529832.	0.3643434	772	1635776.
0.3412180	723	1531950.	0.3648153	773	1637894.
0.3416899	724	1534069.	0.3652873	774	1640013.
0.3421619	725	1536188.	0.3657592	775	1642132.
0.3426338	726	1538307.	0.3662312	776	1644251.
0.3431058	727	1540426.	0.3667031	777	1646370.
0.3435777	728	1542545.	0.3671751	778	1648489.
0.3440497	729	1544664.	0.3676470	779	1650608.
0.3445216	730	1546783.	0.3681190	780	1652727.
0.3449935	731	1548901.	0.3685909	781	1654845.
0.3454655	732	1551020.	0.3690629	782	1656964.
0.3459374	733	1553139.	0.3695348	783	1659083.
0.3464094	734	1555258.	0.3700068	784	1661202.
0.3468813	735	1557377.	0.3704787	785	1663321.
0.3473533	736	1559496.	0.3709507	786	1665440.
0.3478252	737	1561615.	0.3714226	787	1667559.
0.3482972	738	1563734.	0.3718945	788	1669678.
0.3487691	739	1565852.	0.3723665	789	1671797.
0.3492411	740	1567971.	0.3728384	790	1673915.
0.3497130	741	1570090.	0.3733104	791	1676034.
0.3501850	742	1572209.	0.3737823	792	1678153.
0.3506569	743	1574328.	0.3742543	793	1680272.
0.3511289	744	1576447.	0.3747262	794	1682391.
0.3516008	745	1578566.	0.3751982	795	1684510.
0.3520728	746	1580685.	0.3756701	796	1686629.
0.3525447	747	1582804.	0.3761421	797	1688748.
0.3530167	748	1584922.	0.3766140	798	1690866.
0.3534886	749	1587041.	0.3770860	799	1692985.
0.3539605	750	1589160.	0.3775579	800	1695104.

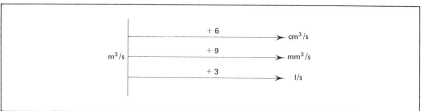

m³/s	REF	ft³/min		m³/s	REF	ft³/min
0.3780299	801	1697223.		0.4016272	851	1803167.
0.3785018	802	1699342.		0.4020992	852	1805286.
0.3789738	803	1701461.		0.4025711	853	1807405.
0.3794457	804	1703580.		0.4030431	854	1809524.
0.3799177	805	1705699.		0.4035150	855	1811643.
0.3803896	806	1707817.		0.4039870	856	1813761.
0.3808616	807	1709936.		0.4044589	857	1815880.
0.3813335	808	1712055.		0.4049309	858	1817999.
0.3818054	809	1714174.		0.4054028	859	1820118.
0.3822774	810	1716293.		0.4058748	860	1822237.
0.3827493	811	1718412.		0.4063467	861	1824356.
0.3832213	812	1720531.		0.4068187	862	1826475.
0.3836932	813	1722650.		0.4072906	863	1828594.
0.3841652	814	1724769.		0.4077626	864	1830713.
0.3846371	815	1726887.		0.4082345	865	1832831.
0.3851091	816	1729006.		0.4087064	866	1834950.
0.3855810	817	1731125.		0.4091784	867	1837069.
0.3860530	818	1733244.		0.4096503	868	1839188.
0.3865249	819	1735363.		0.4101223	869	1841307.
0.3869969	820	1737482.		0.4105942	870	1843426.
0.3874688	821	1739601.		0.4110662	871	1845545.
0.3879408	822	1741720.		0.4115381	872	1847664.
0.3884127	823	1743838.		0.4120101	873	1849782.
0.3888847	824	1745957.		0.4124820	874	1851901.
0.3893566	825	1748076.		0.4129540	875	1854020.
0.3898286	826	1750195.		0.4134259	876	1856139.
0.3903005	827	1752314.		0.4138979	877	1858258.
0.3907724	828	1754433.		0.4143698	878	1860377.
0.3912444	829	1756552.		0.4148418	879	1862496.
0.3917163	830	1758671.		0.4153137	880	1864615.
0.3921883	831	1760789.		0.4157857	881	1866733.
0.3926602	832	1762908.		0.4162576	882	1868852.
0.3931322	833	1765027.		0.4167296	883	1870971.
0.3936041	834	1767146.		0.4172015	884	1873090.
0.3940761	835	1769265.		0.4176734	885	1875209.
0.3945480	836	1771384.		0.4181454	886	1877328.
0.3950200	837	1773503.		0.4186173	887	1879447.
0.3954919	838	1775622.		0.4190893	888	1881566.
0.3959639	839	1777741.		0.4195612	889	1883685.
0.3964358	840	1779859.		0.4200332	890	1885803.
0.3969078	841	1781978.		0.4205051	891	1887922.
0.3973797	842	1784097.		0.4209771	892	1890041.
0.3978517	843	1786216.		0.4214490	893	1892160.
0.3983236	844	1788335.		0.4219210	894	1894279.
0.3987956	845	1790454.		0.4223929	895	1896398.
0.3992675	846	1792573.		0.4228649	896	1898517.
0.3997394	847	1794692.		0.4233368	897	1900636.
0.4002114	848	1796810.		0.4238088	898	1902754.
0.4006833	849	1798929.		0.4242807	899	1904873.
0.4011553	850	1801048.		0.4247527	900	1906992.

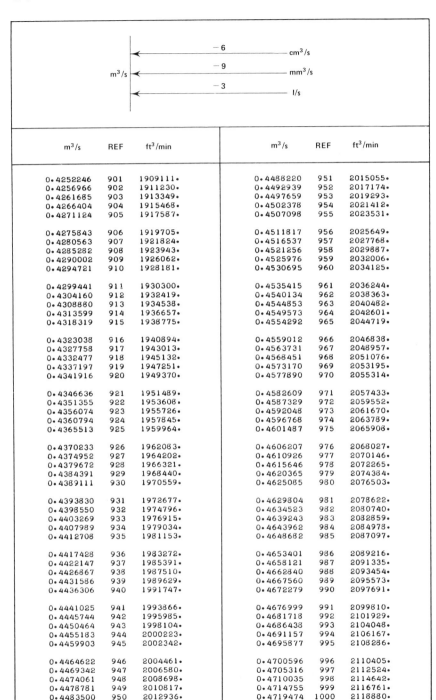

m³/s	REF	ft³/min	m³/s	REF	ft³/min
0.4252246	901	1909111.	0.4488220	951	2015055.
0.4256966	902	1911230.	0.4492939	952	2017174.
0.4261685	903	1913349.	0.4497659	953	2019293.
0.4266404	904	1915468.	0.4502378	954	2021412.
0.4271124	905	1917587.	0.4507098	955	2023531.
0.4275843	906	1919705.	0.4511817	956	2025649.
0.4280563	907	1921824.	0.4516537	957	2027768.
0.4285282	908	1923943.	0.4521256	958	2029887.
0.4290002	909	1926062.	0.4525976	959	2032006.
0.4294721	910	1928181.	0.4530695	960	2034125.
0.4299441	911	1930300.	0.4535415	961	2036244.
0.4304160	912	1932419.	0.4540134	962	2038363.
0.4308880	913	1934538.	0.4544853	963	2040482.
0.4313599	914	1936657.	0.4549573	964	2042601.
0.4318319	915	1938775.	0.4554292	965	2044719.
0.4323038	916	1940894.	0.4559012	966	2046838.
0.4327758	917	1943013.	0.4563731	967	2048957.
0.4332477	918	1945132.	0.4568451	968	2051076.
0.4337197	919	1947251.	0.4573170	969	2053195.
0.4341916	920	1949370.	0.4577890	970	2055314.
0.4346636	921	1951489.	0.4582609	971	2057433.
0.4351355	922	1953608.	0.4587329	972	2059552.
0.4356074	923	1955726.	0.4592048	973	2061670.
0.4360794	924	1957845.	0.4596768	974	2063789.
0.4365513	925	1959964.	0.4601487	975	2065908.
0.4370233	926	1962083.	0.4606207	976	2068027.
0.4374952	927	1964202.	0.4610926	977	2070146.
0.4379672	928	1966321.	0.4615646	978	2072265.
0.4384391	929	1968440.	0.4620365	979	2074384.
0.4389111	930	1970559.	0.4625085	980	2076503.
0.4393830	931	1972677.	0.4629804	981	2078622.
0.4398550	932	1974796.	0.4634523	982	2080740.
0.4403269	933	1976915.	0.4639243	983	2082859.
0.4407989	934	1979034.	0.4643962	984	2084978.
0.4412708	935	1981153.	0.4648682	985	2087097.
0.4417428	936	1983272.	0.4653401	986	2089216.
0.4422147	937	1985391.	0.4658121	987	2091335.
0.4426867	938	1987510.	0.4662840	988	2093454.
0.4431586	939	1989629.	0.4667560	989	2095573.
0.4436306	940	1991747.	0.4672279	990	2097691.
0.4441025	941	1993866.	0.4676999	991	2099810.
0.4445744	942	1995985.	0.4681718	992	2101929.
0.4450464	943	1998104.	0.4686438	993	2104048.
0.4455183	944	2000223.	0.4691157	994	2106167.
0.4459903	945	2002342.	0.4695877	995	2108286.
0.4464622	946	2004461.	0.4700596	996	2110405.
0.4469342	947	2006580.	0.4705316	997	2112524.
0.4474061	948	2008698.	0.4710035	998	2114642.
0.4478781	949	2010817.	0.4714755	999	2116761.
0.4483500	950	2012936.	0.4719474	1000	2118880.

17

VOLUME FLOW RATE

U.S. gallon/Minute — Cubic-metre/Second
(gal/min — m³/s)

U.S. gallon/Minute — Cubic-centimetre/Second
(gal/min — cm³/s)

U.S. gallon/Minute — Litre/Second
(gal/min — l/s)

Conversion Factors:
$$1 \text{ U.S. gallon/Minute (gal/min)} = 0.00006309020 \text{ cubic-metre/second (m}^3\text{/s)}$$
$$(1 \text{ m}^3\text{/s} = 15{,}850.32 \text{ gal/min})$$

In converting m³/s quantities to cm³/s, and l/s, Diagram 17-A, which appears in the headings on alternate pages of this table, will assist the user. It is derived from: $1 \text{ m}^3\text{/s} = (10^6) \text{ cm}^3\text{/s} = (10^3) \text{ l/s}; 1 \text{ l/s} = 1 \text{ dm}^3\text{/s}$.

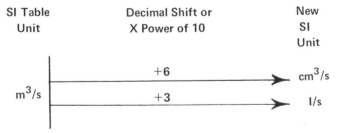

Diagram 17-A

Example (a): Convert 71.5 gal/min to m³/s, cm³/s, l/s.
Since: $71.5 \text{ gal/min} = 715 \times 10^{-1} \text{ gal/min}$:
From table: $715 \text{ gal/min} = 0.04510949 \text{ m}^3\text{/s*} = 451.0949 \times 10^{-4} \text{ m}^3\text{/s}$

---

*All conversion values obtained can be rounded to desired number of significant figures.

Therefore: 715×10^{-1} gal/min $= 0.004510949$ m^3/s $= 451.0949 \times 10^{-5}$
m^3/s $\approx 451 \times 10^{-5}$ m^3/s.

From Diagram 17-A: 1) 0.004510949 m^3/s $= 0.004510949 \, (10^6)$ cm^3/s
$= 4,510.949$ cm^3/s $\approx 4,510$ cm^3/s
(m^3/s to cm^3/s by multiplying m^3/s value by 10^6,
or shifting its decimal 6 places to the right.)

2) 0.004510949 m^3/s $= 0.004510949 \, (10^3)$ l/s
$= 4.510949$ l/s ≈ 4.51 l/s
(m^3/s to l/s by multiplying m^3/s value by 10^3, or
shifting its decimal 3 places to the right.)

Diagram 17-B is also given at the top of alternate table pages:

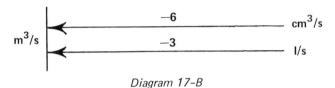

Diagram 17-B

It is the opposite of Diagram 17-A and is used to convert cm^3/s and l/s to m^3/s.
The m^3/s equivalent is then applied to the REF column in the table to obtain
gal/min.

Example (b): Convert 3.15 l/s to gal/min.
From Diagram 17-B: 3.15 l/s $= 3.15 \times 10^{-3}$ m^3/s $= 315 \times 10^{-5}$ m^3/s.
From table: 315 m^3/s $= 4,992,852$ gal/min*
Therefore: 315×10^{-5} m^3/s $= 4,992,852 \times 10^{-5}$ gal/min $= 49.92852$
gal/min or ≈ 49.9 gal/min.

---

*All conversion values obtained can be rounded to desired number of significant figures.

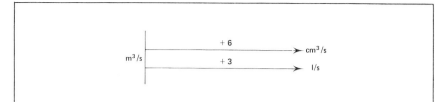

m³/s	REF	gal/min		m³/s	REF	gal/min
0.0000630902	1	15850.32		0.003217600	51	808366.4
0.0001261804	2	31700.64		0.003280690	52	824216.8
0.0001892706	3	47550.97		0.003343781	53	840067.1
0.0002523608	4	63401.29		0.003406871	54	855917.4
0.0003154510	5	79251.61		0.003469961	55	871767.7
0.0003785412	6	95101.93		0.003533051	56	887618.1
0.0004416314	7	110952.3		0.003596141	57	903468.4
0.0005047216	8	126802.6		0.003659232	58	919318.7
0.0005678118	9	142652.9		0.003722322	59	935169.0
0.0006309020	10	158503.2		0.003785412	60	951019.3
0.0006939922	11	174353.5		0.003848502	61	966869.7
0.0007570824	12	190203.9		0.003911592	62	982720.0
0.0008201726	13	206054.2		0.003974683	63	998570.3
0.0008832628	14	221904.5		0.004037773	64	1014421.
0.0009463530	15	237754.8		0.004100863	65	1030271.
0.001009443	16	253605.2		0.004163953	66	1046121.
0.001072533	17	269455.5		0.004227043	67	1061972.
0.001135624	18	285305.8		0.004290134	68	1077822.
0.001198714	19	301156.1		0.004353224	69	1093672.
0.001261804	20	317006.4		0.004416314	70	1109523.
0.001324894	21	332856.8		0.004479404	71	1125373.
0.001387984	22	348707.1		0.004542494	72	1141223.
0.001451075	23	364557.4		0.004605585	73	1157074.
0.001514165	24	380407.7		0.004668675	74	1172924.
0.001577255	25	396258.1		0.004731765	75	1188774.
0.001640345	26	412108.4		0.004794855	76	1204625.
0.001703435	27	427958.7		0.004857945	77	1220475.
0.001766526	28	443809.0		0.004921036	78	1236325.
0.001829616	29	459659.3		0.004984126	79	1252175.
0.001892706	30	475509.7		0.005047216	80	1268026.
0.001955796	31	491360.0		0.005110306	81	1283876.
0.002018886	32	507210.3		0.005173396	82	1299726.
0.002081977	33	523060.6		0.005236487	83	1315577.
0.002145067	34	538911.0		0.005299577	84	1331427.
0.002208157	35	554761.3		0.005362667	85	1347277.
0.002271247	36	570611.6		0.005425757	86	1363128.
0.002334337	37	586461.9		0.005488847	87	1378978.
0.002397428	38	602312.2		0.005551938	88	1394828.
0.002460518	39	618162.6		0.005615028	89	1410679.
0.002523608	40	634012.9		0.005678118	90	1426529.
0.002586698	41	649863.2		0.005741208	91	1442379.
0.002649788	42	665713.5		0.005804298	92	1458230.
0.002712879	43	681563.9		0.005867389	93	1474080.
0.002775969	44	697414.2		0.005930479	94	1489930.
0.002839059	45	713264.5		0.005993569	95	1505781.
0.002902149	46	729114.8		0.006056659	96	1521631.
0.002965239	47	744965.1		0.006119749	97	1537481.
0.003028330	48	760815.5		0.006182840	98	1553332.
0.003091420	49	776665.8		0.006245930	99	1569182.
0.003154510	50	792516.1		0.006309020	100	1585032.

m³/s	REF	gal/min		m³/s	REF	gal/min
0.006372110	101	1600883.		0.009526620	151	2393399.
0.006435200	102	1616733.		0.009589710	152	2409249.
0.006498291	103	1632583.		0.009652801	153	2425099.
0.006561381	104	1648434.		0.009715891	154	2440950.
0.006624471	105	1664284.		0.009778981	155	2456800.
0.006687561	106	1680134.		0.009842071	156	2472650.
0.006750651	107	1695985.		0.009905161	157	2488501.
0.006813742	108	1711835.		0.009968252	158	2504351.
0.006876832	109	1727685.		0.01003134	159	2520201.
0.006939922	110	1743535.		0.01009443	160	2536052.
0.007003012	111	1759386.		0.01015752	161	2551902.
0.007066102	112	1775236.		0.01022061	162	2567752.
0.007129193	113	1791086.		0.01028370	163	2583603.
0.007192283	114	1806937.		0.01034679	164	2599453.
0.007255373	115	1822787.		0.01040988	165	2615303.
0.007318463	116	1838637.		0.01047297	166	2631154.
0.007381553	117	1854488.		0.01053606	167	2647004.
0.007444644	118	1870338.		0.01059915	168	2662854.
0.007507734	119	1886188.		0.01066224	169	2678704.
0.007570824	120	1902039.		0.01072533	170	2694555.
0.007633914	121	1917889.		0.01078842	171	2710405.
0.007697004	122	1933739.		0.01085151	172	2726255.
0.007760095	123	1949590.		0.01091460	173	2742106.
0.007823185	124	1965440.		0.01097769	174	2757956.
0.007886275	125	1981290.		0.01104078	175	2773806.
0.007949365	126	1997141.		0.01110388	176	2789657.
0.008012455	127	2012991.		0.01116697	177	2805507.
0.008075546	128	2028841.		0.01123006	178	2821357.
0.008138636	129	2044692.		0.01129315	179	2837208.
0.008201726	130	2060542.		0.01135624	180	2853058.
0.008264816	131	2076392.		0.01141933	181	2868908.
0.008327906	132	2092243.		0.01148242	182	2884759.
0.008390996	133	2108093.		0.01154551	183	2900609.
0.008454087	134	2123943.		0.01160860	184	2916459.
0.008517177	135	2139794.		0.01167169	185	2932310.
0.008580267	136	2155644.		0.01173478	186	2948160.
0.008643357	137	2171494.		0.01179787	187	2964010.
0.008706448	138	2187345.		0.01186096	188	2979861.
0.008769538	139	2203195.		0.01192405	189	2995711.
0.008832628	140	2219045.		0.01198714	190	3011561.
0.008895718	141	2234895.		0.01205023	191	3027412.
0.008958808	142	2250746.		0.01211332	192	3043262.
0.009021898	143	2266596.		0.01217641	193	3059112.
0.009084989	144	2282446.		0.01223950	194	3074963.
0.009148079	145	2298297.		0.01230259	195	3090813.
0.009211169	146	2314147.		0.01236568	196	3106663.
0.009274259	147	2329997.		0.01242877	197	3122514.
0.009337350	148	2345848.		0.01249186	198	3138364.
0.009400440	149	2361698.		0.01255495	199	3154214.
0.009463530	150	2377548.		0.01261804	200	3170064.

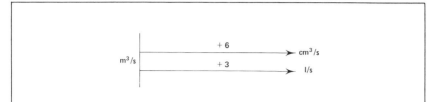

m³/s	REF	gal/min		m³/s	REF	gal/min
0.01268113	201	3185915.		0.01583564	251	3978431.
0.01274422	202	3201765.		0.01589873	252	3994281.
0.01280731	203	3217615.		0.01596182	253	4010132.
0.01287040	204	3233466.		0.01602491	254	4025982.
0.01293349	205	3249316.		0.01608800	255	4041832.
0.01299658	206	3265166.		0.01615109	256	4057683.
0.01305967	207	3281017.		0.01621418	257	4073533.
0.01312276	208	3296867.		0.01627727	258	4089383.
0.01318585	209	3312717.		0.01634036	259	4105234.
0.01324894	210	3328568.		0.01640345	260	4121084.
0.01331203	211	3344418.		0.01646654	261	4136934.
0.01337512	212	3360268.		0.01652963	262	4152784.
0.01343821	213	3376119.		0.01659272	263	4168635.
0.01350130	214	3391969.		0.01665581	264	4184485.
0.01356439	215	3407819.		0.01671890	265	4200335.
0.01362748	216	3423670.		0.01678199	266	4216186.
0.01369057	217	3439520.		0.01684508	267	4232036.
0.01375366	218	3455370.		0.01690817	268	4247886.
0.01381675	219	3471221.		0.01697126	269	4263737.
0.01387984	220	3487071.		0.01703435	270	4279587.
0.01394293	221	3502921.		0.01709744	271	4295437.
0.01400602	222	3518772.		0.01716053	272	4311288.
0.01406911	223	3534622.		0.01722362	273	4327138.
0.01413220	224	3550472.		0.01728671	274	4342988.
0.01419529	225	3566323.		0.01734980	275	4358839.
0.01425839	226	3582173.		0.01741290	276	4374689.
0.01432148	227	3598023.		0.01747599	277	4390539.
0.01438457	228	3613874.		0.01753908	278	4406390.
0.01444766	229	3629724.		0.01760217	279	4422240.
0.01451075	230	3645574.		0.01766526	280	4438090.
0.01457384	231	3661424.		0.01772835	281	4453941.
0.01463693	232	3677275.		0.01779144	282	4469791.
0.01470002	233	3693125.		0.01785453	283	4485641.
0.01476311	234	3708975.		0.01791762	284	4501492.
0.01482620	235	3724826.		0.01798071	285	4517342.
0.01488929	236	3740676.		0.01804380	286	4533192.
0.01495238	237	3756526.		0.01810689	287	4549043.
0.01501547	238	3772377.		0.01816998	288	4564893.
0.01507856	239	3788227.		0.01823307	289	4580743.
0.01514165	240	3804077.		0.01829616	290	4596594.
0.01520474	241	3819928.		0.01835925	291	4612444.
0.01526783	242	3835778.		0.01842234	292	4628294.
0.01533092	243	3851628.		0.01848543	293	4644144.
0.01539401	244	3867479.		0.01854852	294	4659995.
0.01545710	245	3883329.		0.01861161	295	4675845.
0.01552019	246	3899179.		0.01867470	296	4691695.
0.01558328	247	3915030.		0.01873779	297	4707546.
0.01564637	248	3930880.		0.01880088	298	4723396.
0.01570946	249	3946730.		0.01886397	299	4739246.
0.01577255	250	3962581.		0.01892706	300	4755097.

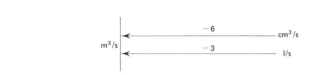

m³/s	REF	gal/min		m³/s	REF	gal/min
0.01899015	301	4770947.		0.02214466	351	5563463.
0.01905324	302	4786797.		0.02220775	352	5579314.
0.01911633	303	4802648.		0.02227084	353	5595164.
0.01917942	304	4818498.		0.02233393	354	5611014.
0.01924251	305	4834348.		0.02239702	355	5626864.
0.01930560	306	4850199.		0.02246011	356	5642715.
0.01936869	307	4866049.		0.02252320	357	5658565.
0.01943178	308	4881899.		0.02258629	358	5674415.
0.01949487	309	4897750.		0.02264938	359	5690266.
0.01955796	310	4913600.		0.02271247	360	5706116.
0.01962105	311	4929450.		0.02277556	361	5721966.
0.01968414	312	4945301.		0.02283865	362	5737817.
0.01974723	313	4961151.		0.02290174	363	5753667.
0.01981032	314	4977001.		0.02296483	364	5769517.
0.01987341	315	4992852.		0.02302792	365	5785368.
0.01993650	316	5008702.		0.02309101	366	5801218.
0.01999959	317	5024552.		0.02315410	367	5817068.
0.02006268	318	5040403.		0.02321719	368	5832919.
0.02012577	319	5056253.		0.02328028	369	5848769.
0.02018886	320	5072103.		0.02334337	370	5864619.
0.02025195	321	5087954.		0.02340646	371	5880470.
0.02031504	322	5103804.		0.02346955	372	5896320.
0.02037813	323	5119654.		0.02353264	373	5912170.
0.02044122	324	5135504.		0.02359573	374	5928021.
0.02050431	325	5151355.		0.02365882	375	5943871.
0.02056741	326	5167205.		0.02372191	376	5959721.
0.02063050	327	5183055.		0.02378501	377	5975572.
0.02069359	328	5198906.		0.02384810	378	5991422.
0.02075668	329	5214756.		0.02391119	379	6007272.
0.02081977	330	5230606.		0.02397428	380	6023123.
0.02088286	331	5246457.		0.02403737	381	6038973.
0.02094595	332	5262307.		0.02410046	382	6054823.
0.02100904	333	5278157.		0.02416355	383	6070674.
0.02107213	334	5294008.		0.02422664	384	6086524.
0.02113522	335	5309858.		0.02428973	385	6102374.
0.02119831	336	5325708.		0.02435282	386	6118224.
0.02126140	337	5341559.		0.02441591	387	6134075.
0.02132449	338	5357409.		0.02447900	388	6149925.
0.02138758	339	5373259.		0.02454209	389	6165775.
0.02145067	340	5389110.		0.02460518	390	6181626.
0.02151376	341	5404960.		0.02466827	391	6197476.
0.02157685	342	5420810.		0.02473136	392	6213326.
0.02163994	343	5436661.		0.02479445	393	6229177.
0.02170303	344	5452511.		0.02485754	394	6245027.
0.02176612	345	5468361.		0.02492063	395	6260877.
0.02182921	346	5484212.		0.02498372	396	6276728.
0.02189230	347	5500062.		0.02504681	397	6292578.
0.02195539	348	5515912.		0.02510990	398	6308428.
0.02201848	349	5531763.		0.02517299	399	6324279.
0.02208157	350	5547613.		0.02523608	400	6340129.

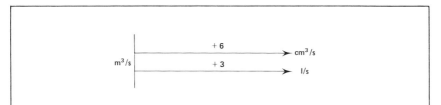

m³/s	REF	gal/min	m³/s	REF	gal/min
0.02529917	401	6355979.	0.02845368	451	7148495.
0.02536226	402	6371830.	0.02851677	452	7164346.
0.02542535	403	6387680.	0.02857986	453	7180196.
0.02548844	404	6403530.	0.02864295	454	7196046.
0.02555153	405	6419381.	0.02870604	455	7211897.
0.02561462	406	6435231.	0.02876913	456	7227747.
0.02567771	407	6451081.	0.02883222	457	7243597.
0.02574080	408	6466932.	0.02889531	458	7259448.
0.02580389	409	6482782.	0.02895840	459	7275298.
0.02586698	410	6498632.	0.02902149	460	7291148.
0.02593007	411	6514483.	0.02908458	461	7306999.
0.02599316	412	6530333.	0.02914767	462	7322849.
0.02605625	413	6546183.	0.02921076	463	7338699.
0.02611934	414	6562033.	0.02927385	464	7354550.
0.02618243	415	6577884.	0.02933694	465	7370400.
0.02624552	416	6593734.	0.02940003	466	7386250.
0.02630861	417	6609584.	0.02946312	467	7402101.
0.02637170	418	6625435.	0.02952621	468	7417951.
0.02643479	419	6641285.	0.02958930	469	7433801.
0.02649788	420	6657135.	0.02965239	470	7449652.
0.02656097	421	6672986.	0.02971548	471	7465502.
0.02662406	422	6688836.	0.02977857	472	7481352.
0.02668715	423	6704686.	0.02984166	473	7497203.
0.02675024	424	6720537.	0.02990475	474	7513053.
0.02681333	425	6736387.	0.02996784	475	7528903.
0.02687642	426	6752237.	0.03003094	476	7544753.
0.02693952	427	6768088.	0.03009403	477	7560604.
0.02700261	428	6783938.	0.03015712	478	7576454.
0.02706570	429	6799788.	0.03022021	479	7592304.
0.02712879	430	6815639.	0.03028330	480	7608155.
0.02719188	431	6831489.	0.03034639	481	7624005.
0.02725497	432	6847339.	0.03040948	482	7639855.
0.02731806	433	6863190.	0.03047257	483	7655706.
0.02738115	434	6879040.	0.03053566	484	7671556.
0.02744424	435	6894890.	0.03059875	485	7687406.
0.02750733	436	6910741.	0.03066184	486	7703257.
0.02757042	437	6926591.	0.03072493	487	7719107.
0.02763351	438	6942441.	0.03078802	488	7734957.
0.02769660	439	6958292.	0.03085111	489	7750808.
0.02775969	440	6974142.	0.03091420	490	7766658.
0.02782278	441	6989992.	0.03097729	491	7782508.
0.02788587	442	7005843.	0.03104038	492	7798359.
0.02794896	443	7021693.	0.03110347	493	7814209.
0.02801205	444	7037543.	0.03116656	494	7830059.
0.02807514	445	7053393.	0.03122965	495	7845910.
0.02813823	446	7069244.	0.03129274	496	7861760.
0.02820132	447	7085094.	0.03135583	497	7877610.
0.02826441	448	7100944.	0.03141892	498	7893461.
0.02832750	449	7116795.	0.03148201	499	7909311.
0.02839059	450	7132645.	0.03154510	500	7925161.

m³/s	REF	gal/min		m³/s	REF	gal/min
0.03160819	501	7941012.		0.03476270	551	8733528.
0.03167128	502	7956862.		0.03482579	552	8749378.
0.03173437	503	7972712.		0.03488888	553	8765228.
0.03179746	504	7988563.		0.03495197	554	8781079.
0.03186055	505	8004413.		0.03501506	555	8796929.
0.03192364	506	8020263.		0.03507815	556	8812779.
0.03198673	507	8036113.		0.03514124	557	8828630.
0.03204982	508	8051964.		0.03520433	558	8844480.
0.03211291	509	8067814.		0.03526742	559	8860330.
0.03217600	510	8083664.		0.03533051	560	8876181.
0.03223909	511	8099515.		0.03539360	561	8892031.
0.03230218	512	8115365.		0.03545669	562	8907881.
0.03236527	513	8131215.		0.03551978	563	8923732.
0.03242836	514	8147066.		0.03558287	564	8939582.
0.03249145	515	8162916.		0.03564596	565	8955432.
0.03255454	516	8178766.		0.03570905	566	8971283.
0.03261763	517	8194617.		0.03577214	567	8987133.
0.03268072	518	8210467.		0.03583523	568	9002983.
0.03274381	519	8226317.		0.03589832	569	9018834.
0.03280690	520	8242168.		0.03596141	570	9034684.
0.03286999	521	8258018.		0.03602450	571	9050534.
0.03293308	522	8273868.		0.03608759	572	9066384.
0.03299617	523	8289719.		0.03615068	573	9082235.
0.03305926	524	8305569.		0.03621377	574	9098085.
0.03312235	525	8321419.		0.03627686	575	9113935.
0.03318544	526	8337270.		0.03633995	576	9129786.
0.03324854	527	8353120.		0.03640305	577	9145636.
0.03331163	528	8368970.		0.03646614	578	9161486.
0.03337472	529	8384821.		0.03652923	579	9177337.
0.03343781	530	8400671.		0.03659232	580	9193187.
0.03350090	531	8416521.		0.03665541	581	9209037.
0.03356399	532	8432372.		0.03671850	582	9224888.
0.03362708	533	8448222.		0.03678159	583	9240738.
0.03369017	534	8464072.		0.03684468	584	9256588.
0.03375326	535	8479923.		0.03690777	585	9272439.
0.03381635	536	8495773.		0.03697086	586	9288289.
0.03387944	537	8511623.		0.03703395	587	9304139.
0.03394253	538	8527474.		0.03709704	588	9319990.
0.03400562	539	8543324.		0.03716013	589	9335840.
0.03406871	540	8559174.		0.03722322	590	9351690.
0.03413180	541	8575024.		0.03728631	591	9367541.
0.03419489	542	8590875.		0.03734940	592	9383391.
0.03425798	543	8606725.		0.03741249	593	9399241.
0.03432107	544	8622575.		0.03747558	594	9415092.
0.03438416	545	8638426.		0.03753867	595	9430942.
0.03444725	546	8654276.		0.03760176	596	9446792.
0.03451034	547	8670126.		0.03766485	597	9462643.
0.03457343	548	8685977.		0.03772794	598	9478493.
0.03463652	549	8701827.		0.03779103	599	9494343.
0.03469961	550	8717677.		0.03785412	600	9510193.

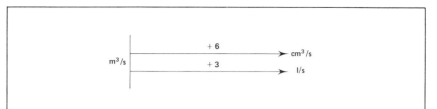

m³/s	REF	gal/min		m³/s	REF	gal/min
0.03791721	601	9526044.		0.04107172	651	10318560.
0.03798030	602	9541894.		0.04113481	652	10334410.
0.03804339	603	9557744.		0.04119790	653	10350260.
0.03810648	604	9573595.		0.04126099	654	10366110.
0.03816957	605	9589445.		0.04132408	655	10381960.
0.03823266	606	9605295.		0.04138717	656	10397810.
0.03829575	607	9621146.		0.04145026	657	10413660.
0.03835884	608	9636996.		0.04151335	658	10429510.
0.03842193	609	9652846.		0.04157644	659	10445360.
0.03848502	610	9668697.		0.04163953	660	10461210.
0.03854811	611	9684547.		0.04170262	661	10477060.
0.03861120	612	9700397.		0.04176571	662	10492910.
0.03867429	613	9716248.		0.04182880	663	10508760.
0.03873738	614	9732098.		0.04189189	664	10524610.
0.03880047	615	9747948.		0.04195498	665	10540460.
0.03886356	616	9763799.		0.04201807	666	10556310.
0.03892665	617	9779649.		0.04208116	667	10572160.
0.03898974	618	9795499.		0.04214425	668	10588020.
0.03905283	619	9811350.		0.04220734	669	10603870.
0.03911592	620	9827200.		0.04227043	670	10619720.
0.03917901	621	9843050.		0.04233352	671	10635570.
0.03924210	622	9858901.		0.04239661	672	10651420.
0.03930519	623	9874751.		0.04245970	673	10667270.
0.03936828	624	9890601.		0.04252279	674	10683120.
0.03943137	625	9906452.		0.04258588	675	10698970.
0.03949447	626	9922302.		0.04264897	676	10714820.
0.03955756	627	9938152.		0.04271206	677	10730670.
0.03962065	628	9954003.		0.04277516	678	10746520.
0.03968374	629	9969853.		0.04283825	679	10762370.
0.03974683	630	9985703.		0.04290134	680	10778220.
0.03980992	631	10001550.		0.04296443	681	10794070.
0.03987301	632	10017400.		0.04302752	682	10809920.
0.03993610	633	10033250.		0.04309061	683	10825770.
0.03999919	634	10049100.		0.04315370	684	10841620.
0.04006228	635	10064950.		0.04321679	685	10857470.
0.04012537	636	10080800.		0.04327988	686	10873320.
0.04018846	637	10096660.		0.04334297	687	10889170.
0.04025155	638	10112510.		0.04340606	688	10905020.
0.04031464	639	10128360.		0.04346915	689	10920870.
0.04037773	640	10144210.		0.04353224	690	10936720.
0.04044082	641	10160060.		0.04359533	691	10952570.
0.04050391	642	10175910.		0.04365842	692	10968420.
0.04056700	643	10191760.		0.04372151	693	10984270.
0.04063009	644	10207610.		0.04378460	694	11000120.
0.04069318	645	10223460.		0.04384769	695	11015970.
0.04075627	646	10239310.		0.04391078	696	11031820.
0.04081936	647	10255160.		0.04397387	697	11047670.
0.04088245	648	10271010.		0.04403696	698	11063520.
0.04094554	649	10286860.		0.04410005	699	11079380.
0.04100863	650	10302710.		0.04416314	700	11095230.

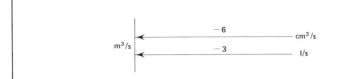

m³/s	REF	gal/min		m³/s	REF	gal/min
0.04422623	701	11111080.		0.04738074	751	11903590.
0.04428932	702	11126930.		0.04744383	752	11919440.
0.04435241	703	11142780.		0.04750692	753	11935290.
0.04441550	704	11158630.		0.04757001	754	11951140.
0.04447859	705	11174480.		0.04763310	755	11966990.
0.04454168	706	11190330.		0.04769619	756	11982840.
0.04460477	707	11206180.		0.04775928	757	11998690.
0.04466786	708	11222030.		0.04782237	758	12014540.
0.04473095	709	11237880.		0.04788546	759	12030390.
0.04479404	710	11253730.		0.04794855	760	12046240.
0.04485713	711	11269580.		0.04801164	761	12062100.
0.04492022	712	11285430.		0.04807473	762	12077950.
0.04498331	713	11301280.		0.04813782	763	12093800.
0.04504640	714	11317130.		0.04820091	764	12109650.
0.04510949	715	11332980.		0.04826400	765	12125500.
0.04517258	716	11348830.		0.04832709	766	12141350.
0.04523567	717	11364680.		0.04839018	767	12157200.
0.04529876	718	11380530.		0.04845327	768	12173050.
0.04536185	719	11396380.		0.04851636	769	12188900.
0.04542494	720	11412230.		0.04857945	770	12204750.
0.04548803	721	11428080.		0.04864254	771	12220600.
0.04555112	722	11443930.		0.04870563	772	12236450.
0.04561421	723	11459780.		0.04876872	773	12252300.
0.04567730	724	11475630.		0.04883181	774	12268150.
0.04574039	725	11491480.		0.04889490	775	12284000.
0.04580348	726	11507330.		0.04895799	776	12299850.
0.04586658	727	11523180.		0.04902109	777	12315700.
0.04592967	728	11539030.		0.04908418	778	12331550.
0.04599276	729	11554880.		0.04914727	779	12347400.
0.04605585	730	11570740.		0.04921036	780	12363250.
0.04611894	731	11586590.		0.04927345	781	12379100.
0.04618203	732	11602440.		0.04933654	782	12394950.
0.04624512	733	11618290.		0.04939963	783	12410800.
0.04630821	734	11634140.		0.04946272	784	12426650.
0.04637130	735	11649990.		0.04952581	785	12442500.
0.04643439	736	11665840.		0.04958890	786	12458350.
0.04649748	737	11681690.		0.04965199	787	12474200.
0.04656057	738	11697540.		0.04971508	788	12490050.
0.04662366	739	11713390.		0.04977817	789	12505900.
0.04668675	740	11729240.		0.04984126	790	12521750.
0.04674984	741	11745090.		0.04990435	791	12537600.
0.04681293	742	11760940.		0.04996744	792	12553460.
0.04687602	743	11776790.		0.05003053	793	12569310.
0.04693911	744	11792640.		0.05009362	794	12585160.
0.04700220	745	11808490.		0.05015671	795	12601010.
0.04706529	746	11824340.		0.05021980	796	12616860.
0.04712838	747	11840190.		0.05028289	797	12632710.
0.04719147	748	11856040.		0.05034598	798	12648560.
0.04725456	749	11871890.		0.05040907	799	12664410.
0.04731765	750	11887740.		0.05047216	800	12680260.

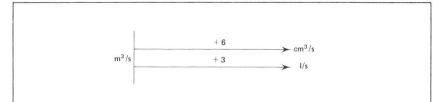

m³/s	REF	gal/min	m³/s	REF	gal/min
0.05053525	801	12696110.	0.05368976	851	13488620.
0.05059834	802	12711960.	0.05375285	852	13504470.
0.05066143	803	12727810.	0.05381594	853	13520320.
0.05072452	804	12743660.	0.05387903	854	13536180.
0.05078761	805	12759510.	0.05394212	855	13552030.
0.05085070	806	12775360.	0.05400521	856	13567880.
0.05091379	807	12791210.	0.05406830	857	13583730.
0.05097688	808	12807060.	0.05413139	858	13599580.
0.05103997	809	12822910.	0.05419448	859	13615430.
0.05110306	810	12838760.	0.05425757	860	13631280.
0.05116615	811	12854610.	0.05432066	861	13647130.
0.05122924	812	12870460.	0.05438375	862	13662980.
0.05129233	813	12886310.	0.05444684	863	13678830.
0.05135542	814	12902160.	0.05450993	864	13694680.
0.05141851	815	12918010.	0.05457302	865	13710530.
0.05148160	816	12933860.	0.05463611	866	13726380.
0.05154469	817	12949710.	0.05469920	867	13742230.
0.05160778	818	12965560.	0.05476229	868	13758080.
0.05167087	819	12981410.	0.05482538	869	13773930.
0.05173396	820	12997260.	0.05488847	870	13789780.
0.05179705	821	13013110.	0.05495156	871	13805630.
0.05186014	822	13028960.	0.05501465	872	13821480.
0.05192323	823	13044820.	0.05507774	873	13837330.
0.05198632	824	13060670.	0.05514083	874	13853180.
0.05204941	825	13076520.	0.05520392	875	13869030.
0.05211250	826	13092370.	0.05526702	876	13884880.
0.05217559	827	13108220.	0.05533011	877	13900730.
0.05223869	828	13124070.	0.05539320	878	13916580.
0.05230178	829	13139920.	0.05545629	879	13932430.
0.05236487	830	13155770.	0.05551938	880	13948280.
0.05242796	831	13171620.	0.05558247	881	13964130.
0.05249105	832	13187470.	0.05564556	882	13979980.
0.05255414	833	13203320.	0.05570865	883	13995830.
0.05261723	834	13219170.	0.05577174	884	14011680.
0.05268032	835	13235020.	0.05583483	885	14027540.
0.05274341	836	13250370.	0.05589792	886	14043390.
0.05280650	837	13266720.	0.05596101	887	14059240.
0.05286959	838	13282570.	0.05602410	888	14075090.
0.05293268	839	13298420.	0.05608719	889	14090940.
0.05299577	840	13314270.	0.05615028	890	14106790.
0.05305886	841	13330120.	0.05621337	891	14122640.
0.05312195	842	13345970.	0.05627646	892	14138490.
0.05318504	843	13361820.	0.05633955	893	14154340.
0.05324813	844	13377670.	0.05640264	894	14170190.
0.05331122	845	13393520.	0.05646573	895	14186040.
0.05337431	846	13409370.	0.05652882	896	14201890.
0.05343740	847	13425220.	0.05659191	897	14217740.
0.05350049	848	13441070.	0.05665500	898	14233590.
0.05356358	849	13456920.	0.05671809	899	14249440.
0.05362667	850	13472770.	0.05678118	900	14265290.

m³/s	REF	gal/min	m³/s	REF	gal/min
0.05684427	901	14281140.	0.05999878	951	15073660.
0.05690736	902	14296990.	0.06006187	952	15089510.
0.05697045	903	14312840.	0.06012496	953	15105360.
0.05703354	904	14328690.	0.06018805	954	15121210.
0.05709663	905	14344540.	0.06025114	955	15137060.
0.05715972	906	14360390.	0.06031423	956	15152910.
0.05722281	907	14376240.	0.06037732	957	15168760.
0.05728590	908	14392090.	0.06044041	958	15184610.
0.05734899	909	14407940.	0.06050350	959	15200460.
0.05741208	910	14423790.	0.06056659	960	15216310.
0.05747517	911	14439640.	0.06062968	961	15232160.
0.05753826	912	14455490.	0.06069277	962	15248010.
0.05760135	913	14471340.	0.06075586	963	15263860.
0.05766444	914	14487190.	0.06081895	964	15279710.
0.05772753	915	14503040.	0.06088204	965	15295560.
0.05779062	916	14518900.	0.06094513	966	15311410.
0.05785371	917	14534750.	0.06100822	967	15327260.
0.05791680	918	14550600.	0.06107131	968	15343110.
0.05797989	919	14566450.	0.06113440	969	15358960.
0.05804298	920	14582300.	0.06119749	970	15374810.
0.05810607	921	14598150.	0.06126058	971	15390660.
0.05816916	922	14614000.	0.06132367	972	15406510.
0.05823225	923	14629850.	0.06138676	973	15422360.
0.05829534	924	14645700.	0.06144985	974	15438210.
0.05835843	925	14661550.	0.06151294	975	15454060.
0.05842152	926	14677400.	0.06157603	976	15469910.
0.05848461	927	14693250.	0.06163912	977	15485760.
0.05854771	928	14709100.	0.06170222	978	15501620.
0.05861080	929	14724950.	0.06176531	979	15517470.
0.05867389	930	14740800.	0.06182840	980	15533320.
0.05873698	931	14756650.	0.06189149	981	15549170.
0.05880007	932	14772500.	0.06195458	982	15565020.
0.05886316	933	14788350.	0.06201767	983	15580870.
0.05892625	934	14804200.	0.06208076	984	15596720.
0.05898934	935	14820050.	0.06214385	985	15612570.
0.05905243	936	14835900.	0.06220694	986	15628420.
0.05911552	937	14851750.	0.06227003	987	15644270.
0.05917861	938	14867600.	0.06233312	988	15660120.
0.05924170	939	14883450.	0.06239621	989	15675970.
0.05930479	940	14899300.	0.06245930	990	15691820.
0.05936788	941	14915150.	0.06252239	991	15707670.
0.05943097	942	14931000.	0.06258548	992	15723520.
0.05949406	943	14946850.	0.06264857	993	15739370.
0.05955715	944	14962700.	0.06271166	994	15755220.
0.05962024	945	14978550.	0.06277475	995	15771070.
0.05968333	946	14994400.	0.06283784	996	15786920.
0.05974642	947	15010260.	0.06290093	997	15802770.
0.05980951	948	15026110.	0.06296402	998	15818620.
0.05987260	949	15041960.	0.06302711	999	15834470.
0.05993569	950	15057810.	0.06309020	1000	15850320.

18

VOLUME FLOW RATE

<u>British Gallon/Minute — Cubic-metre/Second</u>
  (U.K. gal/min — m³/s)
British Gallon/Minute — Cubic-centimetre/Second
  (U.K. gal/min — cm³/s)
British Gallon/Minute — Litre/Second
  (U.K. gal/min — l/s)

Conversion Factors:
  1 British gallon/Minute (U.K. gal/min) = 0.00007576820
  cubic-metre/second (m³/s)
  (1 m³/s = 13,198.15 U.K. gal/min)

In converting m³/s quantities to cm³/s, and l/s, Diagram 18-A, which appears in the headings on alternate pages of this table, will assist the user. It is derived from: 1 m³/s = (10⁶) cm³/s = (10³) l/s; 1 l/s = 1 dm³/s.

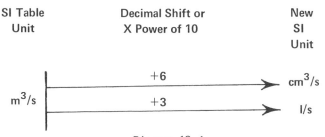

Diagram 18-A

Example (*a*): Convert 67.3 U.K. gal/min to m³/s, cm³/s, l/s
  Since: 67.3 gal/min = 673 × 10⁻¹ gal/min:
  From table: 673 gal/min = 0.05099200 m³/s* = 5.0992 × 10⁻² m³/s

---

*All conversion values obtained can be rounded to desired number of significant figures.

189

Therefore: 673×10^{-1} gal/min $= 5.0992 \times 10^{-3}$ m^3/s $= 0.0050992$ m^3/s
$\approx 5.10 \times 10^{-3}$ m^3/s.

From Diagram 18-A: 1) 0.0050992 m^3/s $= 0.0050992 \, (10^6)$ cm^3/s
$= 5,099.2$ cm^3/s $\approx 5,100$ cm^3/s
(m^3/s to cm^3/s by multiplying m^3/s value by 10^6 or by shifting its decimal 6 places to the right.)

2) 0.0050992 m^3/s $= 0.0050992 \, (10^3)$ l/s $= 5.0992$
l/s ≈ 5.10 l/s
(m^3/s to cm^3/s by multiplying m^3/s value by 10^3 or by shifting its decimal 3 places to the right.)

Diagram 18-B is also given at the top of alternate table pages:

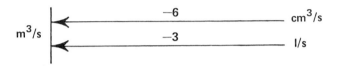

Diagram 18–B

It is the opposite of Diagram 18-A and is used to convert cm^3/s and l/s to m^3/s. The m^3/s equivalent is then applied to the REF column in the table to obtain U.K. gal/min quantities.

Example (*b*): Convert 1.59 l/s to U.K. gal/min.
From Diagram 18-B: 1.59 l/s $= 1.59 \times 10^{-3}$ m^3/s $= 159 \times 10^{-5}$ m^3/s
From table: 159 m^3/s $= 2,098,506$ U.K. gal/min*
Therefore: 159×10^{-5} m^3/s $= 2,098,506 \times 10^{-5} = 20.98506$ U.K. gal/min
≈ 21.0 U.K. gal/min.

---

*All conversion values obtained can be rounded to desired number of significant figures.

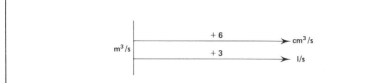

m³/s	REF	U.K. gal/min	m³/s	REF	U.K. gal/min
0.0000757682	1	13198.15	0.003864178	51	673105.6
0.0001515364	2	26396.30	0.003939946	52	686303.7
0.0002273046	3	39594.45	0.004015715	53	699501.9
0.0003030728	4	52792.60	0.004091483	54	712700.0
0.0003788410	5	65990.75	0.004167251	55	725898.2
0.0004546092	6	79188.89	0.004243019	56	739096.3
0.0005303774	7	92387.04	0.004318787	57	752294.5
0.0006061456	8	105585.2	0.004394556	58	765492.6
0.0006819138	9	118783.3	0.004470324	59	778690.8
0.0007576820	10	131981.5	0.004546092	60	791888.9
0.0008334502	11	145179.6	0.004621860	61	805087.1
0.0009092184	12	158377.8	0.004697628	62	818285.2
0.0009849866	13	171575.9	0.004773397	63	831483.4
0.001060755	14	184774.1	0.004849165	64	844681.5
0.001136523	15	197972.2	0.004924933	65	857879.7
0.001212291	16	211170.4	0.005000701	66	871077.8
0.001288059	17	224368.5	0.005076469	67	884276.0
0.001363828	18	237566.7	0.005152238	68	897474.1
0.001439596	19	250764.8	0.005228006	69	910672.3
0.001515364	20	263963.0	0.005303774	70	923870.4
0.001591132	21	277161.1	0.005379542	71	937068.6
0.001666900	22	290359.3	0.005455310	72	950266.7
0.001742669	23	303557.4	0.005531079	73	963464.9
0.001818437	24	316755.6	0.005606847	74	976663.0
0.001894205	25	329953.7	0.005682615	75	989861.2
0.001969973	26	343151.9	0.005758383	76	1003059.
0.002045741	27	356350.0	0.005834151	77	1016257.
0.002121510	28	369548.2	0.005909920	78	1029456.
0.002197278	29	382746.3	0.005985688	79	1042654.
0.002273046	30	395944.5	0.006061456	80	1055852.
0.002348814	31	409142.6	0.006137224	81	1069050.
0.002424582	32	422340.8	0.006212992	82	1082248.
0.002500351	33	435538.9	0.006288761	83	1095446.
0.002576119	34	448737.1	0.006364529	84	1108645.
0.002651887	35	461935.2	0.006440297	85	1121843.
0.002727655	36	475133.4	0.006516065	86	1135041.
0.002803423	37	488331.5	0.006591833	87	1148239.
0.002879192	38	501529.7	0.006667602	88	1161437.
0.002954960	39	514727.8	0.006743370	89	1174635.
0.003030728	40	527926.0	0.006819138	90	1187833.
0.003106496	41	541124.1	0.006894906	91	1201032.
0.003182264	42	554322.3	0.006970674	92	1214230.
0.003258033	43	567520.4	0.007046443	93	1227428.
0.003333801	44	580718.6	0.007122211	94	1240626.
0.003409569	45	593916.7	0.007197979	95	1253824.
0.003485337	46	607114.9	0.007273747	96	1267022.
0.003561105	47	620313.0	0.007349515	97	1280220.
0.003636874	48	633511.2	0.007425284	98	1293419.
0.003712642	49	646709.3	0.007501052	99	1306617.
0.003788410	50	659907.5	0.007576820	100	1319815.

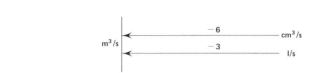

m³/s	REF	U.K. gal/min		m³/s	REF	U.K. gal/min
0.007652588	101	1333013.		0.01144100	151	1992921.
0.007728356	102	1346211.		0.01151677	152	2006119.
0.007804125	103	1359409.		0.01159253	153	2019317.
0.007879893	104	1372608.		0.01166830	154	2032515.
0.007955661	105	1385806.		0.01174407	155	2045713.
0.008031429	106	1399004.		0.01181984	156	2058911.
0.008107197	107	1412202.		0.01189561	157	2072109.
0.008182966	108	1425400.		0.01197138	158	2085308.
0.008258734	109	1438598.		0.01204714	159	2098506.
0.008334502	110	1451796.		0.01212291	160	2111704.
0.008410270	111	1464995.		0.01219868	161	2124902.
0.008486038	112	1478193.		0.01227445	162	2138100.
0.008561807	113	1491391.		0.01235022	163	2151298.
0.008637575	114	1504589.		0.01242598	164	2164496.
0.008713343	115	1517787.		0.01250175	165	2177695.
0.008789111	116	1530985.		0.01257752	166	2190893.
0.008864879	117	1544183.		0.01265329	167	2204091.
0.008940648	118	1557382.		0.01272906	168	2217289.
0.009016416	119	1570580.		0.01280483	169	2230487.
0.009092184	120	1583778.		0.01288059	170	2243685.
0.009167952	121	1596976.		0.01295636	171	2256884.
0.009243720	122	1610174.		0.01303213	172	2270082.
0.009319489	123	1623372.		0.01310790	173	2283280.
0.009395257	124	1636570.		0.01318367	174	2296478.
0.009471025	125	1649769.		0.01325943	175	2309676.
0.009546793	126	1662967.		0.01333520	176	2322874.
0.009622561	127	1676165.		0.01341097	177	2336072.
0.009698330	128	1689363.		0.01348674	178	2349271.
0.009774098	129	1702561.		0.01356251	179	2362469.
0.009849866	130	1715759.		0.01363828	180	2375667.
0.009925634	131	1728958.		0.01371404	181	2388865.
0.01000140	132	1742156.		0.01378981	182	2402063.
0.01007717	133	1755354.		0.01386558	183	2415261.
0.01015294	134	1768552.		0.01394135	184	2428459.
0.01022871	135	1781750.		0.01401712	185	2441658.
0.01030448	136	1794948.		0.01409289	186	2454856.
0.01038024	137	1808146.		0.01416865	187	2468054.
0.01045601	138	1821345.		0.01424442	188	2481252.
0.01053178	139	1834543.		0.01432019	189	2494450.
0.01060755	140	1847741.		0.01439596	190	2507648.
0.01068332	141	1860939.		0.01447173	191	2520846.
0.01075908	142	1874137.		0.01454749	192	2534045.
0.01083485	143	1887335.		0.01462326	193	2547243.
0.01091062	144	1900533.		0.01469903	194	2560441.
0.01098639	145	1913732.		0.01477480	195	2573639.
0.01106216	146	1926930.		0.01485057	196	2586837.
0.01113793	147	1940128.		0.01492634	197	2600035.
0.01121369	148	1953326.		0.01500210	198	2613234.
0.01128946	149	1966524.		0.01507787	199	2626432.
0.01136523	150	1979722.		0.01515364	200	2639630.

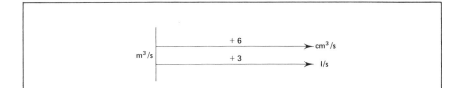

m³/s	REF	U.K. gal/min		m³/s	REF	U.K. gal/min
0.01522941	201	2652828.		0.01901782	251	3312735.
0.01530518	202	2666026.		0.01909359	252	3325934.
0.01538094	203	2679224.		0.01916935	253	3339132.
0.01545671	204	2692422.		0.01924512	254	3352330.
0.01553248	205	2705621.		0.01932089	255	3365528.
0.01560825	206	2718819.		0.01939666	256	3378726.
0.01568402	207	2732017.		0.01947243	257	3391924.
0.01575979	208	2745215.		0.01954820	258	3405122.
0.01583555	209	2758413.		0.01962396	259	3418321.
0.01591132	210	2771611.		0.01969973	260	3431519.
0.01598709	211	2784809.		0.01977550	261	3444717.
0.01606286	212	2798008.		0.01985127	262	3457915.
0.01613863	213	2811206.		0.01992704	263	3471113.
0.01621439	214	2824404.		0.02000281	264	3484311.
0.01629016	215	2837602.		0.02007857	265	3497510.
0.01636593	216	2850800.		0.02015434	266	3510708.
0.01644170	217	2863998.		0.02023011	267	3523906.
0.01651747	218	2877197.		0.02030588	268	3537104.
0.01659324	219	2890395.		0.02038165	269	3550302.
0.01666900	220	2903593.		0.02045741	270	3563500.
0.01674477	221	2916791.		0.02053318	271	3576698.
0.01682054	222	2929989.		0.02060895	272	3589897.
0.01689631	223	2943187.		0.02068472	273	3603095.
0.01697208	224	2956385.		0.02076049	274	3616293.
0.01704785	225	2969584.		0.02083626	275	3629491.
0.01712361	226	2982782.		0.02091202	276	3642689.
0.01719938	227	2995980.		0.02098779	277	3655887.
0.01727515	228	3009178.		0.02106356	278	3669085.
0.01735092	229	3022376.		0.02113933	279	3682284.
0.01742669	230	3035574.		0.02121510	280	3695482.
0.01750245	231	3048772.		0.02129086	281	3708680.
0.01757822	232	3061971.		0.02136663	282	3721878.
0.01765399	233	3075169.		0.02144240	283	3735076.
0.01772976	234	3088367.		0.02151817	284	3748274.
0.01780553	235	3101565.		0.02159394	285	3761472.
0.01788130	236	3114763.		0.02166971	286	3774671.
0.01795706	237	3127961.		0.02174547	287	3787869.
0.01803283	238	3141159.		0.02182124	288	3801067.
0.01810860	239	3154358.		0.02189701	289	3814265.
0.01818437	240	3167556.		0.02197278	290	3827463.
0.01826014	241	3180754.		0.02204855	291	3840661.
0.01833590	242	3193952.		0.02212431	292	3853860.
0.01841167	243	3207150.		0.02220008	293	3867058.
0.01848744	244	3220348.		0.02227585	294	3880256.
0.01856321	245	3233547.		0.02235162	295	3893454.
0.01863898	246	3246745.		0.02242739	296	3906652.
0.01871475	247	3259943.		0.02250316	297	3919850.
0.01879051	248	3273141.		0.02257892	298	3933048.
0.01886628	249	3286339.		0.02265469	299	3946247.
0.01894205	250	3299537.		0.02273046	300	3959445.

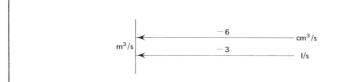

m³/s	REF	U.K. gal/min		m³/s	REF	U.K. gal/min
0.02280623	301	3972643.		0.02659464	351	4632550.
0.02288200	302	3985841.		0.02667041	352	4645748.
0.02295776	303	3999039.		0.02674617	353	4658947.
0.02303353	304	4012237.		0.02682194	354	4672145.
0.02310930	305	4025435.		0.02689771	355	4685343.
0.02318507	306	4038634.		0.02697348	356	4698541.
0.02326084	307	4051832.		0.02704925	357	4711739.
0.02333661	308	4065030.		0.02712502	358	4724937.
0.02341237	309	4078228.		0.02720078	359	4738136.
0.02348814	310	4091426.		0.02727655	360	4751334.
0.02356391	311	4104624.		0.02735232	361	4764532.
0.02363968	312	4117823.		0.02742809	362	4777730.
0.02371545	313	4131021.		0.02750386	363	4790928.
0.02379121	314	4144219.		0.02757962	364	4804126.
0.02386698	315	4157417.		0.02765539	365	4817324.
0.02394275	316	4170615.		0.02773116	366	4830523.
0.02401852	317	4183813.		0.02780693	367	4843721.
0.02409429	318	4197011.		0.02788270	368	4856919.
0.02417006	319	4210210.		0.02795847	369	4870117.
0.02424582	320	4223408.		0.02803423	370	4883315.
0.02432159	321	4236606.		0.02811000	371	4896513.
0.02439736	322	4249804.		0.02818577	372	4909711.
0.02447313	323	4263002.		0.02826154	373	4922910.
0.02454890	324	4276200.		0.02833731	374	4936108.
0.02462467	325	4289398.		0.02841308	375	4949306.
0.02470043	326	4302597.		0.02848884	376	4962504.
0.02477620	327	4315795.		0.02856461	377	4975702.
0.02485197	328	4328993.		0.02864038	378	4988900.
0.02492774	329	4342191.		0.02871615	379	5002099.
0.02500351	330	4355389.		0.02879192	380	5015297.
0.02507927	331	4368587.		0.02886768	381	5028495.
0.02515504	332	4381786.		0.02894345	382	5041693.
0.02523081	333	4394984.		0.02901922	383	5054891.
0.02530658	334	4408182.		0.02909499	384	5068089.
0.02538235	335	4421380.		0.02917076	385	5081287.
0.02545812	336	4434578.		0.02924653	386	5094486.
0.02553388	337	4447776.		0.02932229	387	5107684.
0.02560965	338	4460974.		0.02939806	388	5120882.
0.02568542	339	4474173.		0.02947383	389	5134080.
0.02576119	340	4487371.		0.02954960	390	5147278.
0.02583696	341	4500569.		0.02962537	391	5160476.
0.02591272	342	4513767.		0.02970113	392	5173674.
0.02598849	343	4526965.		0.02977690	393	5186873.
0.02606426	344	4540163.		0.02985267	394	5200071.
0.02614003	345	4553361.		0.02992844	395	5213269.
0.02621580	346	4566560.		0.03000421	396	5226467.
0.02629157	347	4579758.		0.03007998	397	5239665.
0.02636733	348	4592956.		0.03015574	398	5252863.
0.02644310	349	4606154.		0.03023151	399	5266062.
0.02651887	350	4619352.		0.03030728	400	5279260.

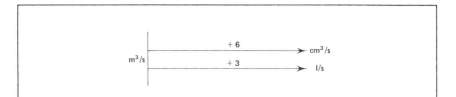

m³/s	REF	U.K. gal/min		m³/s	REF	U.K. gal/min
0.03038305	401	5292458.		0.03417146	451	5952365.
0.03045882	402	5305656.		0.03424723	452	5965563.
0.03053458	403	5318854.		0.03432299	453	5978762.
0.03061035	404	5332052.		0.03439876	454	5991960.
0.03068612	405	5345250.		0.03447453	455	6005158.
0.03076189	406	5358449.		0.03455030	456	6018356.
0.03083766	407	5371647.		0.03462607	457	6031554.
0.03091343	408	5384845.		0.03470184	458	6044752.
0.03098919	409	5398043.		0.03477760	459	6057950.
0.03106496	410	5411241.		0.03485337	460	6071149.
0.03114073	411	5424439.		0.03492914	461	6084347.
0.03121650	412	5437637.		0.03500491	462	6097545.
0.03129227	413	5450836.		0.03508068	463	6110743.
0.03136803	414	5464034.		0.03515644	464	6123941.
0.03144380	415	5477232.		0.03523221	465	6137139.
0.03151957	416	5490430.		0.03530798	466	6150337.
0.03159534	417	5503628.		0.03538375	467	6163536.
0.03167111	418	5516826.		0.03545952	468	6176734.
0.03174688	419	5530024.		0.03553529	469	6189932.
0.03182264	420	5543223.		0.03561105	470	6203130.
0.03189841	421	5556421.		0.03568682	471	6216328.
0.03197418	422	5569619.		0.03576259	472	6229526.
0.03204995	423	5582817.		0.03583836	473	6242725.
0.03212572	424	5596015.		0.03591413	474	6255923.
0.03220149	425	5609213.		0.03598989	475	6269121.
0.03227725	426	5622412.		0.03606566	476	6282319.
0.03235302	427	5635610.		0.03614143	477	6295517.
0.03242879	428	5648808.		0.03621720	478	6308715.
0.03250456	429	5662006.		0.03629297	479	6321913.
0.03258033	430	5675204.		0.03636874	480	6335112.
0.03265609	431	5688402.		0.03644450	481	6348310.
0.03273186	432	5701600.		0.03652027	482	6361508.
0.03280763	433	5714799.		0.03659604	483	6374706.
0.03288340	434	5727997.		0.03667181	484	6387904.
0.03295917	435	5741195.		0.03674758	485	6401102.
0.03303494	436	5754393.		0.03682335	486	6414300.
0.03311070	437	5767591.		0.03689911	487	6427499.
0.03318647	438	5780789.		0.03697488	488	6440697.
0.03326224	439	5793987.		0.03705065	489	6453895.
0.03333801	440	5807186.		0.03712642	490	6467093.
0.03341378	441	5820384.		0.03720219	491	6480291.
0.03348954	442	5833582.		0.03727795	492	6493489.
0.03356531	443	5846780.		0.03735372	493	6506688.
0.03364108	444	5859978.		0.03742949	494	6519886.
0.03371685	445	5873176.		0.03750526	495	6533084.
0.03379262	446	5886375.		0.03758103	496	6546282.
0.03386839	447	5899573.		0.03765680	497	6559480.
0.03394415	448	5912771.		0.03773256	498	6572678.
0.03401992	449	5925969.		0.03780833	499	6585876.
0.03409569	450	5939167.		0.03788410	500	6599075.

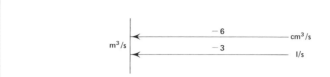

m³/s	REF	U.K. gal/min	m³/s	REF	U.K. gal/min
0.03795987	501	6612273.	0.04174828	551	7272180.
0.03803564	502	6625471.	0.04182405	552	7285378.
0.03811140	503	6638669.	0.04189981	553	7298576.
0.03818717	504	6651867.	0.04197558	554	7311775.
0.03826294	505	6665065.	0.04205135	555	7324973.
0.03833871	506	6678263.	0.04212712	556	7338171.
0.03841448	507	6691462.	0.04220289	557	7351369.
0.03849025	508	6704660.	0.04227866	558	7364567.
0.03856601	509	6717858.	0.04235442	559	7377765.
0.03864178	510	6731056.	0.04243019	560	7390963.
0.03871755	511	6744254.	0.04250596	561	7404162.
0.03879332	512	6757452.	0.04258173	562	7417360.
0.03886909	513	6770650.	0.04265750	563	7430558.
0.03894485	514	6783849.	0.04273326	564	7443756.
0.03902062	515	6797047.	0.04280903	565	7456954.
0.03909639	516	6810245.	0.04288480	566	7470152.
0.03917216	517	6823443.	0.04296057	567	7483351.
0.03924793	518	6836641.	0.04303634	568	7496549.
0.03932370	519	6849839.	0.04311211	569	7509747.
0.03939946	520	6863038.	0.04318787	570	7522945.
0.03947523	521	6876236.	0.04326364	571	7536143.
0.03955100	522	6889434.	0.04333941	572	7549341.
0.03962677	523	6902632.	0.04341518	573	7562539.
0.03970254	524	6915830.	0.04349095	574	7575738.
0.03977831	525	6929028.	0.04356671	575	7588936.
0.03985407	526	6942226.	0.04364248	576	7602134.
0.03992984	527	6955425.	0.04371825	577	7615332.
0.04000561	528	6968623.	0.04379402	578	7628530.
0.04008138	529	6981821.	0.04386979	579	7641728.
0.04015715	530	6995019.	0.04394556	580	7654926.
0.04023291	531	7008217.	0.04402132	581	7668125.
0.04030868	532	7021415.	0.04409709	582	7681323.
0.04038445	533	7034613.	0.04417286	583	7694521.
0.04046022	534	7047812.	0.04424863	584	7707719.
0.04053599	535	7061010.	0.04432440	585	7720917.
0.04061176	536	7074208.	0.04440017	586	7734115.
0.04068752	537	7087406.	0.04447593	587	7747314.
0.04076329	538	7100604.	0.04455170	588	7760512.
0.04083906	539	7113802.	0.04462747	589	7773710.
0.04091483	540	7127001.	0.04470324	590	7786908.
0.04099060	541	7140199.	0.04477901	591	7800106.
0.04106636	542	7153397.	0.04485477	592	7813304.
0.04114213	543	7166595.	0.04493054	593	7826502.
0.04121790	544	7179793.	0.04500631	594	7839701.
0.04129367	545	7192991.	0.04508208	595	7852899.
0.04136944	546	7206189.	0.04515785	596	7866097.
0.04144521	547	7219388.	0.04523362	597	7879295.
0.04152097	548	7232586.	0.04530938	598	7892493.
0.04159674	549	7245784.	0.04538515	599	7905691.
0.04167251	550	7258982.	0.04546092	600	7918889.

m³/s	REF	U.K. gal/min		m³/s	REF	U.K. gal/min
0·04553669	601	7932088·		0·04932510	651	8591995·
0·04561246	602	7945286·		0·04940087	652	8605193·
0·04568822	603	7958484·		0·04947663	653	8618391·
0·04576399	604	7971682·		0·04955240	654	8631590·
0·04583976	605	7984880·		0·04962817	655	8644788·
0·04591553	606	7998078·		0·04970394	656	8657986·
0·04599130	607	8011277·		0·04977971	657	8671184·
0·04606707	608	8024475·		0·04985548	658	8684382·
0·04614283	609	8037673·		0·04993124	659	8697580·
0·04621860	610	8050871·		0·05000701	660	8710778·
0·04629437	611	8064069·		0·05008278	661	8723977·
0·04637014	612	8077267·		0·05015855	662	8737175·
0·04644591	613	8090465·		0·05023432	663	8750373·
0·04652167	614	8103664·		0·05031009	664	8763571·
0·04659744	615	8116862·		0·05038585	665	8776769·
0·04667321	616	8130060·		0·05046162	666	8789967·
0·04674898	617	8143258·		0·05053739	667	8803165·
0·04682475	618	8156456·		0·05061316	668	8816364·
0·04690052	619	8169654·		0·05068893	669	8829562·
0·04705205	621	8196051·		0·05084046	671	8855958·
0·04712782	622	8209249·		0·05091623	672	8869156·
0·04720359	623	8222447·		0·05099200	673	8882354·
0·04727936	624	8235645·		0·05106777	674	8895553·
0·04735513	625	8248843·		0·05114353	675	8908751·
0·04743089	626	8262041·		0·05121930	676	8921949·
0·04750666	627	8275239·		0·05129507	677	8935147·
0·04758243	628	8288438·		0·05137084	678	8948345·
0·04765820	629	8301636·		0·05144661	679	8961543·
0·04773397	630	8314834·		0·05152238	680	8974741·
0·04780973	631	8328032·		0·05159814	681	8987940·
0·04788550	632	8341230·		0·05167391	682	9001138·
0·04796127	633	8354428·		0·05174968	683	9014336·
0·04803704	634	8367627·		0·05182545	684	9027534·
0·04811281	635	8380825·		0·05190122	685	9040732·
0·04818858	636	8394023·		0·05197699	686	9053930·
0·04826434	637	8407221·		0·05205275	687	9067128·
0·04834011	638	8420419·		0·05212852	688	9080327·
0·04841588	639	8433617·		0·05220429	689	9093525·
0·04849165	640	8446815·		0·05228006	690	9106723·
0·04856742	641	8460014·		0·05235583	691	9119921·
0·04864318	642	8473212·		0·05243159	692	9133119·
0·04871895	643	8486410·		0·05250736	693	9146317·
0·04879472	644	8499608·		0·05258313	694	9159516·
0·04887049	645	8512806·		0·05265890	695	9172714·
0·04894626	646	8526004·		0·05273467	696	9185912·
0·04902203	647	8539202·		0·05281044	697	9199110·
0·04909779	648	8552401·		0·05288620	698	9212308·
0·04917356	649	8565599·		0·05296197	699	9225506·
0·04924933	650	8578797·		0·05303774	700	9238704·

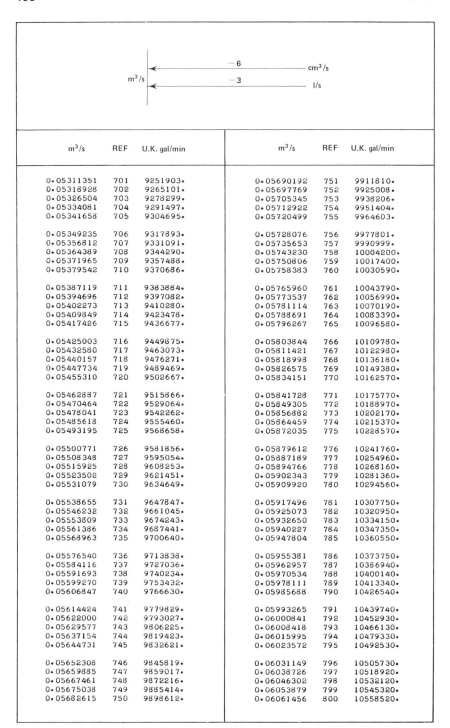

m³/s	REF	U.K. gal/min	m³/s	REF	U.K. gal/min
0.05311351	701	9251903.	0.05690192	751	9911810.
0.05318928	702	9265101.	0.05697769	752	9925008.
0.05326504	703	9278299.	0.05705345	753	9938206.
0.05334081	704	9291497.	0.05712922	754	9951404.
0.05341658	705	9304695.	0.05720499	755	9964603.
0.05349235	706	9317893.	0.05728076	756	9977801.
0.05356812	707	9331091.	0.05735653	757	9990999.
0.05364389	708	9344290.	0.05743230	758	10004200.
0.05371965	709	9357488.	0.05750806	759	10017400.
0.05379542	710	9370686.	0.05758383	760	10030590.
0.05387119	711	9383884.	0.05765960	761	10043790.
0.05394696	712	9397082.	0.05773537	762	10056990.
0.05402273	713	9410280.	0.05781114	763	10070190.
0.05409849	714	9423478.	0.05788691	764	10083390.
0.05417426	715	9436677.	0.05796267	765	10096580.
0.05425003	716	9449875.	0.05803844	766	10109780.
0.05432580	717	9463073.	0.05811421	767	10122980.
0.05440157	718	9476271.	0.05818998	768	10136180.
0.05447734	719	9489469.	0.05826575	769	10149380.
0.05455310	720	9502667.	0.05834151	770	10162570.
0.05462887	721	9515866.	0.05841728	771	10175770.
0.05470464	722	9529064.	0.05849305	772	10188970.
0.05478041	723	9542262.	0.05856882	773	10202170.
0.05485618	724	9555460.	0.05864459	774	10215370.
0.05493195	725	9568658.	0.05872035	775	10228570.
0.05500771	726	9581856.	0.05879612	776	10241760.
0.05508348	727	9595054.	0.05887189	777	10254960.
0.05515925	728	9608253.	0.05894766	778	10268160.
0.05523502	729	9621451.	0.05902343	779	10281360.
0.05531079	730	9634649.	0.05909920	780	10294560.
0.05538655	731	9647847.	0.05917496	781	10307750.
0.05546232	732	9661045.	0.05925073	782	10320950.
0.05553809	733	9674243.	0.05932650	783	10334150.
0.05561386	734	9687441.	0.05940227	784	10347350.
0.05568963	735	9700640.	0.05947804	785	10360550.
0.05576540	736	9713838.	0.05955381	786	10373750.
0.05584116	737	9727036.	0.05962957	787	10386940.
0.05591693	738	9740234.	0.05970534	788	10400140.
0.05599270	739	9753432.	0.05978111	789	10413340.
0.05606847	740	9766630.	0.05985688	790	10426540.
0.05614424	741	9779829.	0.05993265	791	10439740.
0.05622000	742	9793027.	0.06000841	792	10452930.
0.05629577	743	9806225.	0.06008418	793	10466130.
0.05637154	744	9819423.	0.06015995	794	10479330.
0.05644731	745	9832621.	0.06023572	795	10492530.
0.05652308	746	9845819.	0.06031149	796	10505730.
0.05659885	747	9859017.	0.06038726	797	10518920.
0.05667461	748	9872216.	0.06046302	798	10532120.
0.05675038	749	9885414.	0.06053879	799	10545320.
0.05682615	750	9898612.	0.06061456	800	10558520.

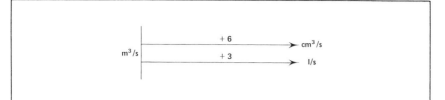

m³/s	REF	U.K. gal/min		m³/s	REF	U.K. gal/min
0.06069033	801	10571720.		0.06447874	851	11231620.
0.06076610	802	10584920.		0.06455451	852	11244820.
0.06084186	803	10598110.		0.06463027	853	11258020.
0.06091763	804	10611310.		0.06470604	854	11271220.
0.06099340	805	10624510.		0.06478181	855	11284420.
0.06106917	806	10637710.		0.06485758	856	11297620.
0.06114494	807	10650910.		0.06493335	857	11310810.
0.06122071	808	10664100.		0.06500912	858	11324010.
0.06129647	809	10677300.		0.06508488	859	11337210.
0.06137224	810	10690500.		0.06516065	860	11350410.
0.06144801	811	10703700.		0.06523642	861	11363610.
0.06152378	812	10716900.		0.06531219	862	11376800.
0.06159955	813	10730100.		0.06538796	863	11390000.
0.06167531	814	10743290.		0.06546372	864	11403200.
0.06175108	815	10756490.		0.06553949	865	11416400.
0.06182685	816	10769690.		0.06561526	866	11429600.
0.06190262	817	10782890.		0.06569103	867	11442800.
0.06197839	818	10796090.		0.06576680	868	11455990.
0.06205416	819	10809280.		0.06584257	869	11469190.
0.06212992	820	10822480.		0.06591833	870	11482390.
0.06220569	821	10835680.		0.06599410	871	11495590.
0.06228146	822	10848880.		0.06606987	872	11508790.
0.06235723	823	10862080.		0.06614564	873	11521980.
0.06243300	824	10875270.		0.06622141	874	11535180.
0.06250876	825	10888470.		0.06629718	875	11548380.
0.06258453	826	10901670.		0.06637294	876	11561580.
0.06266030	827	10914870.		0.06644871	877	11574780.
0.06273607	828	10928070.		0.06652448	878	11587970.
0.06281184	829	10941270.		0.06660025	879	11601170.
0.06288761	830	10954460.		0.06667602	880	11614370.
0.06296337	831	10967660.		0.06675178	881	11627570.
0.06303914	832	10980860.		0.06682755	882	11640770.
0.06311491	833	10994060.		0.06690332	883	11653970.
0.06319068	834	11007260.		0.06697909	884	11667160.
0.06326645	835	11020450.		0.06705486	885	11680360.
0.06334222	836	11033650.		0.06713063	886	11693560.
0.06341798	837	11046850.		0.06720639	887	11706760.
0.06349375	838	11060050.		0.06728216	888	11719960.
0.06356952	839	11073250.		0.06735793	889	11733150.
0.06364529	840	11086450.		0.06743370	890	11746350.
0.06372106	841	11099640.		0.06750947	891	11759550.
0.06379682	842	11112840.		0.06758523	892	11772750.
0.06387259	843	11126040.		0.06766100	893	11785950.
0.06394836	844	11139240.		0.06773677	894	11799150.
0.06402413	845	11152440.		0.06781254	895	11812340.
0.06409990	846	11165630.		0.06788831	896	11825540.
0.06417567	847	11178830.		0.06796408	897	11838740.
0.06425143	848	11192030.		0.06803984	898	11851940.
0.06432720	849	11205230.		0.06811561	899	11865140.
0.06440297	850	11218430.		0.06819138	900	11878330.

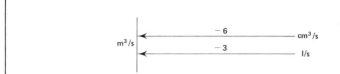

m³/s	REF	U.K. gal/min		m³/s	REF	U.K. gal/min
0.06826715	901	11891530.		0.07205556	951	12551440.
0.06834292	902	11904730.		0.07213133	952	12564640.
0.06841868	903	11917930.		0.07220710	953	12577840.
0.06849445	904	11931130.		0.07228286	954	12591030.
0.06857022	905	11944320.		0.07235863	955	12604230.
0.06864599	906	11957520.		0.07243440	956	12617430.
0.06872176	907	11970720.		0.07251017	957	12630630.
0.06879753	908	11983920.		0.07258594	958	12643830.
0.06887329	909	11997120.		0.07266170	959	12657020.
0.06894906	910	12010320.		0.07273747	960	12670220.
0.06902483	911	12023510.		0.07281324	961	12683420.
0.06910060	912	12036710.		0.07288901	962	12696620.
0.06917637	913	12049910.		0.07296478	963	12709820.
0.06925214	914	12063110.		0.07304054	964	12723020.
0.06932790	915	12076310.		0.07311631	965	12736210.
0.06940367	916	12089500.		0.07319208	966	12749410.
0.06947944	917	12102700.		0.07326785	967	12762610.
0.06955521	918	12115900.		0.07334362	968	12775810.
0.06963098	919	12129100.		0.07341939	969	12789010.
0.06970674	920	12142300.		0.07349515	970	12802200.
0.06978251	921	12155500.		0.07357092	971	12815400.
0.06985828	922	12168690.		0.07364669	972	12828600.
0.06993405	923	12181890.		0.07372246	973	12841800.
0.07000982	924	12195090.		0.07379823	974	12855000.
0.07008559	925	12208290.		0.07387399	975	12868200.
0.07016135	926	12221490.		0.07394976	976	12881390.
0.07023712	927	12234680.		0.07402553	977	12894590.
0.07031289	928	12247880.		0.07410130	978	12907790.
0.07038866	929	12261080.		0.07417707	979	12920990.
0.07046443	930	12274280.		0.07425284	980	12934190.
0.07054019	931	12287480.		0.07432860	981	12947380.
0.07061596	932	12300670.		0.07440437	982	12960580.
0.07069173	933	12313870.		0.07448014	983	12973780.
0.07076750	934	12327070.		0.07455591	984	12986980.
0.07084327	935	12340270.		0.07463168	985	13000180.
0.07091904	936	12353470.		0.07470745	986	13013370.
0.07099480	937	12366670.		0.07478321	987	13026570.
0.07107057	938	12379860.		0.07485898	988	13039770.
0.07114634	939	12393060.		0.07493475	989	13052970.
0.07122211	940	12406260.		0.07501052	990	13066170.
0.07129788	941	12419460.		0.07508629	991	13079370.
0.07137364	942	12432660.		0.07516205	992	13092560.
0.07144941	943	12445850.		0.07523782	993	13105760.
0.07152518	944	12459050.		0.07531359	994	13118960.
0.07160095	945	12472250.		0.07538936	995	13132160.
0.07167672	946	12485450.		0.07546513	996	13145360.
0.07175249	947	12498650.		0.07554090	997	13158550.
0.07182825	948	12511850.		0.07561666	998	13171750.
0.07190402	949	12525040.		0.07569243	999	13184950.
0.07197979	950	12538240.		0.07576820	1000	13198150.

19

MASS

<u>Pound (mass) — Kilogram</u>
         (lbm — kg)
Pound (mass) — Gram
         (lbm — g)
Pound (mass) — Metric ton
         (lbm — t)
Hundredweight — Kilogram
         (cwt — kg)

MASS FLOW RATE

<u>Pound (mass)/Second — Kilogram/Second</u>
         (lbm/s — kg/s)
Pound (mass)/Second — Gram/Second
         (lbm/s — g/s)
<u>Pound (force) — Kilogram (force)</u>   [English to older MKS Force]
         (lbf — kgf)

Conversion Factors:
         1 pound (mass avoirdupois) (lbm) = 0.4535924 kilogram (kg)
                    (1 kg = 2.204622 lbm)

Mass Flow Rate Values

Table 19 can represent mass flow rate values as long as the time units are the same for both English and SI conversion quantities. Besides per second shown above, these consistent time units can also be in per minute (min), per hour (h), etc.

Mass Flow Rate Conversion Factor:

1 lbm/s = 0.4535924 kg/s.

Pound-Kilogram Force Conversions

Table 19 can also be used, when necessary, to convert force values between English and the MKSA (Metre-Kilogram-Second-Ampere) System. The SI system is a modernization of the MKSA system of measurement.

1 Pound-force (lbf) = 0.4535924 Kilogram-force (kgf)

The SI base unit for force is the "newton." Table 28 gives conversion values between English and SI force units. See Introduction for explanation of SI force and mass designations. In SI, the use of the "kilogram" (kg) is confined to mass units.

Hundredweight-kilogram Relation

The short hundredweight (British) is equal to 100 pounds (mass) — avoirdupois.

Therefore its conversion factor is:

1 cwt (short) = 45.35924 kg.

This conversion factor differs from the above only by a power of 10. Table 19 can then be used for conversions of cwt (short) to and from SI by properly shifting decimal places in the values obtained from the table. The following Decimal-shift Diagram 19-A will help:

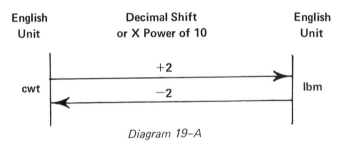

Diagram 19-A

Example (a): Convert 2.67 cwt to kg.
From Diagram 19-A: 2.67 cwt = 2.67 × 10^2 lbm = 267 lbm
From table: 267 lbm = 121.1092 kg* ≈ 121 kg

Example (b): Convert 17,550 kg to cwt.
17,550 kg = 17.55 × 10^3 kg
17.55 kg = (17.00 + 0.55) kg

---

*All conversion values obtained can be rounded to desired number of significant figures.

From table: 17 kg = 37.47858 lbm*
                    55 kg = 121.2542 lbm*

By shifting decimals where necessary and adding:
    17.00 kg = 37.47858   lbm
    $\underline{\text{0.55 kg} = \;\;\underline{1.212542}\;\text{lbm}}$
    17.55 kg = 38.691122 lbm
Therefore: 17.55×10^3 kg = 38.69112×10^3 lbm = 38,691.12 lbm
From Diagram 19-A: 38,691.12 lbm = $38,691.12 \times 10^{-2}$ cwt
                    = 386.9112 cwt ≈ 386.9 cwt

Metric Ton (Tonne) Values

The Metric Ton or tonne (t) is equal to 1000 kilograms or:

$$1 \text{ tonne (t)} = 1 \text{ megagram (Mg)}$$

In converting units containing kg (mass) to g or t, Diagram 19-B, which appears in the headings on alternate pages of this table, will assist the user. It is derived from: $1\text{kg} = (10^3)\,\text{g} = (10^{-3})\,\text{t}$.

Diagram 19-B and Diagram 19-C apply to mass flow rate and kilogram and gram force designations of the older MKSA system of measure.

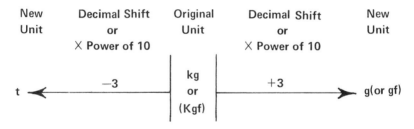

| New Unit | Decimal Shift or X Power of 10 | Original Unit | Decimal Shift or X Power of 10 | New Unit |

Diagram 19-B

Example (*c*): Convert 35,700 lbm to tonnes (metric tons).
    35,700 lbm = 357×10^2 lbm
    From table: 357 lbm = 161.9325 kg*
    Therefore: 357×10^2 lbm = 161.9325×10^2 kg = 16,193.25 kg
    From Diagram 19-B: 16,193.25 kg = $16,193.25\,(10^{-3})$ t = 16.19325 t
                    ≈ 16.2 t
                    (kg to t by multiplying t value by 10^{-3} or shifting its decimal 3 places to the left.)

Diagram 19-C is also given at the top of alternate table pages.

_____

*All conversion values obtained can be rounded to desired number of significant figures.

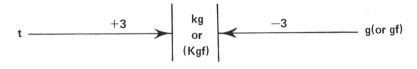

Diagram 19–C

It is the opposite of Diagram 19-B and is used to convert g or gf and t values to kg or kgf. The kg (mass or force) equivalent is then applied to the REF column in the table to obtain lb (mass or force).

Example (*d*): Convert 777 g to lbm.
  From Diagram 19-C: $777 \text{ g} = 777 \times 10^{-3} \text{ kg}$
  From table: 777 kg = 1,712.992 lbm*
  Therefore: $777 \times 10^{-3} \text{ kg} = 1{,}712.992 \times 10^{-3} \text{ lbm} = 1.712992 \text{ lbm}$
       $\approx 1.71 \text{ lbm}$

---

*All conversion values obtained can be rounded to desired number of significant figures.

kg kg/s kgf	REF	lbm lbm/s lbf	kg kg/s kgf	REF	lbm lbm/s lbf
0.4535924	1	2.204622	23.13321	51	112.4357
0.9071848	2	4.409245	23.58680	52	114.6404
1.360777	3	6.613867	24.04040	53	116.8450
1.814370	4	8.818490	24.49399	54	119.0496
2.267962	5	11.02311	24.94758	55	121.2542
2.721554	6	13.22773	25.40117	56	123.4589
3.175147	7	15.43236	25.85477	57	125.6635
3.628739	8	17.63698	26.30836	58	127.8681
4.082332	9	19.84160	26.76195	59	130.0727
4.535924	10	22.04622	27.21554	60	132.2773
4.989516	11	24.25085	27.66914	61	134.4820
5.443109	12	26.45547	28.12273	62	136.6866
5.896701	13	28.66009	28.57632	63	138.8912
6.350294	14	30.86471	29.02991	64	141.0958
6.803886	15	33.06934	29.48351	65	143.3005
7.257478	16	35.27396	29.93710	66	145.5051
7.711071	17	37.47858	30.39069	67	147.7097
8.164663	18	39.68320	30.84428	68	149.9143
8.618256	19	41.88783	31.29788	69	152.1190
9.071848	20	44.09245	31.75147	70	154.3236
9.525440	21	46.29707	32.20506	71	156.5282
9.979033	22	48.50169	32.65865	72	158.7328
10.43263	23	50.70632	33.11225	73	160.9374
10.88622	24	52.91094	33.56584	74	163.1421
11.33981	25	55.11556	34.01943	75	165.3467
11.79340	26	57.32018	34.47302	76	167.5513
12.24699	27	59.52481	34.92661	77	169.7559
12.70059	28	61.72943	35.38021	78	171.9606
13.15418	29	63.93405	35.83380	79	174.1652
13.60777	30	66.13867	36.28739	80	176.3698
14.06136	31	68.34330	36.74098	81	178.5744
14.51496	32	70.54792	37.19458	82	180.7790
14.96855	33	72.75254	37.64817	83	182.9837
15.42214	34	74.95716	38.10176	84	185.1883
15.87573	35	77.16179	38.55535	85	187.3929
16.32933	36	79.36641	39.00895	86	189.5975
16.78292	37	81.57103	39.46254	87	191.8022
17.23651	38	83.77565	39.91613	88	194.0068
17.69010	39	85.98028	40.36972	89	196.2114
18.14370	40	88.18490	40.82332	90	198.4160
18.59729	41	90.38952	41.27691	91	200.6206
19.05088	42	92.59414	41.73050	92	202.8253
19.50447	43	94.79877	42.18409	93	205.0299
19.95807	44	97.00339	42.63769	94	207.2345
20.41166	45	99.20801	43.09128	95	209.4391
20.86525	46	101.4126	43.54487	96	211.6438
21.31884	47	103.6173	43.99846	97	213.8484
21.77244	48	105.8219	44.45206	98	216.0530
22.22603	49	108.0265	44.90565	99	218.2576
22.67962	50	110.2311	45.35924	100	220.4622

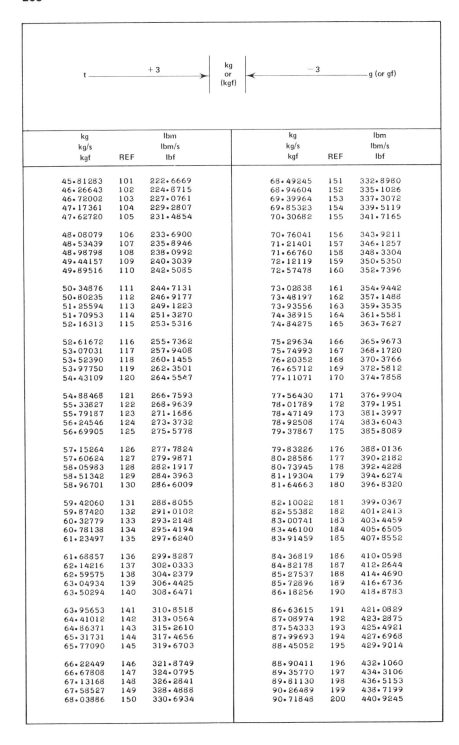

kg kg/s kgf	REF	lbm lbm/s lbf	kg kg/s kgf	REF	lbm lbm/s lbf
45.81283	101	222.6669	68.49245	151	332.8980
46.26643	102	224.8715	68.94604	152	335.1026
46.72002	103	227.0761	69.39964	153	337.3072
47.17361	104	229.2807	69.85323	154	339.5119
47.62720	105	231.4854	70.30682	155	341.7165
48.08079	106	233.6900	70.76041	156	343.9211
48.53439	107	235.8946	71.21401	157	346.1257
48.98798	108	238.0992	71.66760	158	348.3304
49.44157	109	240.3039	72.12119	159	350.5350
49.89516	110	242.5085	72.57478	160	352.7396
50.34876	111	244.7131	73.02838	161	354.9442
50.80235	112	246.9177	73.48197	162	357.1488
51.25594	113	249.1223	73.93556	163	359.3535
51.70953	114	251.3270	74.38915	164	361.5581
52.16313	115	253.5316	74.84275	165	363.7627
52.61672	116	255.7362	75.29634	166	365.9673
53.07031	117	257.9408	75.74993	167	368.1720
53.52390	118	260.1455	76.20352	168	370.3766
53.97750	119	262.3501	76.65712	169	372.5812
54.43109	120	264.5547	77.11071	170	374.7858
54.88468	121	266.7593	77.56430	171	376.9904
55.33827	122	268.9639	78.01789	172	379.1951
55.79187	123	271.1686	78.47149	173	381.3997
56.24546	124	273.3732	78.92508	174	383.6043
56.69905	125	275.5778	79.37867	175	385.8089
57.15264	126	277.7824	79.83226	176	388.0136
57.60624	127	279.9871	80.28586	177	390.2182
58.05983	128	282.1917	80.73945	178	392.4228
58.51342	129	284.3963	81.19304	179	394.6274
58.96701	130	286.6009	81.64663	180	396.8320
59.42060	131	288.8055	82.10022	181	399.0367
59.87420	132	291.0102	82.55382	182	401.2413
60.32779	133	293.2148	83.00741	183	403.4459
60.78138	134	295.4194	83.46100	184	405.6505
61.23497	135	297.6240	83.91459	185	407.8552
61.68857	136	299.8287	84.36819	186	410.0598
62.14216	137	302.0333	84.82178	187	412.2644
62.59575	138	304.2379	85.27537	188	414.4690
63.04934	139	306.4425	85.72896	189	416.6736
63.50294	140	308.6471	86.18256	190	418.8783
63.95653	141	310.8518	86.63615	191	421.0829
64.41012	142	313.0564	87.08974	192	423.2875
64.86371	143	315.2610	87.54333	193	425.4921
65.31731	144	317.4656	87.99693	194	427.6968
65.77090	145	319.6703	88.45052	195	429.9014
66.22449	146	321.8749	88.90411	196	432.1060
66.67808	147	324.0795	89.35770	197	434.3106
67.13168	148	326.2841	89.81130	198	436.5153
67.58527	149	328.4888	90.26489	199	438.7199
68.03886	150	330.6934	90.71848	200	440.9245

t ←————— −3 —————| kg or (kgf) |————— +3 —————→ g (or gf)

kg kg/s kgf	REF	lbm lbm/s lbf		kg kg/s kgf	REF	lbm lbm/s lbf
91•17207	201	443•1291		113•8517	251	553•3602
91•62566	202	445•3337		114•3053	252	555•5649
92•07926	203	447•5384		114•7589	253	557•7695
92•53285	204	449•7430		115•2125	254	559•9741
92•98644	205	451•9476		115•6661	255	562•1787
93•44003	206	454•1522		116•1197	256	564•3834
93•89363	207	456•3569		116•5732	257	566•5880
94•34722	208	458•5615		117•0268	258	568•7926
94•80081	209	460•7661		117•4804	259	570•9972
95•25440	210	462•9707		117•9340	260	573•2018
95•70800	211	465•1753		118•3876	261	575•4065
96•16159	212	467•3800		118•8412	262	577•6111
96•61518	213	469•5846		119•2948	263	579•8157
97•06877	214	471•7892		119•7484	264	582•0203
97•52237	215	473•9938		120•2020	265	584•2250
97•97596	216	476•1985		120•6556	266	586•4296
98•42955	217	478•4031		121•1092	267	588•6342
98•88314	218	480•6077		121•5628	268	590•8388
99•33674	219	482•8123		122•0164	269	593•0434
99•79033	220	485•0169		122•4699	270	595•2481
100•2439	221	487•2216		122•9235	271	597•4527
100•6975	222	489•4262		123•3771	272	599•6573
101•1511	223	491•6308		123•8307	273	601•8619
101•6047	224	493•8354		124•2843	274	604•0666
102•0583	225	496•0401		124•7379	275	606•2712
102•5119	226	498•2447		125•1915	276	608•4758
102•9655	227	500•4493		125•6451	277	610•6804
103•4191	228	502•6539		126•0987	278	612•8850
103•8727	229	504•8585		126•5523	279	615•0897
104•3263	230	507•0632		127•0059	280	617•2943
104•7798	231	509•2678		127•4595	281	619•4989
105•2334	232	511•4724		127•9131	282	621•7035
105•6870	233	513•6770		128•3666	283	623•9082
106•1406	234	515•8817		128•8202	284	626•1128
106•5942	235	518•0863		129•2738	285	628•3174
107•0478	236	520•2909		129•7274	286	630•5220
107•5014	237	522•4955		130•1810	287	632•7267
107•9550	238	524•7001		130•6346	288	634•9313
108•4086	239	526•9048		131•0882	289	637•1359
108•8622	240	529•1094		131•5418	290	639•3405
109•3158	241	531•3140		131•9954	291	641•5451
109•7694	242	533•5186		132•4490	292	643•7498
110•2230	243	535•7233		132•9026	293	645•9544
110•6765	244	537•9279		133•3562	294	648•1590
111•1301	245	540•1325		133•8098	295	650•3636
111•5837	246	542•3371		134•2634	296	652•5683
112•0373	247	544•5418		134•7169	297	654•7729
112•4909	248	546•7464		135•1705	298	656•9775
112•9445	249	548•9510		135•6241	299	659•1821
113•3981	250	551•1556		136•0777	300	661•3867

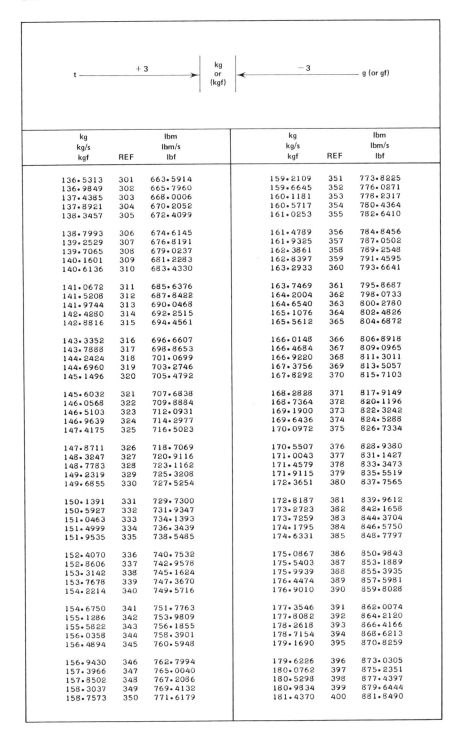

kg kg/s kgf	REF	lbm lbm/s lbf	kg kg/s kgf	REF	lbm lbm/s lbf
136.5313	301	663.5914	159.2109	351	773.8225
136.9849	302	665.7960	159.6645	352	776.0271
137.4385	303	668.0006	160.1181	353	778.2317
137.8921	304	670.2052	160.5717	354	780.4364
138.3457	305	672.4099	161.0253	355	782.6410
138.7993	306	674.6145	161.4789	356	784.8456
139.2529	307	676.8191	161.9325	357	787.0502
139.7065	308	679.0237	162.3861	358	789.2548
140.1601	309	681.2283	162.8397	359	791.4595
140.6136	310	683.4330	163.2933	360	793.6641
141.0672	311	685.6376	163.7469	361	795.8687
141.5208	312	687.8422	164.2004	362	798.0733
141.9744	313	690.0468	164.6540	363	800.2780
142.4280	314	692.2515	165.1076	364	802.4826
142.8816	315	694.4561	165.5612	365	804.6872
143.3352	316	696.6607	166.0148	366	806.8918
143.7888	317	698.8653	166.4684	367	809.0965
144.2424	318	701.0699	166.9220	368	811.3011
144.6960	319	703.2746	167.3756	369	813.5057
145.1496	320	705.4792	167.8292	370	815.7103
145.6032	321	707.6838	168.2828	371	817.9149
146.0568	322	709.8884	168.7364	372	820.1196
146.5103	323	712.0931	169.1900	373	822.3242
146.9639	324	714.2977	169.6436	374	824.5288
147.4175	325	716.5023	170.0972	375	826.7334
147.8711	326	718.7069	170.5507	376	828.9380
148.3247	327	720.9116	171.0043	377	831.1427
148.7783	328	723.1162	171.4579	378	833.3473
149.2319	329	725.3208	171.9115	379	835.5519
149.6855	330	727.5254	172.3651	380	837.7565
150.1391	331	729.7300	172.8187	381	839.9612
150.5927	332	731.9347	173.2723	382	842.1658
151.0463	333	734.1393	173.7259	383	844.3704
151.4999	334	736.3439	174.1795	384	846.5750
151.9535	335	738.5485	174.6331	385	848.7797
152.4070	336	740.7532	175.0867	386	850.9843
152.8606	337	742.9578	175.5403	387	853.1889
153.3142	338	745.1624	175.9939	388	855.3935
153.7678	339	747.3670	176.4474	389	857.5981
154.2214	340	749.5716	176.9010	390	859.8028
154.6750	341	751.7763	177.3546	391	862.0074
155.1286	342	753.9809	177.8082	392	864.2120
155.5822	343	756.1855	178.2618	393	866.4166
156.0358	344	758.3901	178.7154	394	868.6213
156.4894	345	760.5948	179.1690	395	870.8259
156.9430	346	762.7994	179.6226	396	873.0305
157.3966	347	765.0040	180.0762	397	875.2351
157.8502	348	767.2086	180.5298	398	877.4397
158.3037	349	769.4132	180.9834	399	879.6444
158.7573	350	771.6179	181.4370	400	881.8490

t ←	−3	kg or (kgf)	+3	→ g (or gf)

kg kg/s kgf	REF	lbm lbm/s lbf	kg kg/s kgf	REF	lbm lbm/s lbf
181.8906	401	884.0536	204.5702	451	994.2847
182.3441	402	886.2582	205.0238	452	996.4894
182.7977	403	888.4629	205.4774	453	998.6940
183.2513	404	890.6675	205.9310	454	1000.899
183.7049	405	892.8721	206.3845	455	1003.103
184.1585	406	895.0767	206.8381	456	1005.308
184.6121	407	897.2813	207.2917	457	1007.512
185.0657	408	899.4860	207.7453	458	1009.717
185.5193	409	901.6906	208.1989	459	1011.922
185.9729	410	903.8952	208.6525	460	1014.126
186.4265	411	906.0998	209.1061	461	1016.331
186.8801	412	908.3045	209.5597	462	1018.536
187.3337	413	910.5091	210.0133	463	1020.740
187.7873	414	912.7137	210.4669	464	1022.945
188.2408	415	914.9183	210.9205	465	1025.149
188.6944	416	917.1229	211.3741	466	1027.354
189.1480	417	919.3276	211.8277	467	1029.559
189.6016	418	921.5322	212.2812	468	1031.763
190.0552	419	923.7368	212.7348	469	1033.968
190.5088	420	925.9414	213.1884	470	1036.173
190.9624	421	928.1461	213.6420	471	1038.377
191.4160	422	930.3507	214.0956	472	1040.582
191.8696	423	932.5553	214.5492	473	1042.786
192.3232	424	934.7599	215.0028	474	1044.991
192.7768	425	936.9646	215.4564	475	1047.196
193.2304	426	939.1692	215.9100	476	1049.400
193.6840	427	941.3738	216.3636	477	1051.605
194.1375	428	943.5784	216.8172	478	1053.810
194.5911	429	945.7830	217.2708	479	1056.014
195.0447	430	947.9877	217.7244	480	1058.219
195.4983	431	950.1923	218.1779	481	1060.423
195.9519	432	952.3969	218.6315	482	1062.628
196.4055	433	954.6015	219.0851	483	1064.833
196.8591	434	956.8062	219.5387	484	1067.037
197.3127	435	959.0108	219.9923	485	1069.242
197.7663	436	961.2154	220.4459	486	1071.447
198.2199	437	963.4200	220.8995	487	1073.651
198.6735	438	965.6246	221.3531	488	1075.856
199.1271	439	967.8293	221.8067	489	1078.060
199.5807	440	970.0339	222.2603	490	1080.265
200.0342	441	972.2385	222.7139	491	1082.470
200.4878	442	974.4431	223.1675	492	1084.674
200.9414	443	976.6478	223.6211	493	1086.879
201.3950	444	978.8524	224.0746	494	1089.084
201.8486	445	981.0570	224.5282	495	1091.288
202.3022	446	983.2616	224.9818	496	1093.493
202.7558	447	985.4662	225.4354	497	1095.697
203.2094	448	987.6709	225.8890	498	1097.902
203.6630	449	989.8755	226.3426	499	1100.107
204.1166	450	992.0801	226.7962	500	1102.311

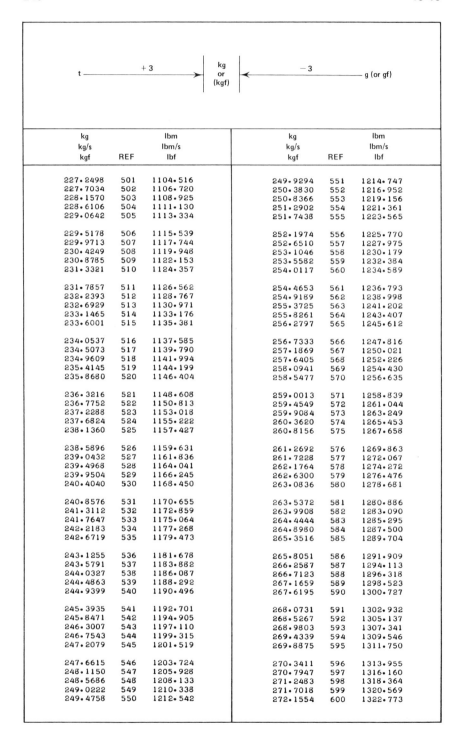

| kg | | lbm | | kg | | lbm |
| kg/s | | lbm/s | | kg/s | | lbm/s |
kgf	REF	lbf		kgf	REF	lbf
227.2498	501	1104.516		249.9294	551	1214.747
227.7034	502	1106.720		250.3830	552	1216.952
228.1570	503	1108.925		250.8366	553	1219.156
228.6106	504	1111.130		251.2902	554	1221.361
229.0642	505	1113.334		251.7438	555	1223.565
229.5178	506	1115.539		252.1974	556	1225.770
229.9713	507	1117.744		252.6510	557	1227.975
230.4249	508	1119.948		253.1046	558	1230.179
230.8785	509	1122.153		253.5582	559	1232.384
231.3321	510	1124.357		254.0117	560	1234.589
231.7857	511	1126.562		254.4653	561	1236.793
232.2393	512	1128.767		254.9189	562	1238.998
232.6929	513	1130.971		255.3725	563	1241.202
233.1465	514	1133.176		255.8261	564	1243.407
233.6001	515	1135.381		256.2797	565	1245.612
234.0537	516	1137.585		256.7333	566	1247.816
234.5073	517	1139.790		257.1869	567	1250.021
234.9609	518	1141.994		257.6405	568	1252.226
235.4145	519	1144.199		258.0941	569	1254.430
235.8680	520	1146.404		258.5477	570	1256.635
236.3216	521	1148.608		259.0013	571	1258.839
236.7752	522	1150.813		259.4549	572	1261.044
237.2288	523	1153.018		259.9084	573	1263.249
237.6824	524	1155.222		260.3620	574	1265.453
238.1360	525	1157.427		260.8156	575	1267.658
238.5896	526	1159.631		261.2692	576	1269.863
239.0432	527	1161.836		261.7228	577	1272.067
239.4968	528	1164.041		262.1764	578	1274.272
239.9504	529	1166.245		262.6300	579	1276.476
240.4040	530	1168.450		263.0836	580	1278.681
240.8576	531	1170.655		263.5372	581	1280.886
241.3112	532	1172.859		263.9908	582	1283.090
241.7647	533	1175.064		264.4444	583	1285.295
242.2183	534	1177.268		264.8980	584	1287.500
242.6719	535	1179.473		265.3516	585	1289.704
243.1255	536	1181.678		265.8051	586	1291.909
243.5791	537	1183.882		266.2587	587	1294.113
244.0327	538	1186.087		266.7123	588	1296.318
244.4863	539	1188.292		267.1659	589	1298.523
244.9399	540	1190.496		267.6195	590	1300.727
245.3935	541	1192.701		268.0731	591	1302.932
245.8471	542	1194.905		268.5267	592	1305.137
246.3007	543	1197.110		268.9803	593	1307.341
246.7543	544	1199.315		269.4339	594	1309.546
247.2079	545	1801.519		269.8875	595	1311.750
247.6615	546	1203.724		270.3411	596	1313.955
248.1150	547	1205.928		270.7947	597	1316.160
248.5686	548	1208.133		271.2483	598	1318.364
249.0222	549	1210.338		271.7018	599	1320.569
249.4758	550	1212.542		272.1554	600	1322.773

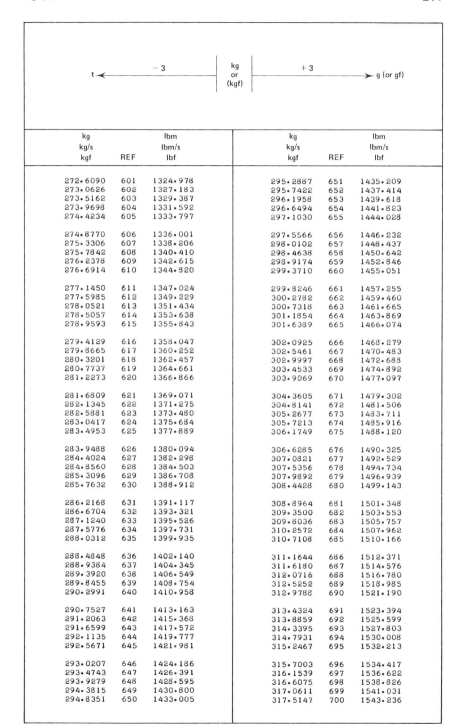

kg kg/s kgf	REF	lbm lbm/s lbf	kg kg/s kgf	REF	lbm lbm/s lbf
272.6090	601	1324.978	295.2887	651	1435.209
273.0626	602	1327.183	295.7422	652	1437.414
273.5162	603	1329.387	296.1958	653	1439.618
273.9698	604	1331.592	296.6494	654	1441.823
274.4234	605	1333.797	297.1030	655	1444.028
274.8770	606	1336.001	297.5566	656	1446.232
275.3306	607	1338.206	298.0102	657	1448.437
275.7842	608	1340.410	298.4638	658	1450.642
276.2378	609	1342.615	298.9174	659	1452.846
276.6914	610	1344.820	299.3710	660	1455.051
277.1450	611	1347.024	299.8246	661	1457.255
277.5985	612	1349.229	300.2782	662	1459.460
278.0521	613	1351.434	300.7318	663	1461.665
278.5057	614	1353.638	301.1854	664	1463.869
278.9593	615	1355.843	301.6389	665	1466.074
279.4129	616	1358.047	302.0925	666	1468.279
279.8665	617	1360.252	302.5461	667	1470.483
280.3201	618	1362.457	302.9997	668	1472.688
280.7737	619	1364.661	303.4533	669	1474.892
281.2273	620	1366.866	303.9069	670	1477.097
281.6809	621	1369.071	304.3605	671	1479.302
282.1345	622	1371.275	304.8141	672	1481.506
282.5881	623	1373.480	305.2677	673	1483.711
283.0417	624	1375.684	305.7213	674	1485.916
283.4953	625	1377.889	306.1749	675	1488.120
283.9488	626	1380.094	306.6285	676	1490.325
284.4024	627	1382.298	307.0821	677	1492.529
284.8560	628	1384.503	307.5356	678	1494.734
285.3096	629	1386.708	307.9892	679	1496.939
285.7632	630	1388.912	308.4428	680	1499.143
286.2168	631	1391.117	308.8964	681	1501.348
286.6704	632	1393.321	309.3500	682	1503.553
287.1240	633	1395.526	309.8036	683	1505.757
287.5776	634	1397.731	310.2572	684	1507.962
288.0312	635	1399.935	310.7108	685	1510.166
288.4848	636	1402.140	311.1644	686	1512.371
288.9384	637	1404.345	311.6180	687	1514.576
289.3920	638	1406.549	312.0716	688	1516.780
289.8455	639	1408.754	312.5252	689	1518.985
290.2991	640	1410.958	312.9788	690	1521.190
290.7527	641	1413.163	313.4324	691	1523.394
291.2063	642	1415.368	313.8859	692	1525.599
291.6599	643	1417.572	314.3395	693	1527.803
292.1135	644	1419.777	314.7931	694	1530.008
292.5671	645	1421.981	315.2467	695	1532.213
293.0207	646	1424.186	315.7003	696	1534.417
293.4743	647	1426.391	316.1539	697	1536.622
293.9279	648	1428.595	316.6075	698	1538.826
294.3815	649	1430.800	317.0611	699	1541.031
294.8351	650	1433.005	317.5147	700	1543.236

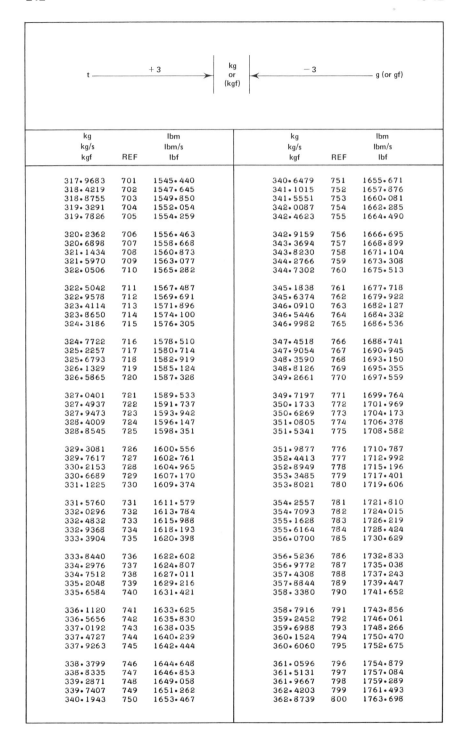

kg kg/s kgf	REF	lbm lbm/s lbf	kg kg/s kgf	REF	lbm lbm/s lbf
317.9683	701	1545.440	340.6479	751	1655.671
318.4219	702	1547.645	341.1015	752	1657.876
318.8755	703	1549.850	341.5551	753	1660.081
319.3291	704	1552.054	342.0087	754	1662.285
319.7826	705	1554.259	342.4623	755	1664.490
320.2362	706	1556.463	342.9159	756	1666.695
320.6898	707	1558.668	343.3694	757	1668.899
321.1434	708	1560.873	343.8230	758	1671.104
321.5970	709	1563.077	344.2766	759	1673.308
322.0506	710	1565.282	344.7302	760	1675.513
322.5042	711	1567.487	345.1838	761	1677.718
322.9578	712	1569.691	345.6374	762	1679.922
323.4114	713	1571.896	346.0910	763	1682.127
323.8650	714	1574.100	346.5446	764	1684.332
324.3186	715	1576.305	346.9982	765	1686.536
324.7722	716	1578.510	347.4518	766	1688.741
325.2257	717	1580.714	347.9054	767	1690.945
325.6793	718	1582.919	348.3590	768	1693.150
326.1329	719	1585.124	348.8126	769	1695.355
326.5865	720	1587.328	349.2661	770	1697.559
327.0401	721	1589.533	349.7197	771	1699.764
327.4937	722	1591.737	350.1733	772	1701.969
327.9473	723	1593.942	350.6269	773	1704.173
328.4009	724	1596.147	351.0805	774	1706.378
328.8545	725	1598.351	351.5341	775	1708.582
329.3081	726	1600.556	351.9877	776	1710.787
329.7617	727	1602.761	352.4413	777	1712.992
330.2153	728	1604.965	352.8949	778	1715.196
330.6689	729	1607.170	353.3485	779	1717.401
331.1225	730	1609.374	353.8021	780	1719.606
331.5760	731	1611.579	354.2557	781	1721.810
332.0296	732	1613.784	354.7093	782	1724.015
332.4832	733	1615.988	355.1628	783	1726.219
332.9368	734	1618.193	355.6164	784	1728.424
333.3904	735	1620.398	356.0700	785	1730.629
333.8440	736	1622.602	356.5236	786	1732.833
334.2976	737	1624.807	356.9772	787	1735.038
334.7512	738	1627.011	357.4308	788	1737.243
335.2048	739	1629.216	357.8844	789	1739.447
335.6584	740	1631.421	358.3380	790	1741.652
336.1120	741	1633.625	358.7916	791	1743.856
336.5656	742	1635.830	359.2452	792	1746.061
337.0192	743	1638.035	359.6988	793	1748.266
337.4727	744	1640.239	360.1524	794	1750.470
337.9263	745	1642.444	360.6060	795	1752.675
338.3799	746	1644.648	361.0596	796	1754.879
338.8335	747	1646.853	361.5131	797	1757.084
339.2871	748	1649.058	361.9667	798	1759.289
339.7407	749	1651.262	362.4203	799	1761.493
340.1943	750	1653.467	362.8739	800	1763.698

kg kg/s kgf	REF	lbm lbm/s lbf	kg kg/s kgf	REF	lbm lbm/s lbf
363.3275	801	1765.903	386.0071	851	1876.134
363.7811	802	1768.107	386.4607	852	1878.338
364.2347	803	1770.312	386.9143	853	1880.543
364.6883	804	1772.516	387.3679	854	1882.748
365.1419	805	1774.721	387.8215	855	1884.952
365.5955	806	1776.926	388.2751	856	1887.157
366.0491	807	1779.130	388.7287	857	1889.361
366.5027	808	1781.335	389.1823	858	1891.566
366.9563	809	1783.540	389.6359	859	1893.771
367.4098	810	1785.744	390.0895	860	1895.975
367.8634	811	1787.949	390.5431	861	1898.180
368.3170	812	1790.153	390.9967	862	1900.385
368.7706	813	1792.358	391.4502	863	1902.589
369.2242	814	1794.563	391.9038	864	1904.794
369.6778	815	1796.767	392.3574	865	1906.998
370.1314	816	1798.972	392.8110	866	1909.203
370.5850	817	1801.177	393.2646	867	1911.408
371.0386	818	1803.381	393.7182	868	1913.612
371.4922	819	1805.586	394.1718	869	1915.817
371.9458	820	1807.790	394.6254	870	1918.022
372.3994	821	1809.995	395.0790	871	1920.226
372.8530	822	1812.200	395.5326	872	1922.431
373.3065	823	1814.404	395.9862	873	1924.635
373.7601	824	1816.609	396.4398	874	1926.840
374.2137	825	1818.814	396.8934	875	1929.045
374.6673	826	1821.018	397.3469	876	1931.249
375.1209	827	1823.223	397.8005	877	1933.454
375.5745	828	1825.427	398.2541	878	1935.659
376.0281	829	1827.632	398.7077	879	1937.863
376.4817	830	1829.837	399.1613	880	1940.068
376.9353	831	1832.041	399.6149	881	1942.272
377.3889	832	1834.246	400.0685	882	1944.477
377.8425	833	1836.451	400.5221	883	1946.682
378.2961	834	1838.655	400.9757	884	1948.886
378.7497	835	1840.860	401.4293	885	1951.091
379.2032	836	1843.064	401.8829	886	1953.296
379.6568	837	1845.269	402.3365	887	1955.500
380.1104	838	1847.474	402.7901	888	1957.705
380.5640	839	1849.678	403.2436	889	1959.909
381.0176	840	1851.883	403.6972	890	1962.114
381.4712	841	1854.088	404.1508	891	1964.319
381.9248	842	1856.292	404.6044	892	1966.523
382.3784	843	1858.497	405.0580	893	1968.728
382.8320	844	1860.701	405.5116	894	1970.932
383.2856	845	1862.906	405.9652	895	1973.137
383.7392	846	1865.111	406.4188	896	1975.342
384.1928	847	1867.315	406.8724	897	1977.546
384.6464	848	1869.520	407.3260	898	1979.751
385.0999	849	1871.724	407.7796	899	1981.956
385.5535	850	1873.929	408.2332	900	1984.160

t ———— +3 ————→ kg or (kgf) ←———— −3 ———— g (or gf)

kg kg/s kgf	REF	lbm lbm/s lbf	kg kg/s kgf	REF	lbm lbm/s lbf
408.6868	901	1986.365	431.3664	951	2096.596
409.1403	902	1988.569	431.8200	952	2098.801
409.5939	903	1990.774	432.2736	953	2101.005
410.0475	904	1992.979	432.7271	954	2103.210
410.5011	905	1995.183	433.1807	955	2105.414
410.9547	906	1997.388	433.6343	956	2107.619
411.4083	907	1999.593	434.0879	957	2109.824
411.8619	908	2001.797	434.5415	958	2112.028
412.3155	909	2004.002	434.9951	959	2114.233
412.7691	910	2006.206	435.4487	960	2116.438
413.2227	911	2008.411	435.9023	961	2118.642
413.6763	912	2010.616	436.3559	962	2120.847
414.1299	913	2012.820	436.8095	963	2123.051
414.5835	914	2015.025	437.2631	964	2125.256
415.0370	915	2017.230	437.7167	965	2127.461
415.4906	916	2019.434	438.1703	966	2129.665
415.9442	917	2021.639	438.6239	967	2131.870
416.3978	918	2023.843	439.0774	968	2134.075
416.8514	919	2026.048	439.5310	969	2136.279
417.3050	920	2028.253	439.9846	970	2138.484
417.7586	921	2030.457	440.4382	971	2140.688
418.2122	922	2032.662	440.8918	972	2142.893
418.6658	923	2034.867	441.3454	973	2145.098
419.1194	924	2037.071	441.7990	974	2147.302
419.5730	925	2039.276	442.2526	975	2149.507
420.0266	926	2041.480	442.7062	976	2151.712
420.4802	927	2043.685	443.1598	977	2153.916
420.9337	928	2045.890	443.6134	978	2156.121
421.3873	929	2048.094	444.0670	979	2158.325
421.8409	930	2050.299	444.5206	980	2160.530
422.2945	931	2052.504	444.9741	981	2162.735
422.7481	932	2054.708	445.4277	982	2164.939
423.2017	933	2056.913	445.8813	983	2167.144
423.6553	934	2059.117	446.3349	984	2169.349
424.1089	935	2061.322	446.7885	985	2171.553
424.5625	936	2063.527	447.2421	986	2173.758
425.0161	937	2065.731	447.6957	987	2175.962
425.4697	938	2067.936	448.1493	988	2178.167
425.9233	939	2070.141	448.6029	989	2180.372
426.3769	940	2072.345	449.0565	990	2182.576
426.8304	941	2074.550	449.5101	991	2184.781
427.2840	942	2076.754	449.9637	992	2186.986
427.7376	943	2078.959	450.4173	993	2189.190
428.1912	944	2081.164	450.8708	994	2191.395
428.6448	945	2083.368	451.3244	995	2193.599
429.0984	946	2085.573	451.7780	996	2195.804
429.5520	947	2087.777	452.2316	997	2198.009
430.0056	948	2089.982	452.6852	998	2200.213
430.4592	949	2092.187	453.1388	999	2202.418
430.9128	950	2094.391	453.5924	1000	2204.622

MASS

Ounce — Gram (Mass)
  (oz — g)
Ounce — Kilogram (Mass)
  (oz — kg)
Ounce — Milligram (Mass)
  (oz — mg)

MASS FLOW RATE

Ounce/Second — Gram/Second
    (oz/s — g/s)
Ounce/Second — Kilogram/Second
    (oz/s — kg/s)
Ounce/Second — Milligram/Second
    (oz/s — mg/s)

Conversion Factors:
  1 ounce (mass-oz) = 28.34952 grams (g)
  (1 g = 0.03527397 oz)

Mass Flow Rate Values

Mass flow rate values are also represented by Table 20, as long as the time units are the same for both English and SI conversion quantities. Besides per second shown above, these consistent time units can also be in per minute (min), per hour (h), etc.

Mass flow rate conversion factor:

1 oz/s (or oz/min, etc.) = 28.34952 g/s (or g/min, etc.)

In converting quantities containing g units to those having kg and mg, Diagram 20-A, which appears in the headings on alternate pages of this table, will assist the user. It is derived from: $1 \text{ g} = (10^{-3}) \text{ kg} = (10^{3}) \text{ mg}$.

New SI Unit	Decimal Shift or X Power of 10	SI Table Unit	Decimal Shift or X Power of 10	New SI Unit

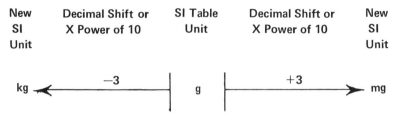

Diagram 20-A

Example (a): Convert 0.25 oz to g, mg.
  Since: $0.25 \text{ oz} = 25 \times 10^{-2}$ oz:
  From table: $25 \text{ oz} = 708.7380 \text{ g}*$
  Therefore: $25 \times 10^{-2} \text{ oz} = 708.7380 \times 10^{-2} \text{ g} = 7.087380 \text{ g} \approx 7.1$ g.
  From Diagram 20-A: $7.087380 \text{ g} = 7.087380 \times 10^{3} \text{ mg} = 7,087.380 \text{ mg}$
                $\approx 7100 \text{ mg}$
                (g to mg by multiplying g value by 10^{3} or by shifting its decimal 3 places to the right.)

Diagram 20-B is also given at the top of alternate table pages.

Diagram 20-B

It is the opposite of Diagram 20-A and is used to convert kg and mg to g values. The g equivalent is then applied to the REF column in the table to obtain oz quantities.

Example (b): Convert 6550 mg/s to oz/s.
  From Diagram 20-B: $6550 \text{ mg/s} = 6550 \times 10^{-3} \text{ g/s} = 655 \times 10^{-2}$ g/s
  From table: $655 \text{ g/s} = 23.10445 \text{ oz/s}*$
  Therefore: $655 \times 10^{-2} \text{ g/s} = 23.10445 \times 10^{-2} \text{ oz/s} = 0.2310445$ oz/s
                $\approx 0.231 \text{ oz/s}.$

---

*All conversion values obtained can be rounded to desired number of significant figures.

g g/s	REF	oz oz/s	g g/s	REF	oz oz/s
28.34952	1	0.03527397	1445.826	51	1.798972
56.69904	2	0.07054793	1474.175	52	1.834246
85.04856	3	0.1058219	1502.525	53	1.869520
113.3981	4	0.1410959	1530.874	54	1.904794
141.7476	5	0.1763698	1559.224	55	1.940068
170.0971	6	0.2116438	1587.573	56	1.975342
198.4466	7	0.2469178	1615.923	57	2.010616
226.7962	8	0.2821917	1644.272	58	2.045890
255.1457	9	0.3174657	1672.622	59	2.081164
283.4952	10	0.3527397	1700.971	60	2.116438
311.8447	11	0.3880136	1729.321	61	2.151712
340.1942	12	0.4232876	1757.670	62	2.186986
368.5438	13	0.4585616	1786.020	63	2.222260
396.8933	14	0.4938355	1814.369	64	2.257534
425.2428	15	0.5291095	1842.719	65	2.292808
453.5923	16	0.5643835	1871.068	66	2.328082
481.9418	17	0.5996574	1899.418	67	2.363356
510.2914	18	0.6349314	1927.767	68	2.398630
538.6409	19	0.6702054	1956.117	69	2.433904
566.9904	20	0.7054793	1984.466	70	2.469178
595.3399	21	0.7407533	2012.816	71	2.504452
623.6894	22	0.7760272	2041.165	72	2.539726
652.0390	23	0.8113012	2069.515	73	2.575000
680.3885	24	0.8465752	2097.864	74	2.610273
708.7380	25	0.8818491	2126.214	75	2.645547
737.0875	26	0.9171231	2154.564	76	2.680821
765.4370	27	0.9523971	2182.913	77	2.716095
793.7866	28	0.9876710	2211.263	78	2.751369
822.1361	29	1.022945	2239.612	79	2.786643
850.4856	30	1.058219	2267.962	80	2.821917
878.8351	31	1.093493	2296.311	81	2.857191
907.1846	32	1.128767	2324.661	82	2.892465
935.5342	33	1.164041	2353.010	83	2.927739
963.8837	34	1.199315	2381.360	84	2.963013
992.2332	35	1.234589	2409.709	85	2.998287
1020.583	36	1.269863	2438.059	86	3.033561
1048.932	37	1.305137	2466.408	87	3.068835
1077.282	38	1.340411	2494.758	88	3.104109
1105.631	39	1.375685	2523.107	89	3.139383
1133.981	40	1.410959	2551.457	90	3.174657
1162.330	41	1.446233	2579.806	91	3.209931
1190.680	42	1.481507	2608.156	92	3.245205
1219.029	43	1.516781	2636.505	93	3.280479
1247.379	44	1.552054	2664.855	94	3.315753
1275.728	45	1.587328	2693.204	95	3.351027
1304.078	46	1.622602	2721.554	96	3.386301
1332.427	47	1.657876	2749.903	97	3.421575
1360.777	48	1.693150	2778.253	98	3.456849
1389.126	49	1.728424	2806.602	99	3.492123
1417.476	50	1.763698	2834.952	100	3.527397

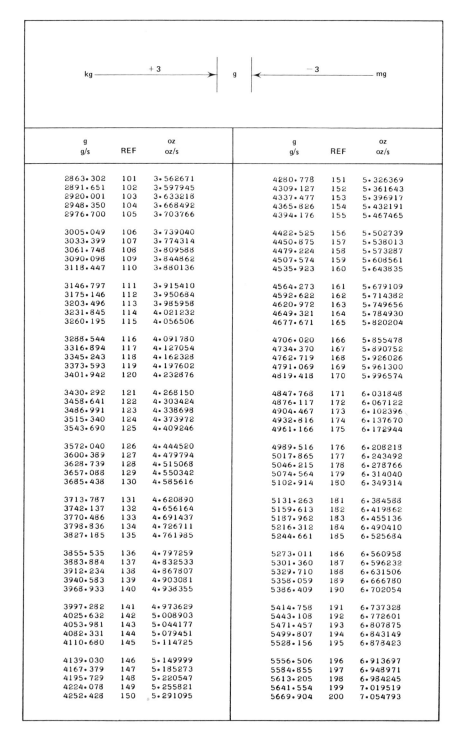

kg ——— +3 ———→ | g | ←——— −3 ——— mg

g g/s	REF	oz oz/s	g g/s	REF	oz oz/s
2863·302	101	3·562671	4280·778	151	5·326369
2891·651	102	3·597945	4309·127	152	5·361643
2920·001	103	3·633218	4337·477	153	5·396917
2948·350	104	3·668492	4365·826	154	5·432191
2976·700	105	3·703766	4394·176	155	5·467465
3005·049	106	3·739040	4422·525	156	5·502739
3033·399	107	3·774314	4450·875	157	5·538013
3061·748	108	3·809588	4479·224	158	5·573287
3090·098	109	3·844862	4507·574	159	5·608561
3118·447	110	3·880136	4535·923	160	5·643835
3146·797	111	3·915410	4564·273	161	5·679109
3175·146	112	3·950684	4592·622	162	5·714382
3203·496	113	3·985958	4620·972	163	5·749656
3231·845	114	4·021232	4649·321	164	5·784930
3260·195	115	4·056506	4677·671	165	5·820204
3288·544	116	4·091780	4706·020	166	5·855478
3316·894	117	4·127054	4734·370	167	5·890752
3345·243	118	4·162328	4762·719	168	5·926026
3373·593	119	4·197602	4791·069	169	5·961300
3401·942	120	4·232876	4819·418	170	5·996574
3430·292	121	4·268150	4847·768	171	6·031848
3458·641	122	4·303424	4876·117	172	6·067122
3486·991	123	4·338698	4904·467	173	6·102396
3515·340	124	4·373972	4932·816	174	6·137670
3543·690	125	4·409246	4961·166	175	6·172944
3572·040	126	4·444520	4989·516	176	6·208218
3600·389	127	4·479794	5017·865	177	6·243492
3628·739	128	4·515068	5046·215	178	6·278766
3657·088	129	4·550342	5074·564	179	6·314040
3685·438	130	4·585616	5102·914	180	6·349314
3713·787	131	4·620890	5131·263	181	6·384588
3742·137	132	4·656164	5159·613	182	6·419862
3770·486	133	4·691437	5187·962	183	6·455136
3798·836	134	4·726711	5216·312	184	6·490410
3827·185	135	4·761985	5244·661	185	6·525684
3855·535	136	4·797259	5273·011	186	6·560958
3883·884	137	4·832533	5301·360	187	6·596232
3912·234	138	4·867807	5329·710	188	6·631506
3940·583	139	4·903081	5358·059	189	6·666780
3968·933	140	4·938355	5386·409	190	6·702054
3997·282	141	4·973629	5414·758	191	6·737328
4025·632	142	5·008903	5443·108	192	6·772601
4053·981	143	5·044177	5471·457	193	6·807875
4082·331	144	5·079451	5499·807	194	6·843149
4110·680	145	5·114725	5528·156	195	6·878423
4139·030	146	5·149999	5556·506	196	6·913697
4167·379	147	5·185273	5584·855	197	6·948971
4195·729	148	5·220547	5613·205	198	6·984245
4224·078	149	5·255821	5641·554	199	7·019519
4252·428	150	5·291095	5669·904	200	7·054793

kg ←————— − 3 ——————| g |—————— + 3 —————→ mg

g g/s	REF	oz oz/s	g g/s	REF	oz oz/s
5698.254	201	7.090067	7115.729	251	8.853765
5726.603	202	7.125341	7144.079	252	8.889039
5754.953	203	7.160615	7172.429	253	8.924313
5783.302	204	7.195889	7200.778	254	8.959587
5811.652	205	7.231163	7229.128	255	8.994861
5840.001	206	7.266437	7257.477	256	9.030135
5868.351	207	7.301711	7285.827	257	9.065409
5896.700	208	7.336985	7314.176	258	9.100683
5925.050	209	7.372259	7342.526	259	9.135957
5953.399	210	7.407533	7370.875	260	9.171231
5981.749	211	7.442807	7399.225	261	9.206505
6010.098	212	7.478081	7427.574	262	9.241779
6038.448	213	7.513355	7455.924	263	9.277053
6066.797	214	7.548629	7484.273	264	9.312327
6095.147	215	7.583903	7512.623	265	9.347601
6123.496	216	7.619177	7540.972	266	9.382875
6151.846	217	7.654451	7569.322	267	9.418149
6180.195	218	7.689725	7597.671	268	9.453423
6208.545	219	7.724999	7626.021	269	9.488697
6236.894	220	7.760273	7654.370	270	9.523971
6265.244	221	7.795546	7682.720	271	9.559245
6293.593	222	7.830820	7711.069	272	9.594519
6321.943	223	7.866094	7739.419	273	9.629793
6350.292	224	7.901368	7767.768	274	9.665067
6378.642	225	7.936642	7796.118	275	9.700341
6406.992	226	7.971916	7824.468	276	9.735615
6435.341	227	8.007190	7852.817	277	9.770889
6463.691	228	8.042464	7881.167	278	9.806162
6492.040	229	8.077738	7909.516	279	9.841437
6520.390	230	8.113012	7937.866	280	9.876710
6548.739	231	8.148286	7966.215	281	9.911984
6577.089	232	8.183560	7994.565	282	9.947258
6605.438	233	8.218834	8022.914	283	9.982532
6633.788	234	8.254108	8051.264	284	10.01781
6662.137	235	8.289382	8079.613	285	10.05308
6690.487	236	8.324656	8107.963	286	10.08835
6718.836	237	8.359930	8136.312	287	10.12363
6747.186	238	8.395204	8164.662	288	10.15890
6775.535	239	8.430478	8193.011	289	10.19418
6803.885	240	8.465752	8221.361	290	10.22945
6832.234	241	8.501026	8249.710	291	10.26472
6860.584	242	8.536300	8278.060	292	10.30000
6888.933	243	8.571574	8306.409	293	10.33527
6917.283	244	8.606848	8334.759	294	10.37055
6945.632	245	8.642122	8363.108	295	10.40582
6973.982	246	8.677396	8391.458	296	10.44109
7002.331	247	8.712670	8419.807	297	10.47637
7030.681	248	8.747944	8448.157	298	10.51164
7059.030	249	8.783218	8476.506	299	10.54692
7087.380	250	8.818491	8504.856	300	10.58219

kg ———— + 3 ————→ g ←———— − 3 ———— mg

g g/s	REF	oz oz/s	g g/s	REF	oz oz/s
8533·206	301	10·61746	9950·682	351	12·38116
8561·555	302	10·65274	9979·031	352	12·41644
8589·905	303	10·68801	10007·38	353	12·45171
8618·254	304	10·72329	10035·73	354	12·48698
8646·604	305	10·75856	10064·08	355	12·52226
8674·953	306	10·79383	10092·43	356	12·55753
8703·303	307	10·82911	10120·78	357	12·59281
8731·652	308	10·86438	10149·13	358	12·62808
8760·002	309	10·89966	10177·48	359	12·66335
8788·351	310	10·93493	10205·83	360	12·69863
8816·701	311	10·97020	10234·18	361	12·73390
8845·050	312	11·00548	10262·53	362	12·76918
8873·400	313	11·04075	10290·88	363	12·80445
8901·749	314	11·07603	10319·23	364	12·83972
8930·099	315	11·11130	10347·57	365	12·87500
8958·448	316	11·14657	10375·92	366	12·91027
8986·798	317	11·18185	10404·27	367	12·94555
9015·147	318	11·21712	10432·62	368	12·98082
9043·497	319	11·25240	10460·97	369	13·01609
9071·846	320	11·28767	10489·32	370	13·05137
9100·196	321	11·32294	10517·67	371	13·08664
9128·545	322	11·35822	10546·02	372	13·12192
9156·895	323	11·39349	10574·37	373	13·15719
9185·245	324	11·42876	10602·72	374	13·19246
9213·594	325	11·46404	10631·07	375	13·22774
9241·943	326	11·49931	10659·42	376	13·26301
9270·293	327	11·53459	10687·77	377	13·29829
9298·643	328	11·56986	10716·12	378	13·33356
9326·992	329	11·60513	10744·47	379	13·36883
9355·342	330	11·64041	10772·82	380	13·40411
9383·691	331	11·67568	10801·17	381	13·43938
9412·041	332	11·71096	10829·52	382	13·47466
9440·390	333	11·74623	10857·87	383	13·50993
9468·740	334	11·78150	10886·22	384	13·54520
9497·089	335	11·81678	10914·57	385	13·58048
9525·439	336	11·85205	10942·91	386	13·61575
9553·788	337	11·88733	10971·26	387	13·65102
9582·138	338	11·92260	10999·61	388	13·68630
9610·487	339	11·95787	11027·96	389	13·72157
9638·837	340	11·99315	11056·31	390	13·75685
9667·186	341	12·02842	11084·66	391	13·79212
9695·536	342	12·06370	11113·01	392	13·82739
9723·885	343	12·09897	11141·36	393	13·86267
9752·235	344	12·13424	11169·71	394	13·89794
9780·584	345	12·16952	11198·06	395	13·93322
9808·934	346	12·20479	11226·41	396	13·96849
9837·283	347	12·24007	11254·76	397	14·00376
9865·633	348	12·27534	11283·11	398	14·03904
9893·982	349	12·31061	11311·46	399	14·07431
9922·332	350	12·34589	11339·81	400	14·10959

g g/s	REF	oz oz/s	g g/s	REF	oz oz/s
11368·16	401	14·14486	12785·63	451	15·90856
11396·51	402	14·18013	12813·98	452	15·94383
11424·86	403	14·21541	12842·33	453	15·97911
11453·21	404	14·25068	12870·68	454	16·01438
11481·56	405	14·28596	12899·03	455	16·04965
11509·91	406	14·32123	12927·38	456	16·08493
11538·25	407	14·35650	12955·73	457	16·12020
11566·60	408	14·39178	12984·08	458	16·15548
11594·95	409	14·42705	13012·43	459	16·19075
11623·30	410	14·46233	13040·78	460	16·22602
11651·65	411	14·49760	13069·13	461	16·26130
11680·00	412	14·53287	13097·48	462	16·29657
11708·35	413	14·56815	13125·83	463	16·33185
11736·70	414	14·60342	13154·18	464	16·36712
11765·05	415	14·63870	13182·53	465	16·40239
11793·40	416	14·67397	13210·88	466	16·43767
11821·75	417	14·70924	13239·23	467	16·47294
11850·10	418	14·74452	13267·58	468	16·50822
11878·45	419	14·77979	13295·92	469	16·54349
11906·80	420	14·81507	13324·27	470	16·57876
11935·15	421	14·85034	13352·62	471	16·61404
11963·50	422	14·88561	13380·97	472	16·64931
11991·85	423	14·92089	13409·32	473	16·68459
12020·20	424	14·95616	13437·67	474	16·71986
12048·55	425	14·99144	13466·02	475	16·75513
12076·90	426	15·02671	13494·37	476	16·79041
12105·24	427	15·06198	13522·72	477	16·82568
12133·59	428	15·09726	13551·07	478	16·86096
12161·94	429	15·13253	13579·42	479	16·89623
12190·29	430	15·16781	13607·77	480	16·93150
12218·64	431	15·20308	13636·12	481	16·96678
12246·99	432	15·23835	13664·47	482	17·00205
12275·34	433	15·27363	13692·82	483	17·03733
12303·69	434	15·30890	13721·17	484	17·07260
12332·04	435	15·34418	13749·52	485	17·10787
12360·39	436	15·37945	13777·87	486	17·14315
12388·74	437	15·41472	13806·22	487	17·17842
12417·09	438	15·45000	13834·57	488	17·21370
12445·44	439	15·48527	13862·92	489	17·24897
12473·79	440	15·52055	13891·26	490	17·28424
12502·14	441	15·55582	13919·61	491	17·31952
12530·49	442	15·59109	13947·96	492	17·35479
12558·84	443	15·62637	13976·31	493	17·39007
12587·19	444	15·66164	14004·66	494	17·42534
12615·54	445	15·69691	14033·01	495	17·46061
12643·89	446	15·73219	14061·36	496	17·49589
12672·24	447	15·76746	14089·71	497	17·53116
12700·58	448	15·80274	14118·06	498	17·56644
12728·93	449	15·83801	14146·41	499	17·60171
12757·28	450	15·87328	14174·76	500	17·63698

kg ——————— + 3 ——————→ g ←————— − 3 ——————— mg

g g/s	REF	oz oz/s	g g/s	REF	oz oz/s
14203.11	501	17.67226	15620.59	551	19.43596
14231.46	502	17.70753	15648.94	552	19.47123
14259.81	503	17.74280	15677.28	553	19.50650
14288.16	504	17.77808	15705.63	554	19.54178
14316.51	505	17.81335	15733.98	555	19.57705
14344.86	506	17.84863	15762.33	556	19.61232
14373.21	507	17.88390	15790.68	557	19.64760
14401.56	508	17.91917	15819.03	558	19.68287
14429.91	509	17.95445	15847.38	559	19.71815
14458.26	510	17.98972	15875.73	560	19.75342
14486.60	511	18.02500	15904.08	561	19.78869
14514.95	512	18.06027	15932.43	562	19.82397
14543.30	513	18.09554	15960.78	563	19.85924
14571.65	514	18.13082	15989.13	564	19.89452
14600.00	515	18.16609	16017.48	565	19.92979
14628.35	516	18.20137	16045.83	566	19.96506
14656.70	517	18.23664	16074.18	567	20.00034
14685.05	518	18.27191	16102.53	568	20.03561
14713.40	519	18.30719	16130.88	569	20.07089
14741.75	520	18.34246	16159.23	570	20.10616
14770.10	521	18.37774	16187.58	571	20.14143
14798.45	522	18.41301	16215.93	572	20.17671
14826.80	523	18.44828	16244.27	573	20.21198
14855.15	524	18.48356	16272.62	574	20.24726
14883.50	525	18.51883	16300.97	575	20.28253
14911.85	526	18.55411	16329.32	576	20.31780
14940.20	527	18.58938	16357.67	577	20.35308
14968.55	528	18.62465	16386.02	578	20.38835
14996.90	529	18.65993	16414.37	579	20.42363
15025.25	530	18.69520	16442.72	580	20.45890
15053.60	531	18.73048	16471.07	581	20.49417
15081.94	532	18.76575	16499.42	582	20.52945
15110.29	533	18.80102	16527.77	583	20.56472
15138.64	534	18.83630	16556.12	584	20.60000
15166.99	535	18.87157	16584.47	585	20.63527
15195.34	536	18.90685	16612.82	586	20.67054
15223.69	537	18.94212	16641.17	587	20.70582
15252.04	538	18.97739	16669.52	588	20.74109
15280.39	539	19.01267	16697.87	589	20.77637
15308.74	540	19.04794	16726.22	590	20.81164
15337.09	541	19.08322	16754.57	591	20.84691
15365.44	542	19.11849	16782.92	592	20.88219
15393.79	543	19.15376	16811.27	593	20.91746
15422.14	544	19.18904	16839.61	594	20.95274
15450.49	545	19.22431	16867.96	595	20.98801
15478.84	546	19.25959	16896.31	596	21.02328
15507.19	547	19.29486	16924.66	597	21.05856
15535.54	548	19.33013	16953.01	598	21.09383
15563.89	549	19.36541	16981.36	599	21.12911
15592.24	550	19.40068	17009.71	600	21.16438

kg ←———— − 3 ————| g |———— + 3 ————→ mg

g g/s	REF	oz oz/s	g g/s	REF	oz oz/s
17038.06	601	21.19965	18455.54	651	22.96335
17066.41	602	21.23493	18483.89	652	22.99863
17094.76	603	21.27020	18512.24	653	23.03390
17123.11	604	21.30548	18540.59	654	23.06917
17151.46	605	21.34075	18568.94	655	23.10445
17179.81	606	21.37602	18597.29	656	23.13972
17208.16	607	21.41130	18625.63	657	23.17500
17236.51	608	21.44657	18653.98	658	23.21027
17264.86	609	21.48185	18682.33	659	23.24554
17293.21	610	21.51712	18710.68	660	23.28082
17321.56	611	21.55239	18739.03	661	23.31609
17349.91	612	21.58767	18767.38	662	23.35137
17378.26	613	21.62294	18795.73	663	23.38664
17406.61	614	21.65822	18824.08	664	23.42191
17434.95	615	21.69349	18852.43	665	23.45719
17463.30	616	21.72876	18880.78	666	23.49246
17491.65	617	21.76404	18909.13	667	23.52774
17520.00	618	21.79931	18937.48	668	23.56301
17548.35	619	21.83458	18965.83	669	23.59828
17576.70	620	21.86986	18994.18	670	23.63356
17605.05	621	21.90513	19022.53	671	23.66883
17633.40	622	21.94041	19050.88	672	23.70411
17661.75	623	21.97568	19079.23	673	23.73938
17690.10	624	22.01095	19107.58	674	23.77465
17718.45	625	22.04623	19135.93	675	23.80993
17746.80	626	22.08150	19164.28	676	23.84520
17775.15	627	22.11678	19192.62	677	23.88047
17803.50	628	22.15205	19220.97	678	23.91575
17831.85	629	22.18732	19249.32	679	23.95102
17860.20	630	22.22260	19277.67	680	23.98630
17888.55	631	22.25787	19306.02	681	24.02157
17916.90	632	22.29315	19334.37	682	24.05684
17945.25	633	22.32842	19362.72	683	24.09212
17973.60	634	22.36369	19391.07	684	24.12739
18001.95	635	22.39897	19419.42	685	24.16267
18030.29	636	22.43424	19447.77	686	24.19794
18058.64	637	22.46952	19476.12	687	24.23321
18086.99	638	22.50479	19504.47	688	24.26849
18115.34	639	22.54006	19532.82	689	24.30376
18143.69	640	22.57534	19561.17	690	24.33904
18172.04	641	22.61061	19589.52	691	24.37431
18200.39	642	22.64589	19617.87	692	24.40958
18228.74	643	22.68116	19646.22	693	24.44486
18257.09	644	22.71643	19674.57	694	24.48013
18285.44	645	22.75171	19702.92	695	24.51541
18313.79	646	22.78698	19731.27	696	24.55068
18342.14	647	22.82226	19759.62	697	24.58595
18370.49	648	22.85753	19787.96	698	24.62123
18398.84	649	22.89280	19816.31	699	24.65650
18427.19	650	22.92808	19844.66	700	24.69178

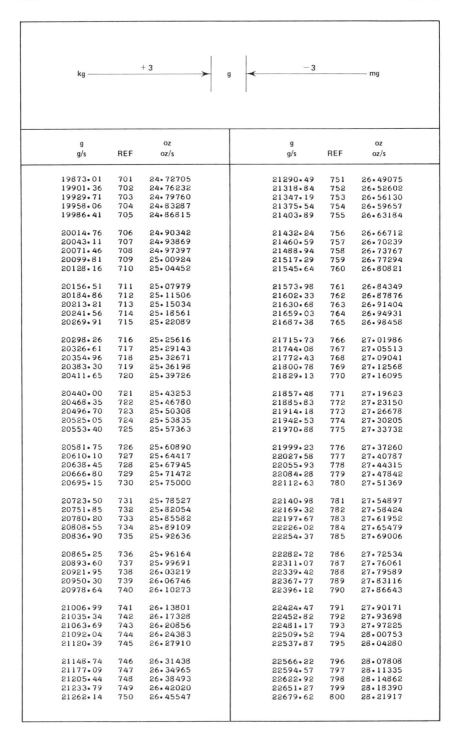

g g/s	REF	oz oz/s	g g/s	REF	oz oz/s
19873.01	701	24.72705	21290.49	751	26.49075
19901.36	702	24.76232	21318.84	752	26.52602
19929.71	703	24.79760	21347.19	753	26.56130
19958.06	704	24.83287	21375.54	754	26.59657
19986.41	705	24.86815	21403.89	755	26.63184
20014.76	706	24.90342	21432.24	756	26.66712
20043.11	707	24.93869	21460.59	757	26.70239
20071.46	708	24.97397	21488.94	758	26.73767
20099.81	709	25.00924	21517.29	759	26.77294
20128.16	710	25.04452	21545.64	760	26.80821
20156.51	711	25.07979	21573.98	761	26.84349
20184.86	712	25.11506	21602.33	762	26.87876
20213.21	713	25.15034	21630.68	763	26.91404
20241.56	714	25.18561	21659.03	764	26.94931
20269.91	715	25.22089	21687.38	765	26.98458
20298.26	716	25.25616	21715.73	766	27.01986
20326.61	717	25.29143	21744.08	767	27.05513
20354.96	718	25.32671	21772.43	768	27.09041
20383.30	719	25.36198	21800.78	769	27.12568
20411.65	720	25.39726	21829.13	770	27.16095
20440.00	721	25.43253	21857.48	771	27.19623
20468.35	722	25.46780	21885.83	772	27.23150
20496.70	723	25.50308	21914.18	773	27.26678
20525.05	724	25.53835	21942.53	774	27.30205
20553.40	725	25.57363	21970.88	775	27.33732
20581.75	726	25.60890	21999.23	776	27.37260
20610.10	727	25.64417	22027.58	777	27.40787
20638.45	728	25.67945	22055.93	778	27.44315
20666.80	729	25.71472	22084.28	779	27.47842
20695.15	730	25.75000	22112.63	780	27.51369
20723.50	731	25.78527	22140.98	781	27.54897
20751.85	732	25.82054	22169.32	782	27.58424
20780.20	733	25.85582	22197.67	783	27.61952
20808.55	734	25.89109	22226.02	784	27.65479
20836.90	735	25.92636	22254.37	785	27.69006
20865.25	736	25.96164	22282.72	786	27.72534
20893.60	737	25.99691	22311.07	787	27.76061
20921.95	738	26.03219	22339.42	788	27.79589
20950.30	739	26.06746	22367.77	789	27.83116
20978.64	740	26.10273	22396.12	790	27.86643
21006.99	741	26.13801	22424.47	791	27.90171
21035.34	742	26.17328	22452.82	792	27.93698
21063.69	743	26.20856	22481.17	793	27.97225
21092.04	744	26.24383	22509.52	794	28.00753
21120.39	745	26.27910	22537.87	795	28.04280
21148.74	746	26.31438	22566.22	796	28.07808
21177.09	747	26.34965	22594.57	797	28.11335
21205.44	748	26.38493	22622.92	798	28.14862
21233.79	749	26.42020	22651.27	799	28.18390
21262.14	750	26.45547	22679.62	800	28.21917

kg ⟵ — 3 | g | + 3 ⟶ mg

g g/s	REF	oz oz/s	g g/s	REF	oz oz/s
22707.97	801	28.25445	24125.44	851	30.01815
22736.31	802	28.28972	24153.79	852	30.05342
22764.66	803	28.32499	24182.14	853	30.08869
22793.01	804	28.36027	24210.49	854	30.12397
22821.36	805	28.39554	24238.84	855	30.15924
22849.71	806	28.43082	24267.19	856	30.19451
22878.06	807	28.46609	24295.54	857	30.22979
22906.41	808	28.50136	24323.89	858	30.26506
22934.76	809	28.53664	24352.24	859	30.30034
22963.11	810	28.57191	24380.59	860	30.33561
22991.46	811	28.60719	24408.94	861	30.37088
23019.81	812	28.64246	24437.29	862	30.40616
23048.16	813	28.67773	24465.64	863	30.44143
23076.51	814	28.71301	24493.99	864	30.47671
23104.86	815	28.74828	24522.33	865	30.51198
23133.21	816	28.78356	24550.68	866	30.54725
23161.56	817	28.81883	24579.03	867	30.58253
23189.91	818	28.85410	24607.38	868	30.61780
23218.26	819	28.88938	24635.73	869	30.65308
23246.61	820	28.92465	24664.08	870	30.68835
23274.96	821	28.95993	24692.43	871	30.72362
23303.31	822	28.99520	24720.78	872	30.75890
23331.66	823	29.03047	24749.13	873	30.79417
23360.00	824	29.06575	24777.48	874	30.82945
23388.35	825	29.10102	24805.83	875	30.86472
23416.70	826	29.13630	24834.18	876	30.89999
23445.05	827	29.17157	24862.53	877	30.93527
23473.40	828	29.20684	24890.88	878	30.97054
23501.75	829	29.24212	24919.23	879	31.00582
23530.10	830	29.27739	24947.58	880	31.04109
23558.45	831	29.31267	24975.93	881	31.07636
23586.80	832	29.34794	25004.28	882	31.11164
23615.15	833	29.38321	25032.63	883	31.14691
23643.50	834	29.41849	25060.98	884	31.18219
23671.85	835	29.45376	25089.33	885	31.21746
23700.20	836	29.48904	25117.67	886	31.25273
23728.55	837	29.52431	25146.02	887	31.28801
23756.90	838	29.55958	25174.37	888	31.32328
23785.25	839	29.59486	25202.72	889	31.35856
23813.60	840	29.63013	25231.07	890	31.39383
23841.95	841	29.66541	25259.42	891	31.42910
23870.30	842	29.70068	25287.77	892	31.46438
23898.65	843	29.73595	25316.12	893	31.49965
23926.99	844	29.77123	25344.47	894	31.53493
23955.34	845	29.80650	25372.82	895	31.57020
23983.69	846	29.84178	25401.17	896	31.60547
24012.04	847	29.87705	25429.52	897	31.64075
24040.39	848	29.91232	25457.87	898	31.67602
24068.74	849	29.94760	25486.22	899	31.71130
24097.09	850	29.98287	25514.57	900	31.74657

kg ———	+ 3 ———→		g	←———	− 3 ——— mg

g g/s	REF	oz oz/s		g g/s	REF	oz oz/s
25542·92	901	31·78184		26960·39	951	33·54554
25571·27	902	31·81712		26988·74	952	33·58082
25599·62	903	31·85239		27017·09	953	33·61609
25627·97	904	31·88767		27045·44	954	33·65136
25656·32	905	31·92294		27073·79	955	33·68664
25684·67	906	31·95821		27102·14	956	33·72191
25713·01	907	31·99349		27130·49	957	33·75719
25741·36	908	32·02876		27158·84	958	33·79246
25769·71	909	32·06403		27187·19	959	33·82773
25798·06	910	32·09931		27215·54	960	33·86301
25826·41	911	32·13458		27243·89	961	33·89828
25854·76	912	32·16986		27272·24	962	33·93356
25883·11	913	32·20513		27300·59	963	33·96883
25911·46	914	32·24041		27328·94	964	34·00410
25939·81	915	32·27568		27357·29	965	34·03938
25968·16	916	32·31095		27385·64	966	34·07465
25996·51	917	32·34623		27413·99	967	34·10993
26024·86	918	32·38150		27442·34	968	34·14520
26053·21	919	32·41677		27470·68	969	34·18047
26081·56	920	32·45205		27499·03	970	34·21575
26109·91	921	32·48732		27527·38	971	34·25102
26138·26	922	32·52260		27555·73	972	34·28629
26166·61	923	32·55787		27584·08	973	34·32157
26194·96	924	32·59314		27612·43	974	34·35684
26223·31	925	32·62842		27640·78	975	34·39212
26251·66	926	32·66369		27669·13	976	34·42739
26280·01	927	32·69897		27697·48	977	34·46266
26308·35	928	32·73424		27725·83	978	34·49794
26336·70	929	32·76951		27754·18	979	34·53321
26365·05	930	32·80479		27782·53	980	34·56849
26393·40	931	32·84006		27810·88	981	34·60376
26421·75	932	32·87534		27839·23	982	34·63903
26450·10	933	32·91061		27867·58	983	34·67431
26478·45	934	32·94588		27895·93	984	34·70958
26506·80	935	32·98116		27924·28	985	34·74486
26535·15	936	33·01643		27952·63	986	34·78013
26563·50	937	33·05171		27980·98	987	34·81540
26591·85	938	33·08698		28009·33	988	34·85068
26620·20	939	33·12225		28037·68	989	34·88595
26648·55	940	33·15753		28066·02	990	34·92123
26676·90	941	33·19280		28094·37	991	34·95650
26705·25	942	33·22808		28122·72	992	34·99177
26733·60	943	33·26335		28151·07	993	35·02705
26761·95	944	33·29862		28179·42	994	35·06232
26790·30	945	33·33390		28207·77	995	35·09760
26818·65	946	33·36917		28236·12	996	35·13287
26847·00	947	33·40445		28264·47	997	35·16814
26875·34	948	33·43972		28292·82	998	35·20342
26903·69	949	33·47499		28321·17	999	35·23869
26932·04	950	33·51027		28349·52	1000	35·27397

21

MASS DENSITY

Pound (mass)/Cubic-foot — Kilogram/Cubic-metre

$$(lbm/ft^3 - kg/m^3)$$

Pound (mass)/Cubic-foot — Gram/Cubic-metre

$$(lbm/ft^3 - g/m^3)$$

Pound (mass)/Cubic-foot — Gram/Cubic-centimetre

$$(lbm/ft^3 - g/cm^3)$$

Conversion Factors:

1 pound (mass)/cubic-foot (lbm/ft^3) = 16.01846 kilogram/cubic-metre (kg/m^3)

$$(1 \ kg/m^3 = 0.06242797 \ lbm/ft^3)$$

In converting kg/m^3 to g/m^3 and g/cm^3, Diagram 21-A, which appears in the headings on alternate pages of this table, will assist the user. It is derived from: $1 \ kg/m^3 = (10^3) \ g/m^3 = (10^{-3}) \ g/cm^3$.

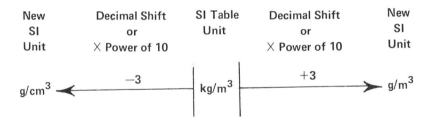

Diagram 21–A

Example (*a*): Convert 415 lb/ft^3 to kg/m^3, g/cm^3.
From table: 415 lb/ft^3 = 6,647.661 kg/m^3 *

227

From Diagram 21-A: $6{,}647.661 \text{ kg/m}^3 = 6{,}647.661 \times 10^{-3} \text{ g/cm}^3$
$= 6.647661 \text{ g/cm}^3 \approx 6.65 \text{ g/cm}^3$
(kg/m^3 to g/cm^3 by multiplying kg/m^3 value by 10^{-3}
or shifting its decimal 3 places to the left.)

Diagram 21-B is also given at the top of alternate table pages:

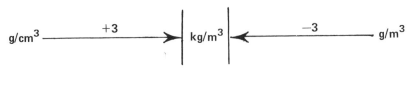

Diagram 21–B

It is the opposite of Diagram 21-A and is used to convert g/m^3 and g/cm^3 to kg/m^3. The kg/m^3 equivalent is then applied to the REF column in the table to obtain lbm/ft^3.

Example (b): Convert 2250 kg/m^3 to lb/ft^3.
Since: $2250 \text{ kg/m}^3 = 225 \times 10 \text{ kg/m}^3$.
From table: $225 \text{ kg/m}^3 = 14.04629 \text{ lb/ft}^3$ *
Therefore: $225 \times 10 \text{ kg/m}^3 = 140.4629 \text{ lb/ft}^3 \approx 140 \text{ lb/ft}^3$.

---

*All conversion values obtained can be rounded to desired number of significant figures.

g/cm³ ←——— −3 ——— | kg/m³ | ——— +3 ———→ g/m³

kg/m³	REF	lbm/ft³	kg/m³	REF	lbm/ft³
16·01846	1	0·06242797	816·9415	51	3·183827
32·03692	2	0·1248559	832·9599	52	3·246255
48·05538	3	0·1872839	848·9784	53	3·308683
64·07384	4	0·2497119	864·9968	54	3·371111
80·09230	5	0·3121399	881·0153	55	3·433539
96·11076	6	0·3745678	897·0338	56	3·495967
112·1292	7	0·4369958	913·0522	57	3·558394
128·1477	8	0·4994238	929·0707	58	3·620822
144·1661	9	0·5618518	945·0891	59	3·683250
160·1846	10	0·6242797	961·1076	60	3·745678
176·2031	11	0·6867077	977·1261	61	3·808106
192·2215	12	0·7491357	993·1445	62	3·870534
208·2400	13	0·8115637	1009·163	63	3·932962
224·2584	14	0·8739916	1025·181	64	3·995390
240·2769	15	0·9364196	1041·200	65	4·057818
256·2954	16	0·9988476	1057·218	66	4·120246
272·3138	17	1·061276	1073·237	67	4·182674
288·3323	18	1·123704	1089·255	68	4·245102
304·3507	19	1·186131	1105·274	69	4·307530
320·3692	20	1·248559	1121·292	70	4·369958
336·3877	21	1·310987	1137·311	71	4·432386
352·4061	22	1·373415	1153·329	72	4·494814
368·4246	23	1·435843	1169·348	73	4·557242
384·4430	24	1·498271	1185·366	74	4·619670
400·4615	25	1·560699	1201·385	75	4·682098
416·4800	26	1·623127	1217·403	76	4·744526
432·4984	27	1·685555	1233·421	77	4·806954
448·5169	28	1·747983	1249·440	78	4·869382
464·5353	29	1·810411	1265·458	79	4·931810
480·5538	30	1·872839	1281·477	80	4·994238
496·5723	31	1·935267	1297·495	81	5·056666
512·5907	32	1·997695	1313·514	82	5·119094
528·6092	33	2·060123	1329·532	83	5·181522
544·6276	34	2·122551	1345·551	84	5·243950
560·6461	35	2·184979	1361·569	85	5·306378
576·6646	36	2·247407	1377·588	86	5·368806
592·6830	37	2·309835	1393·606	87	5·431234
608·7015	38	2·372263	1409·624	88	5·493662
624·7199	39	2·434691	1425·643	89	5·556090
640·7384	40	2·497119	1441·661	90	5·618518
656·7569	41	2·559547	1457·680	91	5·680946
672·7753	42	2·621975	1473·698	92	5·743374
688·7938	43	2·684403	1489·717	93	5·805802
704·8122	44	2·746831	1505·735	94	5·868230
720·8307	45	2·809259	1521·754	95	5·930657
736·8492	46	2·871687	1537·772	96	5·993085
752·8676	47	2·934115	1553·791	97	6·055513
768·8861	48	2·996543	1569·809	98	6·117941
784·9045	49	3·058971	1585·828	99	6·180369
800·9230	50	3·121399	1601·846	100	6·242797

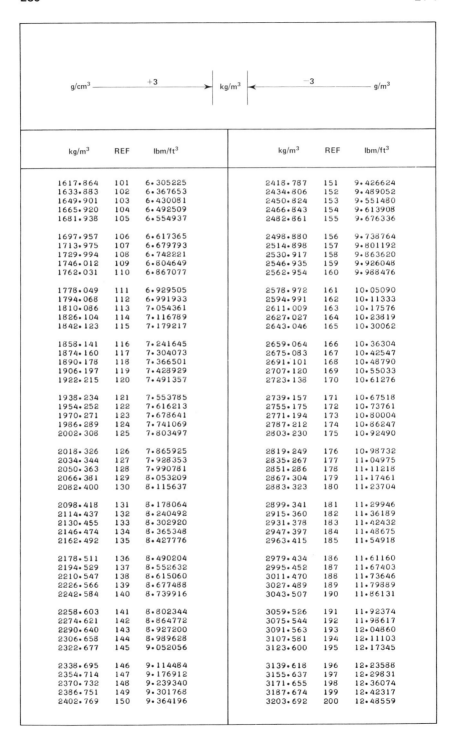

kg/m³	REF	lbm/ft³		kg/m³	REF	lbm/ft³
1617.864	101	6.305225		2418.787	151	9.426624
1633.883	102	6.367653		2434.806	152	9.489052
1649.901	103	6.430081		2450.824	153	9.551480
1665.920	104	6.492509		2466.843	154	9.613908
1681.938	105	6.554937		2482.861	155	9.676336
1697.957	106	6.617365		2498.880	156	9.738764
1713.975	107	6.679793		2514.898	157	9.801192
1729.994	108	6.742221		2530.917	158	9.863620
1746.012	109	6.804649		2546.935	159	9.926048
1762.031	110	6.867077		2562.954	160	9.988476
1778.049	111	6.929505		2578.972	161	10.05090
1794.068	112	6.991933		2594.991	162	10.11333
1810.086	113	7.054361		2611.009	163	10.17576
1826.104	114	7.116789		2627.027	164	10.23819
1842.123	115	7.179217		2643.046	165	10.30062
1858.141	116	7.241645		2659.064	166	10.36304
1874.160	117	7.304073		2675.083	167	10.42547
1890.178	118	7.366501		2691.101	168	10.48790
1906.197	119	7.428929		2707.120	169	10.55033
1922.215	120	7.491357		2723.138	170	10.61276
1938.234	121	7.553785		2739.157	171	10.67518
1954.252	122	7.616213		2755.175	172	10.73761
1970.271	123	7.678641		2771.194	173	10.80004
1986.289	124	7.741069		2787.212	174	10.86247
2002.308	125	7.803497		2803.230	175	10.92490
2018.326	126	7.865925		2819.249	176	10.98732
2034.344	127	7.928353		2835.267	177	11.04975
2050.363	128	7.990781		2851.286	178	11.11218
2066.381	129	8.053209		2867.304	179	11.17461
2082.400	130	8.115637		2883.323	180	11.23704
2098.418	131	8.178064		2899.341	181	11.29946
2114.437	132	8.240492		2915.360	182	11.36189
2130.455	133	8.302920		2931.378	183	11.42432
2146.474	134	8.365348		2947.397	184	11.48675
2162.492	135	8.427776		2963.415	185	11.54918
2178.511	136	8.490204		2979.434	186	11.61160
2194.529	137	8.552632		2995.452	187	11.67403
2210.547	138	8.615060		3011.470	188	11.73646
2226.566	139	8.677488		3027.489	189	11.79889
2242.584	140	8.739916		3043.507	190	11.86131
2258.603	141	8.802344		3059.526	191	11.92374
2274.621	142	8.864772		3075.544	192	11.98617
2290.640	143	8.927200		3091.563	193	12.04860
2306.658	144	8.989628		3107.581	194	12.11103
2322.677	145	9.052056		3123.600	195	12.17345
2338.695	146	9.114484		3139.618	196	12.23588
2354.714	147	9.176912		3155.637	197	12.29831
2370.732	148	9.239340		3171.655	198	12.36074
2386.751	149	9.301768		3187.674	199	12.42317
2402.769	150	9.364196		3203.692	200	12.48559

g/cm³ ←	−3	kg/m³	+3	→ g/m³

kg/m³	REF	lbm/ft³	kg/m³	REF	lbm/ft³
3219.710	201	12.54802	4020.633	251	15.66942
3235.729	202	12.61045	4036.652	252	15.73185
3251.747	203	12.67288	4052.670	253	15.79428
3267.766	204	12.73531	4068.689	254	15.85671
3283.784	205	12.79773	4084.707	255	15.91913
3299.803	206	12.86016	4100.726	256	15.98156
3315.821	207	12.92259	4116.744	257	16.04399
3331.840	208	12.98502	4132.763	258	16.10642
3347.858	209	13.04745	4148.781	259	16.16885
3363.877	210	13.10987	4164.800	260	16.23127
3379.895	211	13.17230	4180.818	261	16.29370
3395.914	212	13.23473	4196.837	262	16.35613
3411.932	213	13.29716	4212.855	263	16.41856
3427.950	214	13.35959	4228.873	264	16.48098
3443.969	215	13.42201	4244.892	265	16.54341
3459.987	216	13.48444	4260.910	266	16.60584
3476.006	217	13.54687	4276.929	267	16.66827
3492.024	218	13.60930	4292.947	268	16.73070
3508.043	219	13.67173	4308.966	269	16.79312
3524.061	220	13.73415	4324.984	270	16.85555
3540.080	221	13.79658	4341.003	271	16.91798
3556.098	222	13.85901	4357.021	272	16.98041
3572.117	223	13.92144	4373.040	273	17.04284
3588.135	224	13.98387	4389.058	274	17.10526
3604.154	225	14.04629	4405.077	275	17.16769
3620.172	226	14.10872	4421.095	276	17.23012
3636.190	227	14.17115	4437.113	277	17.29255
3652.209	228	14.23358	4453.132	278	17.35498
3668.227	229	14.29601	4469.150	279	17.41740
3684.246	230	14.35843	4485.169	280	17.47983
3700.264	231	14.42086	4501.187	281	17.54226
3716.283	232	14.48329	4517.206	282	17.60469
3732.301	233	14.54572	4533.224	283	17.66712
3748.320	234	14.60815	4549.243	284	17.72954
3764.338	235	14.67057	4565.261	285	17.79197
3780.357	236	14.73300	4581.280	286	17.85440
3796.375	237	14.79543	4597.298	287	17.91683
3812.393	238	14.85786	4613.316	288	17.97926
3828.412	239	14.92029	4629.335	289	18.04168
3844.430	240	14.98271	4645.353	290	18.10411
3860.449	241	15.04514	4661.372	291	18.16654
3876.467	242	15.10757	4677.390	292	18.22897
3892.486	243	15.17000	4693.409	293	18.29140
3908.504	244	15.23243	4709.427	294	18.35382
3924.523	245	15.29485	4725.446	295	18.41625
3940.541	246	15.35728	4741.464	296	18.47868
3956.560	247	15.41971	4757.483	297	18.54111
3972.578	248	15.48214	4773.501	298	18.60354
3988.597	249	15.54457	4789.520	299	18.66596
4004.615	250	15.60699	4805.538	300	18.72839

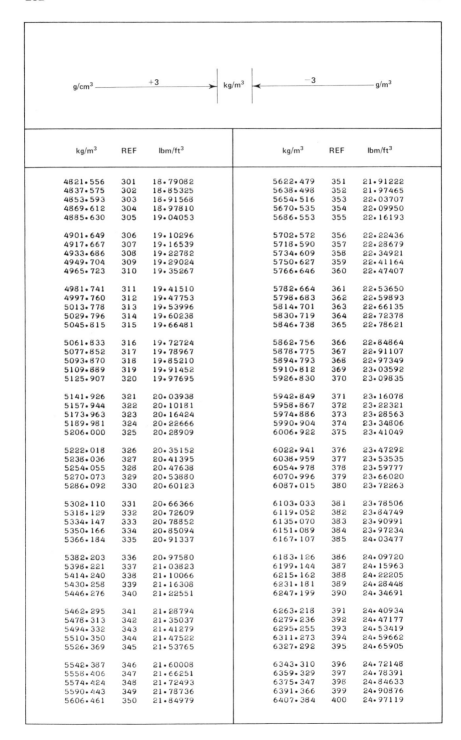

kg/m³	REF	lbm/ft³		kg/m³	REF	lbm/ft³
4821.556	301	18.79082		5622.479	351	21.91222
4837.575	302	18.85325		5638.498	352	21.97465
4853.593	303	18.91568		5654.516	353	22.03707
4869.612	304	18.97810		5670.535	354	22.09950
4885.630	305	19.04053		5686.553	355	22.16193
4901.649	306	19.10296		5702.572	356	22.22436
4917.667	307	19.16539		5718.590	357	22.28679
4933.686	308	19.22782		5734.609	358	22.34921
4949.704	309	19.29024		5750.627	359	22.41164
4965.723	310	19.35267		5766.646	360	22.47407
4981.741	311	19.41510		5782.664	361	22.53650
4997.760	312	19.47753		5798.683	362	22.59893
5013.778	313	19.53996		5814.701	363	22.66135
5029.796	314	19.60238		5830.719	364	22.72378
5045.815	315	19.66481		5846.738	365	22.78621
5061.833	316	19.72724		5862.756	366	22.84864
5077.852	317	19.78967		5878.775	367	22.91107
5093.870	318	19.85210		5894.793	368	22.97349
5109.889	319	19.91452		5910.812	369	23.03592
5125.907	320	19.97695		5926.830	370	23.09835
5141.926	321	20.03938		5942.849	371	23.16078
5157.944	322	20.10181		5958.867	372	23.22321
5173.963	323	20.16424		5974.886	373	23.28563
5189.981	324	20.22666		5990.904	374	23.34806
5206.000	325	20.28909		6006.922	375	23.41049
5222.018	326	20.35152		6022.941	376	23.47292
5238.036	327	20.41395		6038.959	377	23.53535
5254.055	328	20.47638		6054.978	378	23.59777
5270.073	329	20.53880		6070.996	379	23.66020
5286.092	330	20.60123		6087.015	380	23.72263
5302.110	331	20.66366		6103.033	381	23.78506
5318.129	332	20.72609		6119.052	382	23.84749
5334.147	333	20.78852		6135.070	383	23.90991
5350.166	334	20.85094		6151.089	384	23.97234
5366.184	335	20.91337		6167.107	385	24.03477
5382.203	336	20.97580		6183.126	386	24.09720
5398.221	337	21.03823		6199.144	387	24.15963
5414.240	338	21.10066		6215.162	388	24.22205
5430.258	339	21.16308		6231.181	389	24.28448
5446.276	340	21.22551		6247.199	390	24.34691
5462.295	341	21.28794		6263.218	391	24.40934
5478.313	342	21.35037		6279.236	392	24.47177
5494.332	343	21.41279		6295.255	393	24.53419
5510.350	344	21.47522		6311.273	394	24.59662
5526.369	345	21.53765		6327.292	395	24.65905
5542.387	346	21.60008		6343.310	396	24.72148
5558.406	347	21.66251		6359.329	397	24.78391
5574.424	348	21.72493		6375.347	398	24.84633
5590.443	349	21.78736		6391.366	399	24.90876
5606.461	350	21.84979		6407.384	400	24.97119

$$\text{g/cm}^3 \longleftarrow \quad -3 \quad \bigg| \quad \text{kg/m}^3 \quad \bigg| \quad +3 \quad \longrightarrow \text{g/m}^3$$

kg/m³	REF	lbm/ft³	kg/m³	REF	lbm/ft³
6423.402	401	25.03362	7224.326	451	28.15502
6439.421	402	25.09605	7240.344	452	28.21744
6455.439	403	25.15847	7256.362	453	28.27987
6471.458	404	25.22090	7272.381	454	28.34230
6487.476	405	25.28333	7288.399	455	28.40473
6503.495	406	25.34576	7304.418	456	28.46716
6519.513	407	25.40819	7320.436	457	28.52958
6535.532	408	25.47061	7336.455	458	28.59201
6551.550	409	25.53304	7352.473	459	28.65444
6567.569	410	25.59547	7368.492	460	28.71687
6583.587	411	25.65790	7384.510	461	28.77930
6599.606	412	25.72032	7400.529	462	28.84172
6615.624	413	25.78275	7416.547	463	28.90415
6631.642	414	25.84518	7432.565	464	28.96658
6647.661	415	25.90761	7448.584	465	29.02901
6663.679	416	25.97004	7464.602	466	29.09144
6679.698	417	26.03246	7480.621	467	29.15386
6695.716	418	26.09489	7496.639	468	29.21629
6711.735	419	26.15732	7512.658	469	29.27872
6727.753	420	26.21975	7528.676	470	29.34115
6743.772	421	26.28218	7544.695	471	29.40358
6759.790	422	26.34460	7560.713	472	29.46600
6775.809	423	26.40703	7576.732	473	29.52843
6791.827	424	26.46946	7592.750	474	29.59086
6807.846	425	26.53189	7608.768	475	29.65329
6823.864	426	26.59432	7624.787	476	29.71572
6839.882	427	26.65674	7640.805	477	29.77814
6855.901	428	26.71917	7656.824	478	29.84057
6871.919	429	26.78160	7672.842	479	29.90300
6887.938	430	26.84403	7688.861	480	29.96543
6903.956	431	26.90646	7704.879	481	30.02786
6919.975	432	26.96888	7720.898	482	30.09028
6935.993	433	27.03131	7736.916	483	30.15271
6952.012	434	27.09374	7752.935	484	30.21514
6968.030	435	27.15617	7768.953	485	30.27757
6984.049	436	27.21860	7784.972	486	30.34000
7000.067	437	27.28102	7800.990	487	30.40242
7016.086	438	27.34345	7817.008	488	30.46485
7032.104	439	27.40588	7833.027	489	30.52728
7048.122	440	27.46831	7849.045	490	30.58971
7064.141	441	27.53074	7865.064	491	30.65213
7080.159	442	27.59316	7881.082	492	30.71456
7096.178	443	27.65559	7897.101	493	30.77699
7112.196	444	27.71802	7913.119	494	30.83942
7128.215	445	27.78045	7929.138	495	30.90185
7144.233	446	27.84288	7945.156	496	30.96427
7160.252	447	27.90530	7961.175	497	31.02670
7176.270	448	27.96773	7977.193	498	31.08913
7192.289	449	28.03016	7993.212	499	31.15156
7208.307	450	28.09259	8009.230	500	31.21399

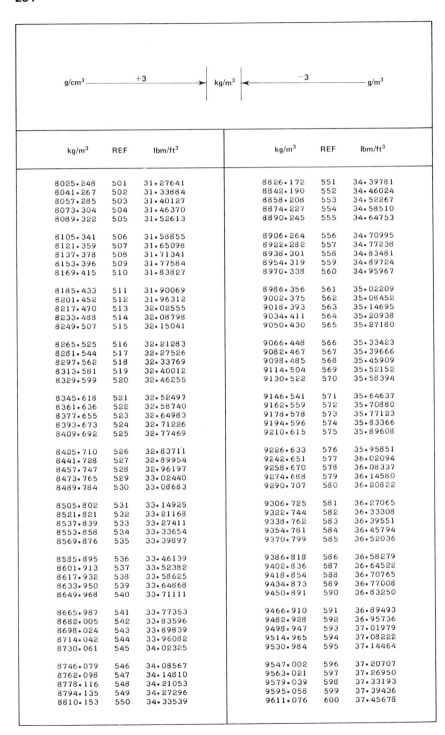

kg/m³	REF	lbm/ft³	kg/m³	REF	lbm/ft³
8025·248	501	31·27641	8826·172	551	34·39781
8041·267	502	31·33884	8842·190	552	34·46024
8057·285	503	31·40127	8858·208	553	34·52267
8073·304	504	31·46370	8874·227	554	34·58510
8089·322	505	31·52613	8890·245	555	34·64753
8105·341	506	31·58855	8906·264	556	34·70995
8121·359	507	31·65098	8922·282	557	34·77238
8137·378	508	31·71341	8938·301	558	34·83481
8153·396	509	31·77584	8954·319	559	34·89724
8169·415	510	31·83827	8970·338	560	34·95967
8185·433	511	31·90069	8986·356	561	35·02209
8201·452	512	31·96312	9002·375	562	35·08452
8217·470	513	32·02555	9018·393	563	35·14695
8233·488	514	32·08798	9034·411	564	35·20938
8249·507	515	32·15041	9050·430	565	35·27180
8265·525	516	32·21283	9066·448	566	35·33423
8281·544	517	32·27526	9082·467	567	35·39666
8297·562	518	32·33769	9098·485	568	35·45909
8313·581	519	32·40012	9114·504	569	35·52152
8329·599	520	32·46255	9130·522	570	35·58394
8345·618	521	32·52497	9146·541	571	35·64637
8361·636	522	32·58740	9162·559	572	35·70880
8377·655	523	32·64983	9178·578	573	35·77123
8393·673	524	32·71226	9194·596	574	35·83366
8409·692	525	32·77469	9210·615	575	35·89608
8425·710	526	32·83711	9226·633	576	35·95851
8441·728	527	32·89954	9242·651	577	36·02094
8457·747	528	32·96197	9258·670	578	36·08337
8473·765	529	33·02440	9274·688	579	36·14580
8489·784	530	33·08683	9290·707	580	36·20822
8505·802	531	33·14925	9306·725	581	36·27065
8521·821	532	33·21168	9322·744	582	36·33308
8537·839	533	33·27411	9338·762	583	36·39551
8553·858	534	33·33654	9354·781	584	36·45794
8569·876	535	33·39897	9370·799	585	36·52036
8585·895	536	33·46139	9386·818	586	36·58279
8601·913	537	33·52382	9402·836	587	36·64522
8617·932	538	33·58625	9418·854	588	36·70765
8633·950	539	33·64868	9434·873	589	36·77008
8649·968	540	33·71111	9450·891	590	36·83250
8665·987	541	33·77353	9466·910	591	36·89493
8682·005	542	33·83596	9482·928	592	36·95736
8698·024	543	33·89839	9498·947	593	37·01979
8714·042	544	33·96082	9514·965	594	37·08222
8730·061	545	34·02325	9530·984	595	37·14464
8746·079	546	34·08567	9547·002	596	37·20707
8762·098	547	34·14810	9563·021	597	37·26950
8778·116	548	34·21053	9579·039	598	37·33193
8794·135	549	34·27296	9595·058	599	37·39436
8810·153	550	34·33539	9611·076	600	37·45678

g/cm³ ← —3 | kg/m³ | +3 → g/m³

kg/m³	REF	lbm/ft³	kg/m³	REF	lbm/ft³
9627.094	601	37.51921	10428.02	651	40.64061
9643.113	602	37.58164	10444.04	652	40.70304
9659.131	603	37.64407	10460.05	653	40.76547
9675.150	604	37.70650	10476.07	654	40.82789
9691.168	605	37.76892	10492.09	655	40.89032
9707.187	606	37.83135	10508.11	656	40.95275
9723.205	607	37.89378	10524.13	657	41.01518
9739.224	608	37.95621	10540.15	658	41.07761
9755.242	609	38.01864	10556.17	659	41.14003
9771.261	610	38.08106	10572.18	660	41.20246
9787.279	611	38.14349	10588.20	661	41.26489
9803.297	612	38.20592	10604.22	662	41.32732
9819.316	613	38.26835	10620.24	663	41.38975
9835.334	614	38.33078	10636.26	664	41.45217
9851.353	615	38.39320	10652.28	665	41.51460
9867.371	616	38.45563	10668.29	666	41.57703
9883.390	617	38.51806	10684.31	667	41.63946
9899.408	618	38.58049	10700.33	668	41.70189
9915.427	619	38.64292	10716.35	669	41.76431
9931.445	620	38.70534	10732.37	670	41.82674
9947.464	621	38.76777	10748.39	671	41.88917
9963.482	622	38.83020	10764.41	672	41.95160
9979.501	623	38.89263	10780.42	673	42.01403
9995.519	624	38.95506	10796.44	674	42.07645
10011.54	625	39.01748	10812.46	675	42.13888
10027.56	626	39.07991	10828.48	676	42.20131
10043.57	627	39.14234	10844.50	677	42.26374
10059.59	628	39.20477	10860.52	678	42.32617
10075.61	629	39.26720	10876.53	679	42.38859
10091.63	630	39.32962	10892.55	680	42.45102
10107.65	631	39.39205	10908.57	681	42.51345
10123.67	632	39.45448	10924.59	682	42.57588
10139.69	633	39.51691	10940.61	683	42.63831
10155.70	634	39.57934	10956.63	684	42.70073
10171.72	635	39.64176	10972.65	685	42.76316
10187.74	636	39.70419	10988.66	686	42.82559
10203.76	637	39.76662	11004.68	687	42.88802
10219.78	638	39.82905	11020.70	688	42.95045
10235.80	639	39.89148	11036.72	689	43.01287
10251.81	640	39.95390	11052.74	690	43.07530
10267.83	641	40.01633	11068.76	691	43.13773
10283.85	642	40.07876	11084.77	692	43.20016
10299.87	643	40.14119	11100.79	693	43.26259
10315.89	644	40.20361	11116.81	694	43.32501
10331.91	645	40.26604	11132.83	695	43.38744
10347.93	646	40.32847	11148.85	696	43.44987
10363.94	647	40.39090	11164.87	697	43.51230
10379.96	648	40.45333	11180.89	698	43.57473
10395.98	649	40.51575	11196.90	699	43.63715
10412.00	650	40.57818	11212.92	700	43.69958

$$g/cm^3 \xrightarrow{\quad +3 \quad} kg/m^3 \xleftarrow{\quad -3 \quad} g/m^3$$

kg/m³	REF	lbm/ft³	kg/m³	REF	lbm/ft³
11228.94	701	43.76201	12029.86	751	46.88341
11244.96	702	43.82444	12045.88	752	46.94584
11260.98	703	43.88687	12061.90	753	47.00826
11277.00	704	43.94929	12077.92	754	47.07069
11293.01	705	44.01172	12093.94	755	47.13312
11309.03	706	44.07415	12109.96	756	47.19555
11325.05	707	44.13658	12125.97	757	47.25798
11341.07	708	44.19901	12141.99	758	47.32040
11357.09	709	44.26143	12158.01	759	47.38283
11373.11	710	44.32386	12174.03	760	47.44526
11389.13	711	44.38629	12190.05	761	47.50769
11405.14	712	44.44872	12206.07	762	47.57012
11421.16	713	44.51115	12222.08	763	47.63254
11437.18	714	44.57357	12238.10	764	47.69497
11453.20	715	44.63600	12254.12	765	47.75740
11469.22	716	44.69843	12270.14	766	47.81983
11485.24	717	44.76086	12286.16	767	47.88226
11501.25	718	44.82329	12302.18	768	47.94468
11517.27	719	44.88571	12318.20	769	48.00711
11533.29	720	44.94814	12334.21	770	48.06954
11549.31	721	45.01057	12350.23	771	48.13197
11565.33	722	45.07300	12366.25	772	48.19440
11581.35	723	45.13542	12382.27	773	48.25682
11597.37	724	45.19785	12398.29	774	48.31925
11613.38	725	45.26028	12414.31	775	48.38168
11629.40	726	45.32271	12430.32	776	48.44411
11645.42	727	45.38514	12446.34	777	48.50654
11661.44	728	45.44756	12462.36	778	48.56896
11677.46	729	45.50999	12478.38	779	48.63139
11693.48	730	45.57242	12494.40	780	48.69382
11709.49	731	45.63485	12510.42	781	48.75625
11725.51	732	45.69728	12526.44	782	48.81868
11741.53	733	45.75970	12542.45	783	48.88110
11757.55	734	45.82213	12558.47	734	48.94353
11773.57	735	45.88456	12574.49	785	49.00596
11789.59	736	45.94699	12590.51	786	49.06839
11805.61	737	46.00942	12606.53	787	49.13082
11821.62	738	46.07184	12622.55	788	49.19324
11837.64	739	46.13427	12638.56	789	49.25567
11853.66	740	46.19670	12654.58	790	49.31810
11869.68	741	46.25913	12670.60	791	49.38053
11885.70	742	46.32156	12686.62	792	49.44296
11901.72	743	46.38398	12702.64	793	49.50538
11917.73	744	46.44641	12718.66	794	49.56781
11933.75	745	46.50884	12734.68	795	49.63024
11949.77	746	46.57127	12750.69	796	49.69267
11965.79	747	46.63370	12766.71	797	49.75510
11981.81	748	46.69612	12782.73	798	49.81752
11997.83	749	46.75855	12798.75	799	49.87995
12013.84	750	46.82098	12814.77	800	49.94238

g/cm³ ◄——— −3 ——— | kg/m³ | ——— +3 ———► g/m³

kg/m³	REF	lbm/ft³	kg/m³	REF	lbm/ft³
12830.79	801	50.00481	13631.71	851	53.12621
12846.80	802	50.06723	13647.73	852	53.18863
12862.82	803	50.12966	13663.75	853	53.25106
12878.84	804	50.19209	13679.76	854	53.31349
12894.86	805	50.25452	13695.78	855	53.37592
12910.88	806	50.31695	13711.80	856	53.43835
12926.90	807	50.37937	13727.82	857	53.50077
12942.92	808	50.44180	13743.84	858	53.56320
12958.93	809	50.50423	13759.86	859	53.62563
12974.95	810	50.56666	13775.88	860	53.68806
12990.97	811	50.62909	13791.89	861	53.75048
13006.99	812	50.69151	13807.91	862	53.81291
13023.01	813	50.75394	13823.93	863	53.87534
13039.03	814	50.81637	13839.95	864	53.93777
13055.04	815	50.87880	13855.97	865	54.00020
13071.06	816	50.94123	13871.99	866	54.06262
13087.08	817	51.00365	13888.00	867	54.12505
13103.10	818	51.06608	13904.02	868	54.18748
13119.12	819	51.12851	13920.04	869	54.24991
13135.14	820	51.19094	13936.06	870	54.31234
13151.16	821	51.25337	13952.08	871	54.37476
13167.17	822	51.31579	13968.10	872	54.43719
13183.19	823	51.37822	13984.12	873	54.49962
13199.21	824	51.44065	14000.13	874	54.56205
13215.23	825	51.50308	14016.15	875	54.62448
13231.25	826	51.56551	14032.17	876	54.68690
13247.27	827	51.62793	14048.19	877	54.74933
13263.28	828	51.69036	14064.21	878	54.81176
13279.30	829	51.75279	14080.23	879	54.87419
13295.32	830	51.81522	14096.24	880	54.93662
13311.34	831	51.87765	14112.26	881	54.99904
13327.36	832	51.94007	14128.28	882	55.06147
13343.38	833	52.00250	14144.30	883	55.12390
13359.40	834	52.06493	14160.32	884	55.18633
13375.41	835	52.12736	14176.34	885	55.24876
13391.43	836	52.18979	14192.36	886	55.31118
13407.45	837	52.25221	14208.37	887	55.37361
13423.47	838	52.31464	14224.39	888	55.43604
13439.49	839	52.37707	14240.41	889	55.49847
13455.51	840	52.43950	14256.43	890	55.56090
13471.52	841	52.50193	14272.45	891	55.62332
13487.54	842	52.56435	14288.47	892	55.68575
13503.56	843	52.62678	14304.48	893	55.74818
13519.58	844	52.68921	14320.50	894	55.81061
13535.60	845	52.75164	14336.52	895	55.87304
13551.62	846	52.81407	14352.54	896	55.93546
13567.64	847	52.87649	14368.56	897	55.99789
13583.65	848	52.93892	14384.58	898	56.06032
13599.67	849	53.00135	14400.60	899	56.12275
13615.69	850	53.06378	14416.61	900	56.18518

g/cm³ ————————— +3 ————————→ | kg/m³ | ←———————— −3 ————————— g/m³

kg/m³	REF	lbm/ft³	kg/m³	REF	lbm/ft³
14432.63	901	56.24760	15233.56	951	59.36900
14448.65	902	56.31003	15249.57	952	59.43143
14464.67	903	56.37246	15265.59	953	59.49386
14480.69	904	56.43489	15281.61	954	59.55629
14496.71	905	56.49732	15297.63	955	59.61871
14512.72	906	56.55974	15313.65	956	59.68114
14528.74	907	56.62217	15329.67	957	59.74357
14544.76	908	56.68460	15345.68	958	59.80600
14560.78	909	56.74703	15361.70	959	59.86843
14576.80	910	56.80946	15377.72	960	59.93085
14592.82	911	56.87188	15393.74	961	59.99328
14608.84	912	56.93431	15409.76	962	60.05571
14624.85	913	56.99674	15425.78	963	60.11814
14640.87	914	57.05917	15441.80	964	60.18057
14656.89	915	57.12160	15457.81	965	60.24299
14672.91	916	57.18402	15473.83	966	60.30542
14688.93	917	57.24645	15489.85	967	60.36785
14704.95	918	57.30888	15505.87	968	60.43028
14720.96	919	57.37131	15521.89	969	60.49271
14736.98	920	57.43374	15537.91	970	60.55513
14753.00	921	57.49616	15553.92	971	60.61756
14769.02	922	57.55859	15569.94	972	60.67999
14785.04	923	57.62102	15585.96	973	60.74242
14801.06	924	57.68345	15601.98	974	60.80485
14817.08	925	57.74588	15618.00	975	60.86727
14833.09	926	57.80830	15634.02	976	60.92970
14849.11	927	57.87073	15650.04	977	60.99213
14865.13	928	57.93316	15666.05	978	61.05456
14881.15	929	57.99559	15682.07	979	61.11699
14897.17	930	58.05802	15698.09	980	61.17941
14913.19	931	58.12044	15714.11	981	61.24184
14929.20	932	58.18287	15730.13	982	61.30427
14945.22	933	58.24530	15746.15	983	61.36670
14961.24	934	58.30773	15762.16	984	61.42913
14977.26	935	58.37016	15778.18	985	61.49155
14993.28	936	58.43258	15794.20	986	61.55398
15009.30	937	58.49501	15810.22	987	61.61641
15025.32	938	58.55744	15826.24	988	61.67884
15041.33	939	58.61987	15842.26	989	61.74127
15057.35	940	58.68229	15858.28	990	61.80369
15073.37	941	58.74472	15874.29	991	61.86612
15089.39	942	58.80715	15890.31	992	61.92855
15105.41	943	58.86958	15906.33	993	61.99098
15121.43	944	58.93201	15922.35	994	62.05341
15137.44	945	58.99443	15938.37	995	62.11583
15153.46	946	59.05686	15954.39	996	62.17826
15169.48	947	59.11929	15970.40	997	62.24069
15185.50	948	59.18172	15986.42	998	62.30312
15201.52	949	59.24415	16002.44	999	62.36555
15217.54	950	59.30657	16018.46	1000	62.42797

22

LIQUID MASS DENSITY (U.S.)

Pound (mass)/U.S. Liquid Gallon — Kilogram/Cubic Metre

$(lbm/gal — kg/m^3)$

Pound (mass)/U.S. Liquid Gallon — Gram/Cubic Centimetre

$(lbm/gal — g/cm^3)$

Pound (mass)/U.S. Liquid Gallon — Kilogram/Litre

$(lbm/gal — kg/l)$

Pound (mass)/U.S. Liquid Gallon — Gram/Litre

$(lbm/gal — g/l)$

Pounds (mass) of substance/U.S. Liquid Gallon of pure water — Milligram of substance/Litre of pure water — Concentration or parts per million of water (ppm)

$(lbm/gal\ (H_2O) — mg/l\ (H_2O) — ppm\ (H_2O))$

Conversion Factors:

1 pound/U.S. gallon (lbm/gal) = 119.8264 kilogram/cubic-metre (kg/m^3)

$(1\ kg/m^3 = 0.008345406\ lbm/gal)$

Concentration in Pure Water (ppm)

Parts per million in water is a measure of proportion by weight that is equivalent to a unit of weight of a substance (solute), in a mixture or solution, per million units of weight of pure water (solvent). This is easily related in SI through the following designation of units:

1 milligram of substance/1 litre of water = 1 part per million of water (ppm)
Where: for most uses, 1 gram of water = $1\ cm^3$ in volume.

Proof: $(mg)_s/(l)_w = (mg)_s/(dm^3)_w = (10^{-3}\ g)_s/(10^3\ cm^3)_w$
$= (10^{-3}\ g)_s/(10^3\ g)_w = 1\ (g)_s/10^6\ (g)_w$

Where: s denotes solute substance, and w denotes water.

The following Diagram 22-A will assist the reader in using Table 22 for concentration conversions and calculations. It is derived from:

$$(kg) \text{ substance}/(m^3) \text{ water} = 1000 \ (mg)_s/(1)_w = 10^3 \text{ ppm}.$$

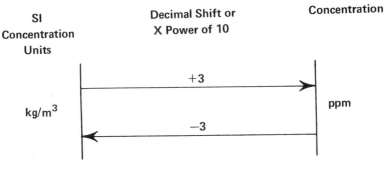

Diagram 22–A

Example (a): Relate in ppm the concentration of a solution of 1/4 oz of dye in 1 gal of H_2O.

Since: $1/4 \ oz \approx 0.016 \ lbm \approx 16 \times 10^{-3} \ lbm$, we therefore have a solution of $16 \times 10^{-3} \ lbm/gal$:

From table: $16 \ lbm/gal = 1917.222 \ kg/m^3$ *

Therefore: $16 \times 10^{-3} \ lbm/gal = 1917.222 \times 10^{-3} \ kg/m^3 = 1.917222 \ kg/m^3$.

From Diagram 22-A: $1.917222 \ kg/m^3 = 1917.222 \ ppm$ or $\approx 1900 \ ppm$

(kg/m^3 to ppm by multiplying kg/m^3 value by 10^3 or by shifting its decimal 3 places to the right.)

Example (b): How much salt must be mixed into a gallon of water to form a 1500 ppm solution?

From Diagram 22-A: $1500 \ ppm = 1500 \times 10^{-3} \ kg/m^3 = 15 \times 10^{-1} \ kg/m^3$

From table: $15 \ kg/m^3 = 0.1251811 \ lbm/gal$*

Therefore: $15 \times 10^{-1} \ kg/m^3 = 0.01251811 \ lbm \ salt/gal \ water \approx 0.20 \ oz/gal$ or $1/5 \ oz \ salt/gal \ water.$

In converting actual density values with Table 22 from kg/m^3 or g/l to g/cm^3 and kg/l, Diagram 22-B, which appears in the headings on alternate pages of this table, will assist the user. It is derived from:

$$1 \ kg/m^3 = 1 \ g/l = (10^{-3}) \ kg/l = (10^{-3}) \ g/cm^3.$$

---

*All conversion values obtained can be rounded to desired number of significant figures.

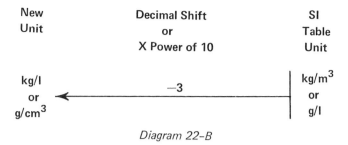

Diagram 22-B

Example (*c*): Convert 15.7 lbm/gal to kg/m^3, g/l, kg/l, g/cm^3.
   Since: 15.7 lbm/gal = 157 × 10^{-1} lbm/gal:
   From table: 157 lbm/gal = 18812.74 kg/m^3 *
   Therefore: 157 × 10^{-1} lbm/gal = 1,881.274 kg/m^3 or g/l
   From Diagram 22-B: 1,881.274 kg/m^3 = 1,881.274 × 10^{-3} kg/l or g/cm^3
                                    = 1.881274 kg/l or g/cm^3 ≈ 1.88 kg/l or g/cm^3

Diagram 22-C is also given at the top of alternate table pages:

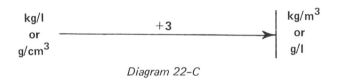

Diagram 22-C

It is the opposite of Diagram 22-B and is used to convert kg/l, g/cm^3 to g/l or kg/m^3. The kg/m^3 or g/l equivalent is then applied to the REF column in the table to obtain lbm/gal.

Example (*d*): Convert 4.25 g/cm^3 to lbm/gal.
   From Diagram 22-C: 4.25 g/cm^3 = 4.25 × 10^3 kg/m^3 = 425 × 10 kg/m^3
   From table: 425 kg/m^3 = 3.546798 lbm/gal*
   Therefore: 425 × 10 kg/m^3 = 35.46798 lbm/gal ≈ 35.5 lbm/gal.

---

*All conversion values obtained can be rounded to desired number of significant figures.

kg/m³ g/l	REF	lbm/gal	kg/m³ g/l	REF	lbm/gal
119.8264	1	0.008345406	6111.146	51	0.4256157
239.6528	2	0.01669081	6230.973	52	0.4339611
359.4792	3	0.02503622	6350.799	53	0.4423065
479.3056	4	0.03338163	6470.626	54	0.4506519
599.1320	5	0.04172703	6590.452	55	0.4589974
718.9584	6	0.05007244	6710.278	56	0.4673428
838.7848	7	0.05841784	6830.105	57	0.4756882
958.6112	8	0.06676325	6949.931	58	0.4840336
1078.438	9	0.07510866	7069.758	59	0.4923790
1198.264	10	0.08345406	7189.584	60	0.5007244
1318.090	11	0.09179947	7309.410	61	0.5090698
1437.917	12	0.1001449	7429.237	62	0.5174152
1557.743	13	0.1084903	7549.063	63	0.5257606
1677.570	14	0.1168357	7668.890	64	0.5341060
1797.396	15	0.1251811	7788.716	65	0.5424514
1917.222	16	0.1335265	7908.542	66	0.5507968
2037.049	17	0.1418719	8028.369	67	0.5591422
2156.875	18	0.1502173	8148.195	68	0.5674876
2276.702	19	0.1585627	8268.022	69	0.5758330
2396.528	20	0.1669081	8387.848	70	0.5841784
2516.354	21	0.1752535	8507.674	71	0.5925239
2636.181	22	0.1835989	8627.501	72	0.6008693
2756.007	23	0.1919443	8747.327	73	0.6092147
2875.834	24	0.2002898	8867.154	74	0.6175601
2995.660	25	0.2086352	8986.980	75	0.6259055
3115.486	26	0.2169806	9106.806	76	0.6342509
3235.313	27	0.2253260	9226.633	77	0.6425963
3355.139	28	0.2336714	9346.459	78	0.6509417
3474.966	29	0.2420168	9466.286	79	0.6592871
3594.792	30	0.2503622	9586.112	80	0.6676325
3714.618	31	0.2587076	9705.938	81	0.6759779
3834.445	32	0.2670530	9825.765	82	0.6843233
3954.271	33	0.2753984	9945.591	83	0.6926687
4074.098	34	0.2837438	10065.42	84	0.7010141
4193.924	35	0.2920892	10185.24	85	0.7093595
4313.750	36	0.3004346	10305.07	86	0.7177050
4433.577	37	0.3087800	10424.90	87	0.7260504
4553.403	38	0.3171254	10544.72	88	0.7343958
4673.230	39	0.3254708	10664.55	89	0.7427412
4793.056	40	0.3338163	10784.38	90	0.7510866
4912.882	41	0.3421617	10904.20	91	0.7594320
5032.709	42	0.3505071	11024.03	92	0.7677774
5152.535	43	0.3588525	11143.86	93	0.7761228
5272.362	44	0.3671979	11263.68	94	0.7844682
5392.188	45	0.3755433	11383.51	95	0.7928136
5512.014	46	0.3838887	11503.33	96	0.8011590
5631.841	47	0.3922341	11623.16	97	0.8095044
5751.667	48	0.4005795	11742.99	98	0.8178498
5871.494	49	0.4089249	11862.81	99	0.8261952
5991.320	50	0.4172703	11982.64	100	0.8345406

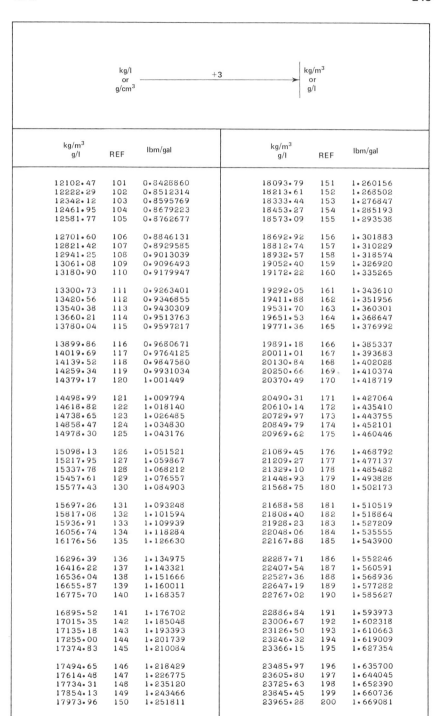

kg/m³ g/l	REF	lbm/gal	kg/m³ g/l	REF	lbm/gal
12102.47	101	0.8428860	18093.79	151	1.260156
12222.29	102	0.8512314	18213.61	152	1.268502
12342.12	103	0.8595769	18333.44	153	1.276847
12461.95	104	0.8679223	18453.27	154	1.285193
12581.77	105	0.8762677	18573.09	155	1.293538
12701.60	106	0.8846131	18692.92	156	1.301883
12821.42	107	0.8929585	18812.74	157	1.310229
12941.25	108	0.9013039	18932.57	158	1.318574
13061.08	109	0.9096493	19052.40	159	1.326920
13180.90	110	0.9179947	19172.22	160	1.335265
13300.73	111	0.9263401	19292.05	161	1.343610
13420.56	112	0.9346855	19411.88	162	1.351956
13540.38	113	0.9430309	19531.70	163	1.360301
13660.21	114	0.9513763	19651.53	164	1.368647
13780.04	115	0.9597217	19771.36	165	1.376992
13899.86	116	0.9680671	19891.18	166	1.385337
14019.69	117	0.9764125	20011.01	167	1.393683
14139.52	118	0.9847580	20130.84	168	1.402028
14259.34	119	0.9931034	20250.66	169	1.410374
14379.17	120	1.001449	20370.49	170	1.418719
14498.99	121	1.009794	20490.31	171	1.427064
14618.82	122	1.018140	20610.14	172	1.435410
14738.65	123	1.026485	20729.97	173	1.443755
14858.47	124	1.034830	20849.79	174	1.452101
14978.30	125	1.043176	20969.62	175	1.460446
15098.13	126	1.051521	21089.45	176	1.468792
15217.95	127	1.059867	21209.27	177	1.477137
15337.78	128	1.068212	21329.10	178	1.485482
15457.61	129	1.076557	21448.93	179	1.493828
15577.43	130	1.084903	21568.75	180	1.502173
15697.26	131	1.093248	21688.58	181	1.510519
15817.08	132	1.101594	21808.40	182	1.518864
15936.91	133	1.109939	21928.23	183	1.527209
16056.74	134	1.118284	22048.06	184	1.535555
16176.56	135	1.126630	22167.88	185	1.543900
16296.39	136	1.134975	22287.71	186	1.552246
16416.22	137	1.143321	22407.54	187	1.560591
16536.04	138	1.151666	22527.36	188	1.568936
16655.87	139	1.160011	22647.19	189	1.577282
16775.70	140	1.168357	22767.02	190	1.585627
16895.52	141	1.176702	22886.84	191	1.593973
17015.35	142	1.185048	23006.67	192	1.602318
17135.18	143	1.193393	23126.50	193	1.610663
17255.00	144	1.201739	23246.32	194	1.619009
17374.83	145	1.210084	23366.15	195	1.627354
17494.65	146	1.218429	23485.97	196	1.635700
17614.48	147	1.226775	23605.80	197	1.644045
17734.31	148	1.235120	23725.63	198	1.652390
17854.13	149	1.243466	23845.45	199	1.660736
17973.96	150	1.251811	23965.28	200	1.669081

kg/m³ g/l	REF	lbm/gal	kg/m³ g/l	REF	lbm/gal
24085.11	201	1.677427	30076.43	251	2.094697
24204.93	202	1.685772	30196.25	252	2.103042
24324.76	203	1.694118	30316.08	253	2.111388
24444.59	204	1.702463	30435.91	254	2.119733
24564.41	205	1.710808	30555.73	255	2.128079
24684.24	206	1.719154	30675.56	256	2.136424
24804.06	207	1.727499	30795.38	257	2.144769
24923.89	208	1.735845	30915.21	258	2.153115
25043.72	209	1.744190	31035.04	259	2.161460
25163.54	210	1.752535	31154.86	260	2.169806
25283.37	211	1.760881	31274.69	261	2.178151
25403.20	212	1.769226	31394.52	262	2.186496
25523.02	213	1.777572	31514.34	263	2.194842
25642.85	214	1.785917	31634.17	264	2.203187
25762.68	215	1.794262	31754.00	265	2.211533
25882.50	216	1.802608	31873.82	266	2.219878
26002.33	217	1.810953	31993.65	267	2.228224
26122.16	218	1.819299	32113.48	268	2.236569
26241.98	219	1.827644	32233.30	269	2.244914
26361.81	220	1.835989	32353.13	270	2.253260
26481.63	221	1.844335	32472.95	271	2.261605
26601.46	222	1.852680	32592.78	272	2.269951
26721.29	223	1.861026	32712.61	273	2.278296
26841.11	224	1.869371	32832.43	274	2.286641
26960.94	225	1.877716	32952.26	275	2.294987
27080.77	226	1.886062	33072.09	276	2.303332
27200.59	227	1.894407	33191.91	277	2.311678
27320.42	228	1.902753	33311.74	278	2.320023
27440.25	229	1.911098	33431.57	279	2.328368
27560.07	230	1.919443	33551.39	280	2.336714
27679.90	231	1.927789	33671.22	281	2.345059
27799.72	232	1.936134	33791.04	282	2.353405
27919.55	233	1.944480	33910.87	283	2.361750
28039.38	234	1.952825	34030.70	284	2.370095
28159.20	235	1.961170	34150.52	285	2.378441
28279.03	236	1.969516	34270.35	286	2.386786
28398.86	237	1.977861	34390.18	287	2.395132
28518.68	238	1.986207	34510.00	288	2.403477
28638.51	239	1.994552	34629.83	289	2.411822
28758.34	240	2.002898	34749.66	290	2.420168
28878.16	241	2.011243	34869.48	291	2.428513
28997.99	242	2.019588	34989.31	292	2.436859
29117.82	243	2.027934	35109.14	293	2.445204
29237.64	244	2.036279	35228.96	294	2.453549
29357.47	245	2.044625	35348.79	295	2.461895
29477.29	246	2.052970	35468.61	296	2.470240
29597.12	247	2.061315	35588.44	297	2.478586
29716.95	248	2.069661	35708.27	298	2.486931
29836.77	249	2.078006	35828.09	299	2.495277
29956.60	250	2.086352	35947.92	300	2.503622

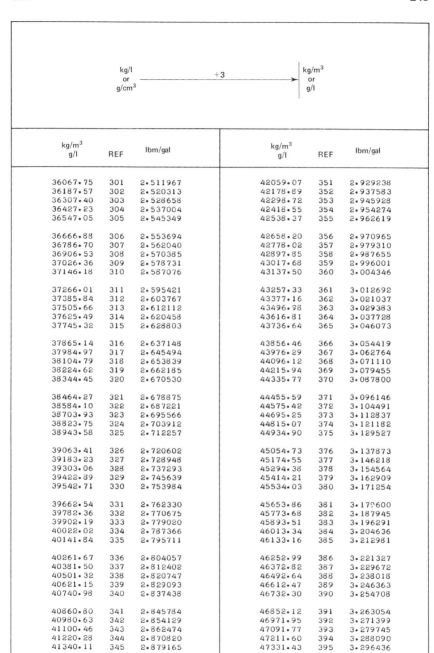

kg/m³ g/l	REF	lbm/gal	kg/m³ g/l	REF	lbm/gal
36067.75	301	2.511967	42059.07	351	2.929238
36187.57	302	2.520313	42178.89	352	2.937583
36307.40	303	2.528658	42298.72	353	2.945928
36427.23	304	2.537004	42418.55	354	2.954274
36547.05	305	2.545349	42538.37	355	2.962619
36666.88	306	2.553694	42658.20	356	2.970965
36786.70	307	2.562040	42778.02	357	2.979310
36906.53	308	2.570385	42897.85	358	2.987655
37026.36	309	2.578731	43017.68	359	2.996001
37146.18	310	2.587076	43137.50	360	3.004346
37266.01	311	2.595421	43257.33	361	3.012692
37385.84	312	2.603767	43377.16	362	3.021037
37505.66	313	2.612112	43496.98	363	3.029383
37625.49	314	2.620458	43616.81	364	3.037728
37745.32	315	2.628803	43736.64	365	3.046073
37865.14	316	2.637148	43856.46	366	3.054419
37984.97	317	2.645494	43976.29	367	3.062764
38104.79	318	2.653839	44096.12	368	3.071110
38224.62	319	2.662185	44215.94	369	3.079455
38344.45	320	2.670530	44335.77	370	3.087800
38464.27	321	2.678875	44455.59	371	3.096146
38584.10	322	2.687221	44575.42	372	3.104491
38703.93	323	2.695566	44695.25	373	3.112837
38823.75	324	2.703912	44815.07	374	3.121182
38943.58	325	2.712257	44934.90	375	3.129527
39063.41	326	2.720602	45054.73	376	3.137873
39183.23	327	2.728948	45174.55	377	3.146218
39303.06	328	2.737293	45294.38	378	3.154564
39422.89	329	2.745639	45414.21	379	3.162909
39542.71	330	2.753984	45534.03	380	3.171254
39662.54	331	2.762330	45653.86	381	3.179600
39782.36	332	2.770675	45773.68	382	3.187945
39902.19	333	2.779020	45893.51	383	3.196291
40022.02	334	2.787366	46013.34	384	3.204636
40141.84	335	2.795711	46133.16	385	3.212981
40261.67	336	2.804057	46252.99	386	3.221327
40381.50	337	2.812402	46372.82	387	3.229672
40501.32	338	2.820747	46492.64	388	3.238018
40621.15	339	2.829093	46612.47	389	3.246363
40740.98	340	2.837438	46732.30	390	3.254708
40860.80	341	2.845784	46852.12	391	3.263054
40980.63	342	2.854129	46971.95	392	3.271399
41100.46	343	2.862474	47091.77	393	3.279745
41220.28	344	2.870820	47211.60	394	3.288090
41340.11	345	2.879165	47331.43	395	3.296436
41459.93	346	2.887511	47451.25	396	3.304781
41579.76	347	2.895856	47571.08	397	3.313126
41699.59	348	2.904201	47690.91	398	3.321472
41819.41	349	2.912547	47810.73	399	3.329817
41939.24	350	2.920892	47930.56	400	3.338163

kg/m³ g/l	REF	lbm/gal	kg/m³ g/l	REF	lbm/gal
48050.39	401	3.346508	54041.71	451	3.763778
48170.21	402	3.354853	54161.53	452	3.772124
48290.04	403	3.363199	54281.36	453	3.780469
48409.87	404	3.371544	54401.19	454	3.788814
48529.69	405	3.379890	54521.01	455	3.797160
48649.52	406	3.388235	54640.84	456	3.805505
48769.34	407	3.396580	54760.66	457	3.813851
48889.17	408	3.404926	54880.49	458	3.822196
49009.00	409	3.413271	55000.32	459	3.830542
49128.82	410	3.421617	55120.14	460	3.838887
49248.65	411	3.429962	55239.97	461	3.847232
49368.48	412	3.438307	55359.80	462	3.855578
49488.30	413	3.446653	55479.62	463	3.863923
49608.13	414	3.454998	55599.45	464	3.872269
49727.96	415	3.463344	55719.28	465	3.880614
49847.78	416	3.471689	55839.10	466	3.888959
49967.61	417	3.480034	55958.93	467	3.897305
50087.44	418	3.488380	56078.75	468	3.905650
50207.26	419	3.496725	56198.58	469	3.913996
50327.09	420	3.505071	56318.41	470	3.922341
50446.91	421	3.513416	56438.23	471	3.930686
50566.74	422	3.521762	56558.06	472	3.939032
50686.57	423	3.530107	56677.89	473	3.947377
50806.39	424	3.538452	56797.71	474	3.955723
50926.22	425	3.546798	56917.54	475	3.964068
51046.05	426	3.555143	57037.37	476	3.972413
51165.87	427	3.563489	57157.19	477	3.980759
51285.70	428	3.571834	57277.02	478	3.989104
51405.53	429	3.580179	57396.85	479	3.997450
51525.35	430	3.588525	57516.67	480	4.005795
51645.18	431	3.596870	57636.50	481	4.014140
51765.00	432	3.605216	57756.32	482	4.022486
51884.83	433	3.613561	57876.15	483	4.030831
52004.66	434	3.621906	57995.98	484	4.039177
52124.48	435	3.630252	58115.80	485	4.047522
52244.31	436	3.638597	58235.63	486	4.055867
52364.14	437	3.646943	58355.46	487	4.064213
52483.96	438	3.655288	58475.28	488	4.072558
52603.79	439	3.663633	58595.11	489	4.080904
52723.62	440	3.671979	58714.94	490	4.089249
52843.44	441	3.680324	58834.76	491	4.097595
52963.27	442	3.688670	58954.59	492	4.105940
53083.10	443	3.697015	59074.42	493	4.114285
53202.92	444	3.705360	59194.24	494	4.122631
53322.75	445	3.713706	59314.07	495	4.130976
53442.57	446	3.722051	59433.89	496	4.139322
53562.40	447	3.730397	59553.72	497	4.147667
53682.23	448	3.738742	59673.55	498	4.156012
53802.05	449	3.747087	59793.37	499	4.164358
53921.88	450	3.755433	59913.20	500	4.172703

| kg/m³ or g/cm³ | +3 → | kg/m³ or g/l |

kg/m³ g/l	REF	lbm/gal	kg/m³ g/l	REF	lbm/gal
60033.03	501	4.181049	66024.35	551	4.598319
60152.85	502	4.189394	66144.17	552	4.606664
60272.68	503	4.197739	66264.00	553	4.615010
60392.51	504	4.206085	66383.83	554	4.623355
60512.33	505	4.214430	66503.65	555	4.631701
60632.16	506	4.222776	66623.48	556	4.640046
60751.98	507	4.231121	66743.30	557	4.648391
60871.81	508	4.239466	66863.13	558	4.656737
60991.64	509	4.247812	66982.96	559	4.665082
61111.46	510	4.256157	67102.78	560	4.673428
61231.29	511	4.264503	67222.61	561	4.681773
61351.12	512	4.272848	67342.44	562	4.690118
61470.94	513	4.281193	67462.26	563	4.698464
61590.77	514	4.289539	67582.09	564	4.706809
61710.60	515	4.297884	67701.92	565	4.715155
61830.42	516	4.306230	67821.74	566	4.723500
61950.25	517	4.314575	67941.57	567	4.731845
62070.08	518	4.322921	68061.40	568	4.740191
62189.90	519	4.331266	68181.22	569	4.748536
62309.73	520	4.339611	68301.05	570	4.756882
62429.55	521	4.347957	68420.87	571	4.765227
62549.38	522	4.356302	68540.70	572	4.773572
62669.21	523	4.364648	68660.53	573	4.781918
62789.03	524	4.372993	68780.35	574	4.790263
62908.86	525	4.381338	68900.18	575	4.798609
63028.69	526	4.389684	69020.01	576	4.806954
63148.51	527	4.398029	69139.83	577	4.815300
63268.34	528	4.406375	69259.66	578	4.823645
63388.17	529	4.414720	69379.49	579	4.831990
63507.99	530	4.423065	69499.31	580	4.840336
63627.82	531	4.431411	69619.14	581	4.848681
63747.64	532	4.439756	69738.96	582	4.857027
63867.47	533	4.448102	69858.79	583	4.865372
63987.30	534	4.456447	69978.62	584	4.873717
64107.12	535	4.464792	70098.44	585	4.882063
64226.95	536	4.473138	70218.27	586	4.890408
64346.78	537	4.481483	70338.10	587	4.898754
64466.60	538	4.489829	70457.92	588	4.907099
64586.43	539	4.498174	70577.75	589	4.915444
64706.26	540	4.506519	70697.58	590	4.923790
64826.08	541	4.514865	70817.40	591	4.932135
64945.91	542	4.523210	70937.23	592	4.940481
65065.73	543	4.531556	71057.05	593	4.948826
65185.56	544	4.539901	71176.88	594	4.957171
65305.39	545	4.548247	71296.71	595	4.965517
65425.21	546	4.556592	71416.53	596	4.973862
65545.04	547	4.564937	71536.36	597	4.982208
65664.87	548	4.573283	71656.19	598	4.990553
65784.69	549	4.581628	71776.01	599	4.998898
65904.52	550	4.589974	71895.84	600	5.007244

kg/m³ g/l	REF	lbm/gal	kg/m³ g/l	REF	lbm/gal
7201 5.67	601	5.015589	7800 6.99	651	5.432860
7213 5.49	602	5.023935	7812 6.81	652	5.441205
7225 5.32	603	5.032280	7824 6.64	653	5.449550
7237 5.15	604	5.040625	7836 6.47	654	5.457896
7249 4.97	605	5.048971	7848 6.29	655	5.466241
7261 4.80	606	5.057316	7860 6.12	656	5.474587
7273 4.62	607	5.065662	7872 5.94	657	5.482932
7285 4.45	608	5.074007	7884 5.77	658	5.491277
7297 4.28	609	5.082352	7896 5.60	659	5.499623
7309 4.10	610	5.090698	7908 5.42	660	5.507968
7321 3.93	611	5.099043	7920 5.25	661	5.516314
7333 3.76	612	5.107389	7932 5.08	662	5.524659
7345 3.58	613	5.115734	7944 4.90	663	5.533004
7357 3.41	614	5.124080	7956 4.73	664	5.541350
7369 3.24	615	5.132425	7968 4.56	665	5.549695
7381 3.06	616	5.140770	7980 4.38	666	5.558041
7393 2.89	617	5.149116	7992 4.21	667	5.566386
7405 2.71	618	5.157461	8004 4.04	668	5.574731
7417 2.54	619	5.165807	8016 3.86	669	5.583077
7429 2.37	620	5.174152	8023 3.69	670	5.591422
7441 2.19	621	5.182497	8040 3.51	671	5.599768
7453 2.02	622	5.190843	8052 3.34	672	5.608113
7465 1.85	623	5.199188	8064 3.17	673	5.616458
7477 1.67	624	5.207534	8076 2.99	674	5.624804
7489 1.50	625	5.215879	8088 2.82	675	5.633149
7501 1.33	626	5.224224	8100 2.65	676	5.641495
7513 1.15	627	5.232570	8112 2.47	677	5.649840
7525 0.98	628	5.240915	8124 2.30	678	5.658186
7537 0.81	629	5.249261	8136 2.12	679	5.666531
7549 0.63	630	5.257606	8148 1.95	680	5.674876
7561 0.46	631	5.265951	8160 1.78	681	5.683222
7573 0.29	632	5.274297	8172 1.60	682	5.691567
7585 0.11	633	5.282642	8184 1.43	683	5.699913
7596 9.94	634	5.290988	8196 1.26	684	5.708258
7608 9.76	635	5.299333	8208 1.08	685	5.716603
7620 9.59	636	5.307678	8220 0.91	686	5.724949
7632 9.42	637	5.316024	8232 0.74	687	5.733294
7644 9.24	638	5.324369	8244 0.56	688	5.741640
7656 9.07	639	5.332715	8256 0.39	689	5.749985
7668 8.90	640	5.341060	8268 0.22	690	5.758330
7680 8.72	641	5.349405	8280 0.04	691	5.766676
7692 8.55	642	5.357751	8291 9.87	692	5.775021
7704 8.37	643	5.366096	8303 9.70	693	5.783367
7716 8.20	644	5.374442	8315 9.52	694	5.791712
7728 8.03	645	5.382787	8327 9.35	695	5.800057
7740 7.85	646	5.391133	8339 9.17	696	5.808403
7752 7.68	647	5.399478	8351 9.00	697	5.816748
7764 7.51	648	5.407823	8363 8.83	698	5.825094
7776 7.33	649	5.416169	8375 8.65	699	5.833439
7788 7.16	650	5.424514	8387 8.48	700	5.841784

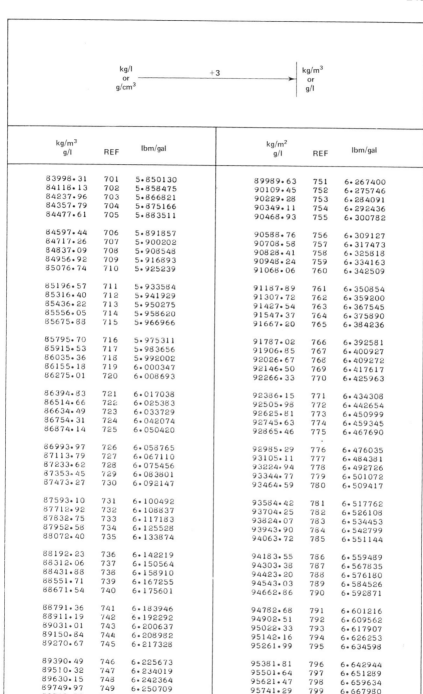

kg/m³ g/l	REF	lbm/gal		kg/m² g/l	REF	lbm/gal
83998.31	701	5.850130		89989.63	751	6.267400
84118.13	702	5.858475		90109.45	752	6.275746
84237.96	703	5.866821		90229.28	753	6.284091
84357.79	704	5.875166		90349.11	754	6.292436
84477.61	705	5.883511		90468.93	755	6.300782
84597.44	706	5.891857		90588.76	756	6.309127
84717.26	707	5.900202		90708.58	757	6.317473
84837.09	708	5.908548		90828.41	758	6.325818
84956.92	709	5.916893		90948.24	759	6.334163
85076.74	710	5.925239		91068.06	760	6.342509
85196.57	711	5.933584		91187.89	761	6.350854
85316.40	712	5.941929		91307.72	762	6.359200
85436.22	713	5.950275		91427.54	763	6.367545
85556.05	714	5.958620		91547.37	764	6.375890
85675.88	715	5.966966		91667.20	765	6.384236
85795.70	716	5.975311		91787.02	766	6.392581
85915.53	717	5.983656		91906.85	767	6.400927
86035.36	718	5.992002		92026.67	768	6.409272
86155.18	719	6.000347		92146.50	769	6.417617
86275.01	720	6.008693		92266.33	770	6.425963
86394.83	721	6.017038		92386.15	771	6.434308
86514.66	722	6.025383		92505.98	772	6.442654
86634.49	723	6.033729		92625.81	773	6.450999
86754.31	724	6.042074		92745.63	774	6.459345
86874.14	725	6.050420		92865.46	775	6.467690
86993.97	726	6.058765		92985.29	776	6.476035
87113.79	727	6.067110		93105.11	777	6.484381
87233.62	728	6.075456		93224.94	778	6.492726
87353.45	729	6.083801		93344.77	779	6.501072
87473.27	730	6.092147		93464.59	780	6.509417
87593.10	731	6.100492		93584.42	781	6.517762
87712.92	732	6.108837		93704.25	782	6.526108
87832.75	733	6.117183		93824.07	783	6.534453
87952.58	734	6.125528		93943.90	784	6.542799
88072.40	735	6.133874		94063.72	785	6.551144
88192.23	736	6.142219		94183.55	786	6.559489
88312.06	737	6.150564		94303.38	787	6.567835
88431.88	738	6.158910		94423.20	788	6.576180
88551.71	739	6.167255		94543.03	789	6.584526
88671.54	740	6.175601		94662.86	790	6.592871
88791.36	741	6.183946		94782.68	791	6.601216
88911.19	742	6.192292		94902.51	792	6.609562
89031.01	743	6.200637		95022.33	793	6.617907
89150.84	744	6.208982		95142.16	794	6.626253
89270.67	745	6.217328		95261.99	795	6.634598
89390.49	746	6.225673		95381.81	796	6.642944
89510.32	747	6.234019		95501.64	797	6.651289
89630.15	748	6.242364		95621.47	798	6.659634
89749.97	749	6.250709		95741.29	799	6.667980
89869.80	750	6.259055		95861.12	800	6.676325

kg/m³ g/l	REF	lbm/gal	kg/m³ g/l	REF	lbm/gal
95980.95	801	6.684671	101972.3	851	7.101941
96100.77	802	6.693016	102092.1	852	7.110286
96220.60	803	6.701361	102211.9	853	7.118632
96340.43	804	6.709707	102331.7	854	7.126977
96460.25	805	6.718052	102451.6	855	7.135322
96580.08	806	6.726398	102571.4	856	7.143668
96699.90	807	6.734743	102691.2	857	7.152013
96819.73	808	6.743088	102811.1	858	7.160359
96939.56	809	6.751434	102930.9	859	7.168704
97059.38	810	6.759779	103050.7	860	7.177050
97179.21	811	6.768125	103170.5	861	7.185395
97299.04	812	6.776470	103290.4	862	7.193740
97418.86	813	6.784815	103410.2	863	7.202086
97538.69	814	6.793161	103530.0	864	7.210431
97658.52	815	6.801506	103649.8	865	7.218777
97778.34	816	6.809852	103769.7	866	7.227122
97898.17	817	6.818197	103889.5	867	7.235467
98018.00	818	6.826542	104009.3	868	7.243813
98137.82	819	6.834888	104129.1	869	7.252158
98257.65	820	6.843233	104249.0	870	7.260504
98377.47	821	6.851579	104368.8	871	7.268849
98497.30	822	6.859924	104488.6	872	7.277194
98617.13	823	6.868269	104608.4	873	7.285540
98736.95	824	6.876615	104728.3	874	7.293885
98856.78	825	6.884960	104848.1	875	7.302231
98976.61	826	6.893306	104967.9	876	7.310576
99096.43	827	6.901651	105087.8	877	7.318921
99216.26	828	6.909997	105207.6	878	7.327267
99336.08	829	6.918342	105327.4	879	7.335612
99455.91	830	6.926687	105447.2	880	7.343958
99575.74	831	6.935033	105567.1	881	7.352303
99695.56	832	6.943378	105686.9	882	7.360648
99815.39	833	6.951724	105806.7	883	7.368994
99935.22	834	6.960069	105926.5	884	7.377339
100055.0	835	6.968414	106046.4	885	7.385685
100174.9	836	6.976760	106166.2	886	7.394030
100294.7	837	6.985105	106286.0	887	7.402375
100414.5	838	6.993451	106405.8	888	7.410721
100534.3	839	7.001796	106525.7	889	7.419066
100654.2	840	7.010141	106645.5	890	7.427412
100774.0	841	7.018487	106765.3	891	7.435757
100893.8	842	7.026832	106885.1	892	7.444103
101013.7	843	7.035178	107005.0	893	7.452448
101133.5	844	7.043523	107124.8	894	7.460793
101253.3	845	7.051868	107244.6	895	7.469139
101373.1	846	7.060214	107364.5	896	7.477484
101493.0	847	7.068559	107484.3	897	7.485830
101612.8	848	7.076905	107604.1	898	7.494175
101732.6	849	7.085250	107723.9	899	7.502520
101852.4	850	7.093595	107843.8	900	7.510866

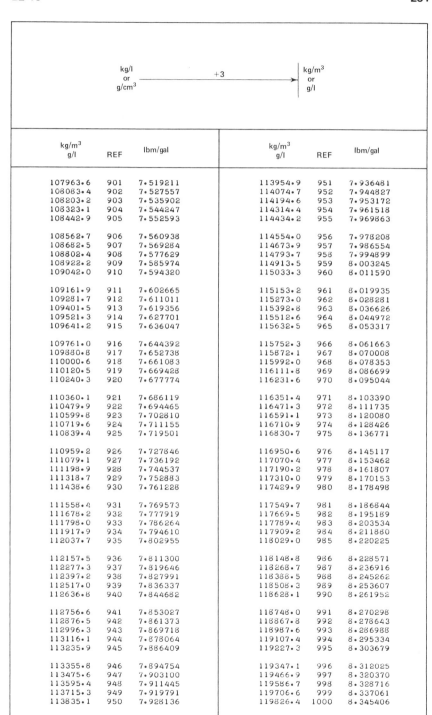

kg/m³ g/l	REF	lbm/gal		kg/m³ g/l	REF	lbm/gal
107963.6	901	7.519211		113954.9	951	7.936481
108083.4	902	7.527557		114074.7	952	7.944827
108203.2	903	7.535902		114194.6	953	7.953172
108323.1	904	7.544247		114314.4	954	7.961518
108442.9	905	7.552593		114434.2	955	7.969863
108562.7	906	7.560938		114554.0	956	7.978208
108682.5	907	7.569284		114673.9	957	7.986554
108802.4	908	7.577629		114793.7	958	7.994899
108922.2	909	7.585974		114913.5	959	8.003245
109042.0	910	7.594320		115033.3	960	8.011590
109161.9	911	7.602665		115153.2	961	8.019935
109281.7	912	7.611011		115273.0	962	8.028281
109401.5	913	7.619356		115392.8	963	8.036626
109521.3	914	7.627701		115512.6	964	8.044972
109641.2	915	7.636047		115632.5	965	8.053317
109761.0	916	7.644392		115752.3	966	8.061663
109880.8	917	7.652738		115872.1	967	8.070008
110000.6	918	7.661083		115992.0	968	8.078353
110120.5	919	7.669428		116111.8	969	8.086699
110240.3	920	7.677774		116231.6	970	8.095044
110360.1	921	7.686119		116351.4	971	8.103390
110479.9	922	7.694465		116471.3	972	8.111735
110599.8	923	7.702810		116591.1	973	8.120080
110719.6	924	7.711155		116710.9	974	8.128426
110839.4	925	7.719501		116830.7	975	8.136771
110959.2	926	7.727846		116950.6	976	8.145117
111079.1	927	7.736192		117070.4	977	8.153462
111198.9	928	7.744537		117190.2	978	8.161807
111318.7	929	7.752883		117310.0	979	8.170153
111438.6	930	7.761228		117429.9	980	8.178498
111558.4	931	7.769573		117549.7	981	8.186844
111678.2	932	7.777919		117669.5	982	8.195189
111798.0	933	7.786264		117789.4	983	8.203534
111917.9	934	7.794610		117909.2	984	8.211880
112037.7	935	7.802955		118029.0	985	8.220225
112157.5	936	7.811300		118148.8	986	8.228571
112277.3	937	7.819646		118268.7	987	8.236916
112397.2	938	7.827991		118388.5	988	8.245262
112517.0	939	7.836337		118508.3	989	8.253607
112636.8	940	7.844682		118628.1	990	8.261952
112756.6	941	7.853027		118748.0	991	8.270298
112876.5	942	7.861373		118867.8	992	8.278643
112996.3	943	7.869718		118987.6	993	8.286988
113116.1	944	7.878064		119107.4	994	8.295334
113235.9	945	7.886409		119227.3	995	8.303679
113355.8	946	7.894754		119347.1	996	8.312025
113475.6	947	7.903100		119466.9	997	8.320370
113595.4	948	7.911445		119586.7	998	8.328716
113715.3	949	7.919791		119706.6	999	8.337061
113835.1	950	7.928136		119826.4	1000	8.345406

23

LIQUID MASS DENSITY (BRITISH)

Pound (mass)/British Liquid Gallon — Kilogram/Cubic-metre
(lbm/U.K. gal — kg/m^3)

Pound (mass)/British Liquid Gallon — Gram/Cubic-centimetre
(lbm/U.K. gal — g/cm^3)

Pound (mass)/British Liquid Gallon — Kilogram/Litre
(lbm/U.K. gal — kg/l)

Pound (mass)/British Liquid Gallon — Gram/Litre
(lbm/U.K. gal — g/l)

Pound (mass) of substance/British Liquid Gallon of pure water — Milligram of
substance/Litre of pure water — Concentration or
parts per million of water (ppm)
(lbm/U.K. gal — mg/l (H$_2$O) — ppm (H$_2$O)

Conversion Factors:

> 1 pound (mass)/British Liquid Gallon (lbm/U. K. gal) = 99.77633
> kilogram/cubic-metre (kg/m^3)
> (1 kg/m^3 = 0.01002242 lbm/U.K. gal)

Concentration in Pure Water (ppm)

Parts per million in water is a measure of proportion by weight that is equivalent to a unit of weight of a substance (solute), in a mixture or solution, per million units of weight of pure water (solvent). This is easily related in SI through the following designation of units:

1 milligram of substance/1 litre of water = 1 part per million of water (ppm)

Where: for most uses, 1 gram of water = 1 cm^3 in volume.

Proof: $(mg)_s/(1)_w = (mg)_s/(dm^3)_w = (10^{-3} g)_s/(10^3 cm^3)_w$
$= (10^{-3} g)_s/(10^3 g)_w = 1 (g)_s/10^6 (g)_w$

Where: s denotes solute substance, and w denotes water.

The following diagram will assist the reader in using Table 23 for concentration conversions and calculations. It is derived from: 1 (kg) substance/(m^3) water $= 1000 \ (mg)_s/(1)_w = 10^3$ ppm.

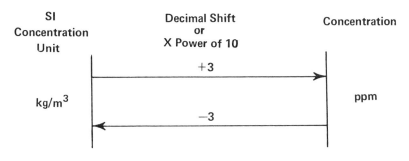

Diagram 23-A

Example (a): How many ppm is 0.25 lbm of substance in 50 U.K. gal of water?
 Since: 0.25 lbm/50 U.K. gal = 0.005 lbm/U.K. gal = 5 × 10^{-3} lbm/U.K. gal:
 From table: 5 lbm/U.K. gal = 498.8816 kg/m^3 *
 Therefore: 5 × 10^{-3} lbm/U.K. gal ≈ 499 × 10^{-3} kg/m^3.
 From Diagram 23-A: 499 × 10^{-3} kg/m^3 = 499 × 10^{-3} (10^3) ppm
                                                         = 499 ppm.

Example (b): How many lbm of substance must be added to 100 U.K. gal of water to form a 1200 ppm mixture?
 From Diagram 23-A: 1200 ppm = 1200 × 10^{-3} kg/m^3 = 120 × 10^{-2} kg/m^3
 From table: 120 kg/m^3 = 1.202690 lbm/U.K. gal*
 Therefore: 120 × 10^{-2} kg/m^3 = 1.202690 × 10^{-2} lbm/U.K. gal.
 Pounds of substance to be added to 100 U.K. gal of water
 = 1.202690 × 10^{-2} (100) = 1.202690 ≈ 1.2 lbm of substance.

In converting actual density values with Table 23 from kg/m^3 and g/l to g/cm^3 and kg/l, Diagram 23-B, which appears in the headings on alternate pages of this table, will assist the user. It is derived from:

$$1 \ kg/m^3 = 1 \ g/l = (10^{-3}) \ kg/l = (10^{-3}) \ g/cm^3.$$

New SI Unit	Decimal Shift or X Power of 10	Original SI Unit
kg/l or g/cm^3	←—— −3 ——	kg/m^3 or g/l

Diagram 23-B

*All conversion values obtained can be rounded to desired number of significant figures.

Example (*c*): Convert 20.3 lbm/U.K. gal to kg/m^3, g/l, kg/l, g/cm^3.
    Since: 20.3 lbm/U.K. gal = 203 X 10^{-1} lbm/U.K. gal:
    From table: 203 lbm/U.K. gal = 20254.59 kg/m^3 *
    Therefore: 20.3 lbm/U.K. gal = 2025.459 kg/m^3 ≈ 2025 kg/m^3.
    From Diagram 23-B: 2025 kg/m^3 = 2025 g/l = 2025 X 10^{-3} kg/l (or g/cm^3)
                            = 2.025 kg/l (or g/cm^3)
                            (kg/m^3 to kg/l by multiplying kg/m^3 value by 10^{-3} or
                            by shifting its decimal 3 places to the left.)

Diagram 23-C is also given at the top of alternate table pages:

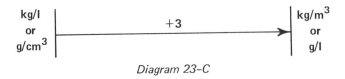

Diagram 23-C

It is the opposite of Diagram 23-B and is used to convert kg/l and g/cm^3 to g/l or
kg/m^3. The kg/m^3 (or g/l) equivalent is then applied to the REF column in
the table to obtain lbm/U.K. gal.

Example (*d*): Convert 1.37 kg/l to lbm/U.K. gal.
    From Diagram 23-C: 1.37 kg/l = 1.37 X 10^3 kg/m^3 = 137 X 10 kg/m^3
    From table: 137 kg/m^3 = 1.373071 lbm/U.K. gal*
    Therefore: 137 X 10 kg/m^3 = 13.73071 ≈ 13.73 lbm/U.K. gal.

---

*All conversion values obtained can be rounded to desired number of significant figures.

kg/m³ g/l	REF	lbm/U.K. gal	kg/m³ g/l	REF	lbm/U.K. gal
99.77633	1	0.01002242	5088.593	51	0.5111433
199.5527	2	0.02004483	5188.369	52	0.5211657
299.3290	3	0.03006725	5288.146	53	0.5311881
399.1053	4	0.04008967	5387.922	54	0.5412105
498.8816	5	0.05011209	5487.698	55	0.5512329
598.6580	6	0.06013450	5587.474	56	0.5612554
698.4343	7	0.07015692	5687.251	57	0.5712778
798.2106	8	0.08017934	5787.027	58	0.5813002
897.9870	9	0.09020176	5886.803	59	0.5913226
997.7633	10	0.1102242	5986.580	60	0.6013450
1097.540	11	0.1102466	6086.356	61	0.6113674
1197.316	12	0.1202690	6186.132	62	0.6213899
1297.092	13	0.1302914	6285.909	63	0.6314123
1396.869	14	0.1403138	6385.685	64	0.6414347
1496.645	15	0.1503363	6485.461	65	0.6514571
1596.421	16	0.1603587	6585.238	66	0.6614795
1696.198	17	0.1703811	6685.014	67	0.6715019
1795.974	18	0.1804035	6784.790	68	0.6815244
1895.750	19	0.1904259	6884.567	69	0.6915468
1995.527	20	0.2004483	6984.343	70	0.7015692
2095.303	21	0.2104708	7084.119	71	0.7115916
2195.079	22	0.2204932	7183.896	72	0.7216140
2294.856	23	0.2305156	7283.672	73	0.7316365
2394.632	24	0.2405380	7383.448	74	0.7416589
2494.408	25	0.2505604	7483.225	75	0.7516813
2594.185	26	0.2605828	7583.001	76	0.7617037
2693.961	27	0.2706053	7682.777	77	0.7717261
2793.737	28	0.2806277	7782.554	78	0.7817485
2893.514	29	0.2906501	7882.330	79	0.7917710
2993.290	30	0.3006725	7982.106	80	0.8017934
3093.066	31	0.3106949	8081.883	81	0.8118158
3192.843	32	0.3207173	8181.659	82	0.8218382
3292.619	33	0.3307398	8281.435	83	0.8318606
3392.395	34	0.3407622	8381.212	84	0.8418830
3492.172	35	0.3507846	8480.988	85	0.8519055
3591.948	36	0.3608070	8580.764	86	0.8619279
3691.724	37	0.3708294	8680.541	87	0.8719503
3791.501	38	0.3808519	8780.317	88	0.8819727
3891.277	39	0.3908743	8880.093	89	0.8919951
3991.053	40	0.4008967	8979.870	90	0.9020175
4090.830	41	0.4109191	9079.646	91	0.9120400
4190.606	42	0.4209415	9179.422	92	0.9220624
4290.382	43	0.4309639	9279.199	93	0.9320848
4390.159	44	0.4409864	9378.975	94	0.9421072
4489.935	45	0.4510088	9478.751	95	0.9521296
4589.711	46	0.4610312	9578.528	96	0.9621520
4689.487	47	0.4710536	9678.304	97	0.9721745
4789.264	48	0.4810760	9778.080	98	0.9821969
4889.040	49	0.4910984	9877.857	99	0.9922193
4988.817	50	0.5011209	9977.633	100	1.002242

kg/m³ g/l	REF	lbm/U.K. gal	kg/m³ g/l	REF	lbm/U.K. gal
10077.41	101	1.012264	15066.23	151	1.513385
10177.19	102	1.022287	15166.00	152	1.523407
10276.96	103	1.032309	15265.78	153	1.533430
10376.74	104	1.042331	15365.55	154	1.543452
10476.51	105	1.052354	15465.33	155	1.553475
10576.29	106	1.062376	15565.11	156	1.563497
10676.07	107	1.072399	15664.88	157	1.573519
10775.84	108	1.082421	15764.66	158	1.583542
10875.62	109	1.092443	15864.44	159	1.593564
10975.40	110	1.102466	15964.21	160	1.603587
11075.17	111	1.112488	16063.99	161	1.613609
11174.95	112	1.122511	16163.77	162	1.623632
11274.73	113	1.132533	16263.54	163	1.633654
11374.50	114	1.142556	16363.32	164	1.643676
11474.28	115	1.152578	16463.09	165	1.653699
11574.05	116	1.162600	16562.87	166	1.663721
11673.83	117	1.172623	16662.65	167	1.673744
11773.61	118	1.182645	16762.42	168	1.683766
11873.38	119	1.192668	16862.20	169	1.693788
11973.16	120	1.202690	16961.98	170	1.703811
12072.94	121	1.212712	17061.75	171	1.713833
12172.71	122	1.222735	17161.53	172	1.723856
12272.49	123	1.232757	17261.31	173	1.733878
12372.26	124	1.242780	17361.08	174	1.743901
12472.04	125	1.252802	17460.86	175	1.753923
12571.82	126	1.262825	17560.63	176	1.763945
12671.59	127	1.272847	17660.41	177	1.773968
12771.37	128	1.282869	17760.19	178	1.783990
12871.15	129	1.292892	17859.96	179	1.794013
12970.92	130	1.302914	17959.74	180	1.804035
13070.70	131	1.312937	18059.52	181	1.814058
13170.48	132	1.322959	18159.29	182	1.824080
13270.25	133	1.332981	18259.07	183	1.834102
13370.03	134	1.343004	18358.84	184	1.844125
13469.80	135	1.353026	18458.62	185	1.854147
13569.58	136	1.363049	18558.40	186	1.864170
13669.36	137	1.373071	18658.17	187	1.874192
13769.13	138	1.383094	18757.95	188	1.884214
13868.91	139	1.393116	18857.73	189	1.894237
13968.69	140	1.403138	18957.50	190	1.904259
14068.46	141	1.413161	19057.28	191	1.914282
14168.24	142	1.423183	19157.06	192	1.924304
14268.02	143	1.433206	19256.83	193	1.934327
14367.79	144	1.443228	19356.61	194	1.944349
14467.57	145	1.453250	19456.38	195	1.954371
14567.34	146	1.463273	19556.16	196	1.964394
14667.12	147	1.473295	19655.94	197	1.974416
14766.90	148	1.483318	19755.71	198	1.984439
14866.67	149	1.493340	19855.49	199	1.994461
14966.45	150	1.503363	19955.27	200	2.004483

kg/m³ g/l	REF	lbm/U.K. gal	kg/m³ g/l	REF	lbm/U.K. gal
20055.04	201	2.014506	25043.86	251	2.515627
20154.82	202	2.024528	25143.64	252	2.525649
20254.59	203	2.034551	25243.41	253	2.535672
20354.37	204	2.044573	25343.19	254	2.545694
20454.15	205	2.054596	25442.96	255	2.555716
20553.92	206	2.064618	25542.74	256	2.565739
20653.70	207	2.074640	25642.52	257	2.575761
20753.48	208	2.084663	25742.29	258	2.585784
20853.25	209	2.094685	25842.07	259	2.595806
20953.03	210	2.104708	25941.85	260	2.605828
21052.81	211	2.114730	26041.62	261	2.615851
21152.58	212	2.124752	26141.40	262	2.625873
21252.36	213	2.134775	26241.17	263	2.635896
21352.13	214	2.144797	26340.95	264	2.645918
21451.91	215	2.154820	26440.73	265	2.655941
21551.69	216	2.164842	26540.50	266	2.665963
21651.46	217	2.174865	26640.28	267	2.675985
21751.24	218	2.184887	26740.06	268	2.686008
21851.02	219	2.194909	26839.83	269	2.696030
21950.79	220	2.204932	26939.61	270	2.706053
22050.57	221	2.214954	27039.39	271	2.716075
22150.35	222	2.224977	27139.16	272	2.726097
22250.12	223	2.234999	27238.94	273	2.736120
22349.90	224	2.245021	27338.71	274	2.746142
22449.67	225	2.255044	27438.49	275	2.756165
22549.45	226	2.265066	27538.27	276	2.766187
22649.23	227	2.275089	27638.04	277	2.776210
22749.00	228	2.285111	27737.82	278	2.786232
22848.78	229	2.295134	27837.60	279	2.796254
22948.56	230	2.305156	27937.37	280	2.806277
23048.33	231	2.315178	28037.15	281	2.816299
23148.11	232	2.325201	28136.93	282	2.826322
23247.89	233	2.335223	28236.70	283	2.836344
23347.66	234	2.345246	28336.48	284	2.846366
23447.44	235	2.355268	28436.25	285	2.856389
23547.21	236	2.365290	28536.03	286	2.866411
23646.99	237	2.375313	28635.81	287	2.876434
23746.77	238	2.385335	28735.58	288	2.886456
23846.54	239	2.395358	28835.36	289	2.896479
23946.32	240	2.405380	28935.14	290	2.906501
24046.10	241	2.415403	29034.91	291	2.916523
24145.87	242	2.425425	29134.69	292	2.926546
24245.65	243	2.435447	29234.46	293	2.936568
24345.42	244	2.445470	29334.24	294	2.946591
24445.20	245	2.455492	29434.02	295	2.956613
24544.98	246	2.465515	29533.79	296	2.966635
24644.75	247	2.475537	29633.57	297	2.976658
24744.53	248	2.485559	29733.35	298	2.986680
24844.31	249	2.495582	29833.12	299	2.996703
24944.08	250	2.505604	29932.90	300	3.006725

kg/m³ · g/l	REF	lbm/U.K. gal	kg/m³ g/l	REF	lbm/U.K. gal
30032.68	301	3.016748	35021.49	351	3.517868
30132.45	302	3.026770	35121.27	352	3.527891
30232.23	303	3.036792	35221.04	353	3.537913
30332.00	304	3.046815	35320.82	354	3.547936
30431.78	305	3.056837	35420.60	355	3.557958
30531.56	306	3.066860	35520.37	356	3.567981
30631.33	307	3.076882	35620.15	357	3.578003
30731.11	308	3.086904	35719.93	358	3.588025
30830.89	309	3.096927	35819.70	359	3.598048
30930.66	310	3.106949	35919.48	360	3.608070
31030.44	311	3.116972	36019.26	361	3.618093
31130.21	312	3.126994	36119.03	362	3.628115
31229.99	313	3.137017	36218.81	363	3.638137
31329.77	314	3.147039	36318.58	364	3.648160
31429.54	315	3.157061	36418.36	365	3.658182
31529.32	316	3.167084	36518.14	366	3.668205
31629.10	317	3.177106	36617.91	367	3.678227
31728.87	318	3.187129	36717.69	368	3.688250
31828.65	319	3.197151	36817.47	369	3.698272
31928.43	320	3.207173	36917.24	370	3.708294
32028.20	321	3.217196	37017.02	371	3.718317
32127.98	322	3.227218	37116.79	372	3.728339
32227.75	323	3.237241	37216.57	373	3.738362
32327.53	324	3.247263	37316.35	374	3.748384
32427.31	325	3.257286	37416.12	375	3.758406
32527.08	326	3.267308	37515.90	376	3.768429
32626.86	327	3.277330	37615.68	377	3.778451
32726.64	328	3.287353	37715.45	378	3.788474
32826.41	329	3.297375	37815.23	379	3.798496
32926.19	330	3.307398	37915.01	380	3.808519
33025.97	331	3.317420	38014.78	381	3.818541
33125.74	332	3.327442	38114.56	382	3.828563
33225.52	333	3.337465	38214.33	383	3.838586
33325.29	334	3.347487	38314.11	384	3.848608
33425.07	335	3.357510	38413.89	385	3.858631
33524.85	336	3.367532	38513.66	386	3.868653
33624.62	337	3.377555	38613.44	387	3.878675
33724.40	338	3.387577	38713.22	388	3.888698
33824.18	339	3.397599	38812.99	389	3.898720
33923.95	340	3.407622	38912.77	390	3.908743
34023.73	341	3.417644	39012.54	391	3.918765
34123.50	342	3.427667	39112.32	392	3.928788
34223.28	343	3.437689	39212.10	393	3.938310
34323.06	344	3.447711	39311.87	394	3.948832
34422.83	345	3.457734	39411.65	395	3.958855
34522.61	346	3.467756	39511.43	396	3.968877
34622.39	347	3.477779	39611.20	397	3.978900
34722.16	348	3.487801	39710.98	398	3.988922
34821.94	349	3.497824	39810.76	399	3.998944
34921.72	350	3.507846	39910.53	400	4.003967

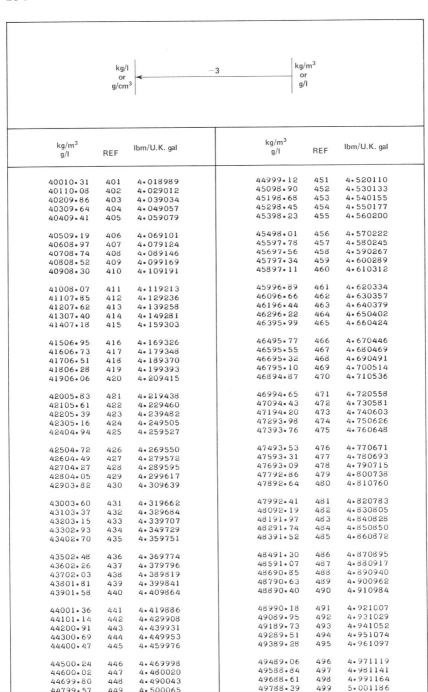

	kg/l or g/cm³	← −3 →	kg/m³ or g/l	

kg/m³ g/l	REF	lbm/U.K. gal	kg/m³ g/l	REF	lbm/U.K. gal
40010.31	401	4.018989	44999.12	451	4.520110
40110.08	402	4.029012	45098.90	452	4.530133
40209.86	403	4.039034	45198.68	453	4.540155
40309.64	404	4.049057	45298.45	454	4.550177
40409.41	405	4.059079	45398.23	455	4.560200
40509.19	406	4.069101	45498.01	456	4.570222
40608.97	407	4.079124	45597.78	457	4.580245
40708.74	408	4.089146	45697.56	458	4.590267
40808.52	409	4.099169	45797.34	459	4.600289
40908.30	410	4.109191	45897.11	460	4.610312
41008.07	411	4.119213	45996.89	461	4.620334
41107.85	412	4.129236	46096.66	462	4.630357
41207.62	413	4.139258	46196.44	463	4.640379
41307.40	414	4.149281	46296.22	464	4.650402
41407.18	415	4.159303	46395.99	465	4.660424
41506.95	416	4.169326	46495.77	466	4.670446
41606.73	417	4.179348	46595.55	467	4.680469
41706.51	418	4.189370	46695.32	468	4.690491
41806.28	419	4.199393	46795.10	469	4.700514
41906.06	420	4.209415	46894.87	470	4.710536
42005.83	421	4.219438	46994.65	471	4.720558
42105.61	422	4.229460	47094.43	472	4.730581
42205.39	423	4.239482	47194.20	473	4.740603
42305.16	424	4.249505	47293.98	474	4.750626
42404.94	425	4.259527	47393.76	475	4.760648
42504.72	426	4.269550	47493.53	476	4.770671
42604.49	427	4.279572	47593.31	477	4.780693
42704.27	428	4.289595	47693.09	478	4.790715
42804.05	429	4.299617	47792.86	479	4.800738
42903.82	430	4.309639	47892.64	480	4.810760
43003.60	431	4.319662	47992.41	481	4.820783
43103.37	432	4.329684	48092.19	482	4.830805
43203.15	433	4.339707	48191.97	483	4.840828
43302.93	434	4.349729	48291.74	484	4.850850
43402.70	435	4.359751	48391.52	485	4.860872
43502.48	436	4.369774	48491.30	486	4.870895
43602.26	437	4.379796	48591.07	487	4.880917
43702.03	438	4.389819	48690.85	488	4.890940
43801.81	439	4.399841	48790.63	489	4.900962
43901.58	440	4.409864	48890.40	490	4.910984
44001.36	441	4.419886	48990.18	491	4.921007
44101.14	442	4.429908	49089.95	492	4.931029
44200.91	443	4.439931	49189.73	493	4.941052
44300.69	444	4.449953	49289.51	494	4.951074
44400.47	445	4.459976	49389.28	495	4.961097
44500.24	446	4.469998	49489.06	496	4.971119
44600.02	447	4.480020	49588.84	497	4.981141
44699.80	448	4.490043	49688.61	498	4.991164
44799.57	449	4.500065	49788.39	499	5.001186
44899.35	450	4.510088	49888.17	500	5.011209

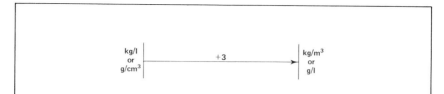

kg/m³ g/l	REF	lbm/U.K. gal	kg/m³ g/l	REF	lbm/U.K. gal
49987.94	501	5.021231	54976.76	551	5.522352
50087.72	502	5.031253	55076.53	552	5.532374
50187.49	503	5.041276	55176.31	553	5.542397
50287.27	504	5.051298	55276.09	554	5.552419
50387.05	505	5.061321	55375.86	555	5.562442
50486.82	506	5.071343	55475.64	556	5.572464
50586.60	507	5.081366	55575.42	557	5.582486
50686.38	508	5.091388	55675.19	558	5.592509
50786.15	509	5.101410	55774.97	559	5.602531
50885.93	510	5.111433	55874.74	560	5.612554
50985.70	511	5.121455	55974.52	561	5.622576
51085.48	512	5.131478	56074.30	562	5.632598
51185.26	513	5.141500	56174.07	563	5.642621
51285.03	514	5.151522	56273.85	564	5.652643
51384.81	515	5.161545	56373.63	565	5.662666
51484.59	516	5.171567	56473.40	566	5.672688
51584.36	517	5.181590	56573.18	567	5.682711
51684.14	518	5.191612	56672.96	568	5.692733
51783.92	519	5.201635	56772.73	569	5.702755
51883.69	520	5.211657	56872.51	570	5.712778
51983.47	521	5.221679	56972.28	571	5.722800
52083.24	522	5.231702	57072.06	572	5.732823
52183.02	523	5.241724	57171.84	573	5.742845
52282.80	524	5.251747	57271.61	574	5.752867
52382.57	525	5.261769	57371.39	575	5.762890
52482.35	526	5.271791	57471.17	576	5.772912
52582.13	527	5.281814	57570.94	577	5.782935
52681.90	528	5.291836	57670.72	578	5.792957
52781.68	529	5.301859	57770.50	579	5.802980
52881.46	530	5.311881	57870.27	580	5.813002
52981.23	531	5.321904	57970.05	581	5.823024
53081.01	532	5.331926	58069.82	582	5.833047
53180.78	533	5.341948	58169.60	583	5.843069
53280.56	534	5.351971	58269.38	584	5.853092
53380.34	535	5.361993	58369.15	585	5.863114
53480.11	536	5.372016	58468.93	586	5.873136
53579.89	537	5.382038	58568.71	587	5.883159
53679.67	538	5.392060	58668.48	588	5.893181
53779.44	539	5.402083	58768.26	589	5.903204
53879.22	540	5.412105	58868.03	590	5.913226
53978.99	541	5.422128	58967.81	591	5.923249
54078.77	542	5.432150	59067.59	592	5.933271
54178.55	543	5.442173	59167.36	593	5.943293
54278.32	544	5.452195	59267.14	594	5.953316
54378.10	545	5.462217	59366.92	595	5.963338
54477.88	546	5.472240	59466.69	596	5.973361
54577.65	547	5.482262	59566.47	597	5.983383
54677.43	548	5.492285	59666.25	598	5.993405
54777.21	549	5.502307	59766.02	599	6.003428
54876.98	550	5.512329	59865.80	600	6.013450

kg/m³ g/l	REF	lbm/U.K. gal	kg/m³ g/l	REF	lbm/U.K. gal
59965·57	601	6·023473	64954·39	651	6·524594
60065·35	602	6·033495	65054·17	652	6·534616
60165·13	603	6·043518	65153·94	653	6·544638
60264·90	604	6·053540	65253·72	654	6·554661
60364·68	605	6·063562	65353·50	655	6·564683
60464·46	606	6·073585	65453·27	656	6·574706
60564·23	607	6·083607	65553·05	657	6·584728
60664·01	608	6·093630	65652·83	658	6·594751
60763·79	609	6·103652	65752·60	659	6·604773
60863·56	610	6·113674	65852·38	660	6·614795
60963·34	611	6·123697	65952·15	661	6·624818
61063·11	612	6·133719	66051·93	662	6·634840
61162·89	613	6·143742	66151·71	663	6·644863
61262·67	614	6·153764	66251·48	664	6·654885
61362·44	615	6·163787	66351·26	665	6·664907
61462·22	616	6·173809	66451·04	666	6·674930
61562·00	617	6·183831	66550·81	667	6·684952
61661·77	618	6·193854	66650·59	668	6·694975
61761·55	619	6·203876	66750·37	669	6·704997
61861·32	620	6·213899	66850·14	670	6·715020
61961·10	621	6·223921	66949·92	671	6·725042
62060·88	622	6·233943	67049·69	672	6·735064
62160·65	623	6·243966	67149·47	673	6·745087
62260·43	624	6·253988	67249·25	674	6·755109
62360·21	625	6·264011	67349·02	675	6·765132
62459·98	626	6·274033	67448·80	676	6·775154
62559·76	627	6·284056	67548·58	677	6·785176
62659·54	628	6·294078	67648·35	678	6·795199
62759·31	629	6·304100	67748·13	679	6·805221
62859·09	630	6·314123	67847·90	680	6·815244
62958·86	631	6·324145	67947·68	681	6·825266
63058·64	632	6·334168	68047·46	682	6·835289
63158·42	633	6·344190	68147·23	683	6·845311
63258·19	634	6·354212	68247·01	684	6·855333
63357·97	635	6·364235	68346·79	685	6·865356
63457·75	636	6·374257	68446·56	686	6·875378
63557·52	637	6·384280	68546·34	687	6·885401
63657·30	638	6·394302	68646·12	688	6·895423
63757·07	639	6·404325	68745·89	689	6·905445
63856·85	640	6·414347	68845·67	690	6·915468
63956·63	641	6·424369	68945·44	691	6·925490
64056·40	642	6·434392	69045·22	692	6·935513
64156·18	643	6·444414	69145·00	693	6·945535
64255·96	644	6·454437	69244·77	694	6·955558
64355·73	645	6·464459	69344·55	695	6·965580
64455·51	646	6·474482	69444·33	696	6·975602
64555·29	647	6·484504	69544·10	697	6·985625
64655·06	648	6·494526	69643·88	698	6·995647
64754·84	649	6·504549	69743·65	699	7·005670
64854·61	650	6·514571	69843·43	700	7·015692

kg/m³ g/l	REF	lbm/U.K. gal	kg/m³ g/l	REF	lbm/U.K. gal
69943.21	701	7.025714	74932.02	751	7.526835
70042.98	702	7.035737	75031.80	752	7.536858
70142.76	703	7.045759	75131.58	753	7.546880
70242.54	704	7.055782	75231.35	754	7.556903
70342.31	705	7.065804	75331.13	755	7.566925
70442.09	706	7.075827	75430.91	756	7.576947
70541.87	707	7.085849	75530.68	757	7.586970
70641.64	708	7.095871	75630.46	758	7.596992
70741.42	709	7.105894	75730.23	759	7.607015
70841.19	710	7.115916	75830.01	760	7.617037
70940.97	711	7.125939	75929.79	761	7.627059
71040.75	712	7.135961	76029.56	762	7.637082
71140.52	713	7.145983	76129.34	763	7.647104
71240.30	714	7.156006	76229.12	764	7.657127
71340.08	715	7.166028	76328.89	765	7.667149
71439.85	716	7.176051	76428.67	766	7.677172
71539.63	717	7.186073	76528.45	767	7.687194
71639.41	718	7.196096	76628.22	768	7.697216
71739.18	719	7.206118	76728.00	769	7.707239
71838.96	720	7.216140	76827.77	770	7.717261
71938.73	721	7.226163	76927.55	771	7.727284
72038.51	722	7.236185	77027.33	772	7.737306
72138.29	723	7.246208	77127.10	773	7.747328
72238.06	724	7.256230	77226.88	774	7.757351
72337.84	725	7.266252	77326.66	775	7.767373
72437.62	726	7.276275	77426.43	776	7.777396
72537.39	727	7.286297	77526.21	777	7.787418
72637.17	728	7.296320	77625.98	778	7.797441
72736.94	729	7.306342	77725.76	779	7.807463
72836.72	730	7.316365	77825.54	780	7.817485
72936.50	731	7.326387	77925.31	781	7.827508
73036.27	732	7.336409	78025.09	782	7.837530
73136.05	733	7.346432	78124.87	783	7.847553
73235.83	734	7.356454	78224.64	784	7.857575
73335.60	735	7.366477	78324.42	785	7.867597
73435.38	736	7.376499	78424.20	786	7.877620
73535.16	737	7.386521	78523.97	787	7.887642
73634.93	738	7.396544	78623.75	788	7.897665
73734.71	739	7.406566	78723.52	789	7.907687
73834.48	740	7.416589	78823.30	790	7.917710
73934.26	741	7.426611	78923.08	791	7.927732
74034.04	742	7.436634	79022.85	792	7.937754
74133.81	743	7.446656	79122.63	793	7.947777
74233.59	744	7.456678	79222.41	794	7.957799
74333.37	745	7.466701	79322.18	795	7.967822
74433.14	746	7.476723	79421.96	796	7.977844
74532.92	747	7.486746	79521.74	797	7.987866
74632.70	748	7.496768	79621.51	798	7.997889
74732.47	749	7.506790	79721.29	799	8.007911
74832.25	750	7.516813	79821.06	800	8.017934

		kg/l or g/cm³	← −3 →	kg/m³ or g/l		

kg/m³ g/l	REF	lbm/U.K. gal	kg/m³ g/l	REF	lbm/U.K. gal
79920.84	801	8.027956	84909.66	851	8.529077
80020.62	802	8.037979	85009.43	852	8.539099
80120.39	803	8.048001	85109.21	853	8.549122
80220.17	804	8.058023	85208.99	854	8.559144
80319.95	805	8.068046	85308.76	855	8.569167
80419.72	806	8.078068	85408.54	856	8.579189
80519.50	807	8.088091	85508.31	857	8.589211
80619.27	808	8.098113	85608.09	858	8.599234
80719.05	809	8.108135	85707.87	859	8.609256
80818.83	810	8.118158	85807.64	860	8.619279
80918.60	811	8.128180	85907.42	861	8.629301
81018.38	812	8.138203	86007.20	862	8.639324
81118.16	813	8.148225	86106.97	863	8.649346
81217.93	814	8.158248	86206.75	864	8.659368
81317.71	815	8.168270	86306.53	865	8.669391
81417.49	816	8.178292	86406.30	866	8.679413
81517.26	817	8.188315	86506.08	867	8.689436
81617.04	818	8.198337	86605.85	868	8.699458
81716.81	819	8.208360	86705.63	869	8.709481
81816.59	820	8.218382	86805.41	870	8.719503
81916.37	821	8.228405	86905.18	871	8.729525
82016.14	822	8.238427	87004.96	872	8.739548
82115.92	823	8.248449	87104.74	873	8.749570
82215.70	824	8.258472	87204.51	874	8.759593
82315.47	825	8.268494	87304.29	875	8.769615
82415.25	826	8.278517	87404.07	876	8.779637
82515.03	827	8.288539	87503.84	877	8.789660
82614.80	828	8.298561	87603.62	878	8.799682
82714.58	829	8.308584	87703.39	879	8.809705
82814.35	830	8.318606	87803.17	880	8.819727
82914.13	831	8.328629	87902.95	881	8.829750
83013.91	832	8.338651	88002.72	882	8.839772
83113.68	833	8.348673	88102.50	883	8.849794
83213.46	834	8.358696	88202.28	884	8.859817
83313.24	835	8.368718	88302.05	885	8.869839
83413.01	836	8.378741	88401.83	886	8.879862
83512.79	837	8.388763	88501.60	887	8.889884
83612.56	838	8.398786	88601.38	888	8.899906
83712.34	839	8.408808	88701.16	889	8.909929
83812.12	840	8.418830	88800.93	890	8.919951
83911.89	841	8.428853	88900.71	891	8.929974
84011.67	842	8.438875	89000.49	892	8.939996
84111.45	843	8.448898	89100.26	893	8.950019
84211.22	844	8.458920	89200.04	894	8.960041
84311.00	845	8.468943	89299.82	895	8.970063
84410.78	846	8.478965	89399.59	896	8.980086
84510.55	847	8.488987	89499.37	897	8.990108
84610.33	848	8.499010	89599.14	898	9.000131
84710.10	849	8.509032	89698.92	899	9.010153
84809.88	850	8.519055	89798.70	900	9.020175

kg/m³ g/l	REF	lbm/U.K. gal	kg/m³ g/l	REF	lbm/U.K. gal
89898.47	901	9.030198	94887.29	951	9.531319
89998.25	902	9.040220	94987.07	952	9.541341
90098.03	903	9.050243	95086.84	953	9.551364
90197.80	904	9.060265	95186.62	954	9.561386
90297.58	905	9.070288	95286.40	955	9.571408
90397.35	906	9.080310	95386.17	956	9.581431
90497.13	907	9.090332	95485.95	957	9.591453
90596.91	908	9.100355	95585.72	958	9.601476
90696.68	909	9.110377	95685.50	959	9.611498
90796.46	910	9.120400	95785.28	960	9.621521
90896.24	911	9.130422	95885.05	961	9.631543
90996.01	912	9.140445	95984.83	962	9.641565
91095.79	913	9.150467	96084.61	963	9.651588
91195.57	914	9.160489	96184.38	964	9.661610
91295.34	915	9.170512	96284.16	965	9.671633
91395.12	916	9.180534	96383.93	966	9.681655
91494.89	917	9.190557	96483.71	967	9.691677
91594.67	918	9.200579	96583.49	968	9.701700
91694.45	919	9.210601	96683.26	969	9.711722
91794.22	920	9.220624	96783.04	970	9.721745
91894.00	921	9.230646	96882.82	971	9.731767
91993.78	922	9.240669	96982.59	972	9.741789
92093.55	923	9.250691	97082.37	973	9.751812
92193.33	924	9.260713	97182.15	974	9.761834
92293.11	925	9.270736	97281.92	975	9.771857
92392.88	926	9.280758	97381.70	976	9.781879
92492.66	927	9.290781	97481.47	977	9.791902
92592.43	928	9.300803	97581.25	978	9.801924
92692.21	929	9.310826	97681.03	979	9.811946
92791.99	930	9.320848	97780.80	980	9.821969
92891.76	931	9.330870	97880.58	981	9.831991
92991.54	932	9.340893	97980.36	982	9.842014
93091.32	933	9.350915	98080.13	983	9.852036
93191.09	934	9.360938	98179.91	984	9.862059
93290.87	935	9.370960	98279.68	985	9.872081
93390.64	936	9.380983	98379.46	986	9.882103
93490.42	937	9.391005	98479.24	987	9.892126
93590.20	938	9.401027	98579.01	988	9.902148
93689.97	939	9.411050	98678.79	989	9.912171
93789.75	940	9.421072	98778.57	990	9.922193
93889.53	941	9.431095	98878.34	991	9.932215
93989.30	942	9.441117	98978.12	992	9.942238
94089.08	943	9.451139	99077.90	993	9.952260
94188.86	944	9.461162	99177.67	994	9.962283
94288.63	945	9.471184	99277.45	995	9.972305
94388.41	946	9.481207	99377.22	996	9.982327
94488.18	947	9.491229	99477.00	997	9.992350
94587.96	948	9.501251	99576.78	998	10.00237
94687.74	949	9.511274	99676.55	999	10.01239
94787.51	950	9.521296	99776.33	1000	10.02242

24

MOMENTUM

Pound (mass)-foot/Second — Kilogram-metre/Second
(lbm-ft/s — kg · m/s)

Pound (mass)-foot/Second — Gram-centimetre/Second
(lbm-ft/s — g · cm/s)

Pound (mass)-foot/Second — Kilogram-centimetre/Second
(lbm-ft/s — kg · cm/s)

FORCE

Poundal — Newton
(pdl — n)

Conversion Factors:
1 pound-foot/second (lbm-ft/s) = 0.1382550 kilogram-metre/second (kg · m/s)
(1 kg · m/s = 7.233011 lbm-ft/s)

Varying Momentum Time Units

Table 24 can convert momentum values in time units other than per second shown above — such as per minute (min). However, the time units must be consistent between the customary and SI quantities used. This reflects no change in the numerical value of the above conversion factor from which Table 24 is derived.

Therefore: 1 lbm-ft/min = 0.1382550 kg · m/min.

Poundal — Newton Values

Table 24 can also be used to convert poundals (pdl) to newtons (N) since the numerical value of the conversion factor is the same as that given above.

Poundal — Newton conversion factor is therefore:

1 pdl = 0.1382550 N
(1N =7.233011 pdl)

The poundal (pdl) is the customary force unit defined as that force acting on a mass of one pound which will impart to it an acceleration of 1 foot per second, squared. The newton (N), the SI base unit for force, is defined in the Introduction, page 5 .

In converting momentum quantities from kg · m/s to g · cm/s and kg · cm/s values, the following Decimal-shift Diagram 24-A, which appears in the headings on alternate pages of this table, will assist the user. It is also applicable with consistent time units other than seconds, and is derived from:

$$1 \text{ kg} \cdot \text{m/s} = (10^5) \text{ g} \cdot \text{cm/s} = (10^2) \text{ kg} \cdot \text{cm/s}.$$

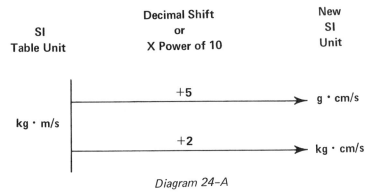

Diagram 24-A

Example (a): Convert 0.478 lbm-ft/s to kg · m/s, g · cm/s, kg · cm/s.
  Since: 0.478 lbm-ft/s = 478 × 10^{-3} lbm-ft/s :
  From table: 478 lbm-ft/s = 66.08589 kg · m/s*
  Therefore: 478 × 10^{-3} lbm-ft/s = 66.08589 × 10^{-3} = 0.06608589 kg · m/s*
      ≈ 6.61 × 10^{-2} kg · m/s
  From Diagram 24-A: 1) 0.06608589 kg · m/s = 0.06608589 (10^5) g · cm/s
                    = 6,608.589 g · cm/s* ≈ 6,610 g · cm/s
                    (kg · m/s to g · cm/s by multiplying kg · m/s value by 10^5 or by shifting its decimal five places to the right.)

                  2) 0.06608589 kg · m/s = 0.06608589 (10^2) kg · cm/s
                    = 6.608589 kg · cm/s* ≈ 6.61 kg · cm/s
                    (kg · m/s to kg · cm/s by multiplying kg · m/s value by 10^2 or by shifting its decimal 2 places to the right.)

---

*All conversion values obtained can be rounded to desired number of significant figures.

Example (*b*): Convert a force of 7,547 pdl to N.
  Since: 7,547 pdl = 75.47 × 10² pdl; and 75.47 = 75.00 + 0.47:
  From table: 75 pdl = 10.36913 N*
         47 pdl = 6.497985 N*

By shifting decimals where necessary and adding:
  75.00 pdl = 10.36913   N
  0.47 pdl = 0.06497985 N
  ―――――――――――――――――――――――
  75.47 pdl = 10.43410985 N

Therefore: 75.47 × 10² pdl = 1,043.311 N ≈ 1,043 N.

Diagram 24-B is also given at the top of alternate table pages:

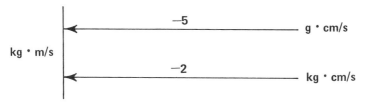

Diagram 24-B

It is the opposite of Diagram 24-A and is used to convert g · cm/s and kg · cm/s to kg · m/s values. The kg · m/s equivalent is then applied to the REF column in the table to obtain the lbm-ft/s quantity.

Example (*c*): Convert 62,900 kg · cm/s to lbm-ft/s.
  From Diagram 24-B: 62,900 kg · cm/s = 62,900 (10⁻²) kg · m/s
                = 629 kg · m/s
  From table: 629 kg · m/s = 4,549.564 lbm-ft/s* ≈ 4,550 lbm-ft/s

Example (*d*): Convert 100.5 N to pdl.
  Since: 100.5 N = (100.0 + 0.5) N:
  From table: 100 N = 723.3011 pdl*
         5 N = 36.16506 pdl*

By proper decimal shift and addition:
  100.0 N = 723.3011   pdl
  0.5 N = 3.616506 pdl
  ――――――――――――――――――――――
  100.5 N = 726.9176   pdl ≈ 726.9 pdl

―――――――――――――――

*All conversion values obtained can be rounded to desired number of significant figures.

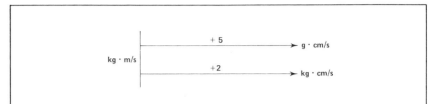

kg · m/s N	REF	lbm–ft/s pdl	kg · m/s N	REF	lbm–ft/s pdl
0·1382550	1	7·233011	7·051005	51	368·8836
0·2765100	2	14·46602	7·189260	52	376·1166
0·4147650	3	21·69903	7·327515	53	383·3496
0·5530200	4	28·93205	7·465770	54	390·5826
0·6912750	5	36·16506	7·604025	55	397·8156
0·8295300	6	43·39807	7·742280	56	405·0486
0·9677850	7	50·63108	7·880535	57	412·2817
1·106040	8	57·86409	8·018790	58	419·5147
1·244295	9	65·09710	8·157045	59	426·7477
1·382550	10	72·33012	8·295300	60	433·9807
1·520805	11	79·56313	8·433555	61	441·2137
1·659060	12	86·79614	8·571810	62	448·4467
1·797315	13	94·02915	8·710065	63	455·6797
1·935570	14	101·2622	8·848320	64	462·9127
2·073825	15	108·4952	8·986575	65	470·1457
2·212080	16	115·7282	9·124830	66	477·3788
2·350335	17	122·9612	9·263085	67	484·6118
2·488590	18	130·1942	9·401340	68	491·8448
2·626845	19	137·4272	9·539595	69	499·0778
2·765100	20	144·6602	9·677850	70	506·3108
2·903355	21	151·8932	9·816105	71	513·5438
3·041610	22	159·1263	9·954360	72	520·7768
3·179865	23	166·3593	10·09262	73	528·0098
3·318120	24	173·5923	10·23087	74	535·2429
3·456375	25	180·8253	10·36913	75	542·4759
3·594630	26	188·0583	10·50738	76	549·7089
3·732885	27	195·2913	10·64564	77	556·9419
3·871140	28	202·5243	10·78389	78	564·1749
4·009395	29	209·7573	10·92215	79	571·4079
4·147650	30	216·9903	11·06040	80	578·6409
4·285905	31	224·2234	11·19866	81	585·8739
4·424160	32	231·4564	11·33691	82	593·1069
4·562415	33	238·6894	11·47517	83	600·3400
4·700670	34	245·9224	11·61342	84	607·5730
4·838925	35	253·1554	11·75168	85	614·8060
4·977180	36	260·3884	11·88993	86	622·0390
5·115435	37	267·6214	12·02819	87	629·2720
5·253690	38	274·8544	12·16644	88	636·5050
5·391945	39	282·0874	12·30470	89	643·7380
5·530200	40	289·3205	12·44295	90	650·9710
5·668455	41	296·5535	12·58121	91	658·2040
5·806710	42	303·7865	12·71946	92	665·4371
5·944965	43	311·0195	12·85772	93	672·6701
6·083220	44	318·2525	12·99597	94	679·9031
6·221475	45	325·4855	13·13423	95	687·1361
6·359730	46	332·7185	13·27248	96	694·3691
6·497985	47	339·9515	13·41074	97	701·6021
6·636240	48	347·1846	13·54899	98	708·8351
6·774495	49	354·4176	13·68725	99	716·0681
6·912750	50	361·6506	13·82550	100	723·3011

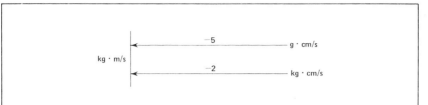

kg · m/s N	REF	lbm-ft/s pdl	kg · m/s N	REF	lbm-ft/s pdl
13.96376	101	730.5342	20.87651	151	1092.185
14.10201	102	737.7672	21.01476	152	1099.418
14.24027	103	745.0002	21.15302	153	1106.651
14.37852	104	752.2332	21.29127	154	1113.884
14.51678	105	759.4662	21.42953	155	1121.117
14.65503	106	766.6992	21.56778	156	1128.350
14.79329	107	773.9322	21.70604	157	1135.583
14.93154	108	781.1652	21.84429	158	1142.816
15.06980	109	788.3983	21.98255	159	1150.049
15.20805	110	795.6313	22.12080	160	1157.282
15.34631	111	802.8643	22.25906	161	1164.515
15.48456	112	810.0973	22.39731	162	1171.748
15.62282	113	817.3303	22.53557	163	1178.981
15.76107	114	824.5633	22.67382	164	1186.214
15.89933	115	831.7963	22.81208	165	1193.447
16.03758	116	839.0293	22.95033	166	1200.680
16.17584	117	846.2623	23.08859	167	1207.913
16.31409	118	853.4954	23.22684	168	1215.146
16.45235	119	860.7284	23.36510	169	1222.379
16.59060	120	867.9614	23.50335	170	1229.612
16.72886	121	875.1944	23.64161	171	1236.845
16.86711	122	882.4274	23.77986	172	1244.078
17.00537	123	889.6604	23.91812	173	1251.311
17.14362	124	896.8934	24.05637	174	1258.544
17.28188	125	904.1264	24.19463	175	1265.777
17.42013	126	911.3594	24.33288	176	1273.010
17.55839	127	918.5925	24.47114	177	1280.243
17.69664	128	925.8255	24.60939	178	1287.476
17.83490	129	933.0585	24.74765	179	1294.709
17.97315	130	940.2915	24.88590	180	1301.942
18.11141	131	947.5245	25.02416	181	1309.175
18.24966	132	954.7575	25.16241	182	1316.408
18.38792	133	961.9905	25.30067	183	1323.641
18.52617	134	969.2235	25.43892	184	1330.874
18.66443	135	976.4566	25.57718	185	1338.107
18.80268	136	983.6896	25.71543	186	1345.340
18.94094	137	990.9226	25.85369	187	1352.573
19.07919	138	998.1556	25.99194	188	1359.806
19.21745	139	1005.389	26.13020	189	1367.039
19.35570	140	1012.622	26.26845	190	1374.272
19.49396	141	1019.855	26.40671	191	1381.505
19.63221	142	1027.088	26.54496	192	1388.738
19.77047	143	1034.321	26.68322	193	1395.971
19.90872	144	1041.554	26.82147	194	1403.204
20.04698	145	1048.787	26.95973	195	1410.437
20.18523	146	1056.020	27.09798	196	1417.670
20.32349	147	1063.253	27.23624	197	1424.903
20.46174	148	1070.486	27.37449	198	1432.136
20.60000	149	1077.719	27.51275	199	1439.369
20.73825	150	1084.952	27.65100	200	1446.602

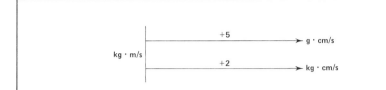

kg· m/s N	REF	lbm–ft/s pdl	kg · m/s N	REF	lbm–ft/s pdl
27.78926	201	1453.835	34.70200	251	1815.486
27.92751	202	1461.068	34.84026	252	1822.719
28.06577	203	1468.301	34.97852	253	1829.952
28.20402	204	1475.534	35.11677	254	1837.185
28.34228	205	1482.767	35.25502	255	1844.418
28.48053	206	1490.000	35.39328	256	1851.651
28.61879	207	1497.233	35.53154	257	1858.884
28.75704	208	1504.466	35.66979	258	1866.117
28.89530	209	1511.699	35.80804	259	1873.350
29.03355	210	1518.932	35.94630	260	1880.583
29.17181	211	1526.165	36.08456	261	1887.816
29.31006	212	1533.398	36.22281	262	1895.049
29.44832	213	1540.631	36.36106	263	1902.282
29.58657	214	1547.864	36.49932	264	1909.515
29.72483	215	1555.097	36.63758	265	1916.748
29.86308	216	1562.330	36.77583	266	1923.981
30.00134	217	1569.563	36.91408	267	1931.214
30.13959	218	1576.797	37.05234	268	1938.447
30.27785	219	1584.030	37.19060	269	1945.680
30.41610	220	1591.263	37.32885	270	1952.913
30.55436	221	1598.496	37.46710	271	1960.146
30.69261	222	1605.729	37.60536	272	1967.379
30.83087	223	1612.962	37.74362	273	1974.612
30.96912	224	1620.195	37.88187	274	1981.845
31.10738	225	1627.428	38.02012	275	1989.078
31.24563	226	1634.661	38.15838	276	1996.311
31.38389	227	1641.894	38.29664	277	2003.544
31.52214	228	1649.127	38.43489	278	2010.777
31.66040	229	1656.360	38.57314	279	2018.010
31.79865	230	1663.593	38.71140	280	2025.243
31.93691	231	1670.826	38.84966	281	2032.476
32.07516	232	1678.059	38.98791	282	2039.709
32.21342	233	1685.292	39.12616	283	2046.942
32.35167	234	1692.525	39.26442	284	2054.175
32.48992	235	1699.758	39.40268	285	2061.408
32.62818	236	1706.991	39.54093	286	2068.641
32.76644	237	1714.224	39.67918	287	2075.874
32.90469	238	1721.457	39.81744	288	2083.107
33.04294	239	1728.690	39.95570	289	2090.340
33.18120	240	1735.923	40.09395	290	2097.573
33.31946	241	1743.156	40.23220	291	2104.806
33.45771	242	1750.389	40.37046	292	2112.039
33.59596	243	1757.622	40.50872	293	2119.272
33.73422	244	1764.855	40.64697	294	2126.505
33.87248	245	1772.088	40.78522	295	2133.738
34.01073	246	1779.321	40.92348	296	2140.971
34.14898	247	1786.554	41.06174	297	2148.204
34.28724	248	1793.787	41.19999	298	2155.437
34.42550	249	1801.020	41.33824	299	2162.670
34.56375	250	1808.253	41.47650	300	2169.903

kg · m/s →−5 g · cm/s

kg · m/s →−2 kg · cm/s

kg · m/s N	REF	lbm-ft/s pdl	kg · m/s N	REF	lbm-ft/s pdl
41.61476	301	2177.136	48.52750	351	2538.787
41.75301	302	2184.369	48.66576	352	2546.020
41.89126	303	2191.602	48.80402	353	2553.253
42.02952	304	2198.835	48.94227	354	2560.486
42.16778	305	2206.069	49.08052	355	2567.719
42.30603	306	2213.302	49.21878	356	2574.952
42.44428	307	2220.535	49.35704	357	2582.185
42.58254	308	2227.768	49.49529	358	2589.418
42.72080	309	2235.001	49.63354	359	2596.651
42.85905	310	2242.234	49.77180	360	2603.884
42.99730	311	2249.467	49.91006	361	2611.117
43.13556	312	2256.700	50.04831	362	2618.350
43.27382	313	2263.933	50.18656	363	2625.583
43.41207	314	2271.166	50.32482	364	2632.816
43.55032	315	2278.399	50.46308	365	2640.049
43.68858	316	2285.632	50.60133	366	2647.282
43.82684	317	2292.865	50.73958	367	2654.515
43.96509	318	2300.098	50.87784	368	2661.748
44.10334	319	2307.331	51.01610	369	2668.981
44.24160	320	2314.564	51.15435	370	2676.214
44.37986	321	2321.797	51.29260	371	2683.447
44.51811	322	2329.030	51.43086	372	2690.680
44.65636	323	2336.263	51.56912	373	2697.913
44.79462	324	2343.496	51.70737	374	2705.146
44.93288	325	2350.729	51.84562	375	2712.379
45.07113	326	2357.962	51.98388	376	2719.612
45.20938	327	2365.195	52.12214	377	2726.845
45.34764	328	2372.428	52.26039	378	2734.078
45.48590	329	2379.661	52.39864	379	2741.311
45.62415	330	2386.894	52.53690	380	2748.544
45.76240	331	2394.127	52.67516	381	2755.777
45.90066	332	2401.360	52.81341	382	2763.010
46.03892	333	2408.593	52.95166	383	2770.243
46.17717	334	2415.826	53.08992	384	2777.476
46.31542	335	2423.059	53.22818	385	2784.709
46.45368	336	2430.292	53.36643	386	2791.942
46.59194	337	2437.525	53.50468	387	2799.175 .
46.73019	338	2444.758	53.64294	388	2806.408
46.86844	339	2451.991	53.78120	389	2813.641
47.00670	340	2459.224	53.91945	390	2820.874
47.14496	341	2466.457	54.05770	391	2828.107
47.28321	342	2473.690	54.19596	392	2835.341
47.42146	343	2480.923	54.33422	393	2842.574
47.55972	344	2488.156	54.47247	394	2849.807
47.69798	345	2495.389	54.61072	395	2857.040
47.83623	346	2502.622	54.74898	396	2864.273
47.97448	347	2509.855	54.88724	397	2871.506
48.11274	348	2517.088	55.02549	398	2878.739
48.25100	349	2524.321	55.16374	399	2885.972
48.38925	350	2531.554	55.30200	400	2893.205

kg · m/s N	REF	lbm-ft/s pdl	kg · m/s N	REF	lbm-ft/s pdl
55·44026	401	2900·438	62·35300	451	3262·088
55·57851	402	2907·671	62·49126	452	3269·321
55·71676	403	2914·904	62·62952	453	3276·554
55·85502	404	2922·137	62·76777	454	3283·787
55·99328	405	2929·370	62·90602	455	3291·020
56·13153	406	2936·603	63·04428	456	3298·253
56·26978	407	2943·836	63·18254	457	3305·486
56·40804	408	2951·069	63·32079	458	3312·719
56·54630	409	2958·302	63·45904	459	3319·952
56·68455	410	2965·535	63·59730	460	3327·185
56·82280	411	2972·768	63·73556	461	3334·418
56·96106	412	2980·001	63·87381	462	3341·651
57·09932	413	2987·234	64·01206	463	3348·884
57·23757	414	2994·467	64·15032	464	3356·117
57·37582	415	3001·700	64·28858	465	3363·350
57·51408	416	3008·933	64·42683	466	3370·583
57·65234	417	3016·166	64·56509	467	3377·816
57·79059	418	3023·399	64·70334	468	3385·049
57·92884	419	3030·632	64·84159	469	3392·282
58·06710	420	3037·865	64·97985	470	3399·515
58·20536	421	3045·098	65·11810	471	3406·748
58·34361	422	3052·331	65·25636	472	3413·981
58·48186	423	3059·564	65·39462	473	3421·214
58·62012	424	3066·797	65·53287	474	3428·447
58·75838	425	3074·030	65·67113	475	3435·680
58·89663	426	3081·263	65·80938	476	3442·913
59·03488	427	3088·496	65·94763	477	3450·146
59·17314	428	3095·729	66·08589	478	3457·379
59·31140	429	3102·962	66·22414	479	3464·612
59·44965	430	3110·195	66·36240	480	3471·846
59·58790	431	3117·428	66·50066	481	3479·079
59·72616	432	3124·661	66·63891	482	3486·312
59·86442	433	3131·894	66·77717	483	3493·545
60·00267	434	3139·127	66·91542	484	3500·778
60·14092	435	3146·360	67·05367	485	3508·011
60·27918	436	3153·593	67·19193	486	3515·244
60·41744	437	3160·826	67·33018	487	3522·477
60·55569	438	3168·059	67·46844	488	3529·710
60·69394	439	3175·292	67·60670	489	3536·943
60·83220	440	3182·525	67·74495	490	3544·176
60·97046	441	3189·758	67·88321	491	3551·409
61·10871	442	3196·991	68·02146	492	3558·642
61·24696	443	3204·224	68·15971	493	3565·875
61·38522	444	3211·457	68·29797	494	3573·108
61·52348	445	3218·690	68·43622	495	3580·341
61·66173	446	3225·923	68·57448	496	3587·574
61·79998	447	3233·156	68·71274	497	3594·807
61·93824	448	3240·389	68·85099	498	3602·040
62·07650	449	3247·622	68·93925	499	3609·273
62·21475	450	3254·855	69·12750	500	3616·506

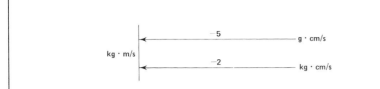

kg · m/s N	REF	lbm-ft/s pdl	kg · m/s N	REF	lbm-ft/s pdl
69.26575	501	3623.739	76.17850	551	3985.389
69.40401	502	3630.972	76.31676	552	3992.622
69.54226	503	3638.205	76.45502	553	3999.855
69.68052	504	3645.438	76.59327	554	4007.088
69.81878	505	3652.671	76.73153	555	4014.321
69.95703	506	3659.904	76.86978	556	4021.554
70.09529	507	3667.137	77.00803	557	4028.787
70.23354	508	3674.370	77.14629	558	4036.020
70.37179	509	3681.603	77.28454	559	4043.253
70.51005	510	3688.836	77.42280	560	4050.486
70.64830	511	3696.069	77.56106	561	4057.719
70.78656	512	3703.302	77.69931	562	4064.952
70.92482	513	3710.535	77.83757	563	4072.185
71.06307	514	3717.768	77.97582	564	4079.418
71.20133	515	3725.001	78.11407	565	4086.651
71.33958	516	3732.234	78.25233	566	4093.884
71.47783	517	3739.467	78.39058	567	4101.117
71.61609	518	3746.700	78.52884	568	4108.351
71.75434	519	3753.933	78.66710	569	4115.584
71.89260	520	3761.166	78.80535	570	4122.817
72.03086	521	3768.399	78.94361	571	4130.050
72.16911	522	3775.632	79.08186	572	4137.283
72.30737	523	3782.865	79.22011	573	4144.516
72.44562	524	3790.098	79.35837	574	4151.749
72.58387	525	3797.331	79.49662	575	4158.982
72.72213	526	3804.564	79.63488	576	4166.215
72.86038	527	3811.797	79.77314	577	4173.448
72.99864	528	3819.030	79.91139	578	4180.681
73.13690	529	3826.263	80.04965	579	4187.914
73.27515	530	3833.496	80.18790	580	4195.147
73.41341	531	3840.729	80.32615	581	4202.380
73.55166	532	3847.962	80.46441	582	4209.613
73.68991	533	3855.195	80.60266	583	4216.846
73.82817	534	3862.428	80.74092	584	4224.079
73.96642	535	3869.661	80.87918	585	4231.312
74.10468	536	3876.894	81.01743	586	4238.545
74.24294	537	3884.127	81.15569	587	4245.778
74.38119	538	3891.360	81.29394	588	4253.011
74.51945	539	3898.593	81.43219	589	4260.244
74.65770	540	3905.826	81.57045	590	4267.477
74.79595	541	3913.059	81.70870	591	4274.710
74.93421	542	3920.292	81.84696	592	4281.943
75.07246	543	3927.525	81.98522	593	4289.176
75.21072	544	3934.758	82.12347	594	4296.409
75.34898	545	3941.991	82.26173	595	4303.642
75.48723	546	3949.224	82.39998	596	4310.875
75.62549	547	3956.457	82.53823	597	4318.108
75.76374	548	3963.690	82.67649	598	4325.341
75.90199	549	3970.923	82.81474	599	4332.574
76.04025	550	3978.156	82.95300	600	4339.807

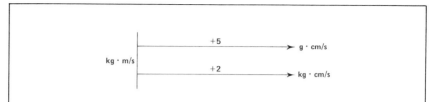

kg · m/s N	REF	lbm–ft/s pdl	kg · m/s N	REF	lbm–ft/s pdl
83.09126	601	4347.040	90.00401	651	4708.690
83.22951	602	4354.273	90.14226	652	4715.923
83.36777	603	4361.506	90.28051	653	4723.156
83.50602	604	4368.739	90.41877	654	4730.390
83.64427	605	4375.972	90.55702	655	4737.622
83.78253	606	4383.205	90.69528	656	4744.856
83.92078	607	4390.438	90.83354	657	4752.089
84.05904	608	4397.671	90.97179	658	4759.322
84.19730	609	4404.904	91.11005	659	4766.555
84.33555	610	4412.137	91.24830	660	4773.788
84.47381	611	4419.370	91.38655	661	4781.021
84.61206	612	4426.603	91.52481	662	4788.254
84.75031	613	4433.836	91.66306	663	4795.487
84.88857	614	4441.069	91.80132	664	4802.720
85.02682	615	4448.302	91.93958	665	4809.953
85.16508	616	4455.535	92.07783	666	4817.186
85.30334	617	4462.768	92.21609	667	4824.419
85.44159	618	4470.001	92.35434	668	4831.652
85.57985	619	4477.234	92.49259	669	4838.885
85.71810	620	4484.467	92.63085	670	4846.118
85.85635	621	4491.700	92.76910	671	4853.351
85.99461	622	4498.933	92.90736	672	4860.584
86.13286	623	4506.166	93.04562	673	4867.817
86.27112	624	4513.399	93.18387	674	4875.050
86.40938	625	4520.632	93.32213	675	4882.283
86.54763	626	4527.865	93.46038	676	4889.516
86.68589	627	4535.098	93.59863	677	4896.749
86.82414	628	4542.331	93.73689	678	4903.982
86.96239	629	4549.564	93.87514	679	4911.215
87.10065	630	4556.797	94.01340	680	4918.448
87.23890	631	4564.030	94.15166	681	4925.681
87.37716	632	4571.263	94.28991	682	4932.914
87.51542	633	4578.496	94.42817	683	4940.147
87.65367	634	4585.729	94.56642	684	4947.380
87.79193	635	4592.962	94.70467	685	4954.613
87.93018	636	4600.195	94.84293	686	4961.846
88.06843	637	4607.428	94.98118	687	4969.079
88.20669	638	4614.661	95.11944	688	4976.312
88.34494	639	4621.894	95.25770	689	4983.545
88.48320	640	4629.127	95.39595	690	4990.778
88.62146	641	4636.360	95.53421	691	4998.011
88.75971	642	4643.593	95.67246	692	5005.244
88.89797	643	4650.826	95.81071	693	5012.477
89.03622	644	4658.059	95.94897	694	5019.710
89.17447	645	4665.292	96.08722	695	5026.943
89.31273	646	4672.525	96.22548	696	5034.176
89.45098	647	4679.758	96.36374	697	5041.409
89.58924	648	4686.991	96.50199	698	5048.642
89.72750	649	4694.224	96.64025	699	5055.875
89.86575	650	4701.457	96.77850	700	5063.108

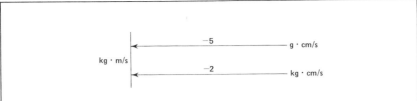

kg · m/s N	REF	lbm–ft/s pdl	kg · m/s N	REF	lbm–ft/s pdl
96.91675	701	5070.341	103.8295	751	5431.992
97.05501	702	5077.574	103.9678	752	5439.225
97.19326	703	5084.807	104.1060	753	5446.458
97.33152	704	5092.040	104.2443	754	5453.691
97.46978	705	5099.273	104.3825	755	5460.924
97.60803	706	5106.506	104.5208	756	5468.157
97.74629	707	5113.739	104.6590	757	5475.390
97.88454	708	5120.972	104.7973	758	5482.623
98.02279	709	5128.205	104.9355	759	5489.856
98.16105	710	5135.438	105.0738	760	5497.089
98.29930	711	5142.671	105.2121	761	5504.322
98.43756	712	5149.904	105.3503	762	5511.555
98.57582	713	5157.137	105.4886	763	5518.788
98.71407	714	5164.370	105.6268	764	5526.021
98.85233	715	5171.603	105.7651	765	5533.254
98.99058	716	5178.836	105.9033	766	5540.487
99.12883	717	5186.069	106.0416	767	5547.720
99.26709	718	5193.302	106.1798	768	5554.953
99.40534	719	5200.535	106.3181	769	5562.186
99.54360	720	5207.768	106.4564	770	5569.419
99.68186	721	5215.001	106.5946	771	5576.652
99.82011	722	5222.234	106.7329	772	5583.885
99.95837	723	5229.467	106.8711	773	5591.118
100.0966	724	5236.700	107.0094	774	5598.351
100.2349	725	5243.933	107.1476	775	5605.584
100.3731	726	5251.166	107.2859	776	5612.817
100.5114	727	5258.399	107.4241	777	5620.050
100.6496	728	5265.632	107.5624	778	5627.283
100.7879	729	5272.865	107.7006	779	5634.516
100.9262	730	5280.098	107.8389	780	5641.749
101.0644	731	5287.331	107.9772	781	5648.982
101.2027	732	5294.564	108.1154	782	5656.215
101.3409	733	5301.797	108.2537	783	5663.448
101.4792	734	5309.030	108.3919	784	5670.681
101.6174	735	5316.263	108.5302	785	5677.914
101.7557	736	5323.496	108.6684	786	5685.147
101.8939	737	5330.729	108.8067	787	5692.380
102.0322	738	5337.962	108.9449	788	5699.613
102.1704	739	5345.195	109.0832	789	5706.846
102.3087	740	5352.429	109.2214	790	5714.079
102.4470	741	5359.661	109.3597	791	5721.312
102.5852	742	5366.895	109.4980	792	5728.545
102.7235	743	5374.128	109.6362	793	5735.778
102.8617	744	5381.361	109.7745	794	5743.011
103.0000	745	5388.594	109.9127	795	5750.244
103.1382	746	5395.827	110.0510	796	5757.477
103.2765	747	5403.060	110.1892	797	5764.710
103.4147	748	5410.293	110.3275	798	5771.943
103.5530	749	5417.526	110.4657	799	5779.176
103.6912	750	5424.759	110.6040	800	5786.409

kg · m/s N	REF	lbm-ft/s pdl		kg · m/s N	REF	lbm-ft/s pdl
110.7423	801	5793.642		117.6550	851	6155.293
110.8805	802	5800.875		117.7933	852	6162.526
111.0188	803	5808.108		117.9315	853	6169.759
111.1570	804	5815.341		118.0698	854	6176.992
111.2953	805	5822.574		118.2080	855	6184.225
111.4335	806	5829.807		118.3463	856	6191.458
111.5718	807	5837.040		118.4845	857	6198.691
111.7100	808	5844.273		118.6228	858	6205.924
111.8483	809	5851.506		118.7610	859	6213.157
111.9866	810	5858.739		118.8993	860	6220.390
112.1248	811	5865.972		119.0376	861	6227.623
112.2631	812	5873.205		119.1758	862	6234.856
112.4013	813	5880.438		119.3141	863	6242.089
112.5396	814	5887.671		119.4523	864	6249.322
112.6778	815	5894.904		119.5906	865	6256.555
112.8161	816	5902.137		119.7288	866	6263.788
112.9543	817	5909.370		119.8671	867	6271.021
113.0926	818	5916.603		120.0053	868	6278.254
113.2308	819	5923.836		120.1436	869	6285.487
113.3691	820	5931.069		120.2818	870	6292.720
113.5074	821	5938.302		120.4201	871	6299.953
113.6456	822	5945.535		120.5584	872	6307.186
113.7839	823	5952.768		120.6966	873	6314.419
113.9221	824	5960.001		120.8349	874	6321.652
114.0604	825	5967.234		120.9731	875	6328.885
114.1986	826	5974.467		121.1114	876	6336.118
114.3369	827	5981.701		121.2496	877	6343.351
114.4751	828	5988.934		121.3879	878	6350.584
114.6134	829	5996.167		121.5261	879	6357.817
114.7516	830	6003.400		121.6644	880	6365.050
114.8899	831	6010.633		121.8027	881	6372.283
115.0282	832	6017.866		121.9409	882	6379.516
115.1664	833	6025.099		122.0792	883	6386.749
115.3047	834	6032.332		122.2174	884	6393.982
115.4429	835	6039.565		122.3557	885	6401.215
115.5812	836	6046.798		122.4939	886	6408.448
115.7194	837	6054.031		122.6322	887	6415.681
115.8577	838	6061.264		122.7704	888	6422.914
115.9959	839	6068.497		122.9087	889	6430.147
116.1342	840	6075.730		123.0470	890	6437.380
116.2725	841	6082.963		123.1852	891	6444.613
116.4107	842	6090.196		123.3235	892	6451.846
116.5490	843	6097.429		123.4617	893	6459.079
116.6872	844	6104.662		123.6000	894	6466.312
116.8255	845	6111.895		123.7382	895	6473.545
116.9637	846	6119.128		123.8765	896	6480.778
117.1020	847	6126.361		124.0147	897	6488.011
117.2402	848	6133.594		124.1530	898	6495.244
117.3785	849	6140.827		124.2912	899	6502.477
117.5168	850	6148.060		124.4295	900	6509.710

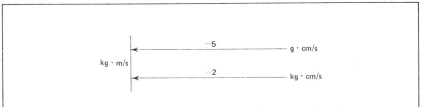

kg · m/s N	REF	lbm-ft/s pdl	kg · m/s N	REF	lbm-ft/s pdl
124.5678	901	6516.943	131.4805	951	6878.594
124.7060	902	6524.176	131.6188	952	6885.827
124.8443	903	6531.409	131.7570	953	6893.060
124.9825	904	6538.642	131.8953	954	6900.293
125.1208	905	6545.875	132.0335	955	6907.526
125.2590	906	6553.108	132.1718	956	6914.759
125.3973	907	6560.341	132.3100	957	6921.992
125.5355	908	6567.574	132.4483	958	6929.225
125.6738	909	6574.807	132.5865	959	6936.458
125.8120	910	6582.040	132.7248	960	6943.691
125.9503	911	6589.273	132.8631	961	6950.924
126.0886	912	6596.506	133.0013	962	6958.157
126.2268	913	6603.740	133.1396	963	6965.390
126.3651	914	6610.972	133.2778	964	6972.623
126.5033	915	6618.206	133.4161	965	6979.856
126.6416	916	6625.439	133.5543	966	6987.089
126.7798	917	6632.672	133.6926	967	6994.322
126.9181	918	6639.905	133.8308	968	7001.555
127.0563	919	6647.138	133.9691	969	7008.788
127.1946	920	6654.371	134.1073	970	7016.021
127.3329	921	6661.604	134.2456	971	7023.254
127.4711	922	6668.837	134.3839	972	7030.487
127.6094	923	6676.070	134.5221	973	7037.720
127.7476	924	6683.303	134.6604	974	7044.953
127.8859	925	6690.536	134.7986	975	7052.186
128.0241	926	6697.769	134.9369	976	7059.419
128.1624	927	6705.002	135.0751	977	7066.652
128.3006	928	6712.235	135.2134	978	7073.885
128.4389	929	6719.468	135.3516	979	7081.118
128.5772	930	6726.701	135.4899	980	7088.351
128.7154	931	6733.934	135.6282	981	7095.584
128.8537	932	6741.167	135.7664	982	7102.817
128.9919	933	6748.400	135.9047	983	7110.050
129.1302	934	6755.633	136.0429	984	7117.283
129.2684	935	6762.866	136.1812	985	7124.516
129.4067	936	6770.099	136.3194	986	7131.749
129.5449	937	6777.332	136.4577	987	7138.982
129.6832	938	6784.565	136.5959	988	7146.215
129.8214	939	6791.793	136.7342	989	7153.448
129.9597	940	6799.031	136.8724	990	7160.681
130.0980	941	6806.264	137.0107	991	7167.914
130.2362	942	6813.497	137.1490	992	7175.147
130.3745	943	6820.730	137.2872	993	7182.380
130.5127	944	6827.963	137.4255	994	7189.613
130.6510	945	6835.196	137.5637	995	7196.846
130.7892	946	6842.429	137.7020	996	7204.079
130.9275	947	6849.662	137.8402	997	7211.312
131.0657	948	6856.895	137.9785	998	7218.545
131.2040	949	6864.128	138.1167	999	7225.778
131.3423	950	6871.361	138.2550	1000	7233.011

MOMENTUM

Pound (mass)-inch/Second — Kilogram-metre/Second
(lbm-in./s — kg · m/s)

Pound (mass)-inch/Second — Kilogram-centimetre/Second
(lbm-in./s — kg · cm/s)

Pound (mass)-inch/Second — Gram-centimetre/Second
(lbm-in./s — g · cm/s)

Conversion Factors:

$\quad$ 1 pound (mass)-inch/second (lbm-in./s) = 0.01152125
$\quad$ kilogram-metre/second (kg · m/s)
$\qquad$ (1 kg · m/s = 86.79614 lbm-in./s)

Varying Momentum Time Units

Table 25 can convert momentum values in time units other than per second shown above — such as per minute (min). However, the time units must be consistent between the customary and SI quantities used. This reflects no change in the numerical value of the above conversion factor from which Table 25 is derived.

Therefore: 1 lbm-in./min = 0.01152125 kg · m/min.

In converting momentum quantities from kg · m/s to g · cm/s and kg · cm/s values, the Decimal-shift Diagram 25-A, which appears in the headings on alternate pages of this table, will assist the user. It is also applicable with consistent time units other than per second, and is derived from:

$$1 \text{ kg} \cdot \text{m/s} = (10^5) \text{ g} \cdot \text{cm/s} = (10^2) \text{ kg} \cdot \text{cm/s}.$$

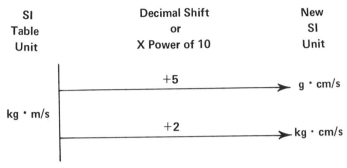

Diagram 25-A

Example (*a*): Convert 1,490 lbm-in./min to kg · m/min, g · cm/min, kg · cm/min.
  Since: 1,490 lbm-in./min = 149 × 10 lbm-in./min:
  From table: 149 lbm-in./min = 1.716666 kg · m/min*
  Therefore: 149 × 10 lbm-in./min = 1.716666 × 10 kg · m/min
      = 17.16666 kg · m/min ≈ 17.17 kg · m/min.
  From Diagram: 1) 17.16666 kg · m/min = 17.16666 (10^2) kg · cm/min
                     = 1,716.666 kg · cm/min ≈ 1,717 kg · cm/min
                     (kg · m/min to kg · cm/min by multiplying kg · m/min
                     value by 10^2 or by shifting its decimal 2 places to the
                     right.)

              2) 17.16666 kg · m/min = 17.16666 × 10^5 g · cm/min
                     = 1,716,666 g · cm/min ≈ 1.717 × 10^6 g · cm/min
                     (kg · m/min to g · cm/min by multiplying kg · m/min
                     value by 10^5 or by shifting its decimal 5 places to the
                     right.)

Diagram 25-B is also given at the top of alternate table pages.

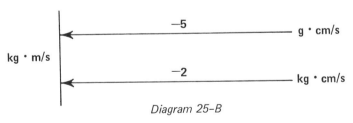

Diagram 25-B

It is the opposite of Diagram 25-A and is used to convert g · cm/s and kg · cm/s
to kg · m/s values. The kg · m/s equivalent is then applied to the REF column
in the table to obtain the lbm-in./s quantity.

---

*All conversion values obtained can be rounded to desired number of significant figures.

Example (*b*): Convert 248.5 g · cm/s to lbm-in./s.

From Diagram 25-B: 248.5 g · cm/s = 248.5 × 10^{-5} kg · m/s

Since: 248.5 kg · m/s = (248.0 + 0.5) kg · m/s:

From table: 248 kg · m/s = 21,525.44 lbm-in./s*

5 kg · m/s = 433.9807 lbm-in./s*

By proper decimal shifting and addition:

248.0 kg · m/s = 21,525.44 lbm-in./s

0.5 kg · m/s = 43.39807 lbm-in./s

248.5 kg · m/s = 21,568.83807 = 21,568.84 lbm-in./s

Therefore: 248.5 × 10^{-5} kg · m/s = 21,568.84 × 10^{-5} lbm-in./s

= 0.2156884 lbm-in./s ≈ 0.2157 lbm-in./s.

---

*All conversion values obtained can be rounded to desired number of significant figures.

kg · m/s	REF	lbm-in/s	kg · m/s	REF	lbm-in/s
0.01152125	1	86.79614	0.5875838	51	4426.603
0.02304250	2	173.5923	0.5991050	52	4513.399
0.03456375	3	260.3884	0.6106263	53	4600.195
0.04608500	4	347.1846	0.6221475	54	4686.991
0.05760625	5	433.9807	0.6336688	55	4773.788
0.06912750	6	520.7768	0.6451900	56	4860.584
0.08064875	7	607.5730	0.6567113	57	4947.380
0.09217000	8	694.3691	0.6682325	58	5034.176
0.1036913	9	781.1652	0.6797538	59	5120.972
0.1152125	10	867.9614	0.6912750	60	5207.768
0.1267338	11	954.7575	0.7027963	61	5294.564
0.1382550	12	1041.554	0.7143175	62	5381.361
0.1497763	13	1128.350	0.7258388	63	5468.157
0.1612975	14	1215.146	0.7373600	64	5554.953
0.1728188	15	1301.942	0.7488813	65	5641.749
0.1843400	16	1388.738	0.7604025	66	5728.545
0.1958613	17	1475.534	0.7719238	67	5815.341
0.2073825	18	1562.330	0.7834450	68	5902.137
0.2189038	19	1649.127	0.7949663	69	5988.934
0.2304250	20	1735.923	0.8064875	70	6075.730
0.2419463	21	1822.719	0.8180088	71	6162.526
0.2534675	22	1909.515	0.8295300	72	6249.322
0.2649888	23	1996.311	0.8410513	73	6336.118
0.2765100	24	2083.107	0.8525725	74	6422.914
0.2880313	25	2169.903	0.8640938	75	6509.710
0.2995525	26	2256.700	0.8756150	76	6596.506
0.3110738	27	2343.496	0.8871363	77	6683.303
0.3225950	28	2430.292	0.8986575	78	6770.099
0.3341163	29	2517.088	0.9101738	79	6856.895
0.3456375	30	2603.884	0.9217000	80	6943.691
0.3571588	31	2690.680	0.9332213	81	7030.487
0.3686800	32	2777.476	0.9447425	82	7117.283
0.3802013	33	2864.273	0.9562638	83	7204.079
0.3917225	34	2951.069	0.9677850	84	7290.876
0.4032438	35	3037.865	0.9793063	85	7377.672
0.4147650	36	3124.661	0.9908275	86	7464.468
0.4262863	37	3211.457	1.002349	87	7551.264
0.4378075	38	3298.253	1.013870	88	7638.060
0.4493288	39	3385.049	1.025391	89	7724.856
0.4608500	40	3471.846	1.036913	90	7811.652
0.4723713	41	3558.642	1.048434	91	7898.449
0.4838925	42	3645.438	1.059955	92	7985.245
0.4954138	43	3732.234	1.071476	93	8072.041
0.5069350	44	3819.030	1.082998	94	8158.837
0.5184563	45	3905.826	1.094519	95	8245.633
0.5299775	46	3992.622	1.106040	96	8332.429
0.5414988	47	4079.418	1.117561	97	8419.225
0.5530200	48	4166.215	1.129083	98	8506.021
0.5645413	49	4253.011	1.140604	99	8592.818
0.5760625	50	4339.807	1.152125	100	8679.614

kg · m/s	REF	lbm-in/s	kg · m/s	REF	lbm-in/s
1.163646	101	8766.410	1.739709	151	13106.22
1.175168	102	8853.206	1.751230	152	13193.01
1.186689	103	8940.002	1.762751	153	13279.81
1.198210	104	9026.798	1.774273	154	13366.61
1.209731	105	9113.594	1.785794	155	13453.40
1.221253	106	9200.391	1.797315	156	13540.20
1.232774	107	9287.187	1.808836	157	13626.99
1.244295	108	9373.983	1.820358	158	13713.79
1.255816	109	9460.779	1.831879	159	13800.59
1.267338	110	9547.575	1.843400	160	13887.38
1.278859	111	9634.371	1.854921	161	13974.18
1.290380	112	9721.167	1.866443	162	14060.97
1.301901	113	9807.964	1.877964	163	14147.77
1.313423	114	9894.760	1.889485	164	14234.57
1.324944	115	9981.556	1.901006	165	14321.36
1.336465	116	10068.35	1.912528	166	14408.16
1.347986	117	10155.15	1.924049	167	14494.95
1.359508	118	10241.94	1.935570	168	14581.75
1.371029	119	10328.74	1.947091	169	14668.55
1.382550	120	10415.54	1.958613	170	14755.34
1.394071	121	10502.33	1.970134	171	14842.14
1.405593	122	10589.13	1.981655	172	14928.94
1.417114	123	10675.92	1.993176	173	15015.73
1.428635	124	10762.72	2.004698	174	15102.53
1.440156	125	10849.52	2.016219	175	15189.32
1.451678	126	10936.31	2.027740	176	15276.12
1.463199	127	11023.11	2.039261	177	15362.92
1.474720	128	11109.91	2.050783	178	15449.71
1.486241	129	11196.70	2.062304	179	15536.51
1.497763	130	11283.50	2.073825	180	15623.30
1.509284	131	11370.29	2.085346	181	15710.10
1.520805	132	11457.09	2.096868	182	15796.90
1.532326	133	11543.89	2.108389	183	15883.69
1.543848	134	11630.68	2.119910	184	15970.49
1.555369	135	11717.48	2.131431	185	16057.29
1.566890	136	11804.27	2.142953	186	16144.08
1.578411	137	11891.07	2.154474	187	16230.88
1.589933	138	11977.87	2.165995	188	16317.67
1.601454	139	12064.66	2.177516	189	16404.47
1.612975	140	12151.46	2.189038	190	16491.27
1.624496	141	12238.26	2.200559	191	16578.06
1.636018	142	12325.05	2.212080	192	16664.86
1.647539	143	12411.85	2.223601	193	16751.65
1.659060	144	12498.64	2.235123	194	16838.45
1.670581	145	12585.44	2.246644	195	16925.25
1.682103	146	12672.24	2.258165	196	17012.04
1.693624	147	12759.03	2.269686	197	17098.84
1.705145	148	12845.83	2.281208	198	17185.64
1.716666	149	12932.62	2.292729	199	17272.43
1.728188	150	13019.42	2.304250	200	17359.23

```
                          +5
                 |————————————————————→ g · cm/s
        kg · m/s |
                 |        +2
                 |————————————————————→ kg · cm/s
```

kg · m/s	REF	lbm–in/s		kg · m/s	REF	lbm–in/s
2.315771	201	17446.02		2.891834	251	21785.83
2.327293	202	17532.82		2.903355	252	21872.63
2.338814	203	17619.62		2.914876	253	21959.42
2.350335	204	17706.41		2.926398	254	22046.22
2.361856	.205	17793.21		2.937919	255	22133.02
2.373378	206	17880.00		2.949440	256	22219.81
2.384899	207	17966.80		2.960961	257	22306.61
2.396420	208	18053.60		2.972483	258	22393.40
2.407941	209	18140.39		2.984004	259	22480.20
2.419463	210	18227.19		2.995525	260	22567.00
2.430984	211	18313.99		3.007046	261	22653.79
2.442505	212	18400.78		3.018568	262	22740.59
2.454026	213	18487.58		3.030089	263	22827.38
2.465548	214	18574.37		3.041610	264	22914.18
2.477069	215	18661.17		3.053131	265	23000.98
2.488590	216	18747.97		3.064653	266	23087.77
2.500111	217	18834.76		3.076174	267	23174.57
2.511633	218	18921.56		3.087695	268	23261.36
2.523154	219	19008.35		3.099216	269	23348.16
2.534675	220	19095.15		3.110738	270	23434.96
2.546196	221	19181.95		3.122259	271	23521.75
2.557718	222	19268.74		3.133780	272	23608.55
2.569239	223	19355.54		3.145301	273	23695.35
2.580760	224	19442.33		3.156823	274	23782.14
2.592281	225	19529.13		3.168344	275	23868.94
2.603803	226	19615.93		3.179865	276	23955.73
2.615324	227	19702.72		3.191386	277	24042.53
2.626845	228	19789.52		3.202908	278	24129.33
2.638366	229	19876.32		3.214429	279	24216.12
2.649888	230	19963.11		3.225950	280	24302.92
2.661409	231	20049.91		3.237471	281	24389.71
2.672930	232	20136.70		3.248993	282	24476.51
2.684451	233	20223.50		3.260514	283	24563.31
2.695973	234	20310.30		3.272035	284	24650.10
2.707494	235	20397.09		3.283556	285	24736.90
2.719015	236	20483.89		3.295078	286	24823.70
2.730536	237	20570.68		3.306599	287	24910.49
2.742058	238	20657.48		3.318120	288	24997.29
2.753579	239	20744.28		3.329641	289	25084.08
2.765100	240	20831.07		3.341163	290	25170.88
2.776621	241	20917.87		3.352684	291	25257.68
2.788143	242	21004.67		3.364205	292	25344.47
2.799664	243	21091.46		3.375726	293	25431.27
2.811185	244	21178.26		3.387248	294	25518.06
2.822706	245	21265.05		3.393769	295	25604.86
2.834228	246	21351.85		3.410290	296	25691.66
2.845749	247	21438.65		3.421811	297	25778.45
2.857270	248	21525.44		3.433333	298	25865.25
2.868791	249	21612.24		3.444854	299	25952.05
2.880313	250	21699.03		3.456375	300	26038.84

kg · m/s ←————— −5 —————→ g · cm/s

kg · m/s ←————— −2 —————→ kg · cm/s

kg · m/s	REF	lbm-in/s	kg · m/s	REF	lbm-in/s
3.467896	301	26125.64	4.043959	351	30465.44
3.479418	302	26212.43	4.055480	352	30552.24
3.490939	303	26299.23	4.067001	353	30639.04
3.502460	304	26386.03	4.078523	354	30725.83
3.513981	305	26472.82	4.090044	355	30812.63
3.525503	306	26559.62	4.101565	356	30899.43
3.537024	307	26646.41	4.113086	357	30986.22
3.548545	308	26733.21	4.124608	358	31073.02
3.560066	309	26820.01	4.136129	359	31159.81
3.571588	310	26906.80	4.147650	360	31246.61
3.583109	311	26993.60	4.159171	361	31333.41
3.594630	312	27080.40	4.170693	362	31420.20
3.606151	313	27167.19	4.182214	363	31507.00
3.617673	314	27253.99	4.193735	364	31593.79
3.629194	315	27340.78	4.205256	365	31680.59
3.640715	316	27427.58	4.216778	366	31767.39
3.652236	317	27514.38	4.228299	367	31854.18
3.663758	318	27601.17	4.239820	368	31940.98
3.675279	319	27687.97	4.251341	369	32027.77
3.686800	320	27774.76	4.262863	370	32114.57
3.698321	321	27861.56	4.274384	371	32201.37
3.709843	322	27948.36	4.285905	372	32288.16
3.721364	323	28035.15	4.297426	373	32374.96
3.732885	324	28121.95	4.308948	374	32461.76
3.744406	325	28208.74	4.320469	375	32548.55
3.755928	326	28295.54	4.331990	376	32635.35
3.767449	327	28382.34	4.343511	377	32722.14
3.778970	328	28469.13	4.355033	378	32808.94
3.790491	329	28555.93	4.366554	379	32895.74
3.802013	330	28642.73	4.378075	380	32982.53
3.813534	331	28729.52	4.389596	381	33069.33
3.825055	332	28816.32	4.401118	382	33156.12
3.836576	333	28903.11	4.412639	383	33242.92
3.848098	334	28989.91	4.424160	384	33329.72
3.859619	335	29076.71	4.435681	385	33416.51
3.871140	336	29163.50	4.447203	386	33503.31
3.882661	337	29250.30	4.458724	387	33590.11
3.894183	338	29337.09	4.470245	388	33676.90
3.905704	339	29423.89	4.481766	389	33763.70
3.917225	340	29510.69	4.493288	390	33850.49
3.928746	341	29597.48	4.504809	391	33937.29
3.940268	342	29684.28	4.516330	392	34024.09
3.951789	343	29771.08	4.527851	393	34110.88
3.963310	344	29857.87	4.539373	394	34197.68
3.974831	345	29944.67	4.550894	395	34284.47
3.986353	346	30031.46	4.562415	396	34371.27
3.997874	347	30118.26	4.573936	397	34458.07
4.009395	348	30205.06	4.585458	398	34544.86
4.020916	349	30291.85	4.596979	399	34631.66
4.032438	350	30378.65	4.608500	400	34718.46

```
                                        +5
                                 ┌──────────────────→ g · cm/s
                        kg · m/s ┤
                                 │       +2
                                 └──────────────────→ kg · cm/s
```

kg · m/s	REF	lbm-in/s		kg · m/s	REF	lbm-in/s
4.620021	401	34805.25		5.196084	451	39145.06
4.631543	402	34892.05		5.207605	452	39231.85
4.643064	403	34978.84		5.219126	453	39318.65
4.654585	404	35065.64		5.230648	454	39405.45
4.666106	405	35152.44		5.242169	455	39492.24
4.677628	406	35239.23		5.253690	456	39579.04
4.689149	407	35326.03		5.265211	457	39665.83
4.700670	408	35412.82		5.276733	458	39752.63
4.712191	409	35499.62		5.288254	459	39839.43
4.723713	410	35586.42		5.299775	460	39926.22
4.735234	411	35673.21		5.311296	461	40013.02
4.746755	412	35760.01		5.322818	462	40099.82
4.758276	413	35846.80		5.334339	463	40186.61
4.769798	414	35933.60		5.345860	464	40273.41
4.781319	415	36020.40		5.357381	465	40360.20
4.792840	416	36107.19		5.368903	466	40447.00
4.804361	417	36193.99		5.380424	467	40533.80
4.815883	418	36280.79		5.391945	468	40620.59
4.827404	419	36367.58		5.403466	469	40707.39
4.838925	420	36454.38		5.414988	470	40794.18
4.850446	421	36541.17		5.426509	471	40880.98
4.861968	422	36627.97		5.438030	472	40967.78
4.873489	423	36714.77		5.449551	473	41054.57
4.885010	424	36801.56		5.461073	474	41141.37
4.896531	425	36888.36		5.472594	475	41228.17
4.908053	426	36975.15		5.484115	476	41314.96
4.919574	427	37061.95		5.495636	477	41401.76
4.931095	428	37148.75		5.507158	478	41488.55
4.942616	429	37235.54		5.518679	479	41575.35
4.954138	430	37322.34		5.530200	480	41662.15
4.965659	431	37409.14		5.541721	481	41748.94
4.977180	432	37495.93		5.553243	482	41835.74
4.988701	433	37582.73		5.564764	483	41922.53
5.000223	434	37669.52		5.576285	484	42009.33
5.011744	435	37756.32		5.587806	485	42096.13
5.023265	436	37843.12		5.599328	486	42182.92
5.034786	437	37929.91		5.610849	487	42269.72
5.046308	438	38016.71		5.622370	488	42356.52
5.057829	439	38103.50		5.633891	489	42443.31
5.069350	440	38190.30		5.645413	490	42530.11
5.080871	441	38277.10		5.656934	491	42616.90
5.092393	442	38363.89		5.668455	492	42703.70
5.103914	443	38450.69		5.679976	493	42790.50
5.115435	444	38537.49		5.691498	494	42877.29
5.126956	445	38624.28		5.703019	495	42964.09
5.133478	446	38711.08		5.714540	496	43050.88
5.149999	447	38797.87		5.726061	497	43137.68
5.161520	448	38884.67		5.737583	498	43224.48
5.173041	449	38971.47		5.749104	499	43311.27
5.184563	450	39058.26		5.760625	500	43398.07

kg · m/s	REF	lbm-in/s	kg · m/s	REF	lbm-in/s
5.772146	501	43484.87	6.348209	551	47824.67
5.783668	502	43571.66	6.359730	552	47911.47
5.795189	503	43658.46	6.371251	553	47998.26
5.806710	504	43745.25	6.382773	554	48085.06
5.818231	505	43832.05	6.394294	555	48171.86
5.829753	506	43918.85	6.405815	556	48258.65
5.841274	507	44005.64	6.417336	557	48345.45
5.852795	508	44092.44	6.428858	558	48432.25
5.864316	509	44179.23	6.440379	559	48519.04
5.875838	510	44266.03	6.451900	560	48605.84
5.887359	511	44352.83	6.463421	561	48692.63
5.898880	512	44439.62	6.474943	562	48779.43
5.910401	513	44526.42	6.486464	563	48866.23
5.921923	514	44613.21	6.497985	564	48953.02
5.933444	515	44700.01	6.509506	565	49039.82
5.944965	516	44786.81	6.521028	566	49126.61
5.956486	517	44873.60	6.532549	567	49213.41
5.968008	518	44960.40	6.544070	568	49300.21
5.979529	519	45047.20	6.555591	569	49387.00
5.991050	520	45133.99	6.567113	570	49473.80
6.002571	521	45220.79	6.578634	571	49560.59
6.014093	522	45307.58	6.590155	572	49647.39
6.025614	523	45394.38	6.601676	573	49734.19
6.037135	524	45481.18	6.613198	574	49820.98
6.048656	525	45567.97	6.624719	575	49907.78
6.060178	526	45654.77	6.636240	576	49994.58
6.071699	527	45741.56	6.647761	577	50081.37
6.083220	528	45828.36	6.659283	578	50168.17
6.094741	529	45915.16	6.670804	579	50254.96
6.106263	530	46001.95	6.682325	580	50341.76
6.117784	531	46088.75	6.693846	581	50428.56
6.129305	532	46175.55	6.705368	582	50515.35
6.140826	533	46262.34	6.716889	583	50602.15
6.152348	534	46349.14	6.728410	584	50688.94
6.163869	535	46435.93	6.739931	585	50775.74
6.175390	536	46522.73	6.751453	586	50862.54
6.186911	537	46609.53	6.762974	587	50949.33
6.198433	538	46696.32	6.774495	588	51036.13
6.209954	539	46783.12	6.786016	589	51122.93
6.221475	540	46869.91	6.797538	590	51209.72
6.232996	541	46956.71	6.809059	591	51296.52
6.244518	542	47043.51	6.820580	592	51383.31
6.256039	543	47130.30	6.832101	593	51470.11
6.267560	544	47217.10	6.843623	594	51556.91
6.279081	545	47303.90	6.855144	595	51643.70
6.290603	546	47390.69	6.866665	596	51730.50
6.302124	547	47477.49	6.878186	597	51817.29
6.313645	548	47564.28	6.889708	598	51904.09
6.325166	549	47651.08	6.901229	599	51990.89
6.336688	550	47737.88	6.912750	600	52077.68

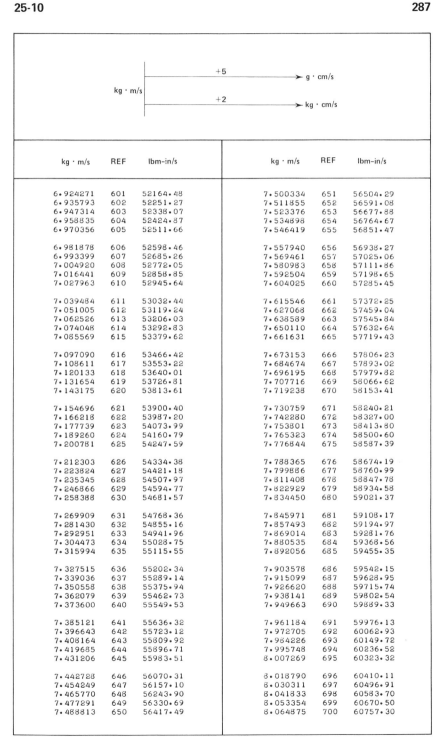

kg · m/s	REF	lbm-in/s		kg · m/s	REF	lbm-in/s
6.924271	601	52164.48		7.500334	651	56504.29
6.935793	602	52251.27		7.511855	652	56591.08
6.947314	603	52338.07		7.523376	653	56677.88
6.958835	604	52424.87		7.534898	654	56764.67
6.970356	605	52511.66		7.546419	655	56851.47
6.981878	606	52598.46		7.557940	656	56938.27
6.993399	607	52685.26		7.569461	657	57025.06
7.004920	608	52772.05		7.580983	658	57111.86
7.016441	609	52858.85		7.592504	659	57198.65
7.027963	610	52945.64		7.604025	660	57285.45
7.039484	611	53032.44		7.615546	661	57372.25
7.051005	612	53119.24		7.627068	662	57459.04
7.062526	613	53206.03		7.638589	663	57545.84
7.074048	614	53292.83		7.650110	664	57632.64
7.085569	615	53379.62		7.661631	665	57719.43
7.097090	616	53466.42		7.673153	666	57806.23
7.108611	617	53553.22		7.684674	667	57893.02
7.120133	618	53640.01		7.696195	668	57979.82
7.131654	619	53726.81		7.707716	669	58066.62
7.143175	620	53813.61		7.719238	670	58153.41
7.154696	621	53900.40		7.730759	671	58240.21
7.166218	622	53987.20		7.742280	672	58327.00
7.177739	623	54073.99		7.753801	673	58413.80
7.189260	624	54160.79		7.765323	674	58500.60
7.200781	625	54247.59		7.776844	675	58587.39
7.212303	626	54334.38		7.788365	676	58674.19
7.223824	627	54421.18		7.799886	677	58760.99
7.235345	628	54507.97		7.811408	678	58847.78
7.246866	629	54594.77		7.822929	679	58934.58
7.258388	630	54681.57		7.834450	680	59021.37
7.269909	631	54768.36		7.845971	681	59108.17
7.281430	632	54855.16		7.857493	682	59194.97
7.292951	633	54941.96		7.869014	683	59281.76
7.304473	634	55028.75		7.880535	684	59368.56
7.315994	635	55115.55		7.892056	685	59455.35
7.327515	636	55202.34		7.903578	686	59542.15
7.339036	637	55289.14		7.915099	687	59628.95
7.350558	638	55375.94		7.926620	688	59715.74
7.362079	639	55462.73		7.938141	689	59802.54
7.373600	640	55549.53		7.949663	690	59889.33
7.385121	641	55636.32		7.961184	691	59976.13
7.396643	642	55723.12		7.972705	692	60062.93
7.408164	643	55809.92		7.984226	693	60149.72
7.419685	644	55896.71		7.995748	694	60236.52
7.431206	645	55983.51		8.007269	695	60323.32
7.442728	646	56070.31		8.018790	696	60410.11
7.454249	647	56157.10		8.030311	697	60496.91
7.465770	648	56243.90		8.041833	698	60583.70
7.477291	649	56330.69		8.053354	699	60670.50
7.488813	650	56417.49		8.064875	700	60757.30

kg · m/s	REF	lbm-in/s	kg · m/s	REF	lbm-in/s
8.076396	701	60844.09	8.652459	751	65183.90
8.087918	702	60930.89	8.663980	752	65270.70
8.099439	703	61017.69	8.675501	753	65357.49
8.110960	704	61104.46	8.687023	754	65444.29
8.122481	705	61191.28	8.698544	755	65531.08
8.134003	706	61278.07	8.710065	756	65617.88
8.145524	707	61364.87	8.721586	757	65704.68
8.157045	708	61451.67	8.733108	758	65791.47
8.168566	709	61538.46	8.744629	759	65878.27
8.180088	710	61625.26	8.756150	760	65965.06
8.191609	711	61712.05	8.767671	761	66051.86
8.203130	712	61798.85	8.779193	762	66138.66
8.214651	713	61885.65	8.790714	763	66225.45
8.226173	714	61972.44	8.802235	764	66312.25
8.237694	715	62059.24	8.813756	765	66399.05
8.249215	716	62146.03	8.825278	766	66485.84
8.260736	717	62232.83	8.836799	767	66572.64
8.272258	718	62319.63	8.848320	768	66659.43
8.283779	719	62406.42	8.859841	769	66746.23
8.295300	720	62493.22	8.871363	770	66833.03
8.306821	721	62580.02	8.882884	771	66919.82
8.318343	722	62666.81	8.894405	772	67006.62
8.329864	723	62753.61	8.905926	773	67093.41
8.341385	724	62840.40	8.917448	774	67180.21
8.352906	725	62927.20	8.928969	775	67267.01
8.364428	726	63014.00	8.940490	776	67353.80
8.375949	727	63100.79	8.952011	777	67440.60
8.387470	728	63187.59	8.963533	778	67527.40
8.398991	729	63274.38	8.975054	779	67614.19
8.410513	730	63361.18	8.986575	780	67700.99
8.422034	731	63447.98	8.998096	781	67787.78
8.433555	732	63534.77	9.009618	782	67874.58
8.445076	733	63621.57	9.021139	783	67961.38
8.456598	734	63708.37	9.032660	784	68048.17
8.468119	735	63795.16	9.044181	785	68134.97
8.479640	736	63881.96	9.055703	786	68221.76
8.491161	737	63968.75	9.067224	787	68308.56
8.502683	738	64055.55	9.078745	788	68395.36
8.514204	739	64142.35	9.090266	789	68482.15
8.525725	740	64229.14	9.101788	790	68568.95
8.537246	741	64315.94	9.113309	791	68655.75
8.548768	742	64402.73	9.124830	792	68742.54
8.560289	743	64489.53	9.136351	793	68829.34
8.571810	744	64576.33	9.147873	794	68916.13
8.583331	745	64663.12	9.159394	795	69002.93
8.594853	746	64749.92	9.170915	796	69089.73
8.606374	747	64836.71	9.182436	797	69176.52
8.617895	748	64923.51	9.193958	798	69263.32
8.629416	749	65010.31	9.205479	799	69350.11
8.640938	750	65097.10	9.217000	800	69436.91

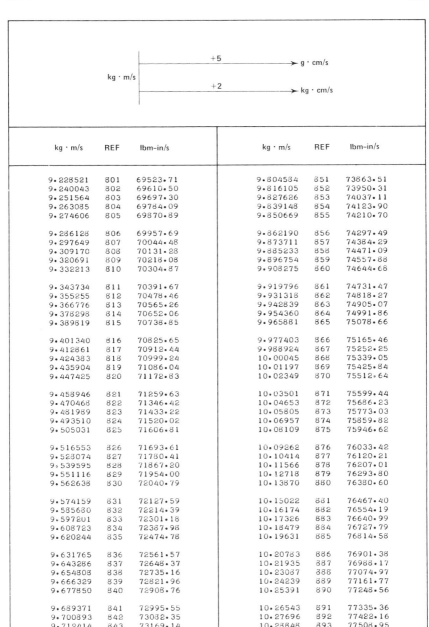

kg · m/s	REF	lbm-in/s		kg · m/s	REF	lbm-in/s
9.228521	801	69523.71		9.804534	851	73863.51
9.240043	802	69610.50		9.816105	852	73950.31
9.251564	803	69697.30		9.827626	853	74037.11
9.263085	804	69784.09		9.839148	854	74123.90
9.274606	805	69870.89		9.850669	855	74210.70
9.286128	806	69957.69		9.862190	856	74297.49
9.297649	807	70044.48		9.873711	857	74384.29
9.309170	808	70131.28		9.885233	858	74471.09
9.320691	809	70218.08		9.896754	859	74557.88
9.332213	810	70304.87		9.908275	860	74644.68
9.343734	811	70391.67		9.919796	861	74731.47
9.355255	812	70478.46		9.931318	862	74818.27
9.366776	813	70565.26		9.942839	863	74905.07
9.378298	814	70652.06		9.954360	864	74991.86
9.389819	815	70738.85		9.965881	865	75078.66
9.401340	816	70825.65		9.977403	866	75165.46
9.412861	817	70912.44		9.988924	867	75252.25
9.424383	818	70999.24		10.00045	868	75339.05
9.435904	819	71086.04		10.01197	869	75425.84
9.447425	820	71172.83		10.02349	870	75512.64
9.458946	821	71259.63		10.03501	871	75599.44
9.470468	822	71346.42		10.04653	872	75686.23
9.481989	823	71433.22		10.05805	873	75773.03
9.493510	824	71520.02		10.06957	874	75859.82
9.505031	825	71606.81		10.08109	875	75946.62
9.516553	826	71693.61		10.09262	876	76033.42
9.528074	827	71780.41		10.10414	877	76120.21
9.539595	828	71867.20		10.11566	878	76207.01
9.551116	829	71954.00		10.12718	879	76293.80
9.562638	830	72040.79		10.13870	880	76380.60
9.574159	831	72127.59		10.15022	881	76467.40
9.585680	832	72214.39		10.16174	882	76554.19
9.597201	833	72301.18		10.17326	883	76640.99
9.608723	834	72387.98		10.18479	884	76727.79
9.620244	835	72474.78		10.19631	885	76814.58
9.631765	836	72561.57		10.20783	886	76901.38
9.643286	837	72648.37		10.21935	887	76988.17
9.654808	838	72735.16		10.23087	888	77074.97
9.666329	839	72821.96		10.24239	889	77161.77
9.677850	840	72908.76		10.25391	890	77248.56
9.689371	841	72995.55		10.26543	891	77335.36
9.700893	842	73082.35		10.27696	892	77422.16
9.712414	843	73169.14		10.28848	893	77508.95
9.723935	844	73255.94		10.30000	894	77595.75
9.735456	845	73342.74		10.31152	895	77682.54
9.746978	846	73429.53		10.32304	896	77769.34
9.758499	847	73516.33		10.33456	897	77856.14
9.770020	848	73603.12		10.34608	898	77942.93
9.781541	849	73689.92		10.35760	899	78029.73
9.793063	850	73776.72		10.36913	900	78116.52

kg · m/s	REF	lbm–in/s	kg · m/s	REF	lbm–in/s
10·38065	901	78203·32	10·95671	951	82543·13
10·39217	902	78290·12	10·96823	952	82629·92
10·40369	903	78376·91	10·97975	953	82716·72
10·41521	904	78463·71	10·99127	954	82803·52
10·42673	905	78550·50	11·00279	955	82890·31
10·43825	906	78637·30	11·01432	956	82977·11
10·44977	907	78724·10	11·02584	957	83063·90
10·46130	908	78810·89	11·03736	958	83150·70
10·47282	909	78897·69	11·04888	959	83237·50
10·48434	910	78984·49	11·06040	960	83324·29
10·49586	911	79071·28	11·07192	961	83411·09
10·50738	912	79158·08	11·08344	962	83497·88
10·51890	913	79244·87	11·09496	963	83584·68
10·53042	914	79331·67	11·10649	964	83671·48
10·54194	915	79418·47	11·11801	965	83758·27
10·55347	916	79505·26	11·12953	966	83845·07
10·56499	917	79592·06	11·14105	967	83931·87
10·57651	918	79678·85	11·15257	968	84018·66
10·58803	919	79765·65	11·16409	969	84105·46
10·59955	920	79852·45	11·17561	970	84192·25
10·61107	921	79939·24	11·18713	971	84279·05
10·62259	922	80026·04	11·19866	972	84365·85
10·63411	923	80112·83	11·21018	973	84452·64
10·64564	924	80199·63	11·22170	974	84539·44
10·65716	925	80286·43	11·23322	975	84626·23
10·66868	926	80373·22	11·24474	976	84713·03
10·68020	927	80460·02	11·25626	977	84799·83
10·69172	928	80546·82	11·26778	978	84886·62
10·70324	929	80633·61	11·27930	979	84973·42
10·71476	930	80720·41	11·29083	980	85060·21
10·72628	931	80807·20	11·30235	981	85147·01
10·73781	932	80894·00	11·31387	982	85233·81
10·74933	933	80980·80	11·32539	983	85320·60
10·76085	934	81067·59	11·33691	984	85407·40
10·77237	935	81154·39	11·34843	985	85494·20
10·78389	936	81241·18	11·35995	986	85580·99
10·79541	937	81327·98	11·37147	987	85667·79
10·80693	938	81414·78	11·38300	988	85754·58
10·81845	939	81501·57	11·39452	989	85841·38
10·82998	940	81588·37	11·40604	990	85928·18
10·84150	941	81675·17	11·41756	991	86014·97
10·85302	942	81761·96	11·42908	992	86101·77
10·86454	943	81848·76	11·44060	993	86188·56
10·87606	944	81935·55	11·45212	994	86275·36
10·88758	945	82022·35	11·46364	995	86362·16
10·89910	946	82109·15	11·47517	996	86448·95
10·91062	947	82195·94	11·48669	997	86535·75
10·92215	948	82282·74	11·49821	998	86622·55
10·93367	949	82369·54	11·50973	999	86709·34
10·94519	950	82456·33	11·52125	1000	86796·14

26

MOMENT of INERTIA

Pound (mass)-foot-square — Kilogram-square-metre
$$(\text{lbm-ft}^2 - \text{kg} \cdot \text{m}^2)$$

Pound (mass)-foot-square — Kilogram-square-centimetre
$$(\text{lbm-ft}^2 - \text{kg} \cdot \text{cm}^2)$$

Pound (mass)-foot-square — Gram-square-centimetre
$$(\text{lbm-ft}^2 - \text{g} \cdot \text{cm}^2)$$

ANGULAR MOMENTUM or MOMENT OF MOMENTUM

Pound (mass)-foot-square/Second — Kilogram-square-metre/Second
$$(\text{lbm-ft}^2/\text{s} - \text{kg} \cdot \text{m}^2/\text{s})$$

Pound (mass)-foot-square/Second — Kilogram-square-centimetre/Second
$$(\text{lbm-ft}^2/\text{s} - \text{kg} \cdot \text{cm}^2/\text{s})$$

Pound (mass)-foot-square/Second — Gram-square-centimetre/Second
$$(\text{lbm-ft}^2/\text{s} - \text{g} \cdot \text{cm}^2/\text{s})$$

WORK AND ENERGY

Foot-poundal — Joule
$$(\text{ft-pdl} - \text{J})$$

Conversion Factors:
  1 Pound-foot square $(\text{lbm-ft}^2) = 0.04214011$ kilogram-square-metre $(\text{kg} \cdot \text{m}^2)$
  $(1 \text{ kg} \cdot \text{m}^2 = 23.73036 \text{ lbm-ft}^2)$

Foot-poundal — Joule Values

Table 26 can be used to convert the work units foot-poundals (ft-pdl) to joules (J) since the numerical value of the conversion factor is the same as that given above.

The foot-poundal to joule conversion factor is therefore:

1 ft-pdl = 0.04214011 J.

The ft-pdl is an English work unit, while in SI, 1 joule equals 1 newton-metre (N · m).

Angular Momentum or Moment of Momentum Values

Table 26 can also be used to convert angular momentum (moment of momentum) values having time units other than per second as above — such as per minute (min). However, the time units must be consistent between the English and SI quantities used. With the satisfaction of this requirement, the numerical value of the above conversion factor — on which Table 26 is based — is the same for SI and English unit relations in angular momentum.

Therefore: 1 lbm-ft^2/ min (or/s) = 0.04214011 kg · m^2/min (or/s).

In converting moment of inertia or angular momentum quantities (with consistent time units) containing kg · m^2 to equivalent quantities having g · cm^2 and kg · cm^2, the Decimal-shift Diagram 26-A, which appears in the headings on alternate pages of this table, will assist the user. It is derived from:

$$1 \text{ kg} \cdot \text{m}^2 = (10^7) \text{ g} \cdot \text{cm}^2 = (10^4) \text{ kg} \cdot \text{cm}^2.$$

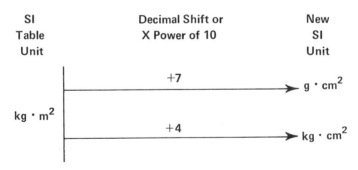

Diagram 26-A

Example (a): Convert 9.75 lbm-ft^2/s to kg · cm^2/s.

Since: 9.75 lbm-ft^2/s = 975 × 10^{-2} lbm-ft^2/s:

From table: 975 lbm-ft^2/s = 41.08661 kg · m^2/s*

Therefore: 975 × 10^{-2} lbm-ft^2/s = 41.08661 × 10^{-2} = 0.4108661 kg · m^2/s.

---

*All conversion values obtained can be rounded to desired number of significant figures.

From Diagram 26-A: $0.4108661 \text{ kg} \cdot \text{m}^2/\text{s} = 0.4108661 (10^4) \text{ kg} \cdot \text{cm}^2/\text{s}$
$= 4{,}108.661 \text{ kg} \cdot \text{cm}^2/\text{s} \approx 4{,}110 \text{ kg} \cdot \text{cm}^2/\text{s}$
(kg $\cdot$ m²/s to kg $\cdot$ cm²/s by multiplying kg $\cdot$ m²/s value by 10^4 or shifting its decimal 4 places to the right.)

Diagram 26-B is also given at the top of alternate table pages:

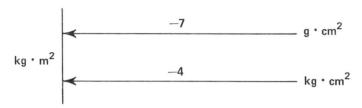

Diagram 26–B

It is the opposite of Diagram 26-A and is used to convert quantities containing g $\cdot$ cm² and kg $\cdot$ cm² to values with kg $\cdot$ m². The equivalents with kg $\cdot$ m² are then applied to the REF column in the table to obtain lbm-ft² relations.

Example (b): Convert 17,270 kg $\cdot$ cm² to lbm-ft².
From Diagram 26-B: $17{,}270 \text{ kg} \cdot \text{cm}^2 = 17{,}270 \times 10^{-4} \text{ kg} \cdot \text{m}^2$
$= 172.7 \times 10^{-2} \text{ kg} \cdot \text{m}^2$
Since: $172.7 \text{ kg} \cdot \text{m}^2 = (172.0 + 0.7) \text{ kg} \cdot \text{m}^2$:
From table: $172 \text{ kg} \cdot \text{m}^2 = 4081.622 \text{ lbm-ft}^2$ *
$7 \text{ kg} \cdot \text{m}^2 = 166.1125 \text{ lbm-ft}^2$ *

By shifting decimals where necessary and adding:
$$172.0 \text{ kg} \cdot \text{m}^2 = 4081.622 \quad \text{lbm-ft}^2$$
$$\underline{\quad 0.7 \text{ kg} \cdot \text{m}^2 = \quad 16.61125 \text{ lbm-ft}^2 }$$
$$172.7 \text{ kg} \cdot \text{m}^2 = 4098.233 \quad \text{lbm-ft}^2$$
Therefore: $172.7 \times 10^{-2} \text{ kg} \cdot \text{m}^2 = 40.98233 \text{ lbm-ft}^2 \approx 40.98 \text{ lbm-ft}^2$.

---

*All conversion values obtained can be rounded to desired number of significant figures.

kg · m² kg · m²/s J	REF	lbm-ft² lbm-ft²/s ft–pdl		kg · m² kg · m²/s J	REF	lbm-ft² lbm-ft²/s ft–pdl
0.04214011	1	23.73036		2.149146	51	1210.248
0.08428022	2	47.46072		2.191286	52	1233.979
0.1264203	3	71.19108		2.233426	53	1257.709
0.1685604	4	94.92144		2.275566	54	1281.439
0.2107005	5	118.6518		2.317706	55	1305.170
0.2528407	6	142.3822		2.359846	56	1328.900
0.2949808	7	166.1125		2.401986	57	1352.631
0.3371209	8	189.8429		2.444126	58	1376.361
0.3792610	9	213.5732		2.486266	59	1400.091
0.4214011	10	237.3036		2.528407	60	1423.822
0.4635412	11	261.0340		2.570547	61	1447.552
0.5056813	12	284.7643		2.612687	62	1471.282
0.5478214	13	308.4947		2.654827	63	1495.013
0.5899615	14	332.2250		2.696967	64	1518.743
0.6321016	15	355.9554		2.739107	65	1542.473
0.6742418	16	379.6858		2.781247	66	1566.204
0.7163819	17	403.4161		2.823387	67	1589.934
0.7585220	18	427.1465		2.865527	68	1613.665
0.8006621	19	450.8769		2.907668	69	1637.395
0.8428022	20	474.6072		2.949808	70	1661.125
0.8849423	21	498.3376		2.991948	71	1684.856
0.9270824	22	522.0679		3.034088	72	1708.586
0.9692225	23	545.7983		3.076228	73	1732.316
1.011363	24	569.5287		3.118368	74	1756.047
1.053503	25	593.2590		3.160508	75	1779.777
1.095643	26	616.9894		3.202648	76	1803.507
1.137783	27	640.7197		3.244788	77	1827.238
1.179923	28	664.4501		3.286929	78	1850.968
1.222063	29	688.1805		3.329069	79	1874.698
1.264203	30	711.9108		3.371209	80	1898.429
1.306343	31	735.6412		3.413349	81	1922.159
1.348484	32	759.3715		3.455489	82	1945.890
1.390624	33	783.1019		3.497629	83	1969.620
1.432764	34	806.8323		3.539769	84	1993.350
1.474904	35	830.5626		3.581909	85	2017.081
1.517044	36	854.2930		3.624049	86	2040.811
1.559184	37	878.0233		3.666190	87	2064.541
1.601324	38	901.7537		3.708330	88	2088.272
1.643464	39	925.4841		3.750470	89	2112.002
1.685604	40	949.2144		3.792610	90	2135.732
1.727745	41	972.9448		3.834750	91	2159.463
1.769885	42	996.6751		3.876890	92	2183.193
1.812025	43	1020.406		3.919030	93	2206.924
1.854165	44	1044.136		3.961170	94	2230.654
1.896305	45	1067.866		4.003310	95	2254.384
1.938445	46	1091.597		4.045451	96	2278.115
1.980585	47	1115.327		4.087591	97	2301.845
2.022725	48	1139.057		4.129731	98	2325.575
2.064865	49	1162.788		4.171871	99	2349.306
2.107006	50	1186.518		4.214011	100	2373.036

kg · m² kg · m²/s J	REF	lbm-ft² lbm-ft²/s ft-pdl	kg · m² kg · m²/s J	REF	lbm-ft² lbm-ft²/s ft-pdl
4·256151	101	2396·766	6·363157	151	3583·284
4·298291	102	2420·497	6·405297	152	3607·015
4·340431	103	2444·227	6·447437	153	3630·745
4·382571	104	2467·957	6·489577	154	3654·476
4·424712	105	2491·688	6·531717	155	3678·206
4·466852	106	2515·418	6·573857	156	3701·936
4·508992	107	2539·149	6·615997	157	3725·667
4·551132	108	2562·879	6·658137	158	3749·397
4·593272	109	2586·609	6·700278	159	3773·127
4·635412	110	2610·340	6·742418	160	3796·858
4·677552	111	2634·070	6·784558	161	3820·588
4·719692	112	2657·800	6·826698	162	3844·318
4·761832	113	2681·531	6·868838	163	3868·049
4·803973	114	2705·261	6·910978	164	3891·779
4·846113	115	2728·991	6·953118	165	3915·509
4·888253	116	2752·722	6·995258	166	3939·240
4·930393	117	2776·452	7·037398	167	3962·970
4·972533	118	2800·183	7·079538	168	3986·701
5·014673	119	2823·913	7·121679	169	4010·431
5·056813	120	2847·643	7·163819	170	4034·161
5·098953	121	2871·374	7·205959	171	4057·892
5·141093	122	2895·104	7·248099	172	4081·622
5·183233	123	2918·834	7·290239	173	4105·352
5·225374	124	2942·565	7·332379	174	4129·083
5·267514	125	2966·295	7·374519	175	4152·813
5·309654	126	2990·025	7·416659	176	4176·543
5·351794	127	3013·756	7·458799	177	4200·274
5·393934	128	3037·486	7·500940	178	4224·004
5·436074	129	3061·216	7·543030	179	4247·735
5·478214	130	3084·947	7·585220	180	4271·465
5·520354	131	3108·677	7·627360	181	4295·195
5·562495	132	3132·408	7·669500	182	4318·926
5·604635	133	3156·133	7·711640	183	4342·656
5·646775	134	3179·868	7·753780	184	4366·386
5·688915	135	3203·599	7·795920	185	4390·117
5·731055	136	3227·329	7·838060	186	4413·847
5·773195	137	3251·059	7·880201	187	4437·577
5·815335	138	3274·790	7·922341	188	4461·308
5·857475	139	3298·520	7·964481	189	4485·038
5·899615	140	3322·250	8·006621	190	4508·768
5·941755	141	3345·981	8·048761	191	4532·499
5·983896	142	3369·711	8·090901	192	4556·229
6·026036	143	3393·442	8·133041	193	4579·960
6·068176	144	3417·172	8·175181	194	4603·690
6·110316	145	3440·902	8·217321	195	4627·420
6·152456	146	3464·633	8·259462	196	4651·151
6·194596	147	3488·363	8·301602	197	4674·881
6·236736	148	3512·093	8·343742	198	4698·611
6·278876	149	3535·824	8·385882	199	4722·342
6·321016	150	3559·554	8·428022	200	4746·072

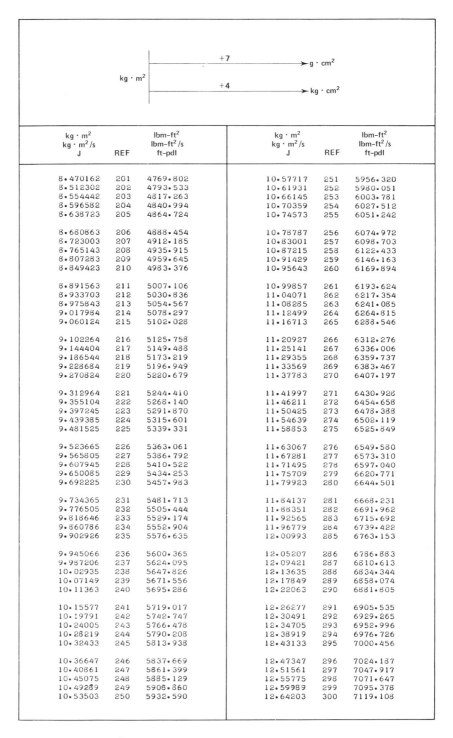

kg · m² kg · m²/s J	REF	lbm-ft² lbm-ft²/s ft–pdl	kg · m² kg · m²/s J	REF	lbm-ft² lbm-ft²/s ft–pdl
8.470162	201	4769.802	10.57717	251	5956.320
8.512302	202	4793.533	10.61931	252	5980.051
8.554442	203	4817.263	10.66145	253	6003.781
8.596582	204	4840.994	10.70359	254	6027.512
8.638723	205	4864.724	10.74573	255	6051.242
8.680863	206	4888.454	10.78787	256	6074.972
8.723003	207	4912.185	10.83001	257	6098.703
8.765143	208	4935.915	10.87215	258	6122.433
8.807283	209	4959.645	10.91429	259	6146.163
8.849423	210	4983.376	10.95643	260	6169.894
8.891563	211	5007.106	10.99857	261	6193.624
8.933703	212	5030.836	11.04071	262	6217.354
8.975843	213	5054.567	11.08285	263	6241.085
9.017984	214	5078.297	11.12499	264	6264.815
9.060124	215	5102.028	11.16713	265	6288.546
9.102264	216	5125.758	11.20927	266	6312.276
9.144404	217	5149.488	11.25141	267	6336.006
9.186544	218	5173.219	11.29355	268	6359.737
9.228684	219	5196.949	11.33569	269	6383.467
9.270824	220	5220.679	11.37783	270	6407.197
9.312964	221	5244.410	11.41997	271	6430.928
9.355104	222	5268.140	11.46211	272	6454.658
9.397245	223	5291.870	11.50425	273	6478.388
9.439385	224	5315.601	11.54639	274	6502.119
9.481525	225	5339.331	11.58853	275	6525.849
9.523665	226	5363.061	11.63067	276	6549.580
9.565805	227	5386.792	11.67281	277	6573.310
9.607945	228	5410.522	11.71495	278	6597.040
9.650085	229	5434.253	11.75709	279	6620.771
9.692225	230	5457.983	11.79923	280	6644.501
9.734365	231	5481.713	11.84137	281	6668.231
9.776505	232	5505.444	11.88351	282	6691.962
9.818646	233	5529.174	11.92565	283	6715.692
9.860786	234	5552.904	11.96779	284	6739.422
9.902926	235	5576.635	12.00993	285	6763.153
9.945066	236	5600.365	12.05207	286	6786.883
9.987206	237	5624.095	12.09421	287	6810.613
10.02935	238	5647.826	12.13635	288	6834.344
10.07149	239	5671.556	12.17849	289	6858.074
10.11363	240	5695.286	12.22063	290	6881.805
10.15577	241	5719.017	12.26277	291	6905.535
10.19791	242	5742.747	12.30491	292	6929.265
10.24005	243	5766.478	12.34705	293	6952.996
10.28219	244	5790.208	12.38919	294	6976.726
10.32433	245	5813.938	12.43133	295	7000.456
10.36647	246	5837.669	12.47347	296	7024.187
10.40861	247	5861.399	12.51561	297	7047.917
10.45075	248	5885.129	12.55775	298	7071.647
10.49289	249	5908.860	12.59989	299	7095.378
10.53503	250	5932.590	12.64203	300	7119.108

kg·m² kg·m²/s J	REF	lbm-ft² lbm-ft²/s ft-pdl	kg·m² kg·m²/s J	REF	lbm-ft² lbm-ft²/s ft-pdl
12.68417	301	7142.839	14.79118	351	8329.357
12.72631	302	7166.569	14.83332	352	8353.087
12.76845	303	7190.299	14.87546	353	8376.817
12.81059	304	7214.030	14.91760	354	8400.548
12.85273	305	7237.760	14.95974	355	8424.278
12.89487	306	7261.490	15.00188	356	8448.008
12.93701	307	7285.221	15.04402	357	8471.739
12.97915	308	7308.951	15.08616	358	8495.469
13.02129	309	7332.681	15.12830	359	8519.199
13.06343	310	7356.412	15.17044	360	8542.930
13.10557	311	7380.142	15.21258	361	8566.660
13.14771	312	7403.872	15.25472	362	8590.391
13.18985	313	7427.603	15.29686	363	8614.121
13.23199	314	7451.333	15.33900	364	8637.851
13.27413	315	7475.064	15.38114	365	8661.582
13.31627	316	7498.794	15.42328	366	8685.312
13.35841	317	7522.524	15.46542	367	8709.042
13.40056	318	7546.255	15.50756	368	8732.773
13.44270	319	7569.985	15.54970	369	8756.503
13.48484	320	7593.715	15.59184	370	8780.233
13.52698	321	7617.446	15.63398	371	8803.964
13.56912	322	7641.176	15.67612	372	8827.694
13.61126	323	7664.906	15.71826	373	8851.424
13.65340	324	7688.637	15.76040	374	8875.155
13.69554	325	7712.367	15.80254	375	8898.885
13.73768	326	7736.098	15.84468	376	8922.616
13.77982	327	7759.828	15.88682	377	8946.346
13.82196	328	7783.558	15.92896	378	8970.076
13.86410	329	7807.289	15.97110	379	8993.807
13.90624	330	7831.019	16.01324	380	9017.537
13.94838	331	7854.749	16.05538	381	9041.267
13.99052	332	7878.480	16.09752	382	9064.998
14.03266	333	7902.210	16.13966	383	9088.728
14.07480	334	7925.940	16.18180	384	9112.458
14.11694	335	7949.671	16.22394	385	9136.189
14.15908	336	7973.401	16.26608	386	9159.919
14.20122	337	7997.131	16.30822	387	9183.650
14.24336	338	8020.862	16.35036	388	9207.380
14.28550	339	8044.592	16.39250	389	9231.110
14.32764	340	8068.323	16.43464	390	9254.841
14.36978	341	8092.053	16.47678	391	9278.571
14.41192	342	8115.783	16.51892	392	9302.301
14.45406	343	8139.514	16.56106	393	9326.032
14.49620	344	8163.244	16.60320	394	9349.762
14.53834	345	8186.974	16.64534	395	9373.492
14.58048	346	8210.705	16.68748	396	9397.223
14.62262	347	8234.435	16.72962	397	9420.953
14.66476	348	8258.165	16.77176	398	9444.683
14.70690	349	8281.896	16.81390	399	9468.414
14.74904	350	8305.626	16.85604	400	9492.144

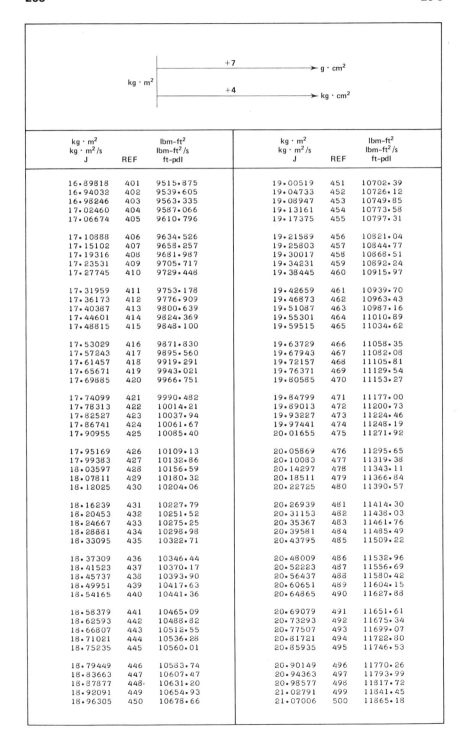

kg · m^2 kg · m^2/s J	REF	lbm-ft^2 lbm-ft^2/s ft–pdl	kg · m^2 kg · m^2/s J	REF	lbm-ft^2 lbm-ft^2/s ft–pdl
16.89818	401	9515.875	19.00519	451	10702.39
16.94032	402	9539.605	19.04733	452	10726.12
16.98246	403	9563.335	19.08947	453	10749.85
17.02460	404	9587.066	19.13161	454	10773.58
17.06674	405	9610.796	19.17375	455	10797.31
17.10888	406	9634.526	19.21589	456	10821.04
17.15102	407	9658.257	19.25803	457	10844.77
17.19316	408	9681.987	19.30017	458	10868.51
17.23531	409	9705.717	19.34231	459	10892.24
17.27745	410	9729.448	19.38445	460	10915.97
17.31959	411	9753.178	19.42659	461	10939.70
17.36173	412	9776.909	19.46873	462	10963.43
17.40387	413	9800.639	19.51087	463	10987.16
17.44601	414	9824.369	19.55301	464	11010.89
17.48815	415	9848.100	19.59515	465	11034.62
17.53029	416	9871.830	19.63729	466	11058.35
17.57243	417	9895.560	19.67943	467	11082.08
17.61457	418	9919.291	19.72157	468	11105.81
17.65671	419	9943.021	19.76371	469	11129.54
17.69885	420	9966.751	19.80585	470	11153.27
17.74099	421	9990.482	19.84799	471	11177.00
17.78313	422	10014.21	19.89013	472	11200.73
17.82527	423	10037.94	19.93227	473	11224.46
17.86741	424	10061.67	19.97441	474	11248.19
17.90955	425	10085.40	20.01655	475	11271.92
17.95169	426	10109.13	20.05869	476	11295.65
17.99383	427	10132.86	20.10083	477	11319.38
18.03597	428	10156.59	20.14297	478	11343.11
18.07811	429	10180.32	20.18511	479	11366.84
18.12025	430	10204.06	20.22725	480	11390.57
18.16239	431	10227.79	20.26939	481	11414.30
18.20453	432	10251.52	20.31153	482	11438.03
18.24667	433	10275.25	20.35367	483	11461.76
18.28881	434	10298.98	20.39581	484	11485.49
18.33095	435	10322.71	20.43795	485	11509.22
18.37309	436	10346.44	20.48009	486	11532.96
18.41523	437	10370.17	20.52223	487	11556.69
18.45737	438	10393.90	20.56437	488	11580.42
18.49951	439	10417.63	20.60651	489	11604.15
18.54165	440	10441.36	20.64865	490	11627.88
18.58379	441	10465.09	20.69079	491	11651.61
18.62593	442	10488.82	20.73293	492	11675.34
18.66807	443	10512.55	20.77507	493	11699.07
18.71021	444	10536.28	20.81721	494	11722.80
18.75235	445	10560.01	20.85935	495	11746.53
18.79449	446	10583.74	20.90149	496	11770.26
18.83663	447	10607.47	20.94363	497	11793.99
18.87877	448	10631.20	20.98577	498	11817.72
18.92091	449	10654.93	21.02791	499	11841.45
18.96305	450	10678.66	21.07006	500	11865.18

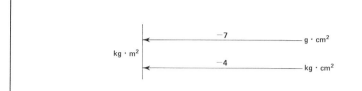

kg · m² kg · m²/s J	REF	lbm-ft² lbm-ft²/s ft-pdl	kg · m² kg · m²/s J	REF	lbm-ft² lbm-ft²/s ft-pdl
21.11220	501	11888.91	23.21920	551	13075.43
21.15434	502	11912.64	23.26134	552	13099.16
21.19648	503	11936.37	23.30348	553	13122.89
21.23862	504	11960.10	23.34562	554	13146.62
21.28076	505	11983.83	23.38776	555	13170.35
21.32290	506	12007.56	23.42990	556	13194.08
21.36504	507	12031.29	23.47204	557	13217.81
21.40718	508	12055.02	23.51418	558	13241.54
21.44932	509	12078.75	23.55632	559	13265.27
21.49146	510	12102.48	23.59846	560	13289.00
21.53360	511	12126.21	23.64060	561	13312.73
21.57574	512	12149.94	23.68274	562	13336.46
21.61788	513	12173.67	23.72488	563	13360.19
21.66002	514	12197.41	23.76702	564	13383.92
21.70216	515	12221.14	23.80916	565	13407.65
21.74430	516	12244.87	23.85130	566	13431.38
21.78644	517	12268.60	23.89344	567	13455.11
21.82858	518	12292.33	23.93558	568	13478.84
21.87072	519	12316.06	23.97772	569	13502.58
21.91286	520	12339.79	24.01986	570	13526.31
21.95500	521	12363.52	24.06200	571	13550.04
21.99714	522	12387.25	24.10414	572	13573.77
22.03928	523	12410.98	24.14628	573	13597.50
22.08142	524	12434.71	24.18842	574	13621.23
22.12356	525	12458.44	24.23056	575	13644.96
22.16570	526	12482.17	24.27270	576	13668.69
22.20784	527	12505.90	24.31484	577	13692.42
22.24998	528	12529.63	24.35698	578	13716.15
22.29212	529	12553.36	24.39912	579	13739.88
22.33426	530	12577.09	24.44126	580	13763.61
22.37640	531	12600.82	24.48340	581	13787.34
22.41854	532	12624.55	24.52554	582	13811.07
22.46068	533	12648.28	24.56768	583	13834.80
22.50282	534	12672.01	24.60982	584	13858.53
22.54496	535	12695.74	24.65196	585	13882.26
22.58710	536	12719.47	24.69410	586	13905.99
22.62924	537	12743.20	24.73624	587	13929.72
22.67138	538	12766.93	24.77838	588	13953.45
22.71352	539	12790.66	24.82052	589	13977.18
22.75566	540	12814.39	24.86266	590	14000.91
22.79780	541	12838.12	24.90480	591	14024.64
22.83994	542	12861.86	24.94695	592	14048.37
22.88208	543	12885.59	24.98909	593	14072.10
22.92422	544	12909.32	25.03123	594	14095.83
22.96636	545	12933.05	25.07337	595	14119.56
23.00850	546	12956.78	25.11551	596	14143.29
23.05064	547	12980.51	25.15765	597	14167.03
23.09278	548	13004.24	25.19979	598	14190.76
23.13492	549	13027.97	25.24193	599	14214.49
23.17706	550	13051.70	25.28407	600	14238.22

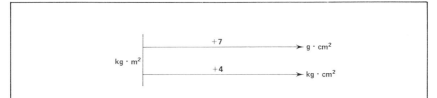

kg · m² kg · m²/s J	REF	lbm-ft² lbm-ft²/s ft-pdl	kg · m² kg · m²/s J	REF	lbm-ft² lbm-ft²/s ft-pdl
25·32621	601	14261·95	27·43321	651	15448·46
25·36835	602	14285·68	27·47535	652	15472·20
25·41049	603	14309·41	27·51749	653	15495·93
25·45263	604	14333·14	27·55963	654	15519·66
25·49477	605	14356·87	27·60177	655	15543·39
25·53691	606	14380·60	27·64391	656	15567·12
25·57905	607	14404·33	27·68605	657	15590·85
25·62119	608	14428·06	27·72819	658	15614·58
25·66333	609	14451·79	27·77033	659	15638·31
25·70547	610	14475·52	27·81247	660	15662·04
25·74761	611	14499·25	27·85461	661	15685·77
25·78975	612	14522·98	27·89675	662	15709·50
25·83189	613	14546·71	27·93889	663	15733·23
25·87403	614	14570·44	27·98103	664	15756·96
25·91617	615	14594·17	28·02317	665	15780·69
25·95831	616	14617·90	28·06531	666	15804·42
26·00045	617	14641·63	28·10745	667	15828·15
26·04259	618	14665·36	28·14959	668	15851·88
26·08473	619	14689·09	28·19173	669	15875·61
26·12687	620	14712·82	28·23387	670	15899·34
26·16901	621	14736·55	28·27601	671	15923·07
26·21115	622	14760·28	28·31815	672	15946·80
26·25329	623	14784·01	28·36029	673	15970·53
26·29543	624	14807·74	28·40243	674	15994·26
26·33757	625	14831·48	28·44457	675	16017·99
26·37971	626	14855·21	28·48671	676	16041·72
26·42185	627	14878·94	28·52885	677	16065·45
26·46399	628	14902·67	28·57099	678	16089·18
26·50613	629	14926·40	28·61313	679	16112·91
26·54827	630	14950·13	28·65527	680	16136·65
26·59041	631	14973·86	28·69741	681	16160·38
26·63255	632	14997·59	28·73955	682	16184·11
26·67469	633	15021·32	28·78170	683	16207·84
26·71683	634	15045·05	28·82384	684	16231·57
26·75897	635	15068·78	28·86598	685	16255·30
26·80111	636	15092·51	28·90812	686	16279·03
26·84325	637	15116·24	28·95026	687	16302·76
26·88539	638	15139·97	28·99240	688	16326·49
26·92753	639	15163·70	29·03454	689	16350·22
26·96967	640	15187·43	29·07668	690	16373·95
27·01181	641	15211·16	29·11882	691	16397·68
27·05395	642	15234·89	29·16096	692	16421·41
27·09609	643	15258·62	29·20310	693	16445·14
27·13823	644	15282·35	29·24524	694	16468·87
27·18037	645	15306·08	29·28738	695	16492·60
27·22251	646	15329·81	29·32952	696	16516·33
27·26465	647	15353·54	29·37166	697	16540·06
27·30679	648	15377·27	29·41380	698	16563·79
27·34893	649	15401·00	29·45594	699	16587·52
27·39107	650	15424·73	29·49808	700	16611·25

kg · m² kg · m²/s J	REF	lbm-ft² lbm-ft²/s ft-pdl	kg · m² kg · m²/s J	REF	lbm-ft² lbm-ft²/s ft-pdl
29.54022	701	16634.98	31.64722	751	17821.50
29.58236	702	16658.71	31.68936	752	17845.23
29.62450	703	16682.44	31.73150	753	17868.96
29.66664	704	16706.17	31.77364	754	17892.69
29.70878	705	16729.90	31.81578	755	17916.42
29.75092	706	16753.63	31.85792	756	17940.15
29.79306	707	16777.36	31.90006	757	17963.88
29.83520	708	16801.10	31.94220	758	17987.61
29.87734	709	16824.83	31.98434	759	18011.34
29.91948	710	16848.56	32.02648	760	18035.07
29.96162	711	16872.29	32.06862	761	18058.80
30.00376	712	16896.02	32.11076	762	18082.53
30.04590	713	16919.75	32.15290	763	18106.27
30.08804	714	16943.48	32.19504	764	18130.00
30.13018	715	16967.21	32.23718	765	18153.73
30.17232	716	16990.94	32.27932	766	18177.46
30.21446	717	17014.67	32.32146	767	18201.19
30.25660	718	17038.40	32.36360	768	18224.92
30.29874	719	17062.13	32.40574	769	18248.65
30.34088	720	17085.86	32.44788	770	18272.38
30.38302	721	17109.59	32.49002	771	18296.11
30.42516	722	17133.32	32.53217	772	18319.84
30.46730	723	17157.05	32.57431	773	18343.57
30.50944	724	17180.78	32.61645	774	18367.30
30.55158	725	17204.51	32.65859	775	18391.03
30.59372	726	17228.24	32.70073	776	18414.76
30.63586	727	17251.97	32.74287	777	18438.49
30.67800	728	17275.70	32.78501	778	18462.22
30.72014	729	17299.43	32.82715	779	18485.95
30.76228	730	17323.16	32.86929	780	18509.68
30.80442	731	17346.89	32.91143	781	18533.41
30.84656	732	17370.62	32.95357	782	18557.14
30.88870	733	17394.35	32.99571	783	18580.87
30.93084	734	17418.08	33.03785	784	18604.60
30.97298	735	17441.81	33.07999	785	18628.33
31.01512	736	17465.55	33.12213	786	18652.06
31.05726	737	17489.28	33.16427	787	18675.79
31.09940	738	17513.01	33.20641	788	18699.52
31.14154	739	17536.74	33.24855	789	18723.25
31.18368	740	17560.47	33.29069	790	18746.98
31.22582	741	17584.20	33.33283	791	18770.72
31.26796	742	17507.93	33.37497	792	18794.45
31.31010	743	17631.66	33.41711	793	18818.18
31.35224	744	17655.39	33.45925	794	18841.91
31.39438	745	17679.12	33.50139	795	18865.64
31.43652	746	17702.85	33.54353	796	18889.37
31.47866	747	17726.58	33.58567	797	18913.10
31.52080	748	17750.31	33.62781	798	18936.83
31.56294	749	17774.04	33.66995	799	18960.56
31.60508	750	17797.77	33.71209	800	18984.29

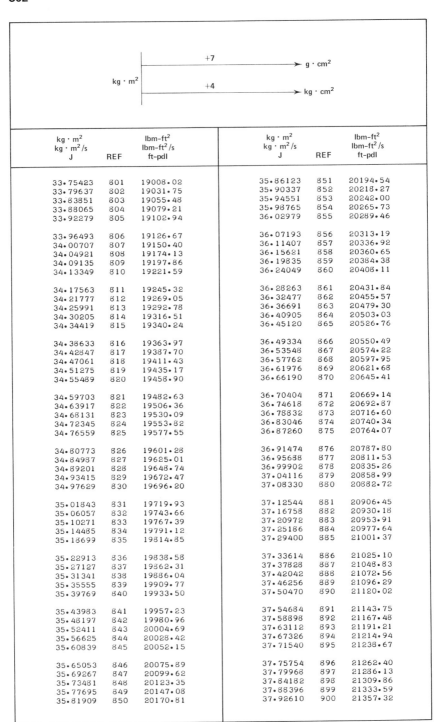

kg · m^2 kg · m^2/s J	REF	lbm-ft^2 lbm-ft^2/s ft-pdl	kg · m^2 kg · m^2/s J	REF	lbm-ft^2 lbm-ft^2/s ft-pdl
33.75423	801	19008.02	35.86123	851	20194.54
33.79637	802	19031.75	35.90337	852	20218.27
33.83851	803	19055.48	35.94551	853	20242.00
33.88065	804	19079.21	35.98765	854	20265.73
33.92279	805	19102.94	36.02979	855	20289.46
33.96493	806	19126.67	36.07193	856	20313.19
34.00707	807	19150.40	36.11407	857	20336.92
34.04921	808	19174.13	36.15621	858	20360.65
34.09135	809	19197.86	36.19835	859	20384.38
34.13349	810	19221.59	36.24049	860	20408.11
34.17563	811	19245.32	36.28263	861	20431.84
34.21777	812	19269.05	36.32477	862	20455.57
34.25991	813	19292.78	36.36691	863	20479.30
34.30205	814	19316.51	36.40905	864	20503.03
34.34419	815	19340.24	36.45120	865	20526.76
34.38633	816	19363.97	36.49334	866	20550.49
34.42847	817	19387.70	36.53548	867	20574.22
34.47061	818	19411.43	36.57762	868	20597.95
34.51275	819	19435.17	36.61976	869	20621.68
34.55489	820	19458.90	36.66190	870	20645.41
34.59703	821	19482.63	36.70404	871	20669.14
34.63917	822	19506.36	36.74618	872	20692.87
34.68131	823	19530.09	36.78832	873	20716.60
34.72345	824	19553.82	36.83046	874	20740.34
34.76559	825	19577.55	36.87260	875	20764.07
34.80773	826	19601.28	36.91474	876	20787.80
34.84987	827	19625.01	36.95688	877	20811.53
34.89201	828	19648.74	36.99902	878	20835.26
34.93415	829	19672.47	37.04116	879	20858.99
34.97629	830	19696.20	37.08330	880	20882.72
35.01843	831	19719.93	37.12544	881	20906.45
35.06057	832	19743.66	37.16758	882	20930.18
35.10271	833	19767.39	37.20972	883	20953.91
35.14485	834	19791.12	37.25186	884	20977.64
35.18699	835	19814.85	37.29400	885	21001.37
35.22913	836	19838.58	37.33614	886	21025.10
35.27127	837	19862.31	37.37828	887	21048.83
35.31341	838	19886.04	37.42042	888	21072.56
35.35555	839	19909.77	37.46256	889	21096.29
35.39769	840	19933.50	37.50470	890	21120.02
35.43983	841	19957.23	37.54684	891	21143.75
35.48197	842	19980.96	37.58898	892	21167.48
35.52411	843	20004.69	37.63112	893	21191.21
35.56625	844	20028.42	37.67326	894	21214.94
35.60839	845	20052.15	37.71540	895	21238.67
35.65053	846	20075.89	37.75754	896	21262.40
35.69267	847	20099.62	37.79968	897	21286.13
35.73481	848	20123.35	37.84182	898	21309.86
35.77695	849	20147.08	37.88396	899	21333.59
35.81909	850	20170.81	37.92610	900	21357.32

Conversion diagram: $kg \cdot m^2$ $\xrightarrow{-7}$ $g \cdot cm^2$; $kg \cdot m^2$ $\xrightarrow{-4}$ $kg \cdot cm^2$

$kg \cdot m^2$ $kg \cdot m^2/s$ J	REF	$lbm\text{-}ft^2$ $lbm\text{-}ft^2/s$ ft-pdl	$kg \cdot m^2$ $kg \cdot m^2/s$ J	REF	$lbm\text{-}ft^2$ $lbm\text{-}ft^2/s$ ft-pdl
37.96824	901	21381.05	40.07524	951	22567.57
38.01038	902	21404.79	40.11738	952	22591.30
38.05252	903	21428.52	40.15952	953	22615.03
38.09466	904	21452.25	40.20166	954	22638.76
38.13680	905	21475.98	40.24380	955	22662.49
38.17894	906	21499.71	40.28594	956	22686.22
38.22108	907	21523.44	40.32809	957	22709.96
38.26322	908	21547.17	40.37023	958	22733.69
38.30536	909	21570.90	40.41237	959	22757.42
38.34750	910	21594.63	40.45451	960	22781.15
38.38964	911	21618.36	40.49665	961	22804.88
38.43178	912	21642.09	40.53879	962	22828.61
38.47392	913	21665.82	40.58093	963	22852.34
38.51606	914	21689.55	40.62307	964	22876.07
38.55820	915	21713.28	40.66521	965	22899.80
38.60034	916	21737.01	40.70735	966	22923.53
38.64248	917	21760.74	40.74949	967	22947.26
38.68462	918	21784.47	40.79163	968	22970.99
38.72676	919	21808.20	40.83377	969	22994.72
38.76890	920	21831.93	40.87591	970	23018.45
38.81104	921	21855.66	40.91805	971	23042.18
38.85318	922	21879.39	40.96019	972	23065.91
38.89532	923	21903.12	41.00233	973	23089.64
38.93746	924	21926.85	41.04447	974	23113.37
38.97960	925	21950.58	41.08661	975	23137.10
39.02174	926	21974.31	41.12875	976	23160.83
39.06388	927	21998.04	41.17089	977	23184.56
39.10602	928	22021.77	41.21303	978	23208.29
39.14816	929	22045.50	41.25517	979	23232.02
39.19030	930	22069.24	41.29731	980	23255.75
39.23244	931	22092.97	41.33945	981	23279.48
39.27458	932	22116.70	41.38159	982	23303.21
39.31672	933	22140.43	41.42373	983	23326.94
39.35886	934	22164.16	41.46587	984	23350.67
39.40100	935	22187.89	41.50801	985	23374.41
39.44314	936	22211.62	41.55015	986	23398.14
39.48528	937	22235.35	41.59229	987	23421.87
39.52742	938	22259.08	41.63443	988	23445.60
39.56956	939	22282.81	41.67657	989	23469.33
39.61170	940	22306.54	41.71871	990	23493.06
39.65384	941	22330.27	41.76085	991	23516.79
39.69598	942	22354.00	41.80299	992	23540.52
39.73812	943	22377.73	41.84513	993	23564.25
39.78026	944	22401.46	41.88727	994	23587.98
39.82240	945	22425.19	41.92941	995	23611.71
39.86454	946	22448.92	41.97155	996	23635.44
39.90668	947	22472.65	42.01369	997	23659.17
39.94882	948	22496.38	42.05583	998	23682.90
39.99096	949	22520.11	42.09797	999	23706.63
40.03310	950	22543.84	42.14011	1000	23730.36

27

MOMENT of INERTIA

Pound (mass)-inch-square — Kilogram-square-metre
$$(lbm\text{-}in.^2 - kg \cdot m^2)$$
Pound (mass)-inch-square — Kilogram-square-centimetre
$$(lbm\text{-}in.^2 - kg \cdot cm^2)$$
Pound (mass)-inch-square — Gram-square-centimetre
$$(lbm\text{-}in.^2 - g \cdot cm^2)$$

ANGULAR MOMENTUM or MOMENT of MOMENTUM

Pound (mass)-inch-square/Second — Kilogram-square-metre/Second
$$(lbm\text{-}in.^2/s - kg \cdot m^2/s)$$
Pound (mass)-inch-square/Second — Kilogram-square-centimetre/Second
$$(lbm\text{-}in.^2/s - kg \cdot cm^2/s)$$
Pound (mass)-inch-square/Second — Gram-square-centimetre/Second
$$(lbm\text{-}in.^2/s - g \cdot cm^2/s)$$

Conversion Factors:
$$1 \text{ Pound (mass)-inch-square } (lbm\text{-}in.^2)$$
$$= 0.0002926397 \text{ Kilogram-square-metre } (kg \cdot m^2)$$
$$(1 \text{ kg} \cdot m^2 = 3417.171 \text{ lbm-in}^2)$$

Angular Momentum or Moment of Momentum Values

Table 27 can be used to convert angular momentum (moment of momentum) values having time units other than per second — such as per minute (min). However, the time units must be consistent between the English and SI quantities used. With the satisfaction of this requirement, the numerical value of the above conversion factor — on which Table 27 is based — is the same for English and SI unit relations in angular momentum.

Therefore: 1 lbm-in.^{2}/s (or/min) = 0.0002926397 kg $\cdot$ in.^{2}/s (or/min).

In converting moment of inertia or angular momentum quantities (with consistent time units) containing kg $\cdot$ m^2 to equivalent quantities having g $\cdot$ cm^2 and kg $\cdot$ cm^2, the Decimal-shift Diagram 27-A, which appears in the headings on alternate pages of this table, will assist the user. It is derived from:

$$1 \text{ kg} \cdot \text{m}^2 = (10^7) \text{ g} \cdot \text{cm}^2 = (10^4) \text{ kg} \cdot \text{cm}^2$$

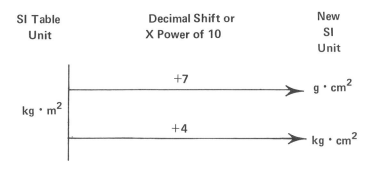

Diagram 27–A

Example (a): Convert 273 lbm-in.^{2} to kg $\cdot$ cm^2
    From table: 273 lbm-in.^{2} = 0.07989064 kg $\cdot$ m^2 *
    From Diagram 27-A: 0.07989064 kg $\cdot$ m^2 = 0.07989064 $\times$ 10^4 kg $\cdot$ cm^2
                = 798.9064 kg $\cdot$ cm^2 or $\approx$ 799 kg $\cdot$ cm^2
                (kg $\cdot$ m^2 to kg $\cdot$ cm^2 by multiplying kg $\cdot$ m^2 value by
                10^4 or by shifting its decimal 4 places to the right.)

Diagram 27-B is also given at the top of alternate table pages:

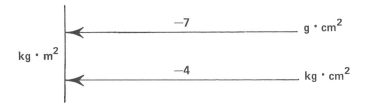

Diagram 27–B

It is the opposite of Diagram 27-A and is used to convert quantities containing g $\cdot$ cm^2 and kg $\cdot$ cm^2 to values with kg $\cdot$ m^2. The equivalents having kg $\cdot$ m^2 are then applied to the REF column in the table to obtain lbm-in.^{2} relations.

---

*All conversion values obtained can be rounded to desired number of significant figures.

Example (*b*): Convert 899,600 kg $\cdot$ cm^2/s to lbm-in.^{2}/s.

From Diagram 27-B: 899,600 kg $\cdot$ cm^2/s $= 899,600 \times 10^{-4}$ kg $\cdot$ m^2/s

$= 89.96$ kg $\cdot$ m^2/s

Since: 89.96 kg $\cdot$ m^2/s $= (89.00 + 0.96)$ kg $\cdot$ m^2/s:

From table: 89 kg $\cdot$ m^2/s $= 304,128.2$ lbm-in.^{2}/s*

96 kg $\cdot$ m^2/s $= 328,048.4$ lbm-in.^{2}/s*

By proper decimalizing and addition:

$$89.00 \text{ kg} \cdot \text{m}^2/\text{s} = 304,128.2 \quad \text{lbm-in.}^2/\text{s}$$

$$\underline{0.96 \text{ kg} \cdot \text{m}^2/\text{s} = \quad 3,280.484 \text{ lbm-in.}^2/\text{s}}$$

$$89.96 \text{ kg} \cdot \text{m}^2/\text{s} = 307,408.\,7 \quad \text{lbm-in.}^2/\text{s} \approx 307,400 \text{ lbm-in.}^2/\text{s}$$

---

*All conversion values obtained can be rounded to desired number of significant figures.

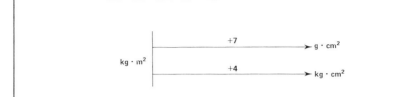

kg · m² kg · m²/s	REF	lbm-in² lbm-in²/s	kg · m² kg · m²/s	REF	lbm-in² lbm-in²/s
0.0002926397	1	3417.171	0.01492462	51	174275.7
0.0005852794	2	6834.343	0.01521726	52	177692.9
0.0008779191	3	10251.51	0.01550990	53	181110.1
0.001170559	4	13668.69	0.01580254	54	184527.3
0.001463198	5	17085.86	0.01609518	55	187944.4
0.001755838	6	20503.03	0.01638782	56	191361.6
0.002048478	7	23920.20	0.01668046	57	194778.8
0.002341118	8	27337.37	0.01697310	58	198195.9
0.002633757	9	30754.54	0.01726574	59	201613.1
0.002926397	10	34171.71	0.01755838	60	205030.3
0.003219037	11	37588.88	0.01785102	61	208447.5
0.003511676	12	41006.06	0.01814366	62	211864.6
0.003804316	13	44423.23	0.01843630	63	215281.8
0.004096956	14	47840.40	0.01872894	64	218699.0
0.004389595	15	51257.57	0.01902158	65	222116.1
0.004682235	16	54674.74	0.01931422	66	225533.3
0.004974875	17	58091.91	0.01960686	67	228950.5
0.005267515	18	61509.08	0.01989950	68	232367.7
0.005560154	19	64926.26	0.02019214	69	235784.8
0.005852794	20	68343.43	0.02048478	70	239202.0
0.006145434	21	71760.60	0.02077742	71	242619.2
0.006438073	22	75177.77	0.02107006	72	246036.3
0.006730713	23	78594.94	0.02136270	73	249453.5
0.007023353	24	82012.11	0.02165534	74	252870.7
0.007315992	25	85429.28	0.02194798	75	256287.9
0.007608632	26	88846.46	0.02224062	76	259705.0
0.007901272	27	92263.63	0.02253326	77	263122.2
0.008193912	28	95680.80	0.02282590	78	266539.4
0.008486551	29	99097.97	0.02311854	79	269956.5
0.008779191	30	102515.1	0.02341118	80	273373.7
0.009071831	31	105932.3	0.02370382	81	276790.9
0.009364470	32	109349.5	0.02399646	82	280208.1
0.009657110	33	112766.7	0.02428910	83	283625.2
0.009949750	34	116183.8	0.02458173	84	287042.4
0.01024239	35	119601.0	0.02487437	85	290459.6
0.01053503	36	123018.2	0.02516701	86	293876.7
0.01082767	37	126435.3	0.02545965	87	297293.9
0.01112031	38	129852.5	0.02575229	88	300711.1
0.01141295	39	133269.7	0.02604493	89	304128.2
0.01170559	40	136686.9	0.02633757	90	307545.4
0.01199823	41	140104.0	0.02663021	91	310962.6
0.01229087	42	143521.2	0.02692285	92	314379.8
0.01258351	43	146938.4	0.02721549	93	317796.9
0.01287615	44	150355.5	0.02750813	94	321214.1
0.01316879	45	153772.7	0.02780077	95	324631.3
0.01346143	46	157189.9	0.02809341	96	328048.4
0.01375407	47	160607.1	0.02838605	97	331465.6
0.01404671	48	164024.2	0.02867869	98	334882.8
0.01433935	49	167441.4	0.02897133	99	338300.0
0.01463198	50	170858.6	0.02926397	100	341717.1

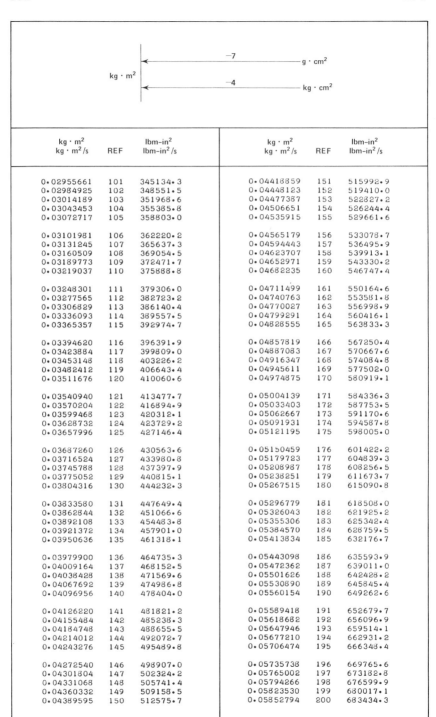

kg · m² kg · m²/s	REF	lbm-in² lbm-in²/s	kg · m² kg · m²/s	REF	lbm-in² lbm-in²/s
0.02955661	101	345134.3	0.04418859	151	515992.9
0.02984925	102	348551.5	0.04448123	152	519410.0
0.03014189	103	351968.6	0.04477387	153	522827.2
0.03043453	104	355385.8	0.04506651	154	526244.4
0.03072717	105	358803.0	0.04535915	155	529661.6
0.03101981	106	362220.2	0.04565179	156	533078.7
0.03131245	107	365637.3	0.04594443	157	536495.9
0.03160509	108	369054.5	0.04623707	158	539913.1
0.03189773	109	372471.7	0.04652971	159	543330.2
0.03219037	110	375888.8	0.04682235	160	546747.4
0.03248301	111	379306.0	0.04711499	161	550164.6
0.03277565	112	382723.2	0.04740763	162	553581.8
0.03306829	113	386140.4	0.04770027	163	556998.9
0.03336093	114	389557.5	0.04799291	164	560416.1
0.03365357	115	392974.7	0.04828555	165	563833.3
0.03394620	116	396391.9	0.04857819	166	567250.4
0.03423884	117	399809.0	0.04887083	167	570667.6
0.03453148	118	403226.2	0.04916347	168	574084.8
0.03482412	119	406643.4	0.04945611	169	577502.0
0.03511676	120	410060.6	0.04974875	170	580919.1
0.03540940	121	413477.7	0.05004139	171	584336.3
0.03570204	122	416894.9	0.05033403	172	587753.5
0.03599468	123	420312.1	0.05062667	173	591170.6
0.03628732	124	423729.2	0.05091931	174	594587.8
0.03657996	125	427146.4	0.05121195	175	598005.0
0.03687260	126	430563.6	0.05150459	176	601422.2
0.03716524	127	433980.8	0.05179723	177	604839.3
0.03745788	128	437397.9	0.05208987	178	608256.5
0.03775052	129	440815.1	0.05238251	179	611673.7
0.03804316	130	444232.3	0.05267515	180	615090.8
0.03833580	131	447649.4	0.05296779	181	618508.0
0.03862844	132	451066.6	0.05326040	182	621925.2
0.03892108	133	454483.8	0.05355306	183	625342.4
0.03921372	134	457901.0	0.05384570	184	628759.5
0.03950636	135	461318.1	0.05413834	185	632176.7
0.03979900	136	464735.3	0.05443098	186	635593.9
0.04009164	137	468152.5	0.05472362	187	639011.0
0.04038428	138	471569.6	0.05501626	188	642428.2
0.04067692	139	474986.8	0.05530890	189	645845.4
0.04096956	140	478404.0	0.05560154	190	649262.6
0.04126220	141	481821.2	0.05589418	191	652679.7
0.04155484	142	485238.3	0.05618682	192	656096.9
0.04184748	143	488655.5	0.05647946	193	659514.1
0.04214012	144	492072.7	0.05677210	194	662931.2
0.04243276	145	495489.8	0.05706474	195	666348.4
0.04272540	146	498907.0	0.05735738	196	669765.6
0.04301804	147	502324.2	0.05765002	197	673182.8
0.04331068	148	505741.4	0.05794266	198	676599.9
0.04360332	149	509158.5	0.05823530	199	680017.1
0.04389595	150	512575.7	0.05852794	200	683434.3

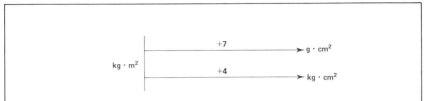

kg · m² kg · m²/s	REF	lbm–in² lbm–in²/s	kg · m² kg · m²/s	REF	lbm–in² lbm–in²/s
0.05882058	201	686851.4	0.07345256	251	857710.0
0.05911322	202	690268.6	0.07374520	252	861127.2
0.05940586	203	693685.8	0.07403784	253	864544.4
0.05969850	204	697103.0	0.07433048	254	867961.5
0.05999114	205	700520.1	0.07462312	255	871378.7
0.06028378	206	703937.3	0.07491576	256	874795.9
0.06057642	207	707354.5	0.07520840	257	878213.0
0.06086906	208	710771.6	0.07550104	258	881630.2
0.06116170	209	714188.8	0.07579368	259	885047.4
0.06145434	210	717606.0	0.07608632	260	888464.6
0.06174698	211	721023.2	0.07637896	261	891881.7
0.06203962	212	724440.3	0.07667160	262	895298.9
0.06233226	213	727857.5	0.07696424	263	898716.1
0.06262490	214	731274.7	0.07725688	264	902133.2
0.06291754	215	734691.8	0.07754952	265	905550.4
0.06321018	216	738109.0	0.07784216	266	908967.6
0.06350281	217	741526.2	0.07813480	267	912384.7
0.06379545	218	744943.4	0.07842744	268	915801.9
0.06408809	219	748360.5	0.07872008	269	919219.1
0.06438073	220	751777.7	0.07901272	270	922636.3
0.06467337	221	755194.9	0.07930536	271	926053.4
0.06496601	222	758612.0	0.07959800	272	929470.6
0.06525865	223	762029.2	0.07989064	273	932887.8
0.06555129	224	765446.4	0.08018328	274	936305.0
0.06584393	225	768863.6	0.08047592	275	939722.1
0.06613657	226	772280.7	0.08076856	276	943139.3
0.06642921	227	775697.9	0.08106120	277	946556.5
0.06672185	228	779115.1	0.08135384	278	949973.6
0.06701449	229	782532.2	0.08164648	279	953390.8
0.06730713	230	785949.4	0.08193912	280	956808.0
0.06759977	231	789366.6	0.08223176	281	960225.1
0.06789241	232	792783.8	0.08252440	282	963642.3
0.06818505	233	796200.9	0.08281703	283	967059.5
0.06847769	234	799618.1	0.08310967	284	970476.7
0.06877033	235	803035.3	0.08340231	285	973893.8
0.06906297	236	806452.4	0.08369495	286	977311.0
0.06935561	237	809869.6	0.08398759	287	980728.2
0.06964825	238	813286.8	0.08428023	288	984145.4
0.06994089	239	816704.0	0.08457287	289	987562.5
0.07023353	240	820121.1	0.08486551	290	990979.7
0.07052617	241	823538.3	0.08515815	291	994396.9
0.07081881	242	826955.5	0.08545079	292	997814.0
0.07111145	243	830372.6	0.08574343	293	1001231.
0.07140409	244	833789.8	0.08603607	294	1004648.
0.07169673	245	837207.0	0.08632871	295	1008066.
0.07198937	246	840624.2	0.08662135	296	1011483.
0.07228201	247	844041.3	0.08691399	297	1014900.
0.07257465	248	847458.5	0.08720663	298	1018317.
0.07286729	249	850875.7	0.08749927	299	1021734.
0.07315992	250	854292.8	0.08779191	300	1025151.

kg · m² kg · m²/s	REF	lbm-in² lbm-in²/s	kg · m² kg · m²/s	REF	lbm-in² lbm-in²/s
0.08808455	301	1028569.	0.1027165	351	1199427.
0.08837719	302	1031986.	0.1030092	352	1202844.
0.08866983	303	1035403.	0.1033018	353	1206261.
0.08896247	304	1038820.	0.1035945	354	1209679.
0.08925511	305	1042237.	0.1038871	355	1213096.
0.08954775	306	1045654.	0.1041797	356	1216513.
0.08984039	307	1049072.	0.1044724	357	1219930.
0.09013303	308	1052489.	0.1047650	358	1223347.
0.09042567	309	1055906.	0.1050577	359	1226765.
0.09071831	310	1059323.	0.1053503	360	1230182.
0.09101095	311	1062740.	0.1056429	361	1233599.
0.09130359	312	1066157.	0.1059356	362	1237016.
0.09159623	313	1069575.	0.1062282	363	1240433.
0.09188887	314	1072992.	0.1065209	364	1243850.
0.09218151	315	1076409.	0.1068135	365	1247268.
0.09247414	316	1079826.	0.1071061	366	1250685.
0.09276679	317	1083243.	0.1073988	367	1254102.
0.09305942	318	1086660.	0.1076914	368	1257519.
0.09335206	319	1090078.	0.1079840	369	1260936.
0.09364470	320	1093495.	0.1082767	370	1264353.
0.09393734	321	1096912.	0.1085693	371	1267771.
0.09422998	322	1100329.	0.1088620	372	1271188.
0.09452262	323	1103746.	0.1091546	373	1274605.
0.09481526	324	1107164.	0.1094472	374	1278022.
0.09510790	325	1110581.	0.1097399	375	1281439.
0.09540054	326	1113998.	0.1100325	376	1284856.
0.09569318	327	1117415.	0.1103252	377	1288274.
0.09598582	328	1120832.	0.1106178	378	1291691.
0.09627846	329	1124249.	0.1109104	379	1295108.
0.09657110	330	1127667.	0.1112031	380	1298525.
0.09686374	331	1131084.	0.1114957	381	1301942.
0.09715638	332	1134501.	0.1117884	382	1305359.
0.09744902	333	1137918.	0.1120810	383	1308777.
0.09774166	334	1141335.	0.1123736	384	1312194.
0.09803430	335	1144752.	0.1126663	385	1315611.
0.09832694	336	1148170.	0.1129589	386	1319028.
0.09861958	337	1151587.	0.1132516	387	1322445.
0.09891222	338	1155004.	0.1135442	388	1325862.
0.09920486	339	1158421.	0.1138368	389	1329280.
0.09949750	340	1161838.	0.1141295	390	1332697.
0.09979014	341	1165255.	0.1144221	391	1336114.
0.1000828	342	1168673.	0.1147148	392	1339531.
0.1003754	343	1172090.	0.1150074	393	1342948.
0.1006681	344	1175507.	0.1153000	394	1346366.
0.1009607	345	1178924.	0.1155927	395	1349783.
0.1012533	346	1182341.	0.1158853	396	1353200.
0.1015460	347	1185758.	0.1161780	397	1356617.
0.1018386	348	1189176.	0.1164706	398	1360034.
0.1021313	349	1192593.	0.1167632	399	1363451.
0.1024239	350	1196010.	0.1170559	400	1366869.

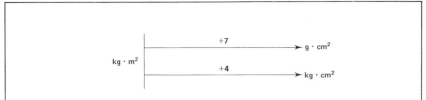

kg · m² kg · m²/s	REF	lbm-in² lbm-in²/s	kg · m² kg · m²/s	REF	lbm-in² lbm-in²/s
0.1173485	401	1370286.	0.1319805	451	1541144.
0.1176412	402	1373703.	0.1322731	452	1544561.
0.1179338	403	1377120.	0.1325658	453	1547979.
0.1182264	404	1380537.	0.1328584	454	1551396.
0.1185191	405	1383954.	0.1331511	455	1554813.
0.1188117	406	1387372.	0.1334437	456	1558230.
0.1191044	407	1390789.	0.1337363	457	1561647.
0.1193970	408	1394206.	0.1340290	458	1565064.
0.1196896	409	1397623.	0.1343216	459	1568482.
0.1199823	410	1401040.	0.1346143	460	1571899.
0.1202749	411	1404457.	0.1349069	461	1575316.
0.1205676	412	1407875.	0.1351995	462	1578733.
0.1208602	413	1411292.	0.1354922	463	1582150.
0.1211528	414	1414709.	0.1357848	464	1585568.
0.1214455	415	1418126.	0.1360775	465	1588985.
0.1217381	416	1421543.	0.1363701	466	1592402.
0.1220308	417	1424960.	0.1366627	467	1595819.
0.1223234	418	1428378.	0.1369554	468	1599236.
0.1226160	419	1431795.	0.1372480	469	1602653.
0.1229087	420	1435212.	0.1375407	470	1606071.
0.1232013	421	1438629.	0.1378333	471	1609488.
0.1234940	422	1442046.	0.1381259	472	1612905.
0.1237866	423	1445463.	0.1384186	473	1616322.
0.1240792	424	1448881.	0.1387112	474	1619739.
0.1243719	425	1452298.	0.1390039	475	1623156.
0.1246645	426	1455715.	0.1392965	476	1626574.
0.1249572	427	1459132.	0.1395891	477	1629991.
0.1252498	428	1462549.	0.1398818	478	1633408.
0.1255424	429	1465967.	0.1401744	479	1636825.
0.1258351	430	1469384.	0.1404671	480	1640242.
0.1261277	431	1472801.	0.1407597	481	1643659.
0.1264204	432	1476218.	0.1410523	482	1647077.
0.1267130	433	1479635.	0.1413450	483	1650494.
0.1270056	434	1483052.	0.1416376	484	1653911.
0.1272983	435	1486470.	0.1419303	485	1657328.
0.1275909	436	1489887.	0.1422229	486	1660745.
0.1278835	437	1493304.	0.1425155	487	1664162.
0.1281762	438	1496721.	0.1428082	488	1667580.
0.1284688	439	1500138.	0.1431008	489	1670997.
0.1287615	440	1503555.	0.1433935	490	1674414.
0.1290541	441	1506973.	0.1436861	491	1677831.
0.1293467	442	1510390.	0.1439787	492	1681248.
0.1296394	443	1513807.	0.1442714	493	1684665.
0.1299320	444	1517224.	0.1445640	494	1688083.
0.1302247	445	1520641.	0.1448567	495	1691500.
0.1305173	446	1524058.	0.1451493	496	1694917.
0.1308099	447	1527476.	0.1454419	497	1698334.
0.1311026	448	1530893.	0.1457346	498	1701751.
0.1313952	449	1534310.	0.1460272	499	1705169.
0.1316879	450	1537727.	0.1463198	500	1708586.

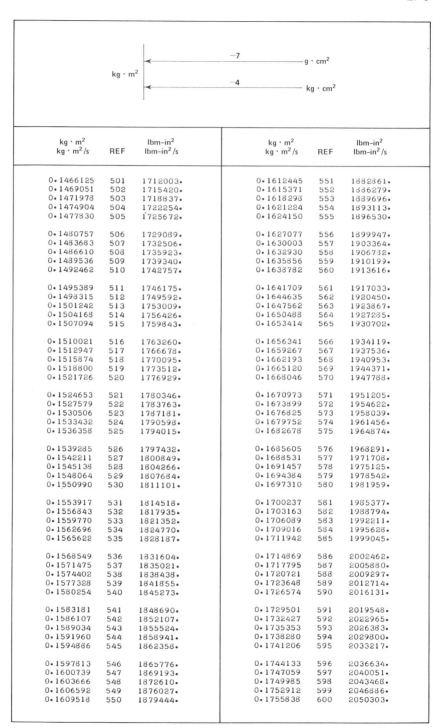

kg · m² kg · m²/s	REF	lbm-in² lbm-in²/s	kg · m² kg · m²/s	REF	lbm-in² lbm-in²/s
0.1466125	501	1712003.	0.1612445	551	1882861.
0.1469051	502	1715420.	0.1615371	552	1886279.
0.1471978	503	1718837.	0.1618298	553	1889696.
0.1474904	504	1722254.	0.1621224	554	1893113.
0.1477830	505	1725672.	0.1624150	555	1896530.
0.1480757	506	1729089.	0.1627077	556	1899947.
0.1483683	507	1732506.	0.1630003	557	1903364.
0.1486610	508	1735923.	0.1632930	558	1906782.
0.1489536	509	1739340.	0.1635856	559	1910199.
0.1492462	510	1742757.	0.1638782	560	1913616.
0.1495389	511	1746175.	0.1641709	561	1917033.
0.1498315	512	1749592.	0.1644635	562	1920450.
0.1501242	513	1753009.	0.1647562	563	1923867.
0.1504168	514	1756426.	0.1650488	564	1927285.
0.1507094	515	1759843.	0.1653414	565	1930702.
0.1510021	516	1763260.	0.1656341	566	1934119.
0.1512947	517	1766678.	0.1659267	567	1937536.
0.1515874	518	1770095.	0.1662193	568	1940953.
0.1518800	519	1773512.	0.1665120	569	1944371.
0.1521726	520	1776929.	0.1668046	570	1947788.
0.1524653	521	1780346.	0.1670973	571	1951205.
0.1527579	522	1783763.	0.1673899	572	1954622.
0.1530506	523	1787181.	0.1676825	573	1958039.
0.1533432	524	1790598.	0.1679752	574	1961456.
0.1536358	525	1794015.	0.1682678	575	1964874.
0.1539285	526	1797432.	0.1685605	576	1968291.
0.1542211	527	1800849.	0.1688531	577	1971708.
0.1545138	528	1804266.	0.1691457	578	1975125.
0.1548064	529	1807684.	0.1694384	579	1978542.
0.1550990	530	1811101.	0.1697310	580	1981959.
0.1553917	531	1814518.	0.1700237	581	1985377.
0.1556843	532	1817935.	0.1703163	582	1988794.
0.1559770	533	1821352.	0.1706089	583	1992211.
0.1562696	534	1824770.	0.1709016	584	1995628.
0.1565622	535	1828187.	0.1711942	585	1999045.
0.1568549	536	1831604.	0.1714869	586	2002462.
0.1571475	537	1835021.	0.1717795	587	2005880.
0.1574402	538	1838438.	0.1720721	588	2009297.
0.1577328	539	1841855.	0.1723648	589	2012714.
0.1580254	540	1845273.	0.1726574	590	2016131.
0.1583181	541	1848690.	0.1729501	591	2019548.
0.1586107	542	1852107.	0.1732427	592	2022965.
0.1589034	543	1855524.	0.1735353	593	2026383.
0.1591960	544	1858941.	0.1738280	594	2029800.
0.1594886	545	1862358.	0.1741206	595	2033217.
0.1597813	546	1865776.	0.1744133	596	2036634.
0.1600739	547	1869193.	0.1747059	597	2040051.
0.1603666	548	1872610.	0.1749985	598	2043468.
0.1606592	549	1876027.	0.1752912	599	2046886.
0.1609518	550	1879444.	0.1755838	600	2050303.

kg · m² kg · m²/s	REF	lbm-in² lbm-in²/s	kg · m² kg · m²/s	REF	lbm-in² lbm-in²/s
0.1758765	601	2053720.	0.1905084	651	2224579.
0.1761691	602	2057137.	0.1908011	652	2227996.
0.1764617	603	2060554.	0.1910937	653	2231413.
0.1767544	604	2063972.	0.1913864	654	2234830.
0.1770470	605	2067389.	0.1916790	655	2238247.
0.1773397	606	2070806.	0.1919716	656	2241664.
0.1776323	607	2074223.	0.1922643	657	2245082.
0.1779249	608	2077640.	0.1925569	658	2248499.
0.1782176	609	2081057.	0.1928496	659	2251916.
0.1785102	610	2084475.	0.1931422	660	2255333.
0.1788029	611	2087892.	0.1934348	661	2258750.
0.1790955	612	2091309.	0.1937275	662	2262167.
0.1793881	613	2094726.	0.1940201	663	2265585.
0.1796808	614	2098143.	0.1943128	664	2269002.
0.1799734	615	2101560.	0.1946054	665	2272419.
0.1802661	616	2104978.	0.1948980	666	2275836.
0.1805587	617	2108395.	0.1951907	667	2279253.
0.1808513	618	2111812.	0.1954833	668	2282670.
0.1811440	619	2115229.	0.1957760	669	2286088.
0.1814366	620	2118646.	0.1960686	670	2289505.
0.1817293	621	2122063.	0.1963612	671	2292922.
0.1820219	622	2125481.	0.1966539	672	2296339.
0.1823145	623	2128898.	0.1969465	673	2299756.
0.1826072	624	2132315.	0.1972392	674	2303174.
0.1828998	625	2135732.	0.1975318	675	2306591.
0.1831925	626	2139149.	0.1978244	676	2310008.
0.1834851	627	2142566.	0.1981171	677	2313425.
0.1837777	628	2145984.	0.1984097	678	2316842.
0.1840704	629	2149401.	0.1987024	679	2320259.
0.1843630	630	2152818.	0.1989950	680	2323677.
0.1846556	631	2156235.	0.1992876	681	2327094.
0.1849483	632	2159652.	0.1995803	682	2330511.
0.1852409	633	2163069.	0.1998729	683	2333928.
0.1855336	634	2166487.	0.2001656	684	2337345.
0.1858262	635	2169904.	0.2004582	685	2340762.
0.1861188	636	2173321.	0.2007508	686	2344180.
0.1864115	637	2176738.	0.2010435	687	2347597.
0.1867041	638	2180155.	0.2013361	688	2351014.
0.1869968	639	2183573.	0.2016288	689	2354431.
0.1872894	640	2186990.	0.2019214	690	2357848.
0.1875820	641	2190407.	0.2022140	691	2361265.
0.1878747	642	2193824.	0.2025067	692	2364683.
0.1881673	643	2197241.	0.2027993	693	2368100.
0.1884600	644	2200658.	0.2030920	694	2371517.
0.1887526	645	2204076.	0.2033846	695	2374934.
0.1890452	646	2207493.	0.2036772	696	2378351.
0.1893379	647	2210910.	0.2039699	697	2381768.
0.1896305	648	2214327.	0.2042625	698	2385186.
0.1899232	649	2217744.	0.2045552	699	2388603.
0.1902158	650	2221161.	0.2048478	700	2392020.

kg · m² kg · m²/s	REF	lbm-in² lbm-in²/s	kg · m² kg · m²/s	REF	lbm-in² lbm-in²/s
0·2051404	701	2395437·	0·2197724	751	2566296·
0·2054331	702	2398854·	0·2200651	752	2569713·
0·2057257	703	2402271·	0·2203577	753	2573130·
0·2060183	704	2405689·	0·2206503	754	2576547·
0·2063110	705	2409106·	0·2209430	755	2579964·
0·2066036	706	2412523·	0·2212356	756	2583382·
0·2068963	707	2415940·	0·2215283	757	2586799·
0·2071889	708	2419357·	0·2218209	758	2590216·
0·2074815	709	2422775·	0·2221135	759	2593633·
0·2077742	710	2426192·	0·2224062	760	2597050·
0·2080668	711	2429609·	0·2226988	761	2600467·
0·2083595	712	2433026·	0·2229915	762	2603885·
0·2086521	713	2436443·	0·2232841	763	2607302·
0·2089447	714	2439860·	0·2235767	764	2610719·
0·2092374	715	2443278·	0·2238694	765	2614136·
0·2095300	716	2446695·	0·2241620	766	2617553·
0·2098227	717	2450112·	0·2244546	767	2620970·
0·2101153	718	2453529·	0·2247473	768	2624388·
0·2104079	719	2456946·	0·2250399	769	2627805·
0·2107006	720	2460363·	0·2253326	770	2631222·
0·2109932	721	2463781·	0·2256252	771	2634639·
0·2112859	722	2467198·	0·2259178	772	2638056·
0·2115785	723	2470615·	0·2262105	773	2641473·
0·2118711	724	2474032·	0·2265031	774	2644891·
0·2121638	·725	2477449·	0·2267958	775	2648308·
0·2124564	726	2480866·	0·2270884	776	2651725·
0·2127491	727	2484284·	0·2273810	777	2655142·
0·2130417	728	2487701·	0·2276737	778	2658559·
0·2133343	729	2491118·	0·2279663	779	2661977·
0·2136270	730	2494535·	0·2282590	780	2665394·
0·2139196	731	2497952·	0·2285516	781	2668811·
0·2142123	732	2501369·	0·2288442	782	2672228·
0·2145049	733	2504787·	0·2291369	783	2675645·
0·2147975	734	2508204·	0·2294295	784	2679062·
0·2150902	735	2511621·	0·2297222	785	2682480·
0·2153828	736	2515038·	0·2300148	786	2685897·
0·2156755	737	2518455·	0·2303074	787	2689314·
0·2159681	738	2521872·	0·2306001	788	2692731·
0·2162607	739	2525290·	0·2308927	789	2696148·
0·2165534	740	2528707·	0·2311854	790	2699565·
0·2168460	741	2532124·	0·2314780	791	2702983·
0·2171387	742	2535541·	0·2317706	792	2706400·
0·2174313	743	2538958·	0·2320633	793	2709817·
0·2177239	744	2542375·	0·2323559	794	2713234·
0·2180166	745	2545793·	0·2326486	795	2716651·
0·2183092	746	2549210·	0·2329412	796	2720068·
0·2186019	747	2552627·	0·2332338	797	2723486·
0·2188945	748	2556044·	0·2335265	798	2726903·
0·2191871	749	2559461·	0·2338191	799	2730320·
0·2194798	750	2562879·	0·2341118	800	2733737·

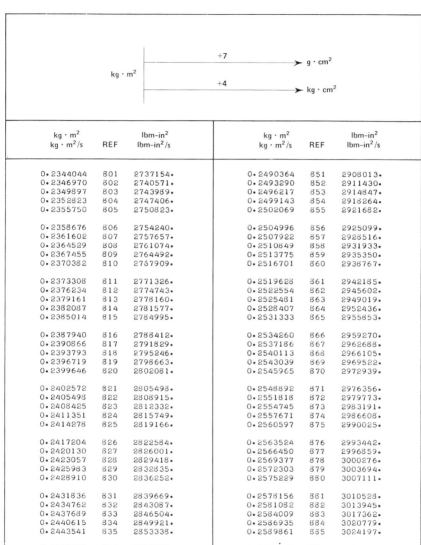

kg · m² kg · m²/s	REF	lbm-in² lbm-in²/s	kg · m² kg · m²/s	REF	lbm-in² lbm-in²/s
0.2344044	801	2737154.	0.2490364	851	2908013.
0.2346970	802	2740571.	0.2493290	852	2911430.
0.2349897	803	2743989.	0.2496217	853	2914847.
0.2352823	804	2747406.	0.2499143	854	2918264.
0.2355750	805	2750823.	0.2502069	855	2921682.
0.2358676	806	2754240.	0.2504996	856	2925099.
0.2361602	807	2757657.	0.2507922	857	2928516.
0.2364529	808	2761074.	0.2510849	858	2931933.
0.2367455	809	2764492.	0.2513775	859	2935350.
0.2370382	810	2767909.	0.2516701	860	2938767.
0.2373308	811	2771326.	0.2519628	861	2942185.
0.2376234	812	2774743.	0.2522554	862	2945602.
0.2379161	813	2778160.	0.2525481	863	2949019.
0.2382087	814	2781577.	0.2528407	864	2952436.
0.2385014	815	2784995.	0.2531333	865	2955853.
0.2387940	816	2788412.	0.2534260	866	2959270.
0.2390866	817	2791829.	0.2537186	867	2962688.
0.2393793	818	2795246.	0.2540113	868	2966105.
0.2396719	819	2798663.	0.2543039	869	2969522.
0.2399646	820	2802081.	0.2545965	870	2972939.
0.2402572	821	2805498.	0.2548892	871	2976356.
0.2405498	822	2808915.	0.2551818	872	2979773.
0.2408425	823	2812332.	0.2554745	873	2983191.
0.2411351	824	2815749.	0.2557671	874	2986608.
0.2414278	825	2819166.	0.2560597	875	2990025.
0.2417204	826	2822584.	0.2563524	876	2993442.
0.2420130	827	2826001.	0.2566450	877	2996859.
0.2423057	828	2829418.	0.2569377	878	3000276.
0.2425983	829	2832835.	0.2572303	879	3003694.
0.2428910	830	2836252.	0.2575229	880	3007111.
0.2431836	831	2839669.	0.2578156	881	3010528.
0.2434762	832	2843087.	0.2581082	882	3013945.
0.2437689	833	2846504.	0.2584009	883	3017362.
0.2440615	834	2849921.	0.2586935	884	3020779.
0.2443541	835	2853338.	0.2589861	885	3024197.
0.2446468	836	2856755.	0.2592788	886	3027614.
0.2449394	837	2860172.	0.2595714	887	3031031.
0.2452321	838	2863590.	0.2598641	888	3034448.
0.2455247	839	2867007.	0.2601567	889	3037865.
0.2458173	840	2870424.	0.2604493	890	3041283.
0.2461100	841	2873841.	0.2607420	891	3044700.
0.2464026	842	2877258.	0.2610346	892	3048117.
0.2466953	843	2880675.	0.2613273	893	3051534.
0.2469879	844	2884093.	0.2616199	894	3054951.
0.2472805	845	2887510.	0.2619125	895	3058368.
0.2475732	846	2890927.	0.2622052	896	3061786.
0.2478658	847	2894344.	0.2624978	897	3065203.
0.2481585	848	2897761.	0.2627904	898	3068620.
0.2484511	849	2901178.	0.2630831	899	3072037.
0.2487437	850	2904596.	0.2633757	900	3075454.

kg · m² kg · m²/s	REF	lbm-in² lbm-in²/s		kg · m² kg · m²/s	REF	lbm-in² lbm-in²/s
0.2636684	901	3078871.		0.2783004	951	3249730.
0.2639610	902	3082289.		0.2785930	952	3253147.
0.2642536	903	3085706.		0.2788856	953	3256564.
0.2645463	904	3089123.		0.2791783	954	3259981.
0.2648389	905	3092540.		0.2794709	955	3263399.
0.2651316	906	3095957.		0.2797636	956	3266816.
0.2654242	907	3099374.		0.2800562	957	3270233.
0.2657168	908	3102792.		0.2803488	958	3273650.
0.2660095	909	3106209.		0.2806415	959	3277067.
0.2663021	910	3109626.		0.2809341	960	3280485.
0.2665948	911	3113043.		0.2812268	961	3283902.
0.2668874	912	3116460.		0.2815194	962	3287319.
0.2671800	913	3119877.		0.2818120	963	3290736.
0.2674727	914	3123295.		0.2821047	964	3294153.
0.2677653	915	3126712.		0.2823973	965	3297570.
0.2680580	916	3130129.		0.2826899	966	3300988.
0.2683506	917	3133546.		0.2829826	967	3304405.
0.2686432	918	3136963.		0.2832752	968	3307822.
0.2689359	919	3140380.		0.2835679	969	3311239.
0.2692285	920	3143798.		0.2838605	970	3314656.
0.2695212	921	3147215.		0.2841531	971	3318073.
0.2698138	922	3150632.		0.2844458	972	3321491.
0.2701064	923	3154049.		0.2847384	973	3324908.
0.2703991	924	3157466.		0.2850311	974	3328325.
0.2706917	925	3160884.		0.2853237	975	3331742.
0.2709844	926	3164301.		0.2856163	976	3335159.
0.2712770	927	3167718.		0.2859090	977	3338576.
0.2715696	928	3171135.		0.2862016	978	3341994.
0.2718623	929	3174552.		0.2864943	979	3345411.
0.2721549	930	3177969.		0.2867869	980	3348828.
0.2724476	931	3181387.		0.2870795	981	3352245.
0.2727402	932	3184804.		0.2873722	982	3355662.
0.2730328	933	3188221.		0.2876648	983	3359079.
0.2733255	934	3191638.		0.2879575	984	3362497.
0.2736181	935	3195055.		0.2882501	985	3365914.
0.2739108	936	3198472.		0.2885427	986	3369331.
0.2742034	937	3201890.		0.2888354	987	3372748.
0.2744960	938	3205307.		0.2891280	988	3376165.
0.2747887	939	3208724.		0.2894207	989	3379582.
0.2750813	940	3212141.		0.2897133	990	3383000.
0.2753740	941	3215558.		0.2900059	991	3386417.
0.2756666	942	3218975.		0.2902986	992	3389834.
0.2759592	943	3222393.		0.2905912	993	3393251.
0.2762519	944	3225810.		0.2908839	994	3396668.
0.2765445	945	3229227.		0.2911765	995	3400086.
0.2768372	946	3232644.		0.2914691	996	3403503.
0.2771298	947	3236061.		0.2917618	997	3406920.
0.2774224	948	3239478.		0.2920544	998	3410337.
0.2777151	949	3242896.		0.2923471	999	3413754.
0.2780077	950	3246313.		0.2926397	1000	3417171.

28

FORCE

Pound (force) — Newton
    (lbf — N)
Kip (kilopound-force) — Kilonewton
    (kip — kN)

Conversion Factors:
    1 pound-force (lbf) = 4.448222 newtons (N)
    (1 N = 0.2248089 lbf)

Kilopound (kip) — Kilonewton (kN) Values

Table 28 can be used directly for kip–kN conversions. Since the kilopound is 1000 pounds and the kilonewton is 1000 newtons, the conversion factors between these units are the same as those above for lbf–N.

Therefore: 1 kip = 4.448222 kN.

In converting force quantities from newton to kilonewton and back, the following Decimal-shift Diagram 28-A, which appears in the headings on alternate pages of this table, will assist the user.

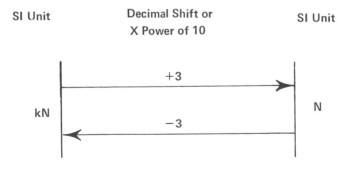

Diagram 28-A

Example (*a*): Convert 754 kips to kN, N.

From table: 754 kips = 3353.959 kN* ≈ 3350 kN

From Diagram 28-A: 3353.959 kN = 3353.959 × 10^3 N = 3,353,959 N

≈ 3,350,000 or 3.35 × 10^6 N

(kN to N by multiplying kN value by 10^3 or shifting its decimal 3 places to the right.)

Example (*b*): Convert 153,000 N to kips.

From Diagram 28-A: 153,000 N = 153,000 × 10^{-3} kN = 153 kN

(N to kN by multiplying N value by 10^{-3} or by shifting its decimal 3 places to the left.)

From table: 153 kN = 34.39577 kip* ≈ 34.4 kip

---

*All conversion values obtained, can be rounded to desired number of significant figures.

N kN	REF	lbf kip	N kN	REF	lbf kip
4.448222	1	0.2248089	226.8593	51	11.46526
8.896444	2	0.4496178	231.3075	52	11.69006
13.34467	3	0.6744268	235.7558	53	11.91487
17.79289	4	0.8992357	240.2040	54	12.13968
22.24111	5	1.124045	244.6522	55	12.36449
26.68933	6	1.348854	249.1004	56	12.58930
31.13755	7	1.573662	253.5487	57	12.81411
35.58578	8	1.798471	257.9969	58	13.03892
40.03400	9	2.023280	262.4451	59	13.26373
44.48222	10	2.248089	266.8933	60	13.48854
48.93044	11	2.472898	271.3415	61	13.71334
53.37866	12	2.697707	275.7898	62	13.93815
57.82689	13	2.922516	280.2380	63	14.16296
62.27511	14	3.147325	284.6862	64	14.38777
66.72333	15	3.372134	289.1344	65	14.61258
71.17155	16	3.596943	293.5826	66	14.83739
75.61977	17	3.821752	298.0309	67	15.06220
80.06800	18	4.046561	302.4791	68	15.28701
84.51622	19	4.271370	306.9273	69	15.51182
88.96444	20	4.496179	311.3755	70	15.73662
93.41266	21	4.720987	315.8238	71	15.96143
97.86088	22	4.945796	320.2720	72	16.18624
102.3091	23	5.170605	324.7202	73	16.41105
106.7573	24	5.395414	329.1684	74	16.63586
111.2055	25	5.620223	333.6166	75	16.86067
115.6538	26	5.845032	338.0649	76	17.08548
120.1020	27	6.069841	342.5131	77	17.31029
124.5502	28	6.294650	346.9613	78	17.53510
128.9984	29	6.519459	351.4095	79	17.75991
133.4467	30	6.744268	355.8578	80	17.98471
137.8949	31	6.969077	360.3060	81	18.20952
142.3431	32	7.193886	364.7542	82	18.43433
146.7913	33	7.418694	369.2024	83	18.65914
151.2395	34	7.643503	373.6506	84	18.88395
155.6878	35	7.868312	378.0989	85	19.10876
160.1360	36	8.093121	382.5471	86	19.33357
164.5842	37	8.317930	386.9953	87	19.55838
169.0324	38	8.542739	391.4435	88	19.78319
173.4807	39	8.767548	395.8918	89	20.00799
177.9289	40	8.992357	400.3400	90	20.23280
182.3771	41	9.217166	404.7882	91	20.45761
186.8253	42	9.441975	409.2364	92	20.68242
191.2735	43	9.666784	413.6846	93	20.90723
195.7218	44	9.891593	418.1329	94	21.13204
200.1700	45	10.11640	422.5811	95	21.35685
204.6182	46	10.34121	427.0293	96	21.58166
209.0664	47	10.56602	431.4775	97	21.80647
213.5147	48	10.79083	435.9258	98	22.03127
217.9629	49	11.01564	440.3740	99	22.25608
222.4111	50	11.24045	444.8222	100	22.48089

N kN	REF	lbf kip		N kN	REF	lbf kip
449.2704	101	22.70570		671.6815	151	33.94615
453.7186	102	22.93051		676.1297	152	34.17096
458.1669	103	23.15532		680.5780	153	34.39577
462.6151	104	23.38013		685.0262	154	34.62057
467.0633	105	23.60494		689.4744	155	34.84538
471.5115	106	23.82975		693.9226	156	35.07019
475.9598	107	24.05455		698.3708	157	35.29500
480.4080	108	24.27936		702.8191	158	35.51981
484.8562	109	24.50417		707.2673	159	35.74462
489.3044	110	24.72898		711.7155	160	35.96943
493.7526	111	24.95379		716.1637	161	36.19424
498.2009	112	25.17860		720.6120	162	36.41905
502.6491	113	25.40341		725.0602	163	36.64385
507.0973	114	25.62822		729.5084	164	36.86866
511.5455	115	25.85303		733.9566	165	37.09347
515.9938	116	26.07784		738.4048	166	37.31828
520.4420	117	26.30264		742.8531	167	37.54309
524.8902	118	26.52745		747.3013	168	37.76790
529.3384	119	26.75226		751.7495	169	37.99271
533.7866	120	26.97707		756.1977	170	38.21752
538.2349	121	27.20188		760.6460	171	38.44233
542.6831	122	27.42669		765.0942	172	38.66713
547.1313	123	27.65150		769.5424	173	38.89194
551.5795	124	27.87631		773.9906	174	39.11675
556.0277	125	28.10112		778.4389	175	39.34156
560.4760	126	28.32592		782.8871	176	39.56637
564.9242	127	28.55073		787.3353	177	39.79118
569.3724	128	28.77554		791.7835	178	40.01599
573.8206	129	29.00035		796.2317	179	40.24080
578.2689	130	29.22516		800.6800	180	40.46561
582.7171	131	29.44997		805.1282	181	40.69042
587.1653	132	29.67478		809.5764	182	40.91522
591.6135	133	29.89959		814.0246	183	41.14003
596.0617	134	30.12440		818.4728	184	41.36484
600.5100	135	30.34920		822.9211	185	41.58965
604.9582	136	30.57401		827.3693	186	41.81446
609.4064	137	30.79882		831.8175	187	42.03927
613.8546	138	31.02363		836.2657	188	42.26408
618.3029	139	31.24844		840.7140	189	42.48889
622.7511	140	31.47325		845.1622	190	42.71370
627.1993	141	31.69806		849.6104	191	42.93850
631.6475	142	31.92287		854.0586	192	43.16331
636.0957	143	32.14768		858.5068	193	43.38812
640.5440	144	32.37249		862.9551	194	43.61293
644.9922	145	32.59729		867.4033	195	43.83774
649.4404	146	32.82210		871.8515	196	44.06255
653.8886	147	33.04691		876.2997	197	44.28736
658.3369	148	33.27172		880.7480	198	44.51217
662.7851	149	33.49653		885.1962	199	44.73698
667.2333	150	33.72134		889.6444	200	44.96178

N kN	REF	lbf kip		N kN	REF	lbf kip
894.0926	201	45.18659		1116.504	251	56.42704
898.5408	202	45.41140		1120.952	252	56.65185
902.9891	203	45.63621		1125.400	253	56.87666
907.4373	204	45.86102		1129.848	254	57.10147
911.8855	205	46.08583		1134.297	255	57.32628
916.3337	206	46.31064		1138.745	256	57.55108
920.7820	207	46.53545		1143.193	257	57.77589
925.2302	208	46.76026		1147.641	258	58.00070
929.6784	209	46.98506		1152.089	259	58.22551
934.1266	210	47.20987		1156.538	260	58.45032
938.5748	211	47.43468		1160.986	261	58.67513
943.0231	212	47.65949		1165.434	262	58.89994
947.4713	213	47.88430		1169.882	263	59.12475
951.9195	214	48.10911		1174.331	264	59.34956
956.3677	215	48.33392		1178.779	265	59.57436
960.8159	216	48.55873		1183.227	266	59.79917
965.2642	217	48.78354		1187.675	267	60.02398
969.7124	218	49.00835		1192.123	268	60.24879
974.1606	219	49.23315		1196.572	269	60.47360
978.6088	220	49.45796		1201.020	270	60.69841
983.0571	221	49.68277		1205.468	271	60.92322
987.5053	222	49.90758		1209.916	272	61.14803
991.9535	223	50.13239		1214.365	273	61.37284
996.4017	224	50.35720		1218.813	274	61.59765
1000.850	225	50.58201		1223.261	275	61.82245
1005.298	226	50.80682		1227.709	276	62.04726
1009.746	227	51.03163		1232.157	277	62.27207
1014.195	228	51.25643		1236.606	278	62.49688
1018.643	229	51.48124		1241.054	279	62.72169
1023.091	230	51.70605		1245.502	280	62.94650
1027.539	231	51.93086		1249.950	281	63.17131
1031.988	232	52.15567		1254.399	282	63.39612
1036.436	233	52.38048		1258.847	283	63.62093
1040.884	234	52.60529		1263.295	284	63.84573
1045.332	235	52.83010		1267.743	285	64.07054
1049.780	236	53.05491		1272.191	286	64.29535
1054.229	237	53.27972		1276.640	287	64.52016
1058.677	238	53.50452		1281.088	288	64.74497
1063.125	239	53.72933		1285.536	289	64.96978
1067.573	240	53.95414		1289.984	290	65.19459
1072.021	241	54.17895		1294.433	291	65.41940
1076.470	242	54.40376		1298.881	292	65.64421
1080.918	243	54.62857		1303.329	293	65.86901
1085.366	244	54.85338		1307.777	294	66.09382
1089.814	245	55.07819		1312.225	295	66.31863
1094.263	246	55.30300		1316.674	296	66.54344
1098.711	247	55.52780		1321.122	297	66.76825
1103.159	248	55.75261		1325.570	298	66.99306
1107.607	249	55.97742		1330.018	299	67.21787
1112.055	250	56.20223		1334.467	300	67.44268

N kN	REF	lbf kip	N kN	REF	lbf kip
1338.915	301	67.66749	1561.326	351	78.90793
1343.363	302	67.89229	1565.774	352	79.13274
1347.811	303	68.11710	1570.222	353	79.35755
1352.259	304	68.34191	1574.671	354	79.58236
1356.708	305	68.56672	1579.119	355	79.80717
1361.156	306	68.79153	1583.567	356	80.03198
1365.604	307	69.01634	1588.015	357	80.25679
1370.052	308	69.24115	1592.463	358	80.48160
1374.501	309	69.46596	1596.912	359	80.70640
1378.949	310	69.69077	1601.360	360	80.93121
1383.397	311	69.91558	1605.808	361	81.15602
1387.845	312	70.14038	1610.256	362	81.38083
1392.293	313	70.36519	1614.705	363	81.60564
1396.742	314	70.59000	1619.153	364	81.83045
1401.190	315	70.81481	1623.601	365	82.05526
1405.638	316	71.03962	1628.049	366	82.28007
1410.086	317	71.26443	1632.497	367	82.50488
1414.535	318	71.48924	1636.946	368	82.72968
1418.983	319	71.71405	1641.394	369	82.95449
1423.431	320	71.93886	1645.842	370	83.17930
1427.879	321	72.16366	1650.290	371	83.40411
1432.327	322	72.38847	1654.739	372	83.62892
1436.776	323	72.61328	1659.187	373	83.85373
1441.224	324	72.83809	1663.635	374	84.07854
1445.672	325	73.06290	1668.083	375	84.30335
1450.120	326	73.28771	1672.531	376	84.52816
1454.569	327	73.51252	1676.980	377	84.75296
1459.017	328	73.73733	1681.428	378	84.97777
1463.465	329	73.96214	1685.876	379	85.20258
1467.913	330	74.18694	1690.324	380	85.42739
1472.361	331	74.41175	1694.773	381	85.65220
1476.810	332	74.63656	1699.221	382	85.87701
1481.258	333	74.86137	1703.669	383	86.10182
1485.706	334	75.08618	1708.117	384	86.32663
1490.154	335	75.31099	1712.565	385	86.55144
1494.603	336	75.53580	1717.014	386	86.77624
1499.051	337	75.76061	1721.462	387	87.00105
1503.499	338	75.98542	1725.910	388	87.22586
1507.947	339	76.21023	1730.358	389	87.45067
1512.395	340	76.43503	1734.807	390	87.67548
1516.844	341	76.65984	1739.255	391	87.90029
1521.292	342	76.88465	1743.703	392	88.12510
1525.740	343	77.10946	1748.151	393	88.34991
1530.188	344	77.33427	1752.599	394	88.57472
1534.637	345	77.55908	1757.048	395	88.79953
1539.085	346	77.78389	1761.496	396	89.02433
1543.533	347	78.00870	1765.944	397	89.24914
1547.981	348	78.23351	1770.392	398	89.47395
1552.429	349	78.45831	1774.841	399	89.69876
1556.878	350	78.68312	1779.289	400	89.92357

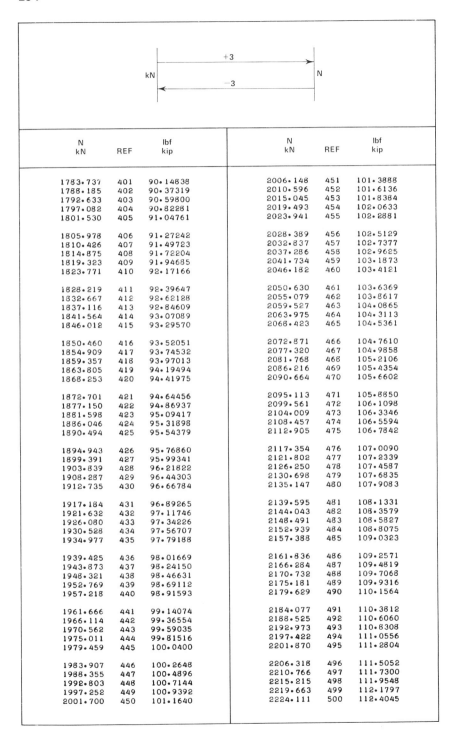

N kN	REF	lbf kip	N kN	REF	lbf kip
1783.737	401	90.14838	2006.148	451	101.3888
1788.185	402	90.37319	2010.596	452	101.6136
1792.633	403	90.59800	2015.045	453	101.8384
1797.082	404	90.82281	2019.493	454	102.0633
1801.530	405	91.04761	2023.941	455	102.2881
1805.978	406	91.27242	2028.389	456	102.5129
1810.426	407	91.49723	2032.837	457	102.7377
1814.875	408	91.72204	2037.286	458	102.9625
1819.323	409	91.94685	2041.734	459	103.1873
1823.771	410	92.17166	2046.182	460	103.4121
1828.219	411	92.39647	2050.630	461	103.6369
1832.667	412	92.62128	2055.079	462	103.8617
1837.116	413	92.84609	2059.527	463	104.0865
1841.564	414	93.07089	2063.975	464	104.3113
1846.012	415	93.29570	2068.423	465	104.5361
1850.460	416	93.52051	2072.871	466	104.7610
1854.909	417	93.74532	2077.320	467	104.9858
1859.357	418	93.97013	2081.768	468	105.2106
1863.805	419	94.19494	2086.216	469	105.4354
1868.253	420	94.41975	2090.664	470	105.6602
1872.701	421	94.64456	2095.113	471	105.8850
1877.150	422	94.86937	2099.561	472	106.1098
1881.598	423	95.09417	2104.009	473	106.3346
1886.046	424	95.31898	2108.457	474	106.5594
1890.494	425	95.54379	2112.905	475	106.7842
1894.943	426	95.76860	2117.354	476	107.0090
1899.391	427	95.99341	2121.802	477	107.2339
1903.839	428	96.21822	2126.250	478	107.4587
1908.287	429	96.44303	2130.698	479	107.6835
1912.735	430	96.66784	2135.147	480	107.9083
1917.184	431	96.89265	2139.595	481	108.1331
1921.632	432	97.11746	2144.043	482	108.3579
1926.080	433	97.34226	2148.491	483	108.5827
1930.528	434	97.56707	2152.939	484	108.8075
1934.977	435	97.79188	2157.388	485	109.0323
1939.425	436	98.01669	2161.836	486	109.2571
1943.873	437	98.24150	2166.284	487	109.4819
1948.321	438	98.46631	2170.732	488	109.7068
1952.769	439	98.69112	2175.181	489	109.9316
1957.218	440	98.91593	2179.629	490	110.1564
1961.666	441	99.14074	2184.077	491	110.3812
1966.114	442	99.36554	2188.525	492	110.6060
1970.562	443	99.59035	2192.973	493	110.8308
1975.011	444	99.81516	2197.422	494	111.0556
1979.459	445	100.0400	2201.870	495	111.2804
1983.907	446	100.2648	2206.318	496	111.5052
1988.355	447	100.4896	2210.766	497	111.7300
1992.803	448	100.7144	2215.215	498	111.9548
1997.252	449	100.9392	2219.663	499	112.1797
2001.700	450	101.1640	2224.111	500	112.4045

N kN	REF	lbf kip	N kN	REF	lbf kip
2228.559	501	112.6293	2450.970	551	123.8697
2233.007	502	112.8541	2455.419	552	124.0945
2237.456	503	113.0789	2459.867	553	124.3193
2241.904	504	113.3037	2464.315	554	124.5441
2246.352	505	113.5285	2468.763	555	124.7690
2250.800	506	113.7533	2473.211	556	124.9938
2255.249	507	113.9781	2477.660	557	125.2186
2259.697	508	114.2029	2482.108	558	125.4434
2264.145	509	114.4277	2486.556	559	125.6682
2268.593	510	114.6526	2491.004	560	125.8930
2273.041	511	114.8774	2495.453	561	126.1178
2277.490	512	115.1022	2499.901	562	126.3426
2281.938	513	115.3270	2504.349	563	126.5674
2286.386	514	115.5518	2508.797	564	126.7922
2290.834	515	115.7766	2513.245	565	127.0170
2295.283	516	116.0014	2517.694	566	127.2419
2299.731	517	116.2262	2522.142	567	127.4667
2304.179	518	116.4510	2526.590	568	127.6915
2308.627	519	116.6758	2531.038	569	127.9163
2313.075	520	116.9006	2535.487	570	128.1411
2317.524	521	117.1254	2539.935	571	128.3659
2321.972	522	117.3503	2544.383	572	128.5907
2326.420	523	117.5751	2548.831	573	128.8155
2330.868	524	117.7999	2553.279	574	129.0403
2335.317	525	118.0247	2557.728	575	129.2651
2339.765	526	118.2495	2562.176	576	129.4899
2344.213	527	118.4743	2566.624	577	129.7147
2348.661	528	118.6991	2571.072	578	129.9396
2353.109	529	118.9239	2575.521	579	130.1644
2357.558	530	119.1487	2579.969	580	130.3892
2362.006	531	119.3735	2584.417	581	130.6140
2366.454	532	119.5983	2588.865	582	130.8388
2370.902	533	119.8232	2593.313	583	131.0636
2375.351	534	120.0480	2597.762	584	131.2884
2379.799	535	120.2728	2602.210	585	131.5132
2384.247	536	120.4976	2606.658	586	131.7380
2388.695	537	120.7224	2611.106	587	131.9628
2393.143	538	120.9472	2615.555	588	132.1876
2397.592	539	121.1720	2620.003	589	132.4125
2402.040	540	121.3968	2624.451	590	132.6373
2406.488	541	121.6216	2628.899	591	132.8621
2410.936	542	121.8464	2633.347	592	133.0869
2415.385	543	122.0712	2637.796	593	133.3117
2419.833	544	122.2961	2642.244	594	133.5365
2424.281	545	122.5209	2646.692	595	133.7613
2428.729	546	122.7457	2651.140	596	133.9861
2433.177	547	122.9705	2655.589	597	134.2109
2437.626	548	123.1953	2660.037	598	134.4357
2442.074	549	123.4201	2664.485	599	134.6605
2446.522	550	123.6449	2668.933	600	134.8854

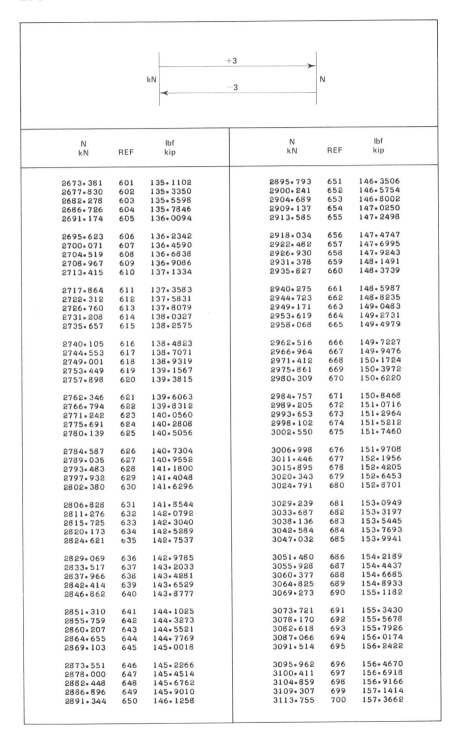

N kN	REF	lbf kip	N kN	REF	lbf kip
2673.381	601	135.1102	2895.793	651	146.3506
2677.830	602	135.3350	2900.241	652	146.5754
2682.278	603	135.5598	2904.689	653	146.8002
2686.726	604	135.7846	2909.137	654	147.0250
2691.174	605	136.0094	2913.585	655	147.2498
2695.623	606	136.2342	2918.034	656	147.4747
2700.071	607	136.4590	2922.482	657	147.6995
2704.519	608	136.6838	2926.930	658	147.9243
2708.967	609	136.9086	2931.378	659	148.1491
2713.415	610	137.1334	2935.827	660	148.3739
2717.864	611	137.3583	2940.275	661	148.5987
2722.312	612	137.5831	2944.723	662	148.8235
2726.760	613	137.8079	2949.171	663	149.0483
2731.208	614	138.0327	2953.619	664	149.2731
2735.657	615	138.2575	2958.068	665	149.4979
2740.105	616	138.4823	2962.516	666	149.7227
2744.553	617	138.7071	2966.964	667	149.9476
2749.001	618	138.9319	2971.412	668	150.1724
2753.449	619	139.1567	2975.861	669	150.3972
2757.898	620	139.3815	2980.309	670	150.6220
2762.346	621	139.6063	2984.757	671	150.8468
2766.794	622	139.8312	2989.205	672	151.0716
2771.242	623	140.0560	2993.653	673	151.2964
2775.691	624	140.2808	2998.102	674	151.5212
2780.139	625	140.5056	3002.550	675	151.7460
2784.587	626	140.7304	3006.998	676	151.9708
2789.035	627	140.9552	3011.446	677	152.1956
2793.483	628	141.1800	3015.895	678	152.4205
2797.932	629	141.4048	3020.343	679	152.6453
2802.380	630	141.6296	3024.791	680	152.8701
2806.828	631	141.8544	3029.239	681	153.0949
2811.276	632	142.0792	3033.687	682	153.3197
2815.725	633	142.3040	3038.136	683	153.5445
2820.173	634	142.5289	3042.584	684	153.7693
2824.621	635	142.7537	3047.032	685	153.9941
2829.069	636	142.9785	3051.480	686	154.2189
2833.517	637	143.2033	3055.928	687	154.4437
2837.966	638	143.4281	3060.377	688	154.6685
2842.414	639	143.6529	3064.825	689	154.8933
2846.862	640	143.8777	3069.273	690	155.1182
2851.310	641	144.1025	3073.721	691	155.3430
2855.759	642	144.3273	3078.170	692	155.5678
2860.207	643	144.5521	3082.618	693	155.7926
2864.655	644	144.7769	3087.066	694	156.0174
2869.103	645	145.0018	3091.514	695	156.2422
2873.551	646	145.2266	3095.962	696	156.4670
2878.000	647	145.4514	3100.411	697	156.6918
2882.448	648	145.6762	3104.859	698	156.9166
2886.896	649	145.9010	3109.307	699	157.1414
2891.344	650	146.1258	3113.755	700	157.3662

N kN	REF	lbf kip	N kN	REF	lbf kip
3118.204	701	157.5911	3340.615	751	168.8315
3122.652	702	157.8159	3345.063	752	169.0563
3127.100	703	158.0407	3349.511	753	169.2811
3131.548	704	158.2655	3353.959	754	169.5059
3135.996	705	158.4903	3358.408	755	169.7307
3140.445	706	158.7151	3362.856	756	169.9555
3144.893	707	158.9399	3367.304	757	170.1804
3149.341	708	159.1647	3371.752	758	170.4052
3153.789	709	159.3895	3376.200	759	170.6300
3158.238	710	159.6143	3380.649	760	170.8548
3162.686	711	159.8391	3385.097	761	171.0796
3167.134	712	160.0640	3389.545	762	171.3044
3171.582	713	160.2888	3393.993	763	171.5292
3176.030	714	160.5136	3398.442	764	171.7540
3180.479	715	160.7384	3402.890	765	171.9788
3184.927	716	160.9632	3407.338	766	172.2036
3189.375	717	161.1880	3411.786	767	172.4284
3193.823	718	161.4128	3416.234	768	172.6533
3198.272	719	161.6376	3420.683	769	172.8781
3202.720	720	161.8624	3425.131	770	173.1029
3207.168	721	162.0872	3429.579	771	173.3277
3211.616	722	162.3120	3434.027	772	173.5525
3216.064	723	162.5369	3438.476	773	173.7773
3220.513	724	162.7617	3442.924	774	174.0021
3224.961	725	162.9865	3447.372	775	174.2269
3229.409	726	163.2113	3451.820	776	174.4517
3233.857	727	163.4361	3456.268	777	174.6765
3238.306	728	163.6609	3460.717	778	174.9013
3242.754	729	163.8857	3465.165	779	175.1262
3247.202	730	164.1105	3469.613	780	175.3510
3251.650	731	164.3353	3474.061	781	175.5758
3256.098	732	164.5601	3478.510	782	175.8006
3260.547	733	164.7849	3482.958	783	176.0254
3264.995	734	165.0098	3487.406	784	176.2502
3269.443	735	165.2346	3491.854	785	176.4750
3273.891	736	165.4594	3496.302	786	176.6998
3278.340	737	165.6842	3500.751	787	176.9246
3282.788	738	165.9090	3505.199	788	177.1494
3287.236	739	166.1338	3509.647	789	177.3742
3291.684	740	166.3586	3514.095	790	177.5991
3296.132	741	166.5834	3518.544	791	177.8239
3300.581	742	166.8082	3522.992	792	178.0487
3305.029	743	167.0330	3527.440	793	178.2735
3309.477	744	167.2578	3531.888	794	178.4983
3313.925	745	167.4826	3536.336	795	178.7231
3318.374	746	167.7075	3540.785	796	178.9479
3322.822	747	167.9323	3545.233	797	179.1727
3327.270	748	168.1571	3549.681	798	179.3975
3331.718	749	168.3819	3554.129	799	179.6223
3336.166	750	168.6067	3558.578	800	179.8471

N kN	REF	lbf kip	N kN	REF	lbf kip
3563.026	801	180.0719	3785.437	851	191.3124
3567.474	802	180.2968	3789.885	852	191.5372
3571.922	803	180.5216	3794.333	853	191.7620
3576.370	804	180.7464	3798.782	854	191.9868
3580.819	805	180.9712	3803.230	855	192.2116
3585.267	806	181.1960	3807.678	856	192.4364
3589.715	807	181.4208	3812.126	857	192.6612
3594.163	808	181.6456	3816.574	858	192.8861
3598.612	809	181.8704	3821.023	859	193.1109
3603.060	810	182.0952	3825.471	860	193.3357
3607.508	811	182.3200	3829.919	861	193.5605
3611.956	812	182.5448	3834.367	862	193.7853
3616.404	813	182.7697	3838.816	863	194.0101
3620.853	814	182.9945	3843.264	864	194.2349
3625.301	815	183.2193	3847.712	865	194.4597
3629.749	816	183.4441	3852.160	866	194.6845
3634.197	817	183.6689	3856.608	867	194.9093
3638.646	818	183.8937	3861.057	868	195.1341
3643.094	819	184.1185	3865.505	869	195.3590
3647.542	820	184.3433	3869.953	870	195.5838
3651.990	821	184.5681	3874.401	871	195.8086
3656.438	822	184.7929	3878.850	872	196.0334
3660.887	823	185.0177	3883.298	873	196.2582
3665.335	824	185.2426	3887.746	874	196.4830
3669.783	825	185.4674	3892.194	875	196.7078
3674.231	826	185.6922	3896.642	876	196.9326
3678.680	827	185.9170	3901.091	877	197.1574
3683.128	828	186.1418	3905.539	878	197.3822
3687.576	829	186.3666	3909.987	879	197.6070
3692.024	830	186.5914	3914.435	880	197.8319
3696.472	831	186.8162	3918.884	881	198.0567
3700.921	832	187.0410	3923.332	882	198.2815
3705.369	833	187.2658	3927.780	883	198.5063
3709.817	834	187.4906	3932.228	884	198.7311
3714.265	835	187.7155	3936.676	885	198.9559
3718.714	836	187.9403	3941.125	886	199.1807
3723.162	837	188.1651	3945.573	887	199.4055
3727.610	838	188.3899	3950.021	888	199.6303
3732.058	839	188.6147	3954.469	889	199.8551
3736.506	840	188.8395	3958.918	890	200.0799
3740.955	841	189.0643	3963.366	891	200.3048
3745.403	842	189.2891	3967.814	892	200.5296
3749.851	843	189.5139	3972.262	893	200.7544
3754.299	844	189.7387	3976.710	894	200.9792
3758.748	845	189.9635	3981.159	895	201.2040
3763.196	846	190.1883	3985.607	896	201.4288
3767.644	847	190.4132	3990.055	897	201.6536
3772.092	848	190.6380	3994.503	898	201.8784
3776.540	849	190.8628	3998.952	899	202.1032
3780.989	850	191.0876	4003.400	900	202.3280

N kN	REF	lbf kip		N kN	REF	lbf kip
4007.848	901	202.5528		4230.259	951	213.7933
4012.296	902	202.7776		4234.707	952	214.0181
4016.744	903	203.0025		4239.156	953	214.2429
4021.193	904	203.2273		4243.604	954	214.4677
4025.641	905	203.4521		4248.052	955	214.6925
4030.089	906	203.6769		4252.500	956	214.9173
4034.537	907	203.9017		4256.948	957	215.1421
4038.986	908	204.1265		4261.397	958	215.3669
4043.434	909	204.3513		4265.845	959	215.5918
4047.882	910	204.5761		4270.293	960	215.8166
4052.330	911	204.8009		4274.741	961	216.0414
4056.778	912	205.0257		4279.190	962	216.2662
4061.227	913	205.2505		4283.638	963	216.4910
4065.675	914	205.4754		4288.086	964	216.7158
4070.123	915	205.7002		4292.534	965	216.9406
4074.571	916	205.9250		4296.982	966	217.1654
4079.020	917	206.1498		4301.431	967	217.3902
4083.468	918	206.3746		4305.879	968	217.6150
4087.916	919	206.5994		4310.327	969	217.8398
4092.364	920	206.8242		4314.775	970	218.0647
4096.812	921	207.0490		4319.224	971	218.2895
4101.261	922	207.2738		4323.672	972	218.5143
4105.709	923	207.4986		4328.120	973	218.7391
4110.157	924	207.7234		4332.568	974	218.9639
4114.605	925	207.9483		4337.016	975	219.1887
4119.054	926	208.1731		4341.465	976	219.4135
4123.502	927	208.3979		4345.913	977	219.6383
4127.950	928	208.6227		4350.361	978	219.8631
4132.398	929	208.8475		4354.809	979	220.0879
4136.846	930	209.0723		4359.258	980	220.3127
4141.295	931	209.2971		4363.706	981	220.5376
4145.743	932	209.5219		4368.154	982	220.7624
4150.191	933	209.7467		4372.602	983	220.9872
4154.639	934	209.9715		4377.050	984	221.2120
4159.088	935	210.1963		4381.499	985	221.4368
4163.536	936	210.4212		4385.947	986	221.6616
4167.984	937	210.6460		4390.395	987	221.8864
4172.432	938	210.8708		4394.843	988	222.1112
4176.880	939	211.0956		4399.292	989	222.3360
4181.329	940	211.3204		4403.740	990	222.5608
4185.777	941	211.5452		4408.188	991	222.7856
4190.225	942	211.7700		4412.636	992	223.0105
4194.673	943	211.9948		4417.084	993	223.2353
4199.122	944	212.2196		4421.533	994	223.4601
4203.570	945	212.4444		4425.981	995	223.6849
4208.018	946	212.6692		4430.429	996	223.9097
4212.466	947	212.8941		4434.877	997	224.1345
4216.914	948	213.1189		4439.326	998	224.3593
4221.363	949	213.3437		4443.774	999	224.5841
4225.811	950	213.5685		4448.222	1000	224.8089

FORCE PER UNIT LENGTH

Pound (force)/Inch — Newton/Metre
    (lbf/in. — N/m)

Pound (force)/Inch — Kilonewton/Metre
    (lbf/in. — kN/m)

Pound (force)/Inch — Newton/Centimetre
    (lbf/in. — N/cm)

Kilopound (force)/Inch — Newton/Metre
    (kip/in. — N/m)

Kilopound (force)/Inch — Kilonewton/Metre
    (kip/in. — kN/m)

Kilopound (force)/Inch — Newton/Centimetre
    (kip/in. — N/cm)

Conversion Factors:
    1 pound (force)/inch (lbf/in.) = 175.1268 Newton/metre (N/m)
    (1 N/m = 0.005710148 lbf/in.)

Using Table 29 for Kilopound-per-inch Values:

Since the kilopound (kip) = 1000 pounds, Table 29, with proper decimal shifting, can be easily used for kip/in. — N/m conversion.

However, the kilonewton (kN) = 1000 Newtons (N). Therefore, conversions between kip/in. and kN/m can be obtained directly from Table 29, without shifting decimals, since: 1 kip/in. = 175.1268 kN/m. The numerical value of this conversion factor is the same as the lbf–N relation above, and on which Table 29 is based.

Diagram 29-A, following, describes decimal-shift relations between kips/in. and lbf/in. when dealing with decimal values:

329

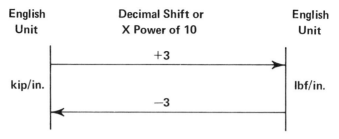

English Unit	Decimal Shift or X Power of 10	English Unit
kip/in.	+3 ⟶ −3 ⟵	lbf/in.

Diagram 29-A

Example (*a*): Convert 0.813 kip/in. to N/m.
From Diagram 29-A: 0.813 kip/in. = 0.813 × 10³ lbf/in. = 813 lbf/in.
(kip/in. to lbf/in. by multiplying kip/in. value by 10³
or by shifting its decimal 3 places to the right.)
From table: 813 lbf/in. = 142,378.1 N/m* ≈ 142,000 N/m.

In converting SI force-per-length quantities from N/m to kN/m, kN/cm, N/cm the Decimal-shift Diagram 29-B, which appears in the headings on alternate pages of this table, will assist table users. It is derived from:

$$1 \text{ N/m} = (10^{-2}) \text{ N/cm} = (10^{-3}) \text{ kN/m} = (10^{-5}) \text{ kN/cm}.$$

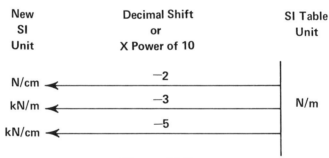

New SI Unit	Decimal Shift or X Power of 10	SI Table Unit
N/cm ⟵ kN/m ⟵ kN/cm ⟵	−2 −3 −5	N/m

Diagram 29-B

Example (*b*): Convert 475 lbf/in. to N/m, kN/m, N/cm, kN/cm.
From table: 475 lbf/in. = 83,185.23 N/m* ≈ 83,200 N/m
From Diagram 29-B: 1) 83,185.23 N/m = 83,185.23 (10⁻²) N/cm
= 831.8523 N/cm ≈ 832 N/cm
(N/m to N/cm by multiplying N/m value by 10⁻²
or by shifting its decimal 2 places to the left.)

2) 83,185.23 N/m = 83,185.23 (10⁻³) kN/m
= 83.18523 kN/m ≈ 83.2 kN/m
(N/m to kN/m by multiplying N/m value by 10⁻³
or by shifting its decimal 3 places to the left.)

---

*All conversion values obtained can be rounded to desired number of significant figures.

3) $83,185.23 \text{ N/m} = 83,185.23 \, (10^{-5}) \text{ kN/cm}$
$= 0.8318523 \text{ kN/cm} \approx 0.832 \text{ kN/cm}$
(N/m to kN/cm by multiplying N/m value by 10^{-5}
or by shifting its decimal 5 places to the left.)

Diagram 29-C is also given at the top of alternate table pages:

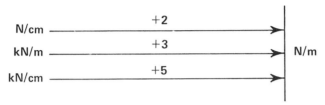

Diagram 29-C

It is the opposite of Diagram 29-B and is used to convert quantities in N/cm, kN/m, kN/cm to N/m. The N/m equivalents are then applied to the REF column in the table to obtain lbf/in. relations. *Please note from what has been shown at the beginning of this introduction, that kN/m can be converted directly to kip/in. through Table 29 without any decimal shifting.*

Example (*c*): Convert 0.177 kN/cm to lbf/in.
   From Diagram 29-C: $0.177 \text{ kN/cm} = 0.177 \times 10^5 \text{ N/m} = 177 \times 10^2 \text{ N/m}$
   From table: $177 \text{ N/m} = 1.010696 \text{ lbf/in.}$*
   Therefore: $177 \times 10^2 \text{ N/m} = 1.010696 \times 10^2 \text{ lbf/in.} \approx 1.01 \times 10^2$ or
            101 lbf/in.

---

*All conversion values obtained can be rounded to desired number of significant figures.

| N/m | | lbf/in | | N/m | | lbf/in |
kN/m	REF	kip/in		kN/m	REF	kip/in
175.1268	1	0.005710148		8931.467	51	0.2912176
350.2536	2	0.01142030		9106.594	52	0.2969277
525.3804	3	0.01713044		9281.720	53	0.3026379
700.5072	4	0.02284059		9456.847	54	0.3083480
875.6340	5	0.02855074		9631.974	55	0.3140582
1050.761	6	0.03426089		9807.101	56	0.3197683
1225.888	7	0.03997104		9982.228	57	0.3254785
1401.014	8	0.04568119		10157.35	58	0.3311886
1576.141	9	0.05139133		10332.48	59	0.3368988
1751.268	10	0.05710148		10507.61	60	0.3426089
1926.395	11	0.06281163		10682.73	61	0.3483190
2101.522	12	0.06852178		10857.86	62	0.3540292
2276.648	13	0.07423193		11032.99	63	0.3597393
2451.775	14	0.07994208		11208.12	64	0.3654495
2626.902	15	0.08565222		11383.24	65	0.3711596
2802.029	16	0.09136237		11558.37	66	0.3768698
2977.156	17	0.09707252		11733.50	67	0.3825799
3152.282	18	0.1027827		11908.62	68	0.3882901
3327.409	19	0.1084928		12083.75	69	0.3940002
3502.536	20	0.1142030		12258.88	70	0.3997104
3677.663	21	0.1199131		12434.00	71	0.4054205
3852.790	22	0.1256233		12609.13	72	0.4111307
4027.916	23	0.1313334		12784.26	73	0.4168408
4203.043	24	0.1370436		12959.38	74	0.4225510
4378.170	25	0.1427537		13134.51	75	0.4282611
4553.297	26	0.1484639		13309.64	76	0.4339713
4728.424	27	0.1541740		13484.76	77	0.4396814
4903.550	28	0.1598842		13659.89	78	0.4453916
5078.677	29	0.1655943		13835.02	79	0.4511017
5253.804	30	0.1713044		14010.14	80	0.4568119
5428.931	31	0.1770146		14185.27	81	0.4625220
5604.058	32	0.1827247		14360.40	82	0.4682322
5779.184	33	0.1884349		14535.52	83	0.4739423
5954.311	34	0.1941450		14710.65	84	0.4796525
6129.438	35	0.1998552		14885.78	85	0.4853626
6304.565	36	0.2055653		15060.90	86	0.4910728
6479.692	37	0.2112755		15236.03	87	0.4967829
6654.818	38	0.2169856		15411.16	88	0.5024931
6829.945	39	0.2226958		15586.29	89	0.5082032
7005.072	40	0.2284059		15761.41	90	0.5139133
7180.199	41	0.2341161		15936.54	91	0.5196235
7355.326	42	0.2398262		16111.67	92	0.5253336
7530.452	43	0.2455364		16286.79	93	0.5310438
7705.579	44	0.2512465		16461.92	94	0.5367539
7880.706	45	0.2569567		16637.05	95	0.5424641
8055.833	46	0.2626668		16812.17	96	0.5481742
8230.960	47	0.2683770		16987.30	97	0.5538844
8406.086	48	0.2740871		17162.43	98	0.5595945
8581.213	49	0.2797973		17337.55	99	0.5653047
8756.340	50	0.2855074		17512.68	100	0.5710148

N/m kN/m	REF	lbf/in kip/in	N/m kN/m	REF	lbf/in kip/in
17687.81	101	0.5767250	26444.15	151	0.8622324
17862.93	102	0.5824351	26619.27	152	0.8679425
18038.06	103	0.5881453	26794.40	153	0.8736527
18213.19	104	0.5938554	26969.53	154	0.8793628
18388.31	105	0.5995656	27144.65	155	0.8850730
18563.44	106	0.6052757	27319.78	156	0.8907831
18738.57	107	0.6109859	27494.91	157	0.8964933
18913.69	108	0.6166960	27670.03	158	0.9022034
19088.82	109	0.6224062	27845.16	159	0.9079136
19263.95	110	0.6281163	28020.29	160	0.9136237
19439.07	111	0.6338265	28195.41	161	0.9193339
19614.20	112	0.6395366	28370.54	162	0.9250440
19789.33	113	0.6452468	28545.67	163	0.9307542
19964.46	114	0.6509569	28720.80	164	0.9364643
20139.58	115	0.6566671	28895.92	165	0.9421745
20314.71	116	0.6623772	29071.05	166	0.9478846
20489.84	117	0.6680874	29246.18	167	0.9535948
20664.96	118	0.6737975	29421.30	168	0.9593049
20840.09	119	0.6795077	29596.43	169	0.9650151
21015.22	120	0.6852178	29771.56	170	0.9707252
21190.34	121	0.6909279	29946.68	171	0.9764354
21365.47	122	0.6966381	30121.81	172	0.9821455
21540.60	123	0.7023482	30296.94	173	0.9878557
21715.72	124	0.7080584	30472.06	174	0.9935658
21890.85	125	0.7137685	30647.19	175	0.9992760
22065.98	126	0.7194787	30822.32	176	1.004986
22241.10	127	0.7251888	30997.44	177	1.010696
22416.23	128	0.7308990	31172.57	178	1.016406
22591.36	129	0.7366091	31347.70	179	1.022117
22766.48	130	0.7423193	31522.82	180	1.027827
22941.61	131	0.7480294	31697.95	181	1.033537
23116.74	132	0.7537396	31873.08	182	1.039247
23291.86	133	0.7594497	32048.20	183	1.044957
23466.99	134	0.7651599	32223.33	184	1.050667
23642.12	135	0.7708700	32398.46	185	1.056377
23817.24	136	0.7765802	32573.58	186	1.062088
23992.37	137	0.7822903	32748.71	187	1.067798
24167.50	138	0.7880005	32923.84	188	1.073508
24342.63	139	0.7937106	33098.97	189	1.079218
24517.75	140	0.7994208	33274.09	190	1.084928
24692.88	141	0.8051309	33449.22	191	1.090638
24868.01	142	0.8108411	33624.35	192	1.096348
25043.13	143	0.8165512	33799.47	193	1.102059
25218.26	144	0.8222614	33974.60	194	1.107769
25393.39	145	0.8279715	34149.73	195	1.113479
25568.51	146	0.8336817	34324.85	196	1.119189
25743.64	147	0.8393918	34499.98	197	1.124899
25918.77	148	0.8451020	34675.11	198	1.130609
26093.89	149	0.8508121	34850.23	199	1.136320
26269.02	150	0.8565222	35025.36	200	1.142030

N/m kN/m	REF	lbf/in kip/in		N/m kN/m	REF	lbf/in kip/in
35200.49	201	1.147740		43956.83	251	1.433247
35375.61	202	1.153450		44131.95	252	1.438957
35550.74	203	1.159160		44307.08	253	1.444668
35725.87	204	1.164870		44482.21	254	1.450378
35900.99	205	1.170580		44657.33	255	1.456088
36076.12	206	1.176291		44832.46	256	1.461798
36251.25	207	1.182001		45007.59	257	1.467508
36426.37	208	1.187711		45182.71	258	1.473218
36601.50	209	1.193421		45357.84	259	1.478928
36776.63	210	1.199131		45532.97	260	1.484639
36951.75	211	1.204841		45708.09	261	1.490349
37126.88	212	1.210551		45883.22	262	1.496059
37302.01	213	1.216262		46058.35	263	1.501769
37477.14	214	1.221972		46233.48	264	1.507479
37652.26	215	1.227682		46408.60	265	1.513189
37827.39	216	1.233392		46583.73	266	1.518899
38002.52	217	1.239102		46758.86	267	1.524610
38177.64	218	1.244812		46933.98	268	1.530320
38352.77	219	1.250522		47109.11	269	1.536030
38527.90	220	1.256233		47284.24	270	1.541740
38703.02	221	1.261943		47459.36	271	1.547450
38878.15	222	1.267653		47634.49	272	1.553160
39053.28	223	1.273363		47809.62	273	1.558870
39228.40	224	1.279073		47984.74	274	1.564581
39403.53	225	1.284783		48159.87	275	1.570291
39578.66	226	1.290494		48335.00	276	1.576001
39753.78	227	1.296204		48510.12	277	1.581711
39928.91	228	1.301914		48685.25	278	1.587421
40104.04	229	1.307624		48860.38	279	1.593131
40279.16	230	1.313334		49035.50	280	1.598842
40454.29	231	1.319044		49210.63	281	1.604552
40629.42	232	1.324754		49385.76	282	1.610262
40804.54	233	1.330465		49560.88	283	1.615972
40979.67	234	1.336175		49736.01	284	1.621682
41154.80	235	1.341885		49911.14	285	1.627392
41329.92	236	1.347595		50086.27	286	1.633102
41505.05	237	1.353305		50261.39	287	1.638813
41680.18	238	1.359015		50436.52	288	1.644523
41855.31	239	1.364725		50611.65	289	1.650233
42030.43	240	1.370436		50786.77	290	1.655943
42205.56	241	1.376146		50961.90	291	1.661653
42380.69	242	1.381856		51137.03	292	1.667363
42555.81	243	1.387566		51312.15	293	1.673073
42730.94	244	1.393276		51487.28	294	1.678784
42906.07	245	1.398986		51662.41	295	1.684494
43081.19	246	1.404696		51837.53	296	1.690204
43256.32	247	1.410407		52012.66	297	1.695914
43431.45	248	1.416117		52187.79	298	1.701624
43606.57	249	1.421827		52362.91	299	1.707334
43781.70	250	1.427537		52538.04	300	1.713044

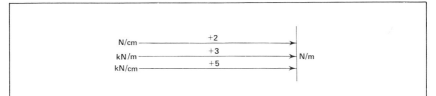

N/m kN/m	REF	lbf/in kip/in		N/m kN/m	REF	lbf/in kip/in
52713.17	301	1.718755		61469.51	351	2.004262
52888.29	302	1.724465		61644.63	352	2.009972
53063.42	303	1.730175		61819.76	353	2.015682
53238.55	304	1.735885		61994.89	354	2.021392
53413.67	305	1.741595		62170.01	355	2.027103
53588.80	306	1.747305		62345.14	356	2.032813
53763.93	307	1.753016		62520.27	357	2.038523
53939.05	308	1.758726		62695.39	358	2.044233
54114.18	309	1.764436		62870.52	359	2.049943
54289.31	310	1.770146		63045.65	360	2.055653
54464.44	311	1.775856		63220.77	361	2.061364
54639.56	312	1.781566		63395.90	362	2.067074
54814.69	313	1.787276		63571.03	363	2.072784
54989.82	314	1.792987		63746.16	364	2.078494
55164.94	315	1.798697		63921.28	365	2.084204
55340.07	316	1.804407		64096.41	366	2.089914
55515.20	317	1.810117		64271.54	367	2.095624
55690.32	318	1.815827		64446.66	368	2.101335
55865.45	319	1.821537		64621.79	369	2.107045
56040.58	320	1.827247		64796.92	370	2.112755
56215.70	321	1.832958		64972.04	371	2.118465
56390.83	322	1.838668		65147.17	372	2.124175
56565.96	323	1.844378		65322.30	373	2.129885
56741.08	324	1.850088		65497.42	374	2.135595
56916.21	325	1.855798		65672.55	375	2.141306
57091.34	326	1.861508		65847.68	376	2.147016
57266.46	327	1.867218		66022.80	377	2.152726
57441.59	328	1.872929		66197.93	378	2.158436
57616.72	329	1.878639		66373.06	379	2.164146
57791.84	330	1.884349		66548.18	380	2.169856
57966.97	331	1.890059		66723.31	381	2.175566
58142.10	332	1.895769		66898.44	382	2.181277
58317.22	333	1.901479		67073.56	383	2.186987
58492.35	334	1.907190		67248.69	384	2.192697
58667.48	335	1.912900		67423.82	385	2.198407
58842.60	336	1.918610		67598.95	386	2.204117
59017.73	337	1.924320		67774.07	387	2.209827
59192.86	338	1.930030		67949.20	388	2.215538
59367.99	339	1.935740		68124.33	389	2.221248
59543.11	340	1.941450		68299.45	390	2.226958
59718.24	341	1.947161		68474.58	391	2.232668
59893.37	342	1.952871		68649.71	392	2.238378
60068.49	343	1.958581		68824.83	393	2.244088
60243.62	344	1.964291		68999.96	394	2.249798
60418.75	345	1.970001		69175.09	395	2.255509
60593.87	346	1.975711		69350.21	396	2.261219
60769.00	347	1.981421		69525.34	397	2.266929
60944.13	348	1.987132		69700.47	398	2.272639
61119.25	349	1.992842		69875.59	399	2.278349
61294.38	350	1.998552		70050.72	400	2.284059

N/cm ← ─2
kN/m ← ─3 → N/m
kN/cm ← ─5

N/m kN/m	REF	lbf/in kip/in	N/m kN/m	REF	lbf/in kip/in
70225.85	401	2.289769	78982.19	451	2.575277
70400.97	402	2.295480	79157.31	452	2.580987
70576.10	403	2.301190	79332.44	453	2.586697
70751.23	404	2.306900	79507.57	454	2.592407
70926.35	405	2.312610	79682.69	455	2.598117
71101.48	406	2.318320	79857.82	456	2.603828
71276.61	407	2.324030	80032.95	457	2.609538
71451.73	408	2.329741	80208.07	458	2.615248
71626.86	409	2.335451	80383.20	459	2.620958
71801.99	410	2.341161	80558.33	460	2.626668
71977.12	411	2.346871	80733.46	461	2.632378
72152.24	412	2.352581	80908.58	462	2.638089
72327.37	413	2.358291	81083.71	463	2.643799
72502.50	414	2.364001	81258.84	464	2.649509
72677.62	415	2.369712	81433.96	465	2.655219
72852.75	416	2.375422	81609.09	466	2.660929
73027.88	417	2.381132	81784.22	467	2.666639
73203.00	418	2.386842	81959.34	468	2.672349
73378.13	419	2.392552	82134.47	469	2.678060
73553.26	420	2.398262	82309.60	470	2.683770
73728.38	421	2.403972	82484.72	471	2.689480
73903.51	422	2.409683	82659.85	472	2.695190
74078.64	423	2.415393	82834.98	473	2.700900
74253.76	424	2.421103	83010.10	474	2.706610
74428.89	425	2.426813	83185.23	475	2.712320
74604.02	426	2.432523	83360.36	476	2.718031
74779.14	427	2.438233	83535.48	477	2.723741
74954.27	428	2.443943	83710.61	478	2.729451
75129.40	429	2.449654	83885.74	479	2.735161
75304.52	430	2.455364	84060.86	480	2.740871
75479.65	431	2.461074	84235.99	481	2.746581
75654.78	432	2.466784	84411.12	482	2.752292
75829.90	433	2.472494	84586.25	483	2.758002
76005.03	434	2.478204	84761.37	484	2.763712
76180.16	435	2.483915	84936.50	485	2.769422
76355.29	436	2.489625	85111.62	486	2.775132
76530.41	437	2.495335	85286.75	487	2.780842
76705.54	438	2.501045	85461.88	488	2.786552
76880.67	439	2.506755	85637.01	489	2.792263
77055.79	440	2.512465	85812.13	490	2.797973
77230.92	441	2.518175	85987.26	491	2.803683
77406.05	442	2.523886	86162.39	492	2.809393
77581.17	443	2.529596	86337.51	493	2.815103
77756.30	444	2.535306	86512.64	494	2.820813
77931.43	445	2.541016	86687.77	495	2.826523
78106.55	446	2.546726	86862.89	496	2.832234
78281.68	447	2.552436	87038.02	497	2.837944
78456.81	448	2.558146	87213.15	498	2.843654
78631.93	449	2.563857	87388.27	499	2.849364
78807.06	450	2.569567	87563.40	500	2.855074

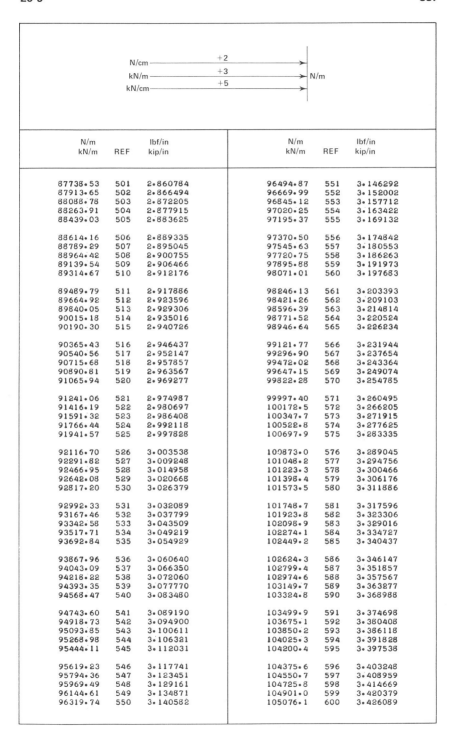

N/m kN/m	REF	lbf/in kip/in	N/m kN/m	REF	lbf/in kip/in
87738.53	501	2.860784	96494.87	551	3.146292
87913.65	502	2.866494	96669.99	552	3.152002
88088.78	503	2.872205	96845.12	553	3.157712
88263.91	504	2.877915	97020.25	554	3.163422
88439.03	505	2.883625	97195.37	555	3.169132
88614.16	506	2.889335	97370.50	556	3.174842
88789.29	507	2.895045	97545.63	557	3.180553
88964.42	508	2.900755	97720.75	558	3.186263
89139.54	509	2.906466	97895.88	559	3.191973
89314.67	510	2.912176	98071.01	560	3.197683
89489.79	511	2.917886	98246.13	561	3.203393
89664.92	512	2.923596	98421.26	562	3.209103
89840.05	513	2.929306	98596.39	563	3.214814
90015.18	514	2.935016	98771.52	564	3.220524
90190.30	515	2.940726	98946.64	565	3.226234
90365.43	516	2.946437	99121.77	566	3.231944
90540.56	517	2.952147	99296.90	567	3.237654
90715.68	518	2.957857	99472.02	568	3.243364
90890.81	519	2.963567	99647.15	569	3.249074
91065.94	520	2.969277	99822.28	570	3.254785
91241.06	521	2.974987	99997.40	571	3.260495
91416.19	522	2.980697	100172.5	572	3.266205
91591.32	523	2.986408	100347.7	573	3.271915
91766.44	524	2.992118	100522.8	574	3.277625
91941.57	525	2.997828	100697.9	575	3.283335
92116.70	526	3.003538	100873.0	576	3.289045
92291.82	527	3.009248	101048.2	577	3.294756
92466.95	528	3.014958	101223.3	578	3.300466
92642.08	529	3.020668	101398.4	579	3.306176
92817.20	530	3.026379	101573.5	580	3.311886
92992.33	531	3.032089	101748.7	581	3.317596
93167.46	532	3.037799	101923.8	582	3.323306
93342.58	533	3.043509	102098.9	583	3.329016
93517.71	534	3.049219	102274.1	584	3.334727
93692.84	535	3.054929	102449.2	585	3.340437
93867.96	536	3.060640	102624.3	586	3.346147
94043.09	537	3.066350	102799.4	587	3.351857
94218.22	538	3.072060	102974.6	588	3.357567
94393.35	539	3.077770	103149.7	589	3.363277
94568.47	540	3.083480	103324.8	590	3.368988
94743.60	541	3.089190	103499.9	591	3.374698
94918.73	542	3.094900	103675.1	592	3.380408
95093.85	543	3.100611	103850.2	593	3.386118
95268.98	544	3.106381	104025.3	594	3.391828
95444.11	545	3.112031	104200.4	595	3.397538
95619.23	546	3.117741	104375.6	596	3.403248
95794.36	547	3.123451	104550.7	597	3.408959
95969.49	548	3.129161	104725.8	598	3.414669
96144.61	549	3.134871	104901.0	599	3.420379
96319.74	550	3.140582	105076.1	600	3.426089

N/m kN/m	REF	lbf/in kip/in	N/m kN/m	REF	lbf/in kip/in
105251.2	601	3.431799	114007.5	651	3.717307
105426.3	602	3.437509	114182.7	652	3.723017
105601.5	603	3.443219	114357.8	653	3.728727
105776.6	604	3.448930	114532.9	654	3.734437
105951.7	605	3.454640	114708.1	655	3.740147
106126.8	606	3.460350	114883.2	656	3.745857
106302.0	607	3.466060	115058.3	657	3.751567
106477.1	608	3.471770	115233.4	658	3.757278
106652.2	609	3.477480	115408.6	659	3.762988
106827.3	610	3.483190	115583.7	660	3.768698
107002.5	611	3.488901	115758.8	661	3.774408
107177.6	612	3.494611	115933.9	662	3.780118
107352.7	613	3.500321	116109.1	663	3.785828
107527.9	614	3.506031	116284.2	664	3.791538
107703.0	615	3.511741	116459.3	665	3.797249
107878.1	616	3.517451	116634.4	666	3.802959
108053.2	617	3.523162	116809.6	667	3.808669
108228.4	618	3.528872	116984.7	668	3.814379
108403.5	619	3.534582	117159.8	669	3.820089
108578.6	620	3.540292	117335.0	670	3.825799
108753.7	621	3.546002	117510.1	671	3.831510
108928.9	622	3.551712	117685.2	672	3.837220
109104.0	623	3.557422	117860.3	673	3.842930
109279.1	624	3.563133	118035.5	674	3.848640
109454.2	625	3.568843	118210.6	675	3.854350
109629.4	626	3.574553	118385.7	676	3.860060
109804.5	627	3.580263	118560.8	677	3.865770
109979.6	628	3.585973	118736.0	678	3.871481
110154.8	629	3.591683	118911.1	679	3.877191
110329.9	630	3.597393	119086.2	680	3.882901
110505.0	631	3.603104	119261.4	681	3.888611
110680.1	632	3.608814	119436.5	682	3.894321
110855.3	633	3.614524	119611.6	683	3.900031
111030.4	634	3.620234	119786.7	684	3.905741
111205.5	635	3.625944	119961.9	685	3.911452
111380.6	636	3.631654	120137.0	686	3.917162
111555.8	637	3.637364	120312.1	687	3.922872
111730.9	638	3.643075	120487.2	688	3.928582
111906.0	639	3.648785	120662.4	689	3.934292
112081.2	640	3.654495	120837.5	690	3.940002
112256.3	641	3.660205	121012.6	691	3.945712
112431.4	642	3.665915	121187.7	692	3.951423
112606.5	643	3.671625	121362.9	693	3.957133
112781.7	644	3.677336	121538.0	694	3.962843
112956.8	645	3.683046	121713.1	695	3.968553
113131.9	646	3.688756	121888.3	696	3.974263
113307.0	647	3.694466	122063.4	697	3.979973
113482.2	648	3.700176	122238.5	698	3.985684
113657.3	649	3.705886	122413.6	699	3.991394
113832.4	650	3.711596	122588.8	700	3.997104

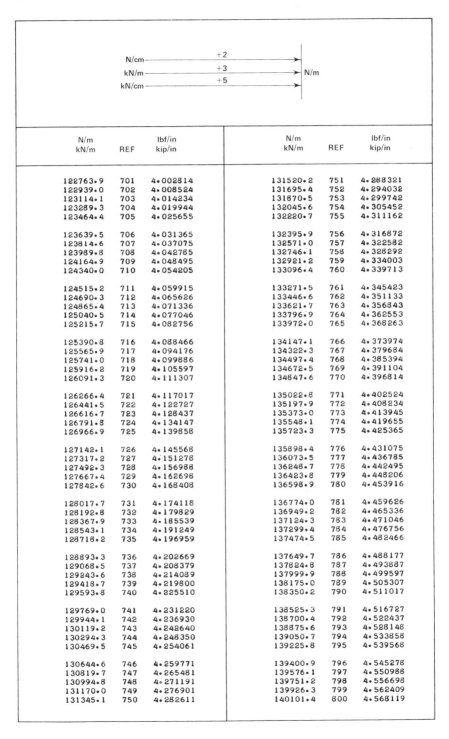

N/m kN/m	REF	lbf/in kip/in	N/m kN/m	REF	lbf/in kip/in
122763.9	701	4.002814	131520.2	751	4.288321
122939.0	702	4.008524	131695.4	752	4.294032
123114.1	703	4.014234	131870.5	753	4.299742
123289.3	704	4.019944	132045.6	754	4.305452
123464.4	705	4.025655	132220.7	755	4.311162
123639.5	706	4.031365	132395.9	756	4.316872
123814.6	707	4.037075	132571.0	757	4.322582
123989.8	708	4.042785	132746.1	758	4.328292
124164.9	709	4.048495	132921.2	759	4.334003
124340.0	710	4.054205	133096.4	760	4.339713
124515.2	711	4.059915	133271.5	761	4.345423
124690.3	712	4.065626	133446.6	762	4.351133
124865.4	713	4.071336	133621.7	763	4.356843
125040.5	714	4.077046	133796.9	764	4.362553
125215.7	715	4.082756	133972.0	765	4.368263
125390.8	716	4.088466	134147.1	766	4.373974
125565.9	717	4.094176	134322.3	767	4.379684
125741.0	718	4.099886	134497.4	768	4.385394
125916.2	719	4.105597	134672.5	769	4.391104
126091.3	720	4.111307	134847.6	770	4.396814
126266.4	721	4.117017	135022.8	771	4.402524
126441.5	722	4.122727	135197.9	772	4.408234
126616.7	723	4.128437	135373.0	773	4.413945
126791.8	724	4.134147	135548.1	774	4.419655
126966.9	725	4.139858	135723.3	775	4.425365
127142.1	726	4.145568	135898.4	776	4.431075
127317.2	727	4.151278	136073.5	777	4.436785
127492.3	728	4.156988	136248.7	778	4.442495
127667.4	729	4.162698	136423.8	779	4.448206
127842.6	730	4.168408	136598.9	780	4.453916
128017.7	731	4.174118	136774.0	781	4.459626
128192.8	732	4.179829	136949.2	782	4.465336
128367.9	733	4.185539	137124.3	783	4.471046
128543.1	734	4.191249	137299.4	784	4.476756
128718.2	735	4.196959	137474.5	785	4.482466
128893.3	736	4.202669	137649.7	786	4.488177
129068.5	737	4.208379	137824.8	787	4.493887
129243.6	738	4.214089	137999.9	788	4.499597
129418.7	739	4.219800	138175.0	789	4.505307
129593.8	740	4.225510	138350.2	790	4.511017
129769.0	741	4.231220	138525.3	791	4.516727
129944.1	742	4.236930	138700.4	792	4.522437
130119.2	743	4.242640	138875.6	793	4.528148
130294.3	744	4.248350	139050.7	794	4.533858
130469.5	745	4.254061	139225.8	795	4.539568
130644.6	746	4.259771	139400.9	796	4.545278
130819.7	747	4.265481	139576.1	797	4.550988
130994.8	748	4.271191	139751.2	798	4.556698
131170.0	749	4.276901	139926.3	799	4.562409
131345.1	750	4.282611	140101.4	800	4.568119

N/cm ← −2
kN/m ← −3
kN/cm ← −5
→ N/m

N/m kN/m	REF	lbf/in kip/in	N/m kN/m	REF	lbf/in kip/in
140276.6	801	4.573829	149032.9	851	4.859336
140451.7	802	4.579539	149208.0	852	4.865046
140626.8	803	4.585249	149383.2	853	4.870757
140801.9	804	4.590959	149558.3	854	4.876467
140977.1	805	4.596669	149733.4	855	4.882177
141152.2	806	4.602380	149908.5	856	4.887887
141327.3	807	4.608090	150083.7	857	4.893597
141502.5	808	4.613800	150258.8	858	4.899307
141677.6	809	4.619510	150433.9	859	4.905017
141852.7	810	4.625220	150609.0	860	4.910728
142027.8	811	4.630930	150784.2	861	4.916438
142203.0	812	4.636640	150959.3	862	4.922148
142378.1	813	4.642351	151134.4	863	4.927858
142553.2	814	4.648061	151309.6	864	4.933568
142728.3	815	4.653771	151484.7	865	4.939278
142903.5	816	4.659481	151659.8	866	4.944988
143078.6	817	4.665191	151834.9	867	4.950699
143253.7	818	4.670901	152010.1	868	4.956409
143428.8	819	4.676611	152185.2	869	4.962119
143604.0	820	4.682322	152360.3	870	4.967829
143779.1	821	4.688032	152535.4	871	4.973539
143954.2	822	4.693742	152710.6	872	4.979249
144129.4	823	4.699452	152885.7	873	4.984959
144304.5	824	4.705162	153060.8	874	4.990670
144479.6	825	4.710872	153236.0	875	4.996380
144654.7	826	4.716582	153411.1	876	5.002090
144829.9	827	4.722293	153586.2	877	5.007800
145005.0	828	4.728003	153761.3	878	5.013510
145180.1	829	4.733713	153936.5	879	5.019220
145355.2	830	4.739423	154111.6	880	5.024931
145530.4	831	4.745133	154286.7	881	5.030641
145705.5	832	4.750843	154461.8	882	5.036351
145880.6	833	4.756554	154637.0	883	5.042061
146055.8	834	4.762264	154812.1	884	5.047771
146230.9	835	4.767974	154987.2	885	5.053481
146406.0	836	4.773684	155162.3	886	5.059191
146581.1	837	4.779394	155337.5	887	5.064902
146756.3	838	4.785104	155512.6	888	5.070612
146931.4	839	4.790814	155687.7	889	5.076322
147106.5	840	4.796525	155862.9	890	5.082032
147281.6	841	4.802235	156038.0	891	5.087742
147456.8	842	4.807945	156213.1	892	5.093452
147631.9	843	4.813655	156388.2	893	5.099162
147807.0	844	4.819365	156563.4	894	5.104873
147982.1	845	4.825075	156738.5	895	5.110583
148157.3	846	4.830785	156913.6	896	5.116293
148332.4	847	4.836496	157088.7	897	5.122003
148507.5	848	4.842206	157263.9	898	5.127713
148682.7	849	4.847916	157439.0	899	5.133423
148857.8	850	4.853626	157614.1	900	5.139133

```
                                                    +2
                        N/cm ──────────────────────────────►
                        kN/m ──────────────────+3──────────►  N/m
                        kN/cm ─────────────────+5──────────►
```

N/m kN/m	REF	lbf/in kip/in	N/m kN/m	REF	lbf/in kip/in
157789.2	901	5.144844	166545.6	951	5.430351
157964.4	902	5.150554	166720.7	952	5.436061
158139.5	903	5.156264	166895.8	953	5.441771
158314.6	904	5.161974	167071.0	954	5.447482
158489.8	905	5.167684	167246.1	955	5.453192
158664.9	906	5.173394	167421.2	956	5.458902
158840.0	907	5.179105	167596.3	957	5.464612
159015.1	908	5.184815	167771.5	958	5.470322
159190.3	909	5.190525	167946.6	959	5.476032
159365.4	910	5.196235	168121.7	960	5.481742
159540.5	911	5.201945	168296.9	961	5.487453
159715.6	912	5.207655	168472.0	962	5.493163
159890.8	913	5.213365	168647.1	963	5.498873
160065.9	914	5.219076	168822.2	964	5.504583
160241.0	915	5.224786	168997.4	965	5.510293
160416.1	916	5.230496	169172.5	966	5.516003
160591.3	917	5.236206	169347.6	967	5.521713
160766.4	918	5.241916	169522.7	968	5.527424
160941.5	919	5.247626	169697.9	969	5.533134
161116.7	920	5.253336	169873.0	970	5.538844
161291.8	921	5.259047	170048.1	971	5.544554
161466.9	922	5.264757	170223.2	972	5.550264
161642.0	923	5.270467	170398.4	973	5.555974
161817.2	924	5.276177	170573.5	974	5.561684
161992.3	925	5.281887	170748.6	975	5.567395
162167.4	926	5.287597	170923.8	976	5.573105
162342.5	927	5.293307	171098.9	977	5.578815
162517.7	928	5.299018	171274.0	978	5.584525
162692.8	929	5.304728	171449.1	979	5.590235
162867.9	930	5.310438	171624.3	980	5.595945
163043.1	931	5.316148	171799.4	981	5.601655
163218.2	932	5.321858	171974.5	982	5.607366
163393.3	933	5.327568	172149.6	983	5.613076
163568.4	934	5.333279	172324.8	984	5.618786
163743.6	935	5.338989	172499.9	985	5.624496
163918.7	936	5.344699	172675.0	986	5.630206
164093.8	937	5.350409	172850.2	987	5.635916
164268.9	938	5.356119	173025.3	988	5.641627
164444.1	939	5.361829	173200.4	989	5.647337
164619.2	940	5.367539	173375.5	990	5.653047
164794.3	941	5.373250	173550.7	991	5.658757
164969.4	942	5.378960	173725.8	992	5.664467
165144.6	943	5.384670	173900.9	993	5.670177
165319.7	944	5.390380	174076.0	994	5.675887
165494.8	945	5.396090	174251.2	995	5.681598
165670.0	946	5.401800	174426.3	996	5.687308
165845.1	947	5.407510	174601.4	997	5.693018
166020.2	948	5.413221	174776.5	998	5.698728
166195.3	949	5.418931	174951.7	999	5.704438
166370.5	950	5.424641	175126.8	1000	5.710148

30

FORCE PER UNIT LENGTH

Pound (force)/Foot — Newton/Metre
$$\text{(lbf/ft — N/m)}$$

Pound (force)/Foot — Kilonewton/Metre
$$\text{(lbf/ft — kN/m)}$$

Pound (force)/Foot — Newton /Centimetre
$$\text{(lbf/ft — N/cm)}$$

Pound (force)/Foot — Kilonewton/Centimetre
$$\text{(lbf/ft — kN/cm)}$$

Kilopound (force)/Foot — Newton/Metre
$$\text{(kip/ft — N/m)}$$

Kilopound (force)/Foot — Kilonewton/Metre
$$\text{(kip/ft — kN/m)}$$

Kilopound (force)/Foot — Newton/Centimetre
$$\text{(kip/ft — N/cm)}$$

Kilopound (force)/Foot — Kilonewton/Centimetre
$$\text{(kip/ft — kN/cm)}$$

MASS

Slug — Kilogram
$$\text{(slug — kg)}$$

Conversion Factors:
$$\text{1 pound (force)/foot (lbf/ft)} = 14.59390 \text{ Newton/metre (N/m)}$$
$$\text{(1 N/m} = 0.06852178 \text{ lbf/ft)}$$

Using Table 30 for Kilopound per Foot Values

Since the kilopound (kip) = 1000 pounds, Table 30, with proper decimal shifting, can be used for kip/ft – N/m conversions.

However, the kilonewton (kN) = 1000 newtons (N). Therefore, conversions between kip/ft and kN/m can be obtained directly from Table 30, without shifting decimals, since 1 kip/ft = 14.53930 kN/m. The numerical value of this conversion factor is the same as the lbf/ft – N/m relation given above, and on which Table 30 is based.

Slug-kilogram Values

Table 30 can be used for conversions between slug (slug) and kilogram (kg) values, since the numerical value of the conversion factor is the same as the lbf/ft–N/m relation on which the derivation of this table is based.

1 slug = 14.59390 kg

The slug is an English unit of mass. It is defined as the mass that will attain an acceleration of 1 ft/sec^2 when acted upon by a force of 1 lb. The numerical value of 1 slug in lbm is equal to that of the customary numerical value for the standard gravitational constant "g": 1 slug = 32.17405 lbm.

Example (a): Convert 119.7 slugs to kg.
  Since: 119.7 slugs = (119.0 + 0.7) slugs:
  From table: 119 slugs = 1736.674 kg*
              7 slugs = 102.1573 kg*

  By shifting decimals where necessary and adding:
     119.0 slugs = 1736.674   kg
       0.7 slugs =   10.21573 kg
     ―――――――――――――――――――――――――――――
     119.7 slugs = 1746.890   kg ≈ 1747 kg

Diagram 30-A describes the decimal-shift relations between kip/ft and lbf/ft when dealing with decimal values:

English Unit	Decimal Shift or X Power of 10	English Unit

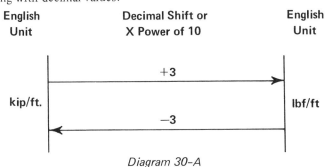

Diagram 30-A

―――――――――――――
*All conversion values obtained can be rounded to desired number of significant figures.

Example (*b*): Convert 10.3 kip/ft to N/m.
  From Diagram 30-A: 10.3 kip/ft = 10.3×10^3 lbf/ft = 103×10^2 lbf/ft
  From table: 103 lbf/ft = 1503.172 N/m*
  Therefore: 103×10^2 lbf/ft = 1503.172×10^2 N/m = 150,317.2 N/m
      $\approx$ 150,000 N/m.

In converting SI force-per-length quantities from N/m to kN/m, kN/cm, N/cm, the Decimal-shift Diagram 30-B, which appears in the headings on alternate table pages, will assist table users. It is derived from:

$$1 \text{ N/m} = (10^{-2}) \text{ N/cm} = (10^{-5}) \text{ kN/cm} = (10^{-3}) \text{ kN/m}.$$

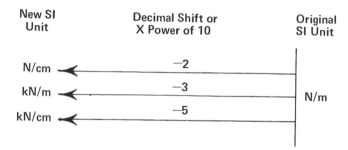

New SI Unit	Decimal Shift or X Power of 10	Original SI Unit
N/cm	−2	
kN/m	−3	N/m
kN/cm	−5	

Diagram 30–B

Example (*c*): Convert 9,340 lbf/ft to N/m, kN/m, N/cm, kN/cm.
  Since: 9,340 lbf/ft = 934×10 lbf/ft:
  From table: 934 lbf/ft = 13,630.70 N/m*
  Therefore: 934×10 lbf/ft = $13,630.70 \times 10$ N/m = 136,307 N/m
      $\approx 1,363 \times 10^2$ N/m
  From Diagram 30-B: 1) 136,307 N/m = $136,307 (10^{-2})$ N/cm
          = 1,363.07 N/cm $\approx$ 1,363 N/cm
          (N/m to N/cm by multiplying N/m value by 10^{-2} or by shifting its decimal 2 places to the left.)

        2) 136,307 N/m = $136,307 (10^{-3})$ kN/m
          = 136.307 kN/m $\approx$ 136.3 kN/m
          (N/m to kN/m by multiplying N/m value by 10^{-3} or by shifting its decimal 3 places to the left.)

        3) 136,307 N/m = $136,307 (10^{-5})$ kN/cm
          = 1.36307 kN/cm $\approx$ 1.363 kN/cm
          (N/m to kN/cm by multiplying N/m value by 10^{-5} or shifting its decimal 5 places to the left.)

*All conversion values obtained can be rounded to desired number of significant figures.

Diagram 30-C is also given at the top of alternate table pages:

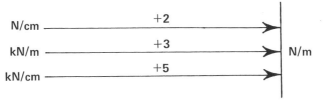

Diagram 30–C

It is the opposite of Diagram 30-B and is used to convert quantities in N/cm, kN/m, kN/cm to N/m. The N/m equivalents are then applied to the REF column in the table to obtain lbf/ft relations. *Please note that from what has been shown in the beginning of this introduction kN/m can be converted directly to kip/ft through Table 30 without any decimal shifting.*

Example (*d*): Convert 0.125 N/cm to lbf/ft.
   From Diagram 30-C: 0.125 N/cm $= 0.125 \times 10^2$ N/m $= 125 \times 10^{-1}$ N/m
   From table: 125 N/m $= 8.565223$ lbf/ft*
   Therefore: 125×10^{-1} N/m $= 8.565223 \times 10^{-1}$ lbf/ft ≈ 0.857 lbf/ft.

---

*All conversion values obtained can be rounded to desired number of significant figures.

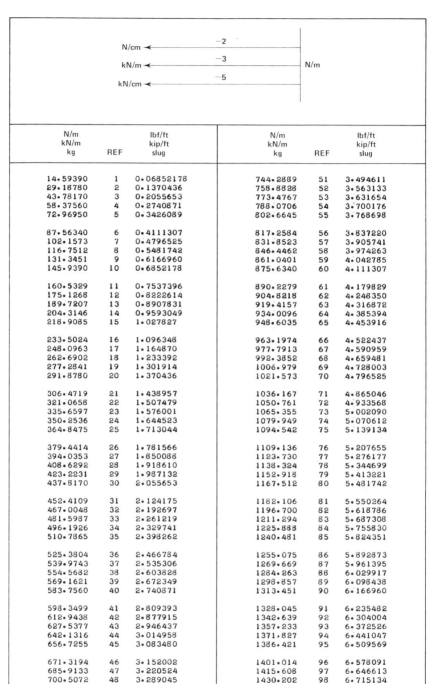

N/m kN/m kg	REF	lbf/ft kip/ft slug	N/m kN/m kg	REF	lbf/ft kip/ft slug
14.59390	1	0.06852178	744.2889	51	3.494611
29.18780	2	0.1370436	758.8828	52	3.563133
43.78170	3	0.2055653	773.4767	53	3.631654
58.37560	4	0.2740871	788.0706	54	3.700176
72.96950	5	0.3426089	802.6645	55	3.768698
87.56340	6	0.4111307	817.2584	56	3.837220
102.1573	7	0.4796525	831.8523	57	3.905741
116.7512	8	0.5481742	846.4462	58	3.974263
131.3451	9	0.6166960	861.0401	59	4.042785
145.9390	10	0.6852178	875.6340	60	4.111307
160.5329	11	0.7537396	890.2279	61	4.179829
175.1268	12	0.8222614	904.8218	62	4.248350
189.7207	13	0.8907831	919.4157	63	4.316872
204.3146	14	0.9593049	934.0096	64	4.385394
218.9085	15	1.027827	948.6035	65	4.453916
233.5024	16	1.096348	963.1974	66	4.522437
248.0963	17	1.164870	977.7913	67	4.590959
262.6902	18	1.233392	992.3852	68	4.659481
277.2841	19	1.301914	1006.979	69	4.728003
291.8780	20	1.370436	1021.573	70	4.796525
306.4719	21	1.438957	1036.167	71	4.865046
321.0658	22	1.507479	1050.761	72	4.933568
335.6597	23	1.576001	1065.355	73	5.002090
350.2536	24	1.644523	1079.949	74	5.070612
364.8475	25	1.713044	1094.542	75	5.139134
379.4414	26	1.781566	1109.136	76	5.207655
394.0353	27	1.850088	1123.730	77	5.276177
408.6292	28	1.918610	1138.324	78	5.344699
423.2231	29	1.987132	1152.918	79	5.413221
437.8170	30	2.055653	1167.512	80	5.481742
452.4109	31	2.124175	1182.106	81	5.550264
467.0048	32	2.192697	1196.700	82	5.618786
481.5987	33	2.261219	1211.294	83	5.687308
496.1926	34	2.329741	1225.888	84	5.755830
510.7865	35	2.398262	1240.481	85	5.824351
525.3804	36	2.466784	1255.075	86	5.892873
539.9743	37	2.535306	1269.669	87	5.961395
554.5682	38	2.603828	1284.263	88	6.029917
569.1621	39	2.672349	1298.857	89	6.098438
583.7560	40	2.740871	1313.451	90	6.166960
598.3499	41	2.809393	1328.045	91	6.235482
612.9438	42	2.877915	1342.639	92	6.304004
627.5377	43	2.946437	1357.233	93	6.372526
642.1316	44	3.014958	1371.827	94	6.441047
656.7255	45	3.083480	1386.421	95	6.509569
671.3194	46	3.152002	1401.014	96	6.578091
685.9133	47	3.220524	1415.608	97	6.646613
700.5072	48	3.289045	1430.202	98	6.715134
715.1011	49	3.357567	1444.796	99	6.783656
729.6950	50	3.426089	1459.390	100	6.852178

N/cm ————————— +2 ————————→
kN/m ————————— +3 ————————→ N/m
kN/cm ———————— +5 ————————→

N/m kN/m kg	REF	lbf/ft kip/ft slug	N/m kN/m kg	REF	lbf/ft kip/ft slug
1473.984	101	6.920700	2203.679	151	10.34679
1488.578	102	6.989222	2218.273	152	10.41531
1503.172	103	7.057743	2232.867	153	10.48383
1517.766	104	7.126265	2247.461	154	10.55235
1532.359	105	7.194787	2262.055	155	10.62088
1546.953	106	7.263309	2276.648	156	10.68940
1561.547	107	7.331830	2291.242	157	10.75792
1576.141	108	7.400352	2305.836	158	10.82644
1590.735	109	7.468874	2320.430	159	10.89496
1605.329	110	7.537396	2335.024	160	10.96348
1619.923	111	7.605918	2349.618	161	11.03201
1634.517	112	7.674439	2364.212	162	11.10053
1649.111	113	7.742961	2378.806	163	11.16905
1663.705	114	7.811483	2393.400	164	11.23757
1678.298	115	7.880005	2407.993	165	11.30609
1692.892	116	7.948527	2422.587	166	11.37462
1707.486	117	8.017048	2437.181	167	11.44314
1722.080	118	8.085570	2451.775	168	11.51166
1736.674	119	8.154092	2466.369	169	11.58018
1751.268	120	8.222614	2480.963	170	11.64870
1765.862	121	8.291135	2495.557	171	11.71722
1780.456	122	8.359657	2510.151	172	11.78575
1795.050	123	8.428179	2524.745	173	11.85427
1809.644	124	8.496701	2539.339	174	11.92279
1824.238	125	8.565223	2553.932	175	11.99131
1838.831	126	8.633744	2568.526	176	12.05983
1853.425	127	8.702266	2583.120	177	12.12836
1868.019	128	8.770788	2597.714	178	12.19688
1882.613	129	8.839310	2612.308	179	12.26540
1897.207	130	8.907831	2626.902	180	12.33392
1911.801	131	8.976353	2641.496	181	12.40244
1926.395	132	9.044875	2656.090	182	12.47096
1940.989	133	9.113397	2670.684	183	12.53949
1955.583	134	9.181919	2685.278	184	12.60801
1970.176	135	9.250440	2699.871	185	12.67653
1984.770	136	9.318962	2714.465	186	12.74505
1999.364	137	9.387484	2729.059	187	12.81357
2013.958	138	9.456006	2743.653	188	12.88209
2028.552	139	9.524527	2758.247	189	12.95062
2043.146	140	9.593049	2772.841	190	13.01914
2057.740	141	9.661571	2787.435	191	13.08766
2072.334	142	9.730093	2802.029	192	13.15618
2086.928	143	9.798615	2816.623	193	13.22470
2101.522	144	9.867136	2831.217	194	13.29323
2116.116	145	9.935658	2845.810	195	13.36175
2130.709	146	10.00418	2860.404	196	13.43027
2145.303	147	10.07270	2874.998	197	13.49879
2159.897	148	10.14122	2889.592	198	13.56731
2174.491	149	10.20975	2904.186	199	13.63583
2189.085	150	10.27827	2918.780	200	13.70436

N/m kN/m kg	REF	lbf/ft kip/ft slug	N/m kN/m kg	REF	lbf/ft kip/ft slug
2933•374	201	13•77288	3663•069	251	17•19897
2947•968	202	13•84140	3677•663	252	17•26749
2962•562	203	13•90992	3692•257	253	17•33601
2977•156	204	13•97844	3706•851	254	17•40453
2991•749	205	14•04696	3721•444	255	17•47305
3006•343	206	14•11549	3736•038	256	17•54158
3020•937	207	14•18401	3750•632	257	17•61010
3035•531	208	14•25253	3765•226	258	17•67862
3050•125	209	14•32105	3779•820	259	17•74714
3064•719	210	14•38957	3794•414	260	17•81566
3079•313	211	14•45810	3809•008	261	17•88418
3093•907	212	14•52662	3823•602	262	17•95271
3108•501	213	14•59514	3838•196	263	18•02123
3123•095	214	14•66366	3852•790	264	18•08975
3137•689	215	14•73218	3867•383	265	18•15827
3152•282	216	14•80070	3881•977	266	18•22679
3166•876	217	14•86923	3896•571	267	18•29532
3181•470	218	14•93775	3911•165	268	18•36384
3196•064	219	15•00627	3925•759	269	18•43236
3210•658	220	15•07479	3940•353	270	18•50088
3225•252	221	15•14331	3954•947	271	18•56940
3239•846	222	15•21184	3969•541	272	18•63792
3254•440	223	15•28036	3984•135	273	18•70645
3269•034	224	15•34888	3998•729	274	18•77497
3283•628	225	15•41740	4013•322	275	18•84349
3298•221	226	15•48592	4027•916	276	18•91201
3312•815	227	15•55444	4042•510	277	18•98053
3327•409	228	15•62297	4057•104	278	19•04905
3342•003	229	15•69149	4071•698	279	19•11758
3356•597	230	15•76001	4086•292	280	19•18610
3371•191	231	15•82853	4100•886	281	19•25462
3385•785	232	15•89705	4115•480	282	19•32314
3400•379	233	15•96557	4130•074	283	19•39166
3414•973	234	16•03410	4144•668	284	19•46019
3429•566	235	16•10262	4159•261	285	19•52871
3444•160	236	16•17114	4173•855	286	19•59723
3458•754	237	16•23966	4188•449	287	19•66575
3473•348	238	16•30818	4203•043	288	19•73427
3487•942	239	16•37671	4217•637	289	19•80279
3502•536	240	16•44523	4232•231	290	19•87132
3517•130	241	16•51375	4246•825	291	19•93984
3531•724	242	16•58227	4261•419	292	20•00836
3546•318	243	16•65079	4276•013	293	20•07688
3560•912	244	16•71931	4290•607	294	20•14540
3575•505	245	16•78784	4305•201	295	20•21393
3590•099	246	16•85636	4319•794	296	20•28245
3604•693	247	16•92488	4334•388	297	20•35097
3619•287	248	16•99340	4348•982	298	20•41949
3633•881	249	17•06192	4363•576	299	20•48801
3648•475	250	17•13045	4378•170	300	20•55653

```
N/cm ————————— +2 ——————————→
kN/m ————————— +3 ——————————→ N/m
kN/cm ———————— +5 ——————————→
```

N/m kN/m kg	REF	lbf/ft kip/ft slug	N/m kN/m kg	REF	lbf/ft kip/ft slug
4392.764	301	20.62506	5122.459	351	24.05114
4407.358	302	20.69358	5137.053	352	24.11967
4421.952	303	20.76210	5151.647	353	24.18819
4436.546	304	20.83062	5166.241	354	24.25671
4451.139	305	20.89914	5180.834	355	24.32523
4465.733	306	20.96766	5195.428	356	24.39375
4480.327	307	21.03619	5210.022	357	24.46228
4494.921	308	21.10471	5224.616	358	24.53080
4509.515	309	21.17323	5239.210	359	24.59932
4524.109	310	21.24175	5253.804	360	24.66784
4538.703	311	21.31027	5268.398	361	24.73636
4553.297	312	21.37880	5282.992	362	24.80488
4567.891	313	21.44732	5297.586	363	24.87341
4582.485	314	21.51584	5312.180	364	24.94193
4597.078	315	21.58436	5326.773	365	25.01045
4611.672	316	21.65288	5341.367	366	25.07897
4626.266	317	21.72140	5355.961	367	25.14749
4640.860	318	21.78993	5370.555	368	25.21602
4655.454	319	21.85845	5385.149	369	25.28454
4670.048	320	21.92697	5399.743	370	25.35306
4684.642	321	21.99549	5414.337	371	25.42158
4699.236	322	22.06401	5428.931	372	25.49010
4713.830	323	22.13253	5443.525	373	25.55862
4728.424	324	22.20106	5458.119	374	25.62715
4743.018	325	22.26958	5472.712	375	25.69567
4757.611	326	22.33810	5487.306	376	25.76419
4772.205	327	22.40662	5501.900	377	25.83271
4786.799	328	22.47514	5516.494	378	25.90123
4801.393	329	22.54367	5531.088	379	25.96975
4815.987	330	22.61219	5545.682	380	26.03828
4830.581	331	22.68071	5560.276	381	26.10680
4845.175	332	22.74923	5574.870	382	26.17532
4859.769	333	22.81775	5589.464	383	26.24384
4874.363	334	22.88627	5604.058	384	26.31236
4888.956	335	22.95480	5618.651	385	26.38089
4903.550	336	23.02332	5633.245	386	26.44941
4918.144	337	23.09184	5647.839	387	26.51793
4932.738	338	23.16036	5662.433	388	26.58645
4947.332	339	23.22888	5677.027	389	26.65497
4961.926	340	23.29741	5691.621	390	26.72349
4976.520	341	23.36593	5706.215	391	26.79202
4991.114	342	23.43445	5720.809	392	26.86054
5005.708	343	23.50297	5735.403	393	26.92906
5020.302	344	23.57149	5749.997	394	26.99758
5034.896	345	23.64001	5764.591	395	27.06610
5049.489	346	23.70854	5779.184	396	27.13462
5064.083	347	23.77706	5793.778	397	27.20315
5078.677	348	23.84558	5808.372	398	27.27167
5093.271	349	23.91410	5822.966	399	27.34019
5107.865	350	23.98262	5837.560	400	27.40871

N/cm ← ——————— −3

kN/m ← ——————— −2 ————— N/m

kN/cm ← ——————— −5

N/m kN/m kg	REF	lbf/ft kip/ft slug		N/m kN/m kg	REF	lbf/ft kip/ft slug
5852.154	401	27.47723		6581.849	451	30.90332
5866.748	402	27.54576		6596.443	452	30.97184
5881.342	403	27.61428		6611.037	453	31.04037
5895.936	404	27.68280		6625.631	454	31.10889
5910.529	405	27.75132		6640.224	455	31.17741
5925.123	406	27.81984		6654.818	456	31.24593
5939.717	407	27.88836		6669.412	457	31.31445
5954.311	408	27.95689		6684.006	458	31.38298
5968.905	409	28.02541		6698.600	459	31.45150
5983.499	410	28.09393		6713.194	460	31.52002
5998.093	411	28.16245		6727.788	461	31.58854
6012.687	412	28.23097		6742.382	462	31.65706
6027.281	413	28.29950		6756.976	463	31.72558
6041.875	414	28.36802		6771.570	464	31.79411
6056.469	415	28.43654		6786.164	465	31.86263
6071.062	416	28.50506		6800.757	466	31.93115
6085.656	417	28.57358		6815.351	467	31.99967
6100.250	418	28.64210		6829.945	468	32.06819
6114.844	419	28.71063		6844.539	469	32.13671
6129.438	420	28.77915		6859.133	470	32.20524
6144.032	421	28.84767		6873.727	471	32.27376
6158.626	422	28.91619		6888.321	472	32.34228
6173.220	423	28.98471		6902.915	473	32.41080
6187.814	424	29.05323		6917.509	474	32.47932
6202.407	425	29.12176		6932.102	475	32.54785
6217.001	426	29.19028		6946.696	476	32.61637
6231.595	427	29.25880		6961.290	477	32.68489
6246.189	428	29.32732		6975.884	478	32.75341
6260.783	429	29.39584		6990.478	479	32.82193
6275.377	430	29.46437		7005.072	480	32.89045
6289.971	431	29.53289		7019.666	481	32.95898
6304.565	432	29.60141		7034.260	482	33.02750
6319.159	433	29.66993		7048.854	483	33.09602
6333.753	434	29.73845		7063.448	484	33.16454
6348.346	435	29.80697		7078.042	485	33.23306
6362.940	436	29.87550		7092.635	486	33.30159
6377.534	437	29.94402		7107.229	487	33.37011
6392.128	438	30.01254		7121.823	488	33.43863
6406.722	439	30.08106		7136.417	489	33.50715
6421.316	440	30.14958		7151.011	490	33.57567
6435.910	441	30.21811		7165.605	491	33.64419
6450.504	442	30.28663		7180.199	492	33.71272
6465.098	443	30.35515		7194.793	493	33.78124
6479.692	444	30.42367		7209.387	494	33.84976
6494.285	445	30.49219		7223.980	495	33.91828
6508.879	446	30.56071		7238.574	496	33.98680
6523.473	447	30.62924		7253.168	497	34.05532
6538.067	448	30.69776		7267.762	498	34.12385
6552.661	449	30.76628		7282.356	499	34.19237
6567.255	450	30.83480		7296.950	500	34.26089

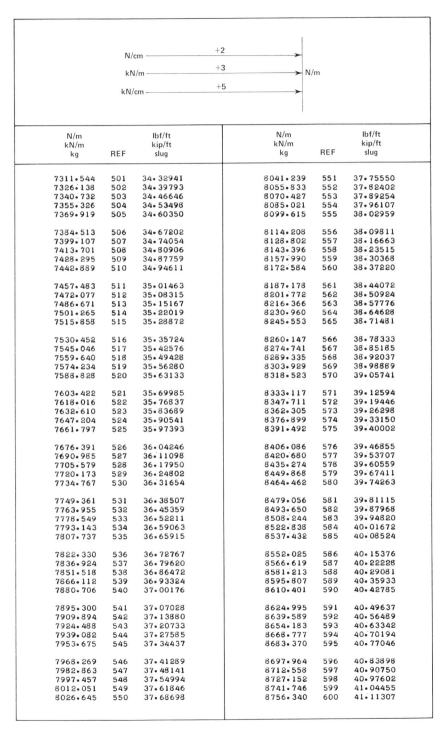

N/m kN/m kg	REF	lbf/ft kip/ft slug		N/m kN/m kg	REF	lbf/ft kip/ft slug
7311.544	501	34.32941		8041.239	551	37.75550
7326.138	502	34.39793		8055.833	552	37.82402
7340.732	503	34.46646		8070.427	553	37.89254
7355.326	504	34.53498		8085.021	554	37.96107
7369.919	505	34.60350		8099.615	555	38.02959
7384.513	506	34.67202		8114.208	556	38.09811
7399.107	507	34.74054		8128.802	557	38.16663
7413.701	508	34.80906		8143.396	558	38.23515
7428.295	509	34.87759		8157.990	559	38.30368
7442.889	510	34.94611		8172.584	560	38.37220
7457.483	511	35.01463		8187.178	561	38.44072
7472.077	512	35.08315		8201.772	562	38.50924
7486.671	513	35.15167		8216.366	563	38.57776
7501.265	514	35.22019		8230.960	564	38.64628
7515.858	515	35.28872		8245.553	565	38.71481
7530.452	516	35.35724		8260.147	566	38.78333
7545.046	517	35.42576		8274.741	567	38.85185
7559.640	518	35.49428		8289.335	568	38.92037
7574.234	519	35.56280		8303.929	569	38.98889
7588.828	520	35.63133		8318.523	570	39.05741
7603.422	521	35.69985		8333.117	571	39.12594
7618.016	522	35.76837		8347.711	572	39.19446
7632.610	523	35.83689		8362.305	573	39.26298
7647.204	524	35.90541		8376.899	574	39.33150
7661.797	525	35.97393		8391.492	575	39.40002
7676.391	526	36.04246		8406.086	576	39.46855
7690.985	527	36.11098		8420.680	577	39.53707
7705.579	528	36.17950		8435.274	578	39.60559
7720.173	529	36.24802		8449.868	579	39.67411
7734.767	530	36.31654		8464.462	580	39.74263
7749.361	531	36.38507		8479.056	581	39.81115
7763.955	532	36.45359		8493.650	582	39.87968
7778.549	533	36.52211		8508.244	583	39.94820
7793.143	534	36.59063		8522.838	584	40.01672
7807.737	535	36.65915		8537.432	585	40.08524
7822.330	536	36.72767		8552.025	586	40.15376
7836.924	537	36.79620		8566.619	587	40.22228
7851.518	538	36.86472		8581.213	588	40.29081
7866.112	539	36.93324		8595.807	589	40.35933
7880.706	540	37.00176		8610.401	590	40.42785
7895.300	541	37.07028		8624.995	591	40.49637
7909.894	542	37.13880		8639.589	592	40.56489
7924.488	543	37.20733		8654.183	593	40.63342
7939.082	544	37.27585		8668.777	594	40.70194
7953.675	545	37.34437		8683.370	595	40.77046
7968.269	546	37.41289		8697.964	596	40.83898
7982.863	547	37.48141		8712.558	597	40.90750
7997.457	548	37.54994		8727.152	598	40.97602
8012.051	549	37.61846		8741.746	599	41.04455
8026.645	550	37.68698		8756.340	600	41.11307

N/cm ◄————————— —2
kN/m ◄————————— —3 ————— N/m
kN/cm ◄———————— —5

N/m kN/m kg	REF	lbf/ft kip/ft slug	N/m kN/m kg	REF	lbf/ft kip/ft slug
8770.934	601	41.18159	9500.629	651	44.60768
8785.528	602	41.25011	9515.223	652	44.67620
8800.122	603	41.31863	9529.817	653	44.74472
8814.716	604	41.38716	9544.411	654	44.81324
8829.309	605	41.45568	9559.005	655	44.88177
8843.903	606	41.52420	9573.598	656	44.95029
8858.497	607	41.59272	9588.192	657	45.01881
8873.091	608	41.66124	9602.786	658	45.08733
8887.685	609	41.72976	9617.380	659	45.15585
8902.279	610	41.79829	9631.974	660	45.22437
8916.873	611	41.86681	9646.568	661	45.29290
8931.467	612	41.93533	9661.162	662	45.36142
8946.061	613	42.00385	9675.756	663	45.42994
8960.655	614	42.07237	9690.350	664	45.49846
8975.249	615	42.14089	9704.943	665	45.56698
8989.842	616	42.20942	9719.537	666	45.63551
9004.436	617	42.27794	9734.131	667	45.70403
9019.030	618	42.34646	9748.725	668	45.77255
9033.624	619	42.41498	9763.319	669	45.84107
9048.218	620	42.48350	9777.913	670	45.90959
9062.812	621	42.55203	9792.507	671	45.97811
9077.406	622	42.62055	9807.101	672	46.04664
9092.000	623	42.68907	9821.695	673	46.11516
9106.594	624	42.75759	9836.289	674	46.18368
9121.188	625	42.82611	9850.882	675	46.25220
9135.781	626	42.89463	9865.476	676	46.32072
9150.375	627	42.96316	9880.070	677	46.38925
9164.969	628	43.03168	9894.664	678	46.45777
9179.563	629	43.10020	9909.258	679	46.52629
9194.157	630	43.16872	9923.852	680	46.59481
9208.751	631	43.23724	9938.446	681	46.66333
9223.345	632	43.30577	9953.040	682	46.73185
9237.939	633	43.37429	9967.634	683	46.80038
9252.533	634	43.44281	9982.228	684	46.86890
9267.126	635	43.51133	9996.822	685	46.93742
9281.720	636	43.57985	10011.42	686	47.00594
9296.314	637	43.64837	10026.01	687	47.07446
9310.908	638	43.71690	10040.60	688	47.14298
9325.502	639	43.78542	10055.20	689	47.21151
9340.096	640	43.85394	10069.79	690	47.28003
9354.690	641	43.92246	10084.38	691	47.34855
9369.284	642	43.99098	10098.98	692	47.41707
9383.878	643	44.05950	10113.57	693	47.48559
9398.472	644	44.12803	10128.17	694	47.55412
9413.065	645	44.19655	10142.76	695	47.62264
9427.659	646	44.26507	10157.35	696	47.69116
9442.253	647	44.33359	10171.95	697	47.75968
9456.847	648	44.40211	10186.54	698	47.82820
9471.441	649	44.47064	10201.14	699	47.89672
9486.035	650	44.53916	10215.73	700	47.96525

N/cm		+2		
kN/m		+3		N/m
kN/cm		+5		

N/m kN/m kg	REF	lbf/ft kip/ft slug		N/m kN/m kg	REF	lbf/ft kip/ft slug
10230.32	701	48.03377		10960.02	751	51.45986
10244.92	702	48.10229		10974.61	752	51.52838
10259.51	703	48.17081		10989.21	753	51.59690
10274.11	704	48.23933		11003.80	754	51.66542
10288.70	705	48.30785		11018.39	755	51.73394
10303.29	706	48.37638		11032.99	756	51.80247
10317.89	707	48.44490		11047.58	757	51.87099
10332.48	708	48.51342		11062.18	758	51.93951
10347.08	709	48.58194		11076.77	759	52.00803
10361.67	710	48.65046		11091.36	760	52.07655
10376.26	711	48.71899		11105.96	761	52.14507
10390.86	712	48.78751		11120.55	762	52.21360
10405.45	713	48.85603		11135.15	763	52.28212
10420.04	714	48.92455		11149.74	764	52.35064
10434.64	715	48.99307		11164.33	765	52.41916
10449.23	716	49.06159		11178.93	766	52.48768
10463.83	717	49.13012		11193.52	767	52.55621
10478.42	718	49.19864		11208.12	768	52.62473
10493.01	719	49.26716		11222.71	769	52.69325
10507.61	720	49.33568		11237.30	770	52.76177
10522.20	721	49.40420		11251.90	771	52.83029
10536.80	722	49.47272		11266.49	772	52.89881
10551.39	723	49.54125		11281.08	773	52.96734
10565.98	724	49.60977		11295.68	774	53.03586
10580.58	725	49.67829		11310.27	775	53.10438
10595.17	726	49.74681		11324.87	776	53.17290
10609.77	727	49.81533		11339.46	777	53.24142
10624.36	728	49.88386		11354.05	778	53.30994
10638.95	729	49.95238		11368.65	779	53.37847
10653.55	730	50.02090		11383.24	780	53.44699
10668.14	731	50.08942		11397.84	781	53.51551
10682.73	732	50.15794		11412.43	782	53.58403
10697.33	733	50.22646		11427.02	783	53.65255
10711.92	734	50.29499		11441.62	784	53.72108
10726.52	735	50.36351		11456.21	785	53.78960
10741.11	736	50.43203		11470.81	786	53.85812
10755.70	737	50.50055		11485.40	787	53.92664
10770.30	738	50.56907		11499.99	788	53.99516
10784.89	739	50.63760		11514.59	789	54.06368
10799.49	740	50.70612		11529.18	790	54.13221
10814.08	741	50.77464		11543.77	791	54.20073
10828.67	742	50.84316		11558.37	792	54.26925
10843.27	743	50.91168		11572.96	793	54.33777
10857.86	744	50.98020		11587.56	794	54.40629
10872.46	745	51.04873		11602.15	795	54.47481
10887.05	746	51.11725		11616.74	796	54.54334
10901.64	747	51.18577		11631.34	797	54.61186
10916.24	748	51.25429		11645.93	798	54.68038
10930.83	749	51.32281		11660.53	799	54.74890
10945.42	750	51.39134		11675.12	800	54.81742

N/cm ← $\quad$ −2
kN/m ← $\quad$ −3 $\quad$ → N/m
kN/cm ← $\quad$ −5

N/m kN/m kg	REF	lbf/ft kip/ft slug	N/m kN/m kg	REF	lbf/ft kip/ft slug
11689.71	801	54.88595	12419.41	851	58.31203
11704.31	802	54.95447	12434.00	852	58.38056
11718.90	803	55.02299	12448.60	853	58.44908
11733.50	804	55.09151	12463.19	854	58.51760
11748.09	805	55.16003	12477.78	855	58.58612
11762.68	806	55.22855	12492.38	856	58.65464
11777.28	807	55.29708	12506.97	857	58.72317
11791.87	808	55.36560	12521.57	858	58.79169
11806.47	809	55.43412	12536.16	859	58.86021
11821.06	810	55.50264	12550.75	860	58.92873
11835.65	811	55.57116	12565.35	861	58.99725
11850.25	812	55.63969	12579.94	862	59.06577
11864.84	813	55.70821	12594.54	863	59.13430
11879.43	814	55.77673	12609.13	864	59.20282
11894.03	815	55.84525	12623.72	865	59.27134
11908.62	816	55.91377	12638.32	866	59.33986
11923.22	817	55.98229	12652.91	867	59.40838
11937.81	818	56.05082	12667.51	868	59.47690
11952.40	819	56.11934	12682.10	869	59.54543
11967.00	820	56.18786	12696.69	870	59.61395
11981.59	821	56.25638	12711.29	871	59.68247
11996.19	822	56.32490	12725.88	872	59.75099
12010.78	823	56.39342	12740.47	873	59.81951
12025.37	824	56.46195	12755.07	874	59.88804
12039.97	825	56.53047	12769.66	875	59.95656
12054.56	826	56.59899	12784.26	876	60.02508
12069.16	827	56.66751	12798.85	877	60.09360
12083.75	828	56.73603	12813.44	878	60.16212
12098.34	829	56.80456	12828.04	879	60.23064
12112.94	830	56.87308	12842.63	880	60.29917
12127.53	831	56.94160	12857.23	881	60.36769
12142.12	832	57.01012	12871.82	882	60.43621
12156.72	833	57.07864	12886.41	883	60.50473
12171.31	834	57.14716	12901.01	884	60.57325
12185.91	835	57.21569	12915.60	885	60.64178
12200.50	836	57.28421	12930.20	886	60.71030
12215.09	837	57.35273	12944.79	887	60.77882
12229.69	838	57.42125	12959.38	888	60.84734
12244.28	839	57.48977	12973.98	889	60.91586
12258.88	840	57.55830	12988.57	890	60.98438
12273.47	841	57.62682	13003.16	891	61.05291
12288.06	842	57.69534	13017.76	892	61.12143
12302.66	843	57.76386	13032.35	893	61.18995
12317.25	844	57.83238	13046.95	894	61.25847
12331.85	845	57.90090	13061.54	895	61.32699
12346.44	846	57.96943	13076.13	896	61.39551
12361.03	847	58.03795	13090.73	897	61.46404
12375.63	848	58.10647	13105.32	898	61.53256
12390.22	849	58.17499	13119.92	899	61.60108
12404.81	850	58.24351	13134.51	900	61.66960

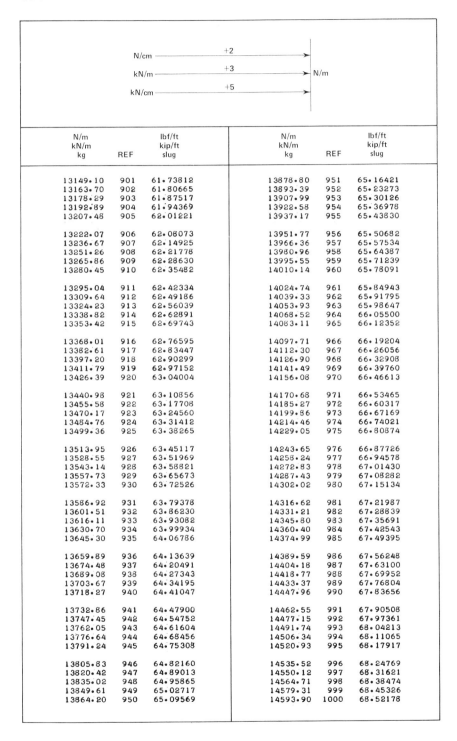

N/m kN/m kg	REF	lbf/ft kip/ft slug	N/m kN/m kg	REF	lbf/ft kip/ft slug
13149.10	901	61.73812	13878.80	951	65.16421
13163.70	902	61.80665	13893.39	952	65.23273
13178.29	903	61.87517	13907.99	953	65.30126
13192.89	904	61.94369	13922.58	954	65.36978
13207.48	905	62.01221	13937.17	955	65.43830
13222.07	906	62.08073	13951.77	956	65.50682
13236.67	907	62.14925	13966.36	957	65.57534
13251.26	908	62.21778	13980.96	958	65.64387
13265.86	909	62.28630	13995.55	959	65.71239
13280.45	910	62.35482	14010.14	960	65.78091
13295.04	911	62.42334	14024.74	961	65.84943
13309.64	912	62.49186	14039.33	962	65.91795
13324.23	913	62.56039	14053.93	963	65.98647
13338.82	914	62.62891	14068.52	964	66.05500
13353.42	915	62.69743	14083.11	965	66.12352
13368.01	916	62.76595	14097.71	966	66.19204
13382.61	917	62.83447	14112.30	967	66.26056
13397.20	918	62.90299	14126.90	968	66.32908
13411.79	919	62.97152	14141.49	969	66.39760
13426.39	920	63.04004	14156.08	970	66.46613
13440.98	921	63.10856	14170.68	971	66.53465
13455.58	922	63.17708	14185.27	972	66.60317
13470.17	923	63.24560	14199.86	973	66.67169
13484.76	924	63.31412	14214.46	974	66.74021
13499.36	925	63.38265	14229.05	975	66.80874
13513.95	926	63.45117	14243.65	976	66.87726
13528.55	927	63.51969	14258.24	977	66.94578
13543.14	928	63.58821	14272.83	978	67.01430
13557.73	929	63.65673	14287.43	979	67.08282
13572.33	930	63.72526	14302.02	980	67.15134
13586.92	931	63.79378	14316.62	981	67.21987
13601.51	932	63.86230	14331.21	982	67.28839
13616.11	933	63.93082	14345.80	983	67.35691
13630.70	934	63.99934	14360.40	984	67.42543
13645.30	935	64.06786	14374.99	985	67.49395
13659.89	936	64.13639	14389.59	986	67.56248
13674.48	937	64.20491	14404.18	987	67.63100
13689.08	938	64.27343	14418.77	988	67.69952
13703.67	939	64.34195	14433.37	989	67.76804
13718.27	940	64.41047	14447.96	990	67.83656
13732.86	941	64.47900	14462.55	991	67.90508
13747.45	942	64.54752	14477.15	992	67.97361
13762.05	943	64.61604	14491.74	993	68.04213
13776.64	944	64.68456	14506.34	994	68.11065
13791.24	945	64.75308	14520.93	995	68.17917
13805.83	946	64.82160	14535.52	996	68.24769
13820.42	947	64.89013	14550.12	997	68.31621
13835.02	948	64.95865	14564.71	998	68.38474
13849.61	949	65.02717	14579.31	999	68.45326
13864.20	950	65.09569	14593.90	1000	68.52178

31

BENDING MOMENT AND TORQUE

Pound (force)–foot — Newton-metre
  (lbf-ft—N·m)

Kilopound (force)–foot — Kilonewton-metre
  (kip-ft — kN-m)

Pound (force)–foot — Newton-centimetre
  (lbf-ft — N·cm)

WORK, ENERGY, QUANTITY of HEAT

Foot-pound (force) — Joule
  (ft-lbf — J)

Foot-kilopound (force) — Kilojoule
  (ft-kip — kJ)

POWER

Foot-pound (force)/Second — Watt

  (ft-lbf/s — W)

Foot-pound (force)/Second — Kilowatt

  (ft-lbf/s — kW)

Conversion Factors:
  1 pound (force)-foot (lbf-ft) = 1.355818 newton-metre (N·m)
  (1 N·m = 0.7375621 lbf-ft)

Torque, Energy, and Power Values

With Table 31, energy (or work or quantity of heat) and power values in ft-lbf and ft-lbf/s can be converted to and from SI.

The SI unit for energy, the joule (J), is equal to 1 newton-metre (N·m). *Note: In symbol arrangement, N·m (or Nm) is preferred to m·N (or mN) to avoid possible confusion of the latter with the SI force submultiple, the milli-newton (mN).*

The numerical values of the conversion factors for energy and power are identical to that given previously for torque, on which Table 31 is based:

$$1 \text{ ft-lbf} = 1.355818 \text{ J}$$
$$1 \text{ ft-lbf/s} = 1.355818 \text{ W}$$

Where the SI derived unit for power, the watt (W), is equal to 1 J/s.

If necessary, Table 31 can also be used for converting power relations given in ft-lbf/min and ft-lbf/h. In these instances maintaining the SI unit watt would be in error. Instead, joule/min and joule/h would be used, respectively:

1 ft-lbf/min = 1.355818 J/min

1 ft-lbf/h = 1.355818 J/h

Also, a kilopound (kip) = 1000 lbf and a kilonewton (kN) = 1000 N. Therefore, Table 31 can be used directly for converting kip-ft and kN·m quantities since: 1 kip-ft = 1.355818 kN·m.

In converting SI torque, energy, and power values from primary compound units to those containing kN, cm, kJ, and kW, the Decimal-shift Diagram 31A, which appears in the headings on alternate table pages, will assist table users. It is derived from: $1 \text{ N} = (10^{-3}) \text{ kN}; 1 \text{ J} = (10^{-3}) \text{ kJ}; 1 \text{ W} = (10^{-3}) \text{ kW}.$

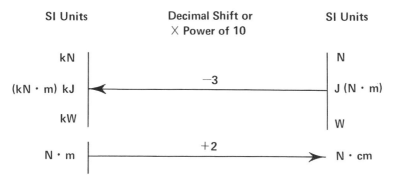

Diagram 31-A

Example (a): Convert 144 kip-ft to kN·m.

Since both English and SI units contain the "kilo" prefix (k), the conversion can be read directly from the table without a decimal shift:

$$144 \text{ kip-ft} = 195.2378 \text{ kN·m*} \approx 195 \text{ kN·m}$$

Example (b): Convert 379,000 ft-lbf/s to kW.

Since: $379{,}000 \text{ ft-lbf/s} = 379 \times 10^3 \text{ ft-lbf/s}$:

From table: 379 ft-lbf/s = 513.8550 W*

Therefore: $379 \times 10^3 \text{ ft-lbf/s} = 513.8550 \times 10^3 \text{ W} = 513{,}855 \text{ W}.$

---

*All conversion values obtained can be rounded to desired number of significant figures.

From Diagram 31-A: 513,855W = 513,855 (10^{-3}) kW = 513.855kW*≈514kW
(W to kW by multiplying W value by 10^{-3} or shifting
its decimal 3 places to the left.)

Diagram 31-B is also given at the top of alternate table pages:

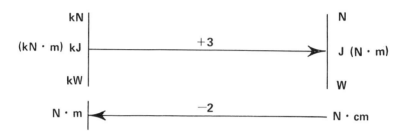

Diagram 31-B

It is the opposite of Diagram 31-A and is used to convert a quantity in kN, kJ, kW,
or N·cm to its primary unit. The primary equivalent is then applied to the REF
column in the table to obtain lbf-ft, ft-lbf, or ft-lbf/s relations.

Example (*c*): Convert 7,170 N·cm to lbf-ft.

From Diagram 31-B: 7,170 N·cm = 7,170 × 10^{-2} N·m = 717 × 10^{-1} N·m
From table: 717 N·m = 528.8320 lbf-ft*
Therefore: 717 × 10^{-1} N·m = 528.8320 lbf-ft × 10^{-1} = 52.88320 lbf-ft
≈52.9 lbf-ft.

---

*All conversion values obtained can be rounded to desired number of significant figures.

```
         kN|                                    N|
(kN · m) kJ| ←————————  −3  ————————  J (N · m)
         kW|                                    W|

              |
    N · m —————————————  + 2  ——————————→ N · cm
              |
```

| N · m | | lbf-ft | N · m | | lbf-ft |
| J | | ft-lbf | J | | ft-lbf |
W	REF	ft-lbf/s	W	REF	ft-lbf/s
1.355818	1	0.7375621	69.14672	51	37.61567
2.711636	2	1.475124	70.50254	52	38.35323
4.067454	3	2.212686	71.85835	53	39.09079
5.423272	4	2.950248	73.21417	54	39.82835
6.779090	5	3.687811	74.56999	55	40.56592
8.134908	6	4.425373	75.92581	56	41.30348
9.490726	7	5.162935	77.28163	57	42.04104
10.84654	8	5.900497	73.63744	58	42.77860
12.20236	9	6.638059	79.99326	59	43.51617
13.55818	10	7.375621	81.34908	60	44.25373
14.91400	11	8.113183	82.70490	61	44.99129
16.26982	12	8.850745	84.06072	62	45.72885
17.62563	13	9.588308	85.41653	63	46.46641
18.98145	14	10.32587	86.77235	64	47.20398
20.33727	15	11.06343	88.12817	65	47.94154
21.69309	16	11.80099	89.48399	66	48.67910
23.04891	17	12.53856	90.83981	67	49.41666
24.40472	18	13.27612	92.19562	68	50.15422
25.76054	19	14.01368	93.55144	69	50.89179
27.11636	20	14.75124	94.90726	70	51.62935
28.47218	21	15.48880	96.26308	71	52.36691
29.82800	22	16.22637	97.61890	72	53.10447
31.18381	23	16.96393	98.97471	73	53.84203
32.53963	24	17.70149	100.3305	74	54.57960
33.89545	25	18.43905	101.6863	75	55.31716
35.25127	26	19.17662	103.0422	76	56.05472
36.60709	27	19.91418	104.3980	77	56.79228
37.96290	28	20.65174	105.7538	78	57.52985
39.31872	29	21.38930	107.1096	79	58.26741
40.67454	30	22.12686	108.4654	80	59.00497
42.03036	31	22.86443	109.8213	81	59.74253
43.38618	32	23.60199	111.1771	82	60.48009
44.74199	33	24.33955	112.5329	83	61.21766
46.09781	34	25.07711	113.8887	84	61.95522
47.45363	35	25.81467	115.2445	85	62.69278
48.80945	36	26.55224	116.6003	86	63.43034
50.16527	37	27.28980	117.9562	87	64.16790
51.52108	38	28.02736	119.3120	88	64.90547
52.87690	39	28.76492	120.6678	89	65.64303
54.23272	40	29.50248	122.0236	90	66.38059
55.58854	41	30.24005	123.3794	91	67.11815
56.94436	42	30.97761	124.7353	92	67.85571
58.30017	43	31.71517	126.0911	93	68.59328
59.65599	44	32.45273	127.4469	94	69.33084
61.01181	45	33.19030	128.8027	95	70.06840
62.36763	46	33.92786	130.1585	96	70.80596
63.72345	47	34.66542	131.5143	97	71.54353
65.07926	48	35.40298	132.8702	98	72.28109
66.43508	49	36.14054	134.2260	99	73.01865
67.79090	50	36.87811	135.5818	100	73.75621

```
                    kN |                          N |
          (kN · m) kJ  |————————+3——————————————▶  J (N · m)
                    kW |                          W |


                       |                              
               N · m ◀—————————————-2—————————————    
                       |                          N · cm
```

N · m J W	REF	lbf-ft ft-lbf ft-lbf/s	N · m J W	REF	lbf-ft ft-lbf ft-lbf/s
136.9376	101	74.49377	204.7285	151	111.3719
138.2934	102	75.23134	206.0843	152	112.1094
139.6493	103	75.96890	207.4402	153	112.8470
141.0051	104	76.70646	208.7960	154	113.5846
142.3609	105	77.44402	210.1518	155	114.3221
143.7167	106	78.18158	211.5076	156	115.0597
145.0725	107	78.91915	212.8634	157	115.7973
146.4283	108	79.65671	214.2192	158	116.5348
147.7842	109	80.39427	215.5751	159	117.2724
149.1400	110	81.13183	216.9309	160	118.0099
150.4958	111	81.86940	218.2867	161	118.7475
151.8516	112	82.60696	219.6425	162	119.4851
153.2074	113	83.34452	220.9983	163	120.2226
154.5633	114	84.08208	222.3542	164	120.9602
155.9191	115	84.81964	223.7100	165	121.6978
157.2749	116	85.55721	225.0658	166	122.4353
158.6307	117	86.29477	226.4216	167	123.1729
159.9865	118	87.03233	227.7774	168	123.9104
161.3423	119	87.76989	229.1332	169	124.6480
162.6982	120	88.50745	230.4891	170	125.3856
164.0540	121	89.24502	231.8449	171	126.1231
165.4098	122	89.98258	233.2007	172	126.8607
166.7656	123	90.72014	234.5565	173	127.5982
168.1214	124	91.45770	235.9123	174	128.3358
169.4773	125	92.19526	237.2682	175	129.0734
170.8331	126	92.93283	238.6240	176	129.8109
172.1889	127	93.67039	239.9798	177	130.5485
173.5447	128	94.40795	241.3356	178	131.2861
174.9005	129	95.14551	242.6914	179	132.0236
176.2563	130	95.88308	244.0472	180	132.7612
177.6122	131	96.62064	245.4031	181	133.4987
178.9680	132	97.35820	246.7589	182	134.2363
180.3238	133	98.09576	248.1147	183	134.9739
181.6796	134	98.83332	249.4705	184	135.7114
183.0354	135	99.57089	250.8263	185	136.4490
184.3912	136	100.3084	252.1821	186	137.1866
185.7471	137	101.0460	253.5380	187	137.9241
187.1029	138	101.7836	254.8938	188	138.6617
188.4587	139	102.5211	256.2496	189	139.3992
189.8145	140	103.2587	257.6054	190	140.1368
191.1703	141	103.9963	258.9612	191	140.8744
192.5262	142	104.7338	260.3171	192	141.6119
193.8820	143	105.4714	261.6729	193	142.3495
195.2378	144	106.2089	263.0287	194	143.0871
196.5936	145	106.9465	264.3845	195	143.8246
197.9494	146	107.6841	265.7403	196	144.5622
199.3052	147	108.4216	267.0961	197	145.2997
200.6611	148	109.1592	268.4520	198	146.0373
202.0169	149	109.8968	269.8078	199	146.7749
203.3727	150	110.6343	271.1636	200	147.5124

```
kN                              N
(kN · m) kJ  ◄──────── −3 ──────┤  J (N · m)
kW                              W


N · m ├─────────── +2 ──────────►  N · cm
```

N · m J W	REF	lbf-ft ft-lbf ft-lbf/s	N · m J W	REF	lbf-ft ft-lbf ft-lbf/s
272.5194	201	148.2500	340.3103	251	185.1281
273.8752	202	148.9875	341.6661	252	185.8657
275.2311	203	149.7251	343.0220	253	186.6032
276.5869	204	150.4627	344.3778	254	187.3408
277.9427	205	151.2002	345.7336	255	188.0783
279.2985	206	151.9378	347.0894	256	188.8159
280.6543	207	152.6754	348.4452	257	189.5535
282.0101	208	153.4129	349.8010	258	190.2910
283.3660	209	154.1505	351.1569	259	191.0286
284.7218	210	154.8880	352.5127	260	191.7662
286.0776	211	155.6256	353.8685	261	192.5037
287.4334	212	156.3632	355.2243	262	193.2413
288.7892	213	157.1007	356.5801	263	193.9788
290.1451	214	157.8383	357.9360	264	194.7164
291.5009	215	158.5759	359.2918	265	195.4540
292.8567	216	159.3134	360.6476	266	196.1915
294.2125	217	160.0510	362.0034	267	196.9291
295.5683	218	160.7885	363.3592	268	197.6666
296.9241	219	161.5261	364.7150	269	198.4042
298.2800	220	162.2637	366.0709	270	199.1418
299.6358	221	163.0012	367.4267	271	199.8793
300.9916	222	163.7388	368.7825	272	200.6169
302.3474	223	164.4764	370.1383	273	201.3545
303.7032	224	165.2139	371.4941	274	202.0920
305.0591	225	165.9515	372.8500	275	202.8296
306.4149	226	166.6890	374.2058	276	203.5671
307.7707	227	167.4266	375.5616	277	204.3047
309.1265	228	168.1642	376.9174	278	205.0423
310.4823	229	168.9017	378.2732	279	205.7798
311.8381	230	169.6393	379.6290	280	206.5174
313.1940	231	170.3769	380.9849	281	207.2550
314.5498	232	171.1144	382.3407	282	207.9925
315.9056	233	171.8520	383.6965	283	208.7301
317.2614	234	172.5895	385.0523	284	209.4676
318.6172	235	173.3271	386.4081	285	210.2052
319.9730	236	174.0647	387.7640	286	210.9428
321.3289	237	174.8022	389.1198	287	211.6803
322.6847	238	175.5398	390.4756	288	212.4179
324.0405	239	176.2773	391.8314	289	213.1555
325.3963	240	177.0149	393.1872	290	213.8930
326.7521	241	177.7525	394.5430	291	214.6306
328.1080	242	178.4900	395.8989	292	215.3681
329.4638	243	179.2276	397.2547	293	216.1057
330.8196	244	179.9652	398.6105	294	216.8433
332.1754	245	180.7027	399.9663	295	217.5808
333.5312	246	181.4403	401.3221	296	218.3184
334.8870	247	182.1778	402.6779	297	219.0560
336.2429	248	182.9154	404.0338	298	219.7935
337.5987	249	183.6530	405.3896	299	220.5311
338.9545	250	184.3905	406.7454	300	221.2686

$$\begin{array}{c} \text{kN} \\ (\text{kN}\cdot\text{m})\ \text{kJ} \\ \text{kW} \end{array} \xrightarrow{\quad +3 \quad} \begin{array}{c} \text{N} \\ \text{J (N}\cdot\text{m)} \\ \text{W} \end{array}$$

$$\text{N}\cdot\text{m} \xleftarrow{\quad -2 \quad} \text{N}\cdot\text{cm}$$

N · m J W	REF	lbf-ft ft-lbf ft-lbf/s	N · m J W	REF	lbf-ft ft-lbf ft-lbf/s
408.1012	301	222.0062	475.8921	351	258.8843
409.4570	302	222.7438	477.2479	352	259.6219
410.8129	303	223.4813	478.6038	353	260.3594
412.1687	304	224.2189	479.9596	354	261.0970
413.5245	305	224.9564	481.3154	355	261.8346
414.8803	306	225.6940	482.6712	356	262.5721
416.2361	307	226.4316	484.0270	357	263.3097
417.5919	308	227.1691	485.3828	358	264.0472
418.9478	309	227.9067	486.7387	359	264.7848
420.3036	310	228.6443	488.0945	360	265.5224
421.6594	311	229.3818	489.4503	361	266.2599
423.0152	312	230.1194	490.8061	362	266.9975
424.3710	313	230.8569	492.1619	363	267.7351
425.7269	314	231.5945	493.5178	364	268.4726
427.0827	315	232.3321	494.8736	365	269.2102
428.4385	316	233.0696	496.2294	366	269.9477
429.7943	317	233.8072	497.5852	367	270.6853
431.1501	318	234.5448	498.9410	368	271.4229
432.5059	319	235.2823	500.2968	369	272.1604
433.8618	320	236.0199	501.6527	370	272.8980
435.2176	321	236.7574	503.0085	371	273.6355
436.5734	322	237.4950	504.3643	372	274.3731
437.9292	323	238.2326	505.7201	373	275.1107
439.2850	324	238.9701	507.0759	374	275.8482
440.6409	325	239.7077	508.4318	375	276.5858
441.9967	326	240.4453	509.7876	376	277.3234
443.3525	327	241.1828	511.1434	377	278.0609
444.7083	328	241.9204	512.4992	378	278.7985
446.0641	329	242.6579	513.8550	379	279.5360
447.4199	330	243.3955	515.2108	380	280.2736
448.7758	331	244.1331	516.5667	381	281.0112
450.1316	332	244.8706	517.9225	382	281.7487
451.4874	333	245.6082	519.2783	383	282.4863
452.8432	334	246.3457	520.6341	384	283.2239
454.1990	335	247.0833	521.9899	385	283.9614
455.5548	336	247.8209	523.3457	386	284.6990
456.9107	337	248.5584	524.7016	387	285.4365
458.2665	338	249.2960	526.0574	388	286.1741
459.6223	339	250.0336	527.4132	389	286.9117
460.9781	340	250.7711	528.7690	390	287.6492
462.3339	341	251.5087	530.1248	391	288.3868
463.6898	342	252.2462	531.4807	392	289.1244
465.0456	343	252.9838	532.8365	393	289.8619
466.4014	344	253.7214	534.1923	394	290.5995
467.7572	345	254.4589	535.5481	395	291.3370
469.1130	346	255.1965	536.9039	396	292.0746
470.4688	347	255.9341	538.2598	397	292.8122
471.8247	348	256.6716	539.6156	398	293.5497
473.1805	349	257.4092	540.9714	399	294.2873
474.5363	350	258.1467	542.3272	400	295.0248

kN
(kN · m) kJ ←————— −3 ————— N
kW   J (N · m)
                                W

N · m ——————— +2 ——————→ N · cm

N · m J W	REF	lbf-ft ft-lbf ft-lbf/s	N · m J W	REF	lbf-ft ft-lbf ft-lbf/s
543·6830	401	295·7624	611·4739	451	332·6405
545·0388	402	296·5000	612·8297	452	333·3781
546·3947	403	297·2375	614·1856	453	334·1156
547·7505	404	297·9751	615·5414	454	334·8532
549·1063	405	298·7127	616·8972	455	335·5908
550·4621	406	299·4502	618·2530	456	336·3283
551·8179	407	300·1878	619·6088	457	337·0659
553·1737	408	300·9253	620·9646	458	337·8035
554·5296	409	301·6629	622·3205	459	338·5410
555·8854	410	302·4005	623·6763	460	339·2786
557·2412	411	303·1380	625·0321	461	340·0161
558·5970	412	303·8756	626·3879	462	340·7537
559·9528	413	304·6132	627·7437	463	341·4913
561·3087	414	305·3507	629·0996	464	342·2288
562·6645	415	306·0883	630·4554	465	342·9664
564·0203	416	306·8258	631·8112	466	343·7039
565·3761	417	307·5634	633·1670	467	344·4415
566·7319	418	308·3010	634·5228	468	345·1791
568·0877	419	309·0385	635·8786	469	345·9166
569·4436	420	309·7761	637·2345	470	346·6542
570·7994	421	310·5137	638·5903	471	347·3918
572·1552	422	311·2512	639·9461	472	348·1293
573·5110	423	311·9888	641·3019	473	348·8669
574·8668	424	312·7263	642·6577	474	349·6044
576·2226	425	313·4639	644·0135	475	350·3420
577·5785	426	314·2015	645·3694	476	351·0796
578·9343	427	314·9390	646·7252	477	351·8171
580·2901	428	315·6766	648·0810	478	352·5547
581·6459	429	316·4142	649·4368	479	353·2923
583·0017	430	317·1517	650·7926	480	354·0298
584·3576	431	317·8893	652·1485	481	354·7674
585·7134	432	318·6268	653·5043	482	355·5049
587·0692	433	319·3644	654·8601	483	356·2425
588·4250	434	320·1020	656·2159	484	356·9801
589·7808	435	320·8395	657·5717	485	357·7176
591·1367	436	321·5771	658·9276	486	358·4552
592·4925	437	322·3146	660·2834	487	359·1928
593·8483	438	323·0522	661·6392	488	359·9303
595·2041	439	323·7898	662·9950	489	360·6679
596·5599	440	324·5273	664·3508	490	361·4054
597·9157	441	325·2649	665·7066	491	362·1430
599·2716	442	326·0025	667·0625	492	362·8806
600·6274	443	326·7400	668·4183	493	363·6181
601·9832	444	327·4776	669·7741	494	364·3557
603·3390	445	328·2151	671·1299	495	365·0933
604·6948	446	328·9527	672·4857	496	365·8308
606·0506	447	329·6903	673·8415	497	366·5684
607·4065	448	330·4278	675·1974	498	367·3059
608·7623	449	331·1654	676·5532	499	368·0435
610·1181	450	331·9030	677·9090	500	368·7811

```
                    kN                              N
        (kN · m) kJ ─────────── +3 ──────────────▶ J (N · m)
                    kW                              W


            N · m ◀───────────── −2 ────────────── N · cm
```

N · m J W	REF	lbf-ft ft-lbf ft-lbf/s	N · m J W	REF	lbf-ft ft-lbf ft-lbf/s
679·2648	501	369·5186	747·0557	551	406·3967
680·6206	502	370·2562	748·4115	552	407·1343
681·9765	503	370·9937	749·7674	553	407·8719
683·3323	504	371·7313	751·1232	554	408·6094
684·6881	505	372·4689	752·4790	555	409·3470
686·0439	506	373·2064	753·8348	556	410·0845
687·3997	507	373·9440	755·1906	557	410·8221
688·7555	508	374·6816	756·5464	558	411·5597
690·1114	509	375·4191	757·9023	559	412·2972
691·4672	510	376·1567	759·2581	560	413·0348
692·8230	511	376·8942	760·6139	561	413·7724
694·1788	512	377·6318	761·9697	562	414·5099
695·5346	513	378·3694	763·3255	563	415·2475
696·8905	514	379·1069	764·6814	564	415·9850
698·2463	515	379·8445	766·0372	565	416·7226
699·6021	516	380·5821	767·3930	566	417·4602
700·9579	517	381·3196	768·7488	567	418·1977
702·3137	518	382·0572	770·1046	568	418·9353
703·6695	519	382·7947	771·4604	569	419·6728
705·0254	520	383·5323	772·8163	570	420·4104
706·3812	521	384·2699	774·1721	571	421·1480
707·7370	522	385·0074	775·5279	572	421·8855
709·0928	523	385·7450	776·8837	573	422·6231
710·4486	524	386·4826	778·2395	574	423·3607
711·8045	525	387·2201	779·5954	575	424·0982
713·1603	526	387·9577	780·9512	576	424·8358
714·5161	527	388·6952	782·3070	577	425·5733
715·8719	528	389·4328	783·6628	578	426·3109
717·2277	529	390·1704	785·0186	579	427·0485
718·5835	530	390·9079	786·3744	580	427·7860
719·9394	531	391·6455	787·7303	581	428·5236
721·2952	532	392·3830	789·0861	582	429·2612
722·6510	533	393·1206	790·4419	583	429·9987
724·0068	534	393·8582	791·7977	584	430·7363
725·3626	535	394·5957	793·1535	585	431·4738
726·7185	536	395·3333	794·5094	586	432·2114
728·0743	537	396·0709	795·8652	587	432·9490
729·4301	538	396·8084	797·2210	588	433·6865
730·7859	539	397·5460	798·5768	589	434·4241
732·1417	540	398·2835	799·9326	590	435·1617
733·4975	541	399·0211	801·2884	591	435·8992
734·8534	542	399·7587	802·6443	592	436·6368
736·2092	543	400·4962	804·0001	593	437·3743
737·5650	544	401·2338	805·3559	594	438·1119
738·9208	545	401·9714	806·7117	595	438·8495
740·2766	546	402·7089	808·0675	596	439·5870
741·6324	547	403·4465	809·4233	597	440·3246
742·9883	548	404·1840	810·7792	598	441·0621
744·3441	549	404·9216	812·1350	599	441·7997
745·6999	550	405·6592	813·4908	600	442·5373

```
              kN                              N
    (kN · m) kJ ◄──────── −3 ──────── J (N · m)
              kW                              W


      N · m ─────────── +2 ──────────► N · cm
```

N · m J W	REF	lbf·ft ft·lbf ft·lbf/s	N · m J W	REF	lbf·ft ft·lbf ft·lbf/s
814.8466	601	443.2748	882.6375	651	480.1529
816.2024	602	444.0124	883.9933	652	480.8905
817.5583	603	444.7500	885.3492	653	481.6281
818.9141	604	445.4875	886.7050	654	482.3656
820.2699	605	446.2251	888.0608	655	483.1032
821.6257	606	446.9626	889.4166	656	483.8408
822.9815	607	447.7002	890.7724	657	484.5783
824.3373	608	448.4378	892.1282	658	485.3159
825.6932	609	449.1753	893.4841	659	486.0534
827.0490	610	449.9129	894.8399	660	486.7910
828.4048	611	450.6505	896.1957	661	487.5286
829.7606	612	451.3880	897.5515	662	488.2661
831.1164	613	452.1256	898.9073	663	489.0037
832.4723	614	452.8631	900.2632	664	489.7412
833.8281	615	453.6007	901.6190	665	490.4788
835.1839	616	454.3383	902.9748	666	491.2164
836.5397	617	455.0758	904.3306	667	491.9539
837.8955	618	455.8134	905.6864	668	492.6915
839.2513	619	456.5510	907.0422	669	493.4291
840.6072	620	457.2885	908.3981	670	494.1666
841.9630	621	458.0261	909.7539	671	494.9042
843.3188	622	458.7636	911.1097	672	495.6417
844.6746	623	459.5012	912.4655	673	496.3793
846.0304	624	460.2388	913.8213	674	497.1169
847.3863	625	460.9763	915.1772	675	497.8544
848.7421	626	461.7139	916.5330	676	498.5920
850.0979	627	462.4515	917.8888	677	499.3296
851.4537	628	463.1890	919.2446	678	500.0671
852.8095	629	463.9266	920.6004	679	500.8047
854.1653	630	464.6641	921.9562	680	501.5422
855.5212	631	465.4017	923.3121	681	502.2798
856.8770	632	466.1393	924.6679	682	503.0174
858.2328	633	466.8768	926.0237	683	503.7549
859.5886	634	467.6144	927.3795	684	504.4925
860.9444	635	468.3519	928.7353	685	505.2301
862.3002	636	469.0895	930.0911	686	505.9676
863.6561	637	469.8271	931.4470	687	506.7052
865.0119	638	470.5646	932.8028	688	507.4427
866.3677	639	471.3022	934.1586	689	508.1803
867.7235	640	472.0398	935.5144	690	508.9179
869.0793	641	472.7773	936.8702	691	509.6554
870.4352	642	473.5149	938.2261	692	510.3930
871.7910	643	474.2524	939.5819	693	511.1306
873.1468	644	474.9900	940.9377	694	511.8681
874.5026	645	475.7276	942.2935	695	512.6057
875.8584	646	476.4651	943.6493	696	513.3432
877.2142	647	477.2027	945.0051	697	514.0808
878.5701	648	477.9403	946.3610	698	514.8184
879.9259	649	478.6778	947.7168	699	515.5559
881.2817	650	479.4154	949.0726	700	516.2935

```
          kN                                        N
(kN · m)  kJ  ─────────── +3 ───────────────→  J (N · m)
          kW                                        W


              N · m  ←─────────── −2 ─────────── N · cm
```

N · m J W	REF	lbf-ft ft-lbf ft-lbf/s	N · m J W	REF	lbf-ft ft-lbf ft-lbf/s
950.4284	701	517.0310	1018.219	751	553.9091
951.7842	702	517.7686	1019.575	752	554.6467
953.1401	703	518.5062	1020.931	753	555.3843
954.4959	704	519.2437	1022.287	754	556.1218
955.8517	705	519.9813	1023.643	755	556.8594
957.2075	706	520.7189	1024.998	756	557.5970
958.5633	707	521.4564	1026.354	757	558.3345
959.9191	708	522.1940	1027.710	758	559.0721
961.2750	709	522.9315	1029.066	759	559.8096
962.6308	710	523.6691	1030.422	760	560.5472
963.9866	711	524.4067	1031.777	761	561.2848
965.3424	712	525.1442	1033.133	762	562.0223
966.6982	713	525.8818	1034.489	763	562.7599
968.0541	714	526.6194	1035.845	764	563.4975
969.4099	715	527.3569	1037.201	765	564.2350
970.7657	716	528.0945	1038.557	766	564.9726
972.1215	717	528.8320	1039.912	767	565.7101
973.4773	718	529.5696	1041.268	768	566.4477
974.8331	719	530.3072	1042.624	769	567.1853
976.1890	720	531.0447	1043.980	770	567.9228
977.5448	721	531.7823	1045.336	771	568.6604
978.9006	722	532.5199	1046.691	772	569.3980
980.2564	723	533.2574	1048.047	773	570.1355
981.6122	724	533.9950	1049.403	774	570.8731
982.9681	725	534.7325	1050.759	775	571.6106
984.3239	726	535.4701	1052.115	776	572.3482
985.6797	727	536.2077	1053.471	777	573.0858
987.0355	728	536.9452	1054.826	778	573.8233
988.3913	729	537.6828	1056.182	779	574.5609
989.7471	730	538.4203	1057.538	780	575.2985
991.1030	731	539.1579	1058.894	781	576.0360
992.4588	732	539.8955	1060.250	782	576.7736
993.8146	733	540.6330	1061.605	783	577.5111
995.1704	734	541.3706	1062.961	784	578.2487
996.5262	735	542.1082	1064.317	785	578.9863
997.8820	736	542.8457	1065.673	786	579.7238
999.2379	737	543.5833	1067.029	787	580.4614
1000.594	738	544.3208	1068.385	788	581.1990
1001.950	739	545.0584	1069.740	789	581.9365
1003.305	740	545.7960	1071.096	790	582.6741
1004.661	741	546.5335	1072.452	791	583.4116
1006.017	742	547.2711	1073.808	792	584.1492
1007.373	743	548.0087	1075.164	793	584.8868
1008.729	744	548.7462	1076.520	794	585.6243
1010.084	745	549.4838	1077.875	795	586.3619
1011.440	746	550.2213	1079.231	796	587.0994
1012.796	747	550.9589	1080.587	797	587.8370
1014.152	748	551.6965	1081.943	798	588.5746
1015.508	749	552.4340	1083.299	799	589.3121
1016.864	750	553.1716	1084.654	800	590.0497

```
                    kN|                              |N
               (kN · m) kJ ◄─────────── −3 ──────────| J (N · m)
                    kW|                              |W


                  N · m |────────────── +2 ──────────────► N · cm
```

N · m J W	REF	lbf-ft ft-lbf ft-lbf/s	N · m J W	REF	lbf-ft ft-lbf ft-lbf/s
1086.010	801	590.7873	1153.801	851	627.6654
1087.366	802	591.5248	1155.157	852	628.4029
1088.722	803	592.2624	1156.513	853	629.1405
1090.078	804	592.9999	1157.869	854	629.8781
1091.433	805	593.7375	1159.224	855	630.6156
1092.789	806	594.4751	1160.580	856	631.3532
1094.145	807	595.2126	1161.936	857	632.0907
1095.501	808	595.9502	1163.292	858	632.8283
1096.857	809	596.6878	1164.648	859	633.5659
1098.213	810	597.4253	1166.003	860	634.3034
1099.568	811	598.1629	1167.359	861	635.0410
1100.924	812	598.9004	1168.715	862	635.7785
1102.280	813	599.6380	1170.071	863	636.5161
1103.636	814	600.3756	1171.427	864	637.2537
1104.992	815	601.1131	1172.783	865	637.9912
1106.347	816	601.8507	1174.138	866	638.7288
1107.703	817	602.5882	1175.494	867	639.4664
1109.059	818	603.3258	1176.850	868	640.2039
1110.415	819	604.0634	1178.206	869	640.9415
1111.771	820	604.8009	1179.562	870	641.6790
1113.127	821	605.5385	1180.917	871	642.4166
1114.482	822	606.2761	1182.273	872	643.1542
1115.838	823	607.0136	1183.629	873	643.8917
1117.194	824	607.7512	1184.985	874	644.6293
1118.550	825	608.4887	1186.341	875	645.3669
1119.906	826	609.2263	1187.697	876	646.1044
1121.261	827	609.9639	1189.052	877	646.8420
1122.617	828	610.7014	1190.408	878	647.5795
1123.973	829	611.4390	1191.764	879	648.3171
1125.329	830	612.1766	1193.120	880	649.0547
1126.685	831	612.9141	1194.476	881	649.7922
1128.041	832	613.6517	1195.831	882	650.5298
1129.396	833	614.3892	1197.187	883	651.2673
1130.752	834	615.1268	1198.543	884	652.0049
1132.108	835	615.8644	1199.899	885	652.7425
1133.464	836	616.6019	1201.255	886	653.4800
1134.820	837	617.3395	1202.611	887	654.2176
1136.175	838	618.0771	1203.966	888	654.9552
1137.531	839	618.8146	1205.322	889	655.6927
1138.887	840	619.5522	1206.678	890	656.4303
1140.243	841	620.2897	1208.034	891	657.1678
1141.599	842	621.0273	1209.390	892	657.9054
1142.955	843	621.7649	1210.745	893	658.6430
1144.310	844	622.5024	1212.101	894	659.3805
1145.666	845	623.2400	1213.457	895	660.1181
1147.022	846	623.9776	1214.813	896	660.8557
1148.378	847	624.7151	1216.169	897	661.5932
1149.734	848	625.4527	1217.525	898	662.3308
1151.089	849	626.1902	1218.880	899	663.0683
1152.445	850	626.9278	1220.236	900	663.8059

```
        kN                                      N
(kN · m) kJ ─────────── +3 ──────────▶ J (N · m)
        kW                                      W


         N · m ◀────────── −2 ──────────── N · cm
```

N · m J W	REF	lbf-ft ft-lbf ft-lbf/s	N · m J W	REF	lbf-ft ft-lbf ft-lbf/s
1221.592	901	664.5435	1289.383	951	701.4216
1222.948	902	665.2810	1290.739	952	702.1591
1224.304	903	666.0186	1292.095	953	702.8967
1225.659	904	666.7562	1293.450	954	703.6343
1227.015	905	667.4937	1294.806	955	704.3718
1228.371	906	668.2313	1296.162	956	705.1094
1229.727	907	668.9688	1297.518	957	705.8469
1231.083	908	669.7064	1298.874	958	706.5845
1232.439	909	670.4440	1300.229	959	707.3221
1233.794	910	671.1815	1301.585	960	708.0596
1235.150	911	671.9191	1302.941	961	708.7972
1236.506	912	672.6567	1304.297	962	709.5348
1237.862	913	673.3942	1305.653	963	710.2723
1239.218	914	674.1318	1307.009	964	711.0099
1240.573	915	674.8693	1308.364	965	711.7474
1241.929	916	675.6069	1309.720	966	712.4850
1243.285	917	676.3445	1311.076	967	713.2226
1244.641	918	677.0820	1312.432	968	713.9601
1245.997	919	677.8196	1313.788	969	714.6977
1247.353	920	678.5572	1315.143	970	715.4353
1248.708	921	679.2947	1316.499	971	716.1728
1250.064	922	680.0323	1317.855	972	716.9104
1251.420	923	680.7698	1319.211	973	717.6479
1252.776	924	681.5074	1320.567	974	718.3855
1254.132	925	682.2450	1321.923	975	719.1231
1255.487	926	682.9825	1323.278	976	719.8606
1256.843	927	683.7201	1324.634	977	720.5982
1258.199	928	684.4576	1325.990	978	721.3358
1259.555	929	685.1952	1327.346	979	722.0733
1260.911	930	685.9328	1328.702	980	722.8109
1262.267	931	686.6703	1330.057	981	723.5484
1263.622	932	687.4079	1331.413	982	724.2860
1264.978	933	688.1455	1332.769	983	725.0236
1266.334	934	688.8830	1334.125	984	725.7611
1267.690	935	689.6206	1335.481	985	726.4987
1269.046	936	690.3581	1336.837	986	727.2363
1270.401	937	691.0957	1338.192	987	727.9738
1271.757	938	691.8333	1339.548	988	728.7114
1273.113	939	692.5708	1340.904	989	729.4489
1274.469	940	693.3084	1342.260	990	730.1865
1275.825	941	694.0460	1343.616	991	730.9241
1277.181	942	694.7835	1344.971	992	731.6616
1278.536	943	695.5211	1346.327	993	732.3992
1279.892	944	696.2586	1347.683	994	733.1367
1281.248	945	696.9962	1349.039	995	733.8743
1282.604	946	697.7338	1350.395	996	734.6119
1283.960	947	698.4713	1351.751	997	735.3494
1285.315	948	699.2089	1353.106	998	736.0870
1286.671	949	699.9464	1354.462	999	736.8246
1288.027	950	700.6840	1355.818	1000	737.5621

32

PRESSURE AND STRESS

Pound (force)/Square-inch — Kilopascal (or Kilonewton/Square-metre)

$\quad$ (lbf/in.^{2} — kPa — kN/m^2)

Pound (force)/Square-inch — Pascal (or Newton/Square-metre)

$\quad$ (lbf/in.^{2} — Pa — N/m^2)

Pound (force)/Square-inch — Newton/Square-centimetre

$\quad$ (lbf/in.^{2} — N/cm^2)

Kilopound (force)/Square-inch — Kilopascal (or Kilonewton/Square-metre)

$\quad$ (kip/in.^{2} — kPa — kN/m^2)

Kilopound (force)/Square-inch — Pascal (or Kilonewton/Square-metre)

$\quad$ (kip/in.^{2} — Pa — N/m^2)

Kilopound (force)/Square-inch — Newton/Square-centimetre

$\quad$ (kip/in.^{2} — N/cm^2)

Conversion Factors:

$\quad$ 1 pound (force)/square-inch (lbf/in.^{2} or psi) = 6.894757 kilopascal (kPa)

$\quad$ (1 kPa = 0.1450377 lbf/in.^{2})

Pound/Square-inch (psi) and Kilopound/Square-inch (kpsi) Relation

The kilopound per square inch equals 1000 pounds per square inch. The following Diagram 32-A, describes the decimal-shift relations between psi and kpsi when dealing with decimal values.

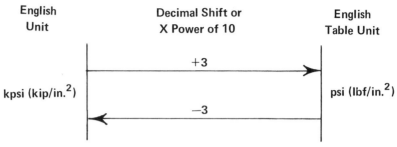

Diagram 32-A

Example (a): Convert 48 kpsi to kPa.
    From Diagram 32-A: 48 kpsi = 48 × 10³ psi
    From table: 48 psi = 330.9483 kPa*
    Therefore: 48 × 10³ psi = 330.9483 × 10³ kPa ≈ 331,000 kPa

In converting kPa values to Pa and N/cm², the Decimal-shift Diagram 32-B, which appears in the headings on alternate table pages, will assist table users. It is derived from: 1 kPa = (10³) Pa = (10⁻¹) N/cm³.

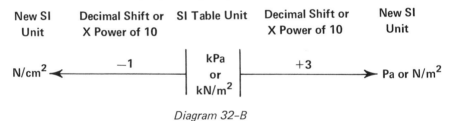

Diagram 32-B

Example (b): Convert 275 psi to kPa, Pa, and N/cm².
    From table: 275 psi = 1,896.058 kPa* ≈ 1,900 kPa
    From Diagram 32-B: 1) 1,896.058 kPa = 1,896.058 × 10³ Pa
                                = 1,896,058 Pa ≈ 1,900 × 10³ Pa
                                (kPa to Pa by multiplying kPa value by 10³ or by shifting its decimal 3 places to the right.)

                        2) 1,896,058 kPa = 1,896.058 × 10⁻¹ N/cm²
                                = 189.6058 N/cm² ≈ 190 N/cm²
                                (kPa to N/cm² by multiplying kPa value by 10⁻¹ or by shifting its decimal 1 place to the left.)

---

*All conversion values obtained can be rounded to desired number of significant figures.

Diagram 32-C is also given at the top of alternate table pages:

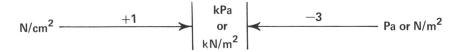

Diagram 32-C

It is the opposite of Diagram 32-B and is used to convert a quantity in N/cm^2 or Pa to kPa. The kPa equivalent is then applied to the REF column in the table to obtain psi.

Example (*c*): Convert 17,300 Pa to psi.
    From Diagram 32-C: 17,300 Pa $= 17,300 \times 10^{-3}$ kPa $= 173 \times 10^{-1}$ kPa
    From table: 173 kPa $= 25.09153$ psi*
    Therefore: 173×10^{-1} kPa $= 25.09153 \times 10^{-1}$ psi $= 2.509153$ psi
                 ≈ 2.51 psi.

---

*All conversion values obtained can be rounded to desired number of significant figures.

N/cm² ←――――― −1 | kPa or kN/m² | +3 ―――――→ Pa or N/m²

kPa kN/m²	REF	lbf/in²	kPa kN/m²	REF	lbf/in²
6.894757	1	0.1450377	351.6326	51	7.396925
13.78951	2	0.2900755	358.5274	52	7.541963
20.68427	3	0.4351132	365.4221	53	7.687000
27.57903	4	0.5801510	372.3169	54	7.832038
34.47378	5	0.7251887	379.2116	55	7.977076
41.36854	6	0.8702265	386.1064	56	8.122114
48.26330	7	1.015264	393.0011	57	8.267151
55.15806	8	1.160302	399.8959	58	8.412189
62.05281	9	1.305340	406.7907	59	8.557227
68.94757	10	1.450377	413.6854	60	8.702265
75.84233	11	1.595415	420.5802	61	8.847302
82.73708	12	1.740453	427.4749	62	8.992340
89.63184	13	1.885491	434.3697	63	9.137378
96.52660	14	2.030528	441.2644	64	9.282416
103.4214	15	2.175566	448.1592	65	9.427453
110.3161	16	2.320604	455.0540	66	9.572491
117.2109	17	2.465642	461.9487	67	9.717529
124.1056	18	2.610679	468.8435	68	9.862567
131.0004	19	2.755717	475.7382	69	10.00760
137.8951	20	2.900755	482.6330	70	10.15264
144.7899	21	3.045793	489.5277	71	10.29768
151.6847	22	3.190830	496.4225	72	10.44272
158.5794	23	3.335868	503.3173	73	10.58776
165.4742	24	3.480906	510.2120	74	10.73279
172.3689	25	3.625944	517.1068	75	10.87783
179.2637	26	3.770981	524.0015	76	11.02287
186.1584	27	3.916019	530.8963	77	11.16791
193.0532	28	4.061057	537.7910	78	11.31294
199.9480	29	4.206095	544.6858	79	11.45798
206.8427	30	4.351132	551.5806	80	11.60302
213.7375	31	4.496170	558.4753	81	11.74806
220.6322	32	4.641208	565.3701	82	11.89310
227.5270	33	4.786246	572.2648	83	12.03813
234.4217	34	4.931283	579.1596	84	12.18317
241.3165	35	5.076321	586.0543	85	12.32821
248.2113	36	5.221359	592.9491	86	12.47325
255.1060	37	5.366397	599.8439	87	12.61828
262.0008	38	5.511434	606.7386	88	12.76332
268.8955	39	5.656472	613.6334	89	12.90836
275.7903	40	5.801510	620.5281	90	13.05340
282.6850	41	5.946548	627.4229	91	13.19843
289.5798	42	6.091585	634.3176	92	13.34347
296.4745	43	6.236623	641.2124	93	13.48851
303.3693	44	6.381661	648.1072	94	13.63355
310.2641	45	6.526698	655.0019	95	13.77859
317.1588	46	6.671736	661.8967	96	13.92362
324.0536	47	6.816774	668.7914	97	14.06866
330.9483	48	6.961812	675.6862	98	14.21370
337.8431	49	7.106849	682.5809	99	14.35874
344.7379	50	7.251887	689.4757	100	14.50377

$$N/cm^2 \xrightarrow{\quad +1 \quad} \begin{array}{c} kPa \\ or \\ kN/m^2 \end{array} \xleftarrow{\quad -3 \quad} Pa \text{ or } N/m^2$$

kPa kN/m²	REF	lbf/in²	kPa kN/m²	REF	lbf/in²
696.3705	101	14.64881	1041.108	151	21.90070
703.2652	102	14.79385	1048.003	152	22.04574
710.1600	103	14.93889	1054.898	153	22.19077
717.0547	104	15.08393	1061.793	154	22.33581
723.9495	105	15.22896	1068.687	155	22.48085
730.8442	106	15.37400	1075.582	156	22.62589
737.7390	107	15.51904	1082.477	157	22.77093
744.6338	108	15.66408	1089.372	158	22.91596
751.5285	109	15.80911	1096.266	159	23.06100
758.4233	110	15.95415	1103.161	160	23.20604
765.3180	111	16.09919	1110.056	161	23.35108
772.2128	112	16.24423	1116.951	162	23.49611
779.1075	113	16.38927	1123.845	163	23.64115
786.0023	114	16.53430	1130.740	164	23.78619
792.8970	115	16.67934	1137.635	165	23.93123
799.7918	116	16.82438	1144.530	166	24.07627
806.6866	117	16.96942	1151.424	167	24.22130
813.5813	118	17.11445	1158.319	168	24.36634
820.4761	119	17.25949	1165.214	169	24.51138
827.3708	120	17.40453	1172.109	170	24.65642
834.2656	121	17.54957	1179.003	171	24.80145
841.1603	122	17.69460	1185.898	172	24.94649
848.0551	123	17.83964	1192.793	173	25.09153
854.9499	124	17.98468	1199.688	174	25.23657
861.8446	125	18.12972	1206.582	175	25.38161
868.7394	126	18.27476	1213.477	176	25.52664
875.6341	127	18.41979	1220.372	177	25.67168
882.5289	128	18.56483	1227.267	178	25.81672
889.4237	129	18.70987	1234.161	179	25.96176
896.3184	130	18.85491	1241.056	180	26.10679
903.2132	131	18.99994	1247.951	181	26.25183
910.1079	132	19.14498	1254.846	182	26.39687
917.0027	133	19.29002	1261.741	183	26.54191
923.8974	134	19.43506	1268.635	184	26.68694
930.7922	135	19.58010	1275.530	185	26.83198
937.6870	136	19.72513	1282.425	186	26.97702
944.5817	137	19.87017	1289.320	187	27.12206
951.4765	138	20.01521	1296.214	188	27.26710
958.3712	139	20.16025	1303.109	189	27.41213
965.2660	140	20.30528	1310.004	190	27.55717
972.1607	141	20.45032	1316.899	191	27.70221
979.0555	142	20.59536	1323.793	192	27.84725
985.9502	143	20.74040	1330.688	193	27.99228
992.8450	144	20.88544	1337.583	194	28.13732
999.7398	145	21.03047	1344.478	195	28.28236
1006.635	146	21.17551	1351.372	196	28.42740
1013.529	147	21.32055	1358.267	197	28.57244
1020.424	148	21.46559	1365.162	198	28.71747
1027.319	149	21.61062	1372.057	199	28.86251
1034.214	150	21.75566	1378.951	200	29.00755

| N/cm² | ← | −1 | | kPa or kN/m² | | +3 | → | Pa or N/m² |

kPa kN/m²	REF	lbf/in²		kPa kN/m²	REF	lbf/in²
1385.846	201	29.15259		1730.584	251	36.40447
1392.741	202	29.29762		1737.479	252	36.54951
1399.636	203	29.44266		1744.374	253	36.69455
1406.530	204	29.58770		1751.268	254	36.83959
1413.425	205	29.73274		1758.163	255	36.98462
1420.320	206	29.87778		1765.058	256	37.12966
1427.215	207	30.02281		1771.953	257	37.27470
1434.109	208	30.16785		1778.847	258	37.41974
1441.004	209	30.31289		1785.742	259	37.56478
1447.899	210	30.45793		1792.637	260	37.70981
1454.794	211	30.60296		1799.532	261	37.85485
1461.688	212	30.74800		1806.426	262	37.99989
1468.583	213	30.89304		1813.321	263	38.14493
1475.478	214	31.03808		1820.216	264	38.28996
1482.373	215	31.18312		1827.111	265	38.43500
1489.268	216	31.32815		1834.005	266	38.58004
1496.162	217	31.47319		1840.900	267	38.72508
1503.057	218	31.61823		1847.795	268	38.87012
1509.952	219	31.76327		1854.690	269	39.01515
1516.847	220	31.90830		1861.584	270	39.16019
1523.741	221	32.05334		1868.479	271	39.30523
1530.636	222	32.19838		1875.374	272	39.45027
1537.531	223	32.34342		1882.269	273	39.59530
1544.426	224	32.48845		1889.163	274	39.74034
1551.320	225	32.63349		1896.058	275	39.88538
1558.215	226	32.77853		1902.953	276	40.03042
1565.110	227	32.92357		1909.848	277	40.17546
1572.005	228	33.06861		1916.742	278	40.32049
1578.899	229	33.21364		1923.637	279	40.46553
1585.794	230	33.35868		1930.532	280	40.61057
1592.689	231	33.50372		1937.427	281	40.75561
1599.584	232	33.64876		1944.321	282	40.90064
1606.478	233	33.79379		1951.216	283	41.04568
1613.373	234	33.93883		1958.111	284	41.19072
1620.268	235	34.08387		1965.006	285	41.33576
1627.163	236	34.22891		1971.900	286	41.48079
1634.057	237	34.37395		1978.795	287	41.62583
1640.952	238	34.51898		1985.690	288	41.77087
1647.847	239	34.66402		1992.585	289	41.91591
1654.742	240	34.80906		1999.480	290	42.06095
1661.636	241	34.95410		2006.374	291	42.20598
1668.531	242	35.09913		2013.269	292	42.35102
1675.426	243	35.24417		2020.164	293	42.49606
1682.321	244	35.38921		2027.059	294	42.64110
1689.215	245	35.53425		2033.953	295	42.78613
1696.110	246	35.67929		2040.848	296	42.93117
1703.005	247	35.82432		2047.743	297	43.07621
1709.900	248	35.96936		2054.638	298	43.22125
1716.794	249	36.11440		2061.532	299	43.36629
1723.689	250	36.25944		2068.427	300	43.51132

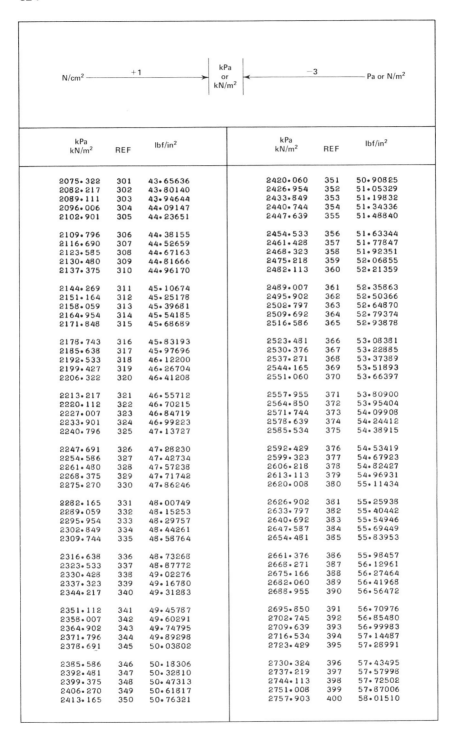

$N/cm^2 \xrightarrow{\quad +1 \quad} \begin{array}{c} kPa \\ or \\ kN/m^2 \end{array} \xleftarrow{\quad -3 \quad} Pa\ or\ N/m^2$

kPa kN/m²	REF	lbf/in²	kPa kN/m²	REF	lbf/in²
2075.322	301	43.65636	2420.060	351	50.90825
2082.217	302	43.80140	2426.954	352	51.05329
2089.111	303	43.94644	2433.849	353	51.19832
2096.006	304	44.09147	2440.744	354	51.34336
2102.901	305	44.23651	2447.639	355	51.48840
2109.796	306	44.38155	2454.533	356	51.63344
2116.690	307	44.52659	2461.428	357	51.77847
2123.585	308	44.67163	2468.323	358	51.92351
2130.480	309	44.81666	2475.218	359	52.06855
2137.375	310	44.96170	2482.113	360	52.21359
2144.269	311	45.10674	2489.007	361	52.35863
2151.164	312	45.25178	2495.902	362	52.50366
2158.059	313	45.39681	2502.797	363	52.64870
2164.954	314	45.54185	2509.692	364	52.79374
2171.848	315	45.68689	2516.586	365	52.93878
2178.743	316	45.83193	2523.481	366	53.08381
2185.638	317	45.97696	2530.376	367	53.22885
2192.533	318	46.12200	2537.271	368	53.37389
2199.427	319	46.26704	2544.165	369	53.51893
2206.322	320	46.41208	2551.060	370	53.66397
2213.217	321	46.55712	2557.955	371	53.80900
2220.112	322	46.70215	2564.850	372	53.95404
2227.007	323	46.84719	2571.744	373	54.09908
2233.901	324	46.99223	2578.639	374	54.24412
2240.796	325	47.13727	2585.534	375	54.38915
2247.691	326	47.28230	2592.429	376	54.53419
2254.586	327	47.42734	2599.323	377	54.67923
2261.480	328	47.57238	2606.218	378	54.82427
2268.375	329	47.71742	2613.113	379	54.96931
2275.270	330	47.86246	2620.008	380	55.11434
2282.165	331	48.00749	2626.902	381	55.25938
2289.059	332	48.15253	2633.797	382	55.40442
2295.954	333	48.29757	2640.692	383	55.54946
2302.849	334	48.44261	2647.587	384	55.69449
2309.744	335	48.58764	2654.481	385	55.83953
2316.638	336	48.73268	2661.376	386	55.98457
2323.533	337	48.87772	2668.271	387	56.12961
2330.428	338	49.02276	2675.166	388	56.27464
2337.323	339	49.16780	2682.060	389	56.41968
2344.217	340	49.31283	2688.955	390	56.56472
2351.112	341	49.45787	2695.850	391	56.70976
2358.007	342	49.60291	2702.745	392	56.85480
2364.902	343	49.74795	2709.639	393	56.99983
2371.796	344	49.89298	2716.534	394	57.14487
2378.691	345	50.03802	2723.429	395	57.28991
2385.586	346	50.18306	2730.324	396	57.43495
2392.481	347	50.32810	2737.219	397	57.57998
2399.375	348	50.47313	2744.113	398	57.72502
2406.270	349	50.61817	2751.008	399	57.87006
2413.165	350	50.76321	2757.903	400	58.01510

N/cm² ← ——— −1 ——— | kPa or kN/m² | ——— +3 ——→ Pa or N/m²

kPa kN/m²	REF	lbf/in²	kPa kN/m²	REF	lbf/in²
2764.798	401	58.16014	3109.535	451	65.41202
2771.692	402	58.30517	3116.430	452	65.55706
2778.587	403	58.45021	3123.325	453	65.70210
2785.482	404	58.59525	3130.220	454	65.84714
2792.377	405	58.74029	3137.114	455	65.99217
2799.271	406	58.88532	3144.009	456	66.13721
2806.166	407	59.03036	3150.904	457	66.28225
2813.061	408	59.17540	3157.799	458	66.42729
2819.956	409	59.32044	3164.693	459	66.57232
2826.850	410	59.46548	3171.588	460	66.71736
2833.745	411	59.61051	3178.483	461	66.86240
2840.640	412	59.75555	3185.378	462	67.00744
2847.535	413	59.90059	3192.272	463	67.15248
2854.429	414	60.04563	3199.167	464	67.29751
2861.324	415	60.19066	3206.062	465	67.44255
2868.219	416	60.33570	3212.957	466	67.58759
2875.114	417	60.48074	3219.852	467	67.73263
2882.008	418	60.62578	3226.746	468	67.87766
2888.903	419	60.77081	3233.641	469	68.02270
2895.798	420	60.91585	3240.536	470	68.16774
2902.693	421	61.06089	3247.431	471	68.31278
2909.587	422	61.20593	3254.325	472	68.45782
2916.482	423	61.35097	3261.220	473	68.60285
2923.377	424	61.49600	3268.115	474	68.74789
2930.272	425	61.64104	3275.010	475	68.89293
2937.166	426	61.78608	3281.904	476	69.03797
2944.061	427	61.93112	3288.799	477	69.18300
2950.956	428	62.07615	3295.694	478	69.32804
2957.851	429	62.22119	3302.589	479	69.47308
2964.745	430	62.36623	3309.483	480	69.61812
2971.640	431	62.51127	3316.378	481	69.76315
2978.535	432	62.65631	3323.273	482	69.90819
2985.430	433	62.80134	3330.168	483	70.05323
2992.325	434	62.94638	3337.062	484	70.19827
2999.219	435	63.09142	3343.957	485	70.34331
3006.114	436	63.23646	3350.852	486	70.48834
3013.009	437	63.38149	3357.747	487	70.63338
3019.904	438	63.52653	3364.641	488	70.77842
3026.798	439	63.67157	3371.536	489	70.92346
3033.693	440	63.81661	3378.431	490	71.06849
3040.588	441	63.96165	3385.326	491	71.21353
3047.483	442	64.10668	3392.220	492	71.35857
3054.377	443	64.25172	3399.115	493	71.50361
3061.272	444	64.39676	3406.010	494	71.64865
3068.167	445	64.54180	3412.905	495	71.79368
3075.062	446	64.68683	3419.799	496	71.93872
3081.956	447	64.83187	3426.694	497	72.08376
3088.851	448	64.97691	3433.589	498	72.22880
3095.746	449	65.12195	3440.484	499	72.37383
3102.641	450	65.26698	3447.378	500	72.51887

N/cm² ——————— +1 —————→ | kPa or kN/m² | ←————— −3 ————— Pa or N/m²

kPa kN/m²	REF	lbf/in²	kPa kN/m²	REF	lbf/in²
3454•273	501	72•66391	3799•011	551	79•91580
3461•168	502	72•80895	3805•906	552	80•06083
3468•063	503	72•95399	3812•801	553	80•20587
3474•958	504	73•09902	3819•695	554	80•35091
3481•852	505	73•24406	3826•590	555	80•49595
3488•747	506	73•38910	3833•485	556	80•64099
3495•642	507	73•53414	3840•380	557	80•78602
3502•537	508	73•67917	3847•274	558	80•93106
3509•431	509	73•82421	3854•169	559	81•07610
3516•326	510	73•96925	3861•064	560	81•22114
3523•221	511	74•11429	3867•959	561	81•36617
3530•116	512	74•25933	3874•853	562	81•51121
3537•010	513	74•40436	3881•748	563	81•65625
3543•905	514	74•54940	3888•643	564	81•80129
3550•800	515	74•69444	3895•538	565	81•94633
3557•695	516	74•83948	3902•432	566	82•09136
3564•589	517	74•98451	3909•327	567	82•23640
3571•484	518	75•12955	3916•222	568	82•38144
3578•379	519	75•27459	3923•117	569	82•52648
3585•274	520	75•41963	3930•011	570	82•67151
3592•168	521	75•56466	3936•906	571	82•81655
3599•063	522	75•70970	3943•801	572	82•96159
3605•958	523	75•85474	3950•696	573	83•10663
3612•853	524	75•99978	3957•591	574	83•25167
3619•747	525	76•14482	3964•485	575	83•39670
3626•642	526	76•28985	3971•380	576	83•54174
3633•537	527	76•43489	3978•275	577	83•68678
3640•432	528	76•57993	3985•170	578	83•83182
3647•326	529	76•72497	3992•064	579	83•97685
3654•221	530	76•87000	3998•959	580	84•12189
3661•116	531	77•01504	4005•854	581	84•26693
3668•011	532	77•16008	4012•749	582	84•41197
3674•905	533	77•30512	4019•643	583	84•55700
3681•800	534	77•45016	4026•538	584	84•70204
3688•695	535	77•59519	4033•433	585	84•84708
3695•590	536	77•74023	4040•328	586	84•99212
3702•484	537	77•88527	4047•222	587	85•13716
3709•379	538	78•03031	4054•117	588	85•28219
3716•274	539	78•17534	4061•012	589	85•42723
3723•169	540	78•32038	4067•907	590	85•57227
3730•064	541	78•46542	4074•801	591	85•71731
3736•958	542	78•61046	4081•696	592	85•86234
3743•853	543	78•75550	4088•591	593	86•00738
3750•748	544	78•90053	4095•486	594	86•15242
3757•643	545	79•04557	4102•380	595	86•29746
3764•537	546	79•19061	4109•275	596	86•44250
3771•432	547	79•33565	4116•170	597	86•58753
3778•327	548	79•48068	4123•065	598	86•73257
3785•222	549	79•62572	4129•959	599	86•87761
3792•116	550	79•77076	4136•854	600	87•02265

kPa kN/m²	REF	lbf/in²	kPa kN/m²	REF	lbf/in²
4143.749	601	87.16768	4488.487	651	94.41957
4150.644	602	87.31272	4495.382	652	94.56461
4157.538	603	87.45776	4502.276	653	94.70965
4164.433	604	87.60280	4509.171	654	94.85468
4171.328	605	87.74784	4516.066	655	94.99972
4178.223	606	87.89287	4522.961	656	95.14476
4185.117	607	88.03791	4529.855	657	95.28980
4192.012	608	88.18295	4536.750	658	95.43484
4198.907	609	88.32799	4543.645	659	95.57987
4205.802	610	88.47302	4550.540	660	95.72491
4212.697	611	88.61806	4557.434	661	95.86995
4219.591	612	88.76310	4564.329	662	96.01499
4226.486	613	88.90814	4571.224	663	96.16002
4233.381	614	89.05317	4578.119	664	96.30506
4240.276	615	89.19821	4585.013	665	96.45010
4247.170	616	89.34325	4591.908	666	96.59514
4254.065	617	89.48829	4598.803	667	96.74018
4260.960	618	89.63333	4605.698	668	96.88521
4267.855	619	89.77836	4612.592	669	97.03025
4274.749	620	89.92340	4619.487	670	97.17529
4281.644	621	90.06844	4626.382	671	97.32033
4288.539	622	90.21348	4633.277	672	97.46536
4295.434	623	90.35851	4640.171	673	97.61040
4302.328	624	90.50355	4647.066	674	97.75544
4309.223	625	90.64859	4653.961	675	97.90048
4316.118	626	90.79363	4660.856	676	98.04552
4323.013	627	90.93867	4667.750	677	98.19055
4329.907	628	91.08370	4674.645	678	98.33559
4336.802	629	91.22874	4681.540	679	98.48063
4343.697	630	91.37378	4688.435	680	98.62567
4350.592	631	91.51882	4695.330	681	98.77070
4357.486	632	91.66385	4702.224	682	98.91574
4364.381	633	91.80889	4709.119	683	99.06078
4371.276	634	91.95393	4716.014	684	99.20582
4378.171	635	92.09897	4722.909	685	99.35085
4385.065	636	92.24401	4729.803	686	99.49589
4391.960	637	92.38904	4736.698	687	99.64093
4398.855	638	92.53408	4743.593	688	99.78597
4405.750	639	92.67912	4750.488	689	99.93101
4412.644	640	92.82416	4757.382	690	100.0760
4419.539	641	92.96919	4764.277	691	100.2211
4426.434	642	93.11423	4771.172	692	100.3661
4433.329	643	93.25927	4778.067	693	100.5112
4440.224	644	93.40431	4784.961	694	100.6562
4447.118	645	93.54935	4791.856	695	100.8012
4454.013	646	93.69438	4798.751	696	100.9463
4460.908	647	93.83942	4805.646	697	101.0913
4467.802	648	93.98446	4812.540	698	101.2363
4474.697	649	94.12950	4819.435	699	101.3814
4481.592	650	94.27453	4826.330	700	101.5264

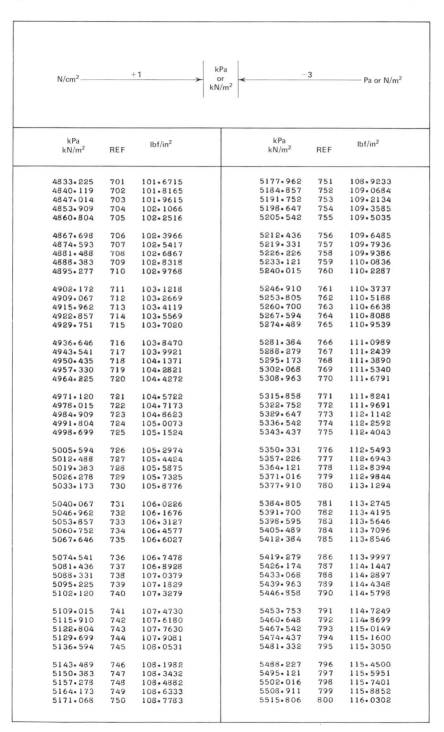

kPa kN/m²	REF	lbf/in²	kPa kN/m²	REF	lbf/in²
4833.225	701	101.6715	5177.962	751	108.9233
4840.119	702	101.8165	5184.857	752	109.0684
4847.014	703	101.9615	5191.752	753	109.2134
4853.909	704	102.1066	5198.647	754	109.3585
4860.804	705	102.2516	5205.542	755	109.5035
4867.698	706	102.3966	5212.436	756	109.6485
4874.593	707	102.5417	5219.331	757	109.7936
4881.488	708	102.6867	5226.226	758	109.9386
4888.383	709	102.8318	5233.121	759	110.0836
4895.277	710	102.9768	5240.015	760	110.2287
4902.172	711	103.1218	5246.910	761	110.3737
4909.067	712	103.2669	5253.805	762	110.5188
4915.962	713	103.4119	5260.700	763	110.6638
4922.857	714	103.5569	5267.594	764	110.8088
4929.751	715	103.7020	5274.489	765	110.9539
4936.646	716	103.8470	5281.384	766	111.0989
4943.541	717	103.9921	5288.279	767	111.2439
4950.435	718	104.1371	5295.173	768	111.3890
4957.330	719	104.2821	5302.068	769	111.5340
4964.225	720	104.4272	5308.963	770	111.6791
4971.120	721	104.5722	5315.858	771	111.8241
4978.015	722	104.7173	5322.752	772	111.9691
4984.909	723	104.8623	5329.647	773	112.1142
4991.804	724	105.0073	5336.542	774	112.2592
4998.699	725	105.1524	5343.437	775	112.4043
5005.594	726	105.2974	5350.331	776	112.5493
5012.488	727	105.4424	5357.226	777	112.6943
5019.383	728	105.5875	5364.121	778	112.8394
5026.278	729	105.7325	5371.016	779	112.9844
5033.173	730	105.8776	5377.910	780	113.1294
5040.067	731	106.0226	5384.805	781	113.2745
5046.962	732	106.1676	5391.700	782	113.4195
5053.857	733	106.3127	5398.595	783	113.5646
5060.752	734	106.4577	5405.489	784	113.7096
5067.646	735	106.6027	5412.384	785	113.8546
5074.541	736	106.7478	5419.279	786	113.9997
5081.436	737	106.8928	5426.174	787	114.1447
5088.331	738	107.0379	5433.068	788	114.2897
5095.225	739	107.1829	5439.963	789	114.4348
5102.120	740	107.3279	5446.858	790	114.5798
5109.015	741	107.4730	5453.753	791	114.7249
5115.910	742	107.6180	5460.648	792	114.8699
5122.804	743	107.7630	5467.542	793	115.0149
5129.699	744	107.9081	5474.437	794	115.1600
5136.594	745	108.0531	5481.332	795	115.3050
5143.489	746	108.1982	5488.227	796	115.4500
5150.383	747	108.3432	5495.121	797	115.5951
5157.278	748	108.4882	5502.016	798	115.7401
5164.173	749	108.6333	5508.911	799	115.8852
5171.068	750	108.7783	5515.806	800	116.0302

| N/cm² | ← | −1 | kPa or kN/m² | +3 | → | Pa or N/m² |

kPa kN/m²	REF	lbf/in²	kPa kN/m²	REF	lbf/in²
5522.700	801	116.1752	5867.438	851	123.4271
5529.595	802	116.3203	5874.333	852	123.5722
5536.490	803	116.4653	5881.228	853	123.7172
5543.385	804	116.6103	5888.122	854	123.8622
5550.279	805	116.7554	5895.017	855	124.0073
5557.174	806	116.9004	5901.912	856	124.1523
5564.069	807	117.0455	5908.807	857	124.2973
5570.964	808	117.1905	5915.701	858	124.4424
5577.858	809	117.3355	5922.596	859	124.5874
5584.753	810	117.4806	5929.491	860	124.7325
5591.648	811	117.6256	5936.386	861	124.8775
5598.543	812	117.7706	5943.281	862	125.0225
5605.437	813	117.9157	5950.175	863	125.1676
5612.332	814	118.0607	5957.070	864	125.3126
5619.227	815	118.2058	5963.965	865	125.4576
5626.122	816	118.3508	5970.860	866	125.6027
5633.016	817	118.4958	5977.754	867	125.7477
5639.911	818	118.6409	5984.649	868	125.8928
5646.806	819	118.7859	5991.544	869	126.0378
5653.701	820	118.9310	5998.439	870	126.1828
5660.595	821	119.0760	6005.333	871	126.3279
5667.490	822	119.2210	6012.228	872	126.4729
5674.385	823	119.3661	6019.123	873	126.6180
5681.280	824	119.5111	6026.018	874	126.7630
5688.174	825	119.6561	6032.912	875	126.9080
5695.069	826	119.8012	6039.807	876	127.0531
5701.964	827	119.9462	6046.702	877	127.1981
5708.859	828	120.0913	6053.597	878	127.3431
5715.754	829	120.2363	6060.491	879	127.4882
5722.648	830	120.3813	6067.386	880	127.6332
5729.543	831	120.5264	6074.281	881	127.7783
5736.438	832	120.6714	6081.176	882	127.9233
5743.333	833	120.8164	6088.070	883	128.0683
5750.227	834	120.9615	6094.965	884	128.2134
5757.122	835	121.1065	6101.860	885	128.3584
5764.017	836	121.2516	6108.755	886	128.5034
5770.912	837	121.3966	6115.649	887	128.6485
5777.806	838	121.5416	6122.544	888	128.7935
5784.701	839	121.6867	6129.439	889	128.9386
5791.596	840	121.8317	6136.334	890	129.0836
5798.491	841	121.9767	6143.228	891	129.2286
5805.385	842	122.1218	6150.123	892	129.3737
5812.280	843	122.2668	6157.018	893	129.5187
5819.175	844	122.4119	6163.913	894	129.6637
5826.070	845	122.5569	6170.807	895	129.8088
5832.964	846	122.7019	6177.702	896	129.9538
5839.859	847	122.8470	6184.597	897	130.0989
5846.754	848	122.9920	6191.492	898	130.2439
5853.649	849	123.1370	6198.387	899	130.3889
5860.543	850	123.2821	6205.281	900	130.5340

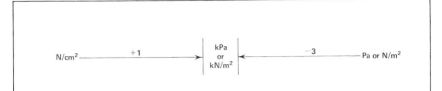

kPa kN/m²	REF	lbf/in²	kPa kN/m²	REF	lbf/in²
6212.176	901	130.6790	6556.914	951	137.9309
6219.071	902	130.8240	6563.809	952	138.0759
6225.966	903	130.9691	6570.703	953	138.2210
6232.860	904	131.1141	6577.598	954	138.3660
6239.755	905	131.2592	6584.493	955	138.5110
6246.650	906	131.4042	6591.388	956	138.6561
6253.545	907	131.5492	6598.282	957	138.8011
6260.439	908	131.6943	6605.177	958	138.9462
6267.334	909	131.8393	6612.072	959	139.0912
6274.229	910	131.9843	6618.967	960	139.2362
6281.124	911	132.1294	6625.861	961	139.3813
6288.018	912	132.2744	6632.756	962	139.5263
6294.913	913	132.4195	6639.651	963	139.6713
6301.808	914	132.5645	6646.546	964	139.8164
6308.703	915	132.7095	6653.440	965	139.9614
6315.597	916	132.8546	6660.335	966	140.1065
6322.492	917	132.9996	6667.230	967	140.2515
6329.387	918	133.1446	6674.125	968	140.3965
6336.282	919	133.2897	6681.020	969	140.5416
6343.176	920	133.4347	6687.914	970	140.6866
6350.071	921	133.5798	6694.809	971	140.8316
6356.966	922	133.7248	6701.704	972	140.9767
6363.861	923	133.8698	6708.599	973	141.1217
6370.755	924	134.0149	6715.493	974	141.2668
6377.650	925	134.1599	6722.388	975	141.4118
6384.545	926	134.3050	6729.283	976	141.5568
6391.440	927	134.4500	6736.178	977	141.7019
6398.334	928	134.5950	6743.072	978	141.8469
6405.229	929	134.7401	6749.967	979	141.9920
6412.124	930	134.8851	6756.862	980	142.1370
6419.019	931	135.0301	6763.757	981	142.2820
6425.914	932	135.1752	6770.651	982	142.4271
6432.808	933	135.3202	6777.546	983	142.5721
6439.703	934	135.4653	6784.441	984	142.7171
6446.598	935	135.6103	6791.336	985	142.8622
6453.493	936	135.7553	6798.230	986	143.0072
6460.387	937	135.9004	6805.125	987	143.1523
6467.282	938	136.0454	6812.020	988	143.2973
6474.177	939	136.1904	6818.915	989	143.4423
6481.072	940	136.3355	6825.809	990	143.5874
6487.966	941	136.4805	6832.704	991	143.7324
6494.861	942	136.6256	6839.599	992	143.8774
6501.756	943	136.7706	6846.494	993	144.0225
6508.651	944	136.9156	6853.388	994	144.1675
6515.545	945	137.0607	6860.283	995	144.3126
6522.440	946	137.2057	6867.178	996	144.4576
6529.335	947	137.3507	6874.073	997	144.6026
6536.230	948	137.4958	6880.967	998	144.7477
6543.124	949	137.6408	6887.862	999	144.8927
6550.019	950	137.7859	6894.757	1000	145.0377

33

PRESSURE AND STRESS

Pound (force)/Square-foot — Pascal (or Newton/Square-metre)

$(lbf/ft^2 — Pa — N/m^2)$

Pound (force)/Square-foot — Kilopascal (or Kilonewton/Square-metre)

$(lbf/ft^2 — kPa — kN/m^2)$

Pound (force)/Square-foot — Newton/Square-centimetre

$(lbf/ft^2 — N/cm^2)$

Kilopound (force)/Square-foot — Pascal (or Newton/Square-metre)

$(kip/ft^2 — Pa — N/m^2)$

Kilopound (force)/Square-foot — Kilopascal (or Kilonewton/Square-metre)

$(kip/ft^2 — kPa — kN/m^2)$

Kilopound (force)/Square-foot — Newton/Square-centimetre

$(kip/ft^2 — N/cm^2)$

DYNAMIC or ABSOLUTE VISCOSITY

Pound (force)-second/Square-foot — Pascal-second (or Newton-second/Metre^2)

$(lbf\text{-}s/ft^2 — Pa \cdot s — N \cdot s/m^2)$

Pound (force)-second/Square-foot — Centipoise (or Millipascal-second)

$(lbf\text{-}s/ft^2 — cP — mPa \cdot s)$

Conversion Factors:

1 Pound (force)/square-foot (or lbf/ft^2 or psf) = 47.88026 Pascals (Pa)

(1 Pa = 0.02088543 lbf/ft^2)

Pound (force)/Square-foot (lbf/ft² or psf) —
Kilopound/Square-foot (kip/ft² or kpsf) Relation

Since the kilopound per square foot equals 1000 pounds per square foot, Table 33 can be used with proper decimal shifting for converting relations to and from kpsf. Diagram 33-A describes the decimal-shift relation between kpsf and psf:

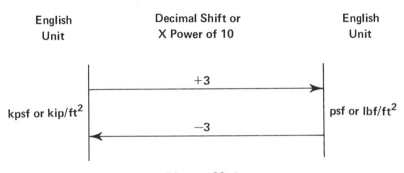

English Unit	Decimal Shift or X Power of 10	English Unit

Diagram 33–A

Also, 1 kilopascal (kPa) = 1000 pascals (Pa). Therefore, conversions between kpsf and kPa can be made directly through Table 33 without any decimal shifting, since: 1 kpsf = 47.88026 kPa.

The numerical value of this conversion factor is exactly the same as that given above for psf to Pa, and from which the values of this table are derived.

Example (a): Convert 3.05 kpsf to Pa.
  From Diagram 33-A: 3.05 kpsf = 3.05×10^3 psf = 305×10 psf.
  From table: 305 psf = 14,603.48 Pa*
  Therefore: 305×10 psf = $14,603.48 \times 10$ Pa = 146,034.8 Pa
        $\approx 1.46 \times 10^5$ Pa.

Dynamic or Absolute Viscosity Values

Dynamic (or absolute) viscosity quantities having the English and SI units shown above can be converted by means of Table 33, since:

1 pound (force)-second/square-foot ($lbf\text{-}s/ft^2$) = 47.88026 pascal-seconds ($Pa \cdot s$ or $N \cdot s/m^2$)

Again, the numerical value of this conversion factor is the same as that from which Table 33 is derived. Also, 1 centipoise (cP) = (10^3) Pa · s. Therefore, Table 33 can be used for converting $lbf\text{-}s/ft^2$ –cP relations, with appropriate decimal shifting.

---

*All conversion values obtained can be rounded to desired number of significant figures.

In converting Pa values to kPa and N/cm^2, and $Pa \cdot s$ to cP, the Decimal-shift Diagram 33-B, which appears in the headings on alternate table pages, will assist table users. It is derived from: $1 \, Pa = 1 \, N/m^2 = (10^{-3}) \, kPa = (10^{-4}) \, N/cm^2$; and $1 \, Pa \cdot s = (10^{-3}) \, cP$.

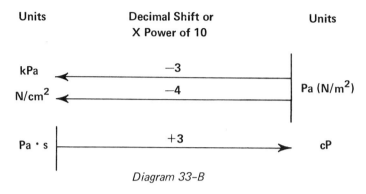

| Units | Decimal Shift or X Power of 10 | Units |

Diagram 33-B

Example (b): Convert 49.7 psf to Pa, kPa, N/cm^2.
  Since: $49.7 \, psf = 497 \times 10^{-1} \, psf$:
  From table: $497 \, psf = 23{,}796.49 \, Pa$*
  Therefore: $497 \times 10^{-1} \, psf = 23{,}796.49 \times 10^{-1} \, Pa = 2{,}379.649 \, Pa$
      $\approx 2{,}380 \, Pa$.
  From Diagram 33-B: 1) $2{,}379.649 \, Pa = 2{,}379.649 \times 10^{-3} \, kPa$
              $= 2.379649 \, kPa \approx 2.38 \, kPa$
              (Pa to kPa by multiplying Pa value by 10^{-3} or by shifting its decimal 3 places to the left.)

          2) $2{,}379.649 \, Pa = 2{,}379.649 \times 10^{-4} \, N/cm^2$
              $= 0.2379649 \, N/cm^2 \approx 0.238 \, N/cm^2$
              (Pa to N/cm^2 by multiplying Pa value by 10^{-4} or by shifting its decimal 4 places to the left.)

Example (c): Convert $0.940 \times 10^{-3} \, lbf\text{-}s/ft^2$ to cP.
  Since: $0.940 \times 10^{-3} \, lbf\text{-}s/ft^2 = 940 \times 10^{-6} \, lbf\text{-}s/ft^2$:
  From table: $940 \, lbf\text{-}s/ft^2 = 45{,}007.44 \, Pa \cdot s$*
  Therefore: $940 \times 10^{-6} \, lbf\text{-}s/ft^2 = 45{,}007.44 \times 10^{-6} \, Pa \cdot s$
      $= 0.04500744 \, Pa \cdot s$.
  From Diagram: $0.04500744 \, Pa \cdot s = 0.04500744 \times 10^3 \, cP$
          $= 45.00744 \, cP \approx 45 \, cP$
          (Pa $\cdot$ s to cP by multiplying Pa $\cdot$ s value by 10^3 or shifting its decimal 3 places to the right.)

---

*All conversion values obtained can be rounded to desired number of significant figures.

Diagram 33-C is also given at the top of alternate table pages:

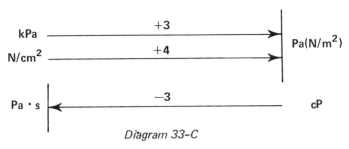

Diagram 33-C

It is the opposite of Diagram 33-B and is used to convert a quantity in N/cm^2 or kPa to Pa; and cP to Pa $\cdot$ s. The Pa or Pa $\cdot$ s equivalent is then applied to the REF column in the table to obtain psf or $lbf\text{-}s/ft^2$.

Example (d): Convert 7.31 N/cm^2 to psf.
    From Diagram 33-C: 7.31 $N/cm^2 = 7.31 \times 10^4$ Pa $= 731 \times 10^2$ Pa
    From table: 731 Pa = 15.26725 psf*
    Therefore: 731×10^2 Pa $= 15.26725 \times 10^2$ psf $= 1526.725$ psf $\approx 1,530$ psf.

Example (e): Convert 661 cP to $lbf\text{-}s/ft^2$.
    From Diagram 33-C: 661 cP $= 661 \times 10^{-3}$ Pa $\cdot$ s
    From table: 661 Pa $\cdot$ s $= 13.80527$ $lbf\text{-}s/ft^2$ *
    Therefore: 661×10^{-3} Pa $\cdot$ s $= 13.80527 \times 10^{-3}$ $lbf\text{-}s/ft^2$
                $= 0.01380527$ $lbf\text{-}s/ft^2$ ≈ 0.0138 $lbf\text{-}s/ft^2$.

---

*All conversion values obtained can be rounded to desired number of significant figures.

kPa	−3	Pa(N/m²)
N/cm²	−4	
Pa · s	+3	cP

Pa N/m² kPa kN/m² Pa · s N · s/m²	REF	lbf/ft² kip/ft² lbf–s/ft²	Pa N/m² kPa kN/m² Pa · s N · s/m²	REF	lbf/ft² kip/ft² lbf–s/ft²
47.88026	1	0.02088543	2441.893	51	1.065157
95.76052	2	0.04177087	2489.774	52	1.086043
143.6408	3	0.06265630	2537.654	53	1.106928
191.5210	4	0.08354174	2585.534	54	1.127813
239.4013	5	0.1044272	2633.414	55	1.148699
287.2816	6	0.1253126	2681.295	56	1.169584
335.1618	7	0.1461980	2729.175	57	1.190470
383.0421	8	0.1670835	2777.055	58	1.211355
430.9223	9	0.1879689	2824.935	59	1.232241
478.8026	10	0.2088543	2872.816	60	1.253126
526.6829	11	0.2297398	2920.696	61	1.274011
574.5631	12	0.2506252	2968.576	62	1.294897
622.4434	13	0.2715106	3016.456	63	1.315782
670.3236	14	0.2923961	3064.337	64	1.336668
718.2039	15	0.3132815	3112.217	65	1.357553
766.0842	16	0.3341669	3160.097	66	1.378439
813.9644	17	0.3550524	3207.977	67	1.399324
861.8447	18	0.3759378	3255.858	68	1.420209
909.7249	19	0.3968232	3303.738	69	1.441095
957.6052	20	0.4177087	3351.618	70	1.461980
1005.485	21	0.4385941	3399.498	71	1.482866
1053.366	22	0.4594795	3447.379	72	1.503751
1101.246	23	0.4803650	3495.259	73	1.524637
1149.126	24	0.5012504	3543.139	74	1.545522
1197.007	25	0.5221358	3591.020	75	1.566408
1244.887	26	0.5430213	3638.900	76	1.587293
1292.767	27	0.5639067	3686.780	77	1.608178
1340.647	28	0.5847921	3734.660	78	1.629064
1388.528	29	0.6056776	3782.541	79	1.649949
1436.408	30	0.6265630	3830.421	80	1.670835
1484.288	31	0.6474485	3878.301	81	1.691720
1532.168	32	0.6683339	3926.181	82	1.712606
1580.049	33	0.6892193	3974.062	83	1.733491
1627.929	34	0.7101047	4021.942	84	1.754376
1675.809	35	0.7309902	4069.822	85	1.775262
1723.689	36	0.7518756	4117.702	86	1.796147
1771.570	37	0.7727610	4165.583	87	1.817033
1819.450	38	0.7936465	4213.463	88	1.837918
1867.330	39	0.8145319	4261.343	89	1.858804
1915.210	40	0.8354174	4309.223	90	1.879689
1963.091	41	0.8563028	4357.104	91	1.900574
2010.971	42	0.8771882	4404.984	92	1.921460
2058.851	43	0.8980737	4452.864	93	1.942345
2106.731	44	0.9189591	4500.744	94	1.963231
2154.612	45	0.9398445	4548.625	95	1.984116
2202.492	46	0.9607300	4596.505	96	2.005002
2250.372	47	0.9816154	4644.385	97	2.025887
2298.252	48	1.002501	4692.266	98	2.046773
2346.133	49	1.023386	4740.146	99	2.067658
2394.013	50	1.044272	4788.026	100	2.088543

kPa ———— +3 ———→	Pa(N/m²)
N/cm² ———— +4 ———→	
Pa · s ←———— −3 ———— cP	

Pa N/m² kPa kN/m² Pa · s N · s/m²	REF	lbf/ft² kip/ft² lbf-s/ft²	Pa N/m² kPa kN/m² Pa · s N · s/m²	REF	lbf/ft² kip/ft² lbf-s/ft²
4835.906	101	2.109429	7229.919	151	3.153701
4883.786	102	2.130314	7277.799	152	3.174586
4931.667	103	2.151200	7325.680	153	3.195471
4979.547	104	2.172085	7373.560	154	3.216357
5027.427	105	2.192971	7421.440	155	3.237242
5075.308	106	2.213856	7469.321	156	3.258128
5123.188	107	2.234741	7517.201	157	3.279013
5171.068	108	2.255627	7565.081	158	3.299899
5218.948	109	2.276512	7612.961	159	3.320784
5266.829	110	2.297398	7660.842	160	3.341669
5314.709	111	2.318283	7708.722	161	3.362555
5362.589	112	2.339169	7756.602	162	3.383440
5410.469	113	2.360054	7804.482	163	3.404326
5458.350	114	2.380939	7852.363	164	3.425211
5506.230	115	2.401825	7900.243	165	3.446097
5554.110	116	2.422710	7948.123	166	3.466982
5601.990	117	2.443596	7996.003	167	3.487867
5649.871	118	2.464481	8043.884	168	3.508753
5697.751	119	2.485367	8091.764	169	3.529638
5745.631	120	2.506252	8139.644	170	3.550524
5793.511	121	2.527137	8187.524	171	3.571409
5841.392	122	2.548023	8235.405	172	3.592295
5889.272	123	2.568908	8283.285	173	3.613180
5937.152	124	2.589794	8331.165	174	3.634065
5985.032	125	2.610679	8379.046	175	3.654951
6032.913	126	2.631565	8426.926	176	3.675836
6080.793	127	2.652450	8474.806	177	3.696722
6128.673	128	2.673336	8522.686	178	3.717607
6176.554	129	2.694221	8570.567	179	3.738493
6224.434	130	2.715106	8618.447	180	3.759378
6272.314	131	2.735992	8666.327	181	3.780264
6320.194	132	2.756877	8714.207	182	3.801149
6368.075	133	2.777763	8762.088	183	3.822034
6415.955	134	2.798648	8809.968	184	3.842920
6463.835	135	2.819534	8857.848	185	3.863805
6511.715	136	2.840419	8905.728	186	3.884691
6559.596	137	2.861304	8953.609	187	3.905576
6607.476	138	2.882190	9001.489	188	3.926462
6655.356	139	2.903075	9049.369	189	3.947347
6703.236	140	2.923961	9097.249	190	3.968232
6751.117	141	2.944846	9145.130	191	3.989118
6798.997	142	2.965732	9193.010	192	4.010003
6846.877	143	2.986617	9240.890	193	4.030889
6894.757	144	3.007502	9288.770	194	4.051774
6942.638	145	3.028388	9336.651	195	4.072660
6990.518	146	3.049273	9384.531	196	4.093545
7038.398	147	3.070159	9432.411	197	4.114430
7086.279	148	3.091044	9480.292	198	4.135316
7134.159	149	3.111930	9528.172	199	4.156201
7182.039	150	3.132815	9576.052	200	4.177087

Pa N/m²		lbf/ft²		Pa N/m²		lbf/ft²
kPa kN/m²		kip/ft²		kPa kN/m²		kip/ft²
Pa · s N · s/m²	REF	lbf–s/ft²		Pa · s N · s/m²	REF	lbf–s/ft²
9623·932	201	4·197972		12017·95	251	5·242244
9671·812	202	4·218858		12065·83	252	5·263129
9719·693	203	4·239743		12113·71	253	5·284015
9767·573	204	4·260628		12161·59	254	5·304900
9815·453	205	4·281514		12209·47	255	5·325786
9863·334	206	4·302399		12257·35	256	5·346671
9911·214	207	4·323285		12305·23	257	5·367556
9959·094	208	4·344170		12353·11	258	5·388442
10006·97	209	4·365056		12400·99	259	5·409327
10054·85	210	4·385941		12448·87	260	5·430213
10102·73	211	4·406826		12496·75	261	5·451098
10150·62	212	4·427712		12544·63	262	5·471984
10198·50	213	4·448597		12592·51	263	5·492869
10246·38	214	4·469483		12640·39	264	5·513754
10294·26	215	4·490368		12688·27	265	5·534640
10342·14	216	4·511254		12736·15	266	5·555525
10390·02	217	4·532139		12784·03	267	5·576411
10437·90	218	4·553025		12831·91	268	5·597296
10485·78	219	4·573910		12879·79	269	5·618182
10533·66	220	4·594795		12927·67	270	5·639067
10581·54	221	4·615681		12975·55	271	5·659953
10629·42	222	4·636566		13023·43	272	5·680838
10677·30	223	4·657452		13071·31	273	5·701723
10725·18	224	4·678337		13119·19	274	5·722609
10773·06	225	4·699223		13167·07	275	5·743494
10820·94	226	4·720108		13214·95	276	5·764380
10868·82	227	4·740993		13262·83	277	5·785265
10916·70	228	4·761879		13310·71	278	5·806151
10964·58	229	4·782764		13358·59	279	5·827036
11012·46	230	4·803650		13406·47	280	5·847921
11060·34	231	4·824535		13454·35	281	5·868807
11108·22	232	4·845421		13502·23	282	5·889692
11156·10	233	4·866306		13550·11	283	5·910578
11203·98	234	4·887191		13597·99	284	5·931463
11251·86	235	4·908077		13645·87	285	5·952349
11299·74	236	4·928962		13693·75	286	5·973234
11347·62	237	4·949848		13741·63	287	5·994119
11395·50	238	4·970733		13789·51	288	6·015005
11443·38	239	4·991619		13837·40	289	6·035890
11491·26	240	5·012504		13885·28	290	6·056776
11539·14	241	5·033390		13933·16	291	6·077661
11587·02	242	5·054275		13981·04	292	6·098547
11634·90	243	5·075160		14028·92	293	6·119432
11682·78	244	5·096046		14076·80	294	6·140317
11730·66	245	5·116931		14124·68	295	6·161203
11778·54	246	5·137817		14172·56	296	6·182088
11826·42	247	5·158702		14220·44	297	6·202974
11874·30	248	5·179588		14268·32	298	6·223859
11922·18	249	5·200473		14316·20	299	6·244745
11970·06	250	5·221358		14364·08	300	6·265630

Pa N/m² kPa kN/m² Pa · s N · s/m²	REF	lbf/ft² kip/ft² lbf-s/ft²	Pa N/m² kPa kN/m² Pa · s N · s/m²	REF	lbf/ft² kip/ft² lbf-s/ft²
14411.96	301	6.286516	16805.97	351	7.330787
14459.84	302	6.307401	16853.85	352	7.351673
14507.72	303	6.328286	16901.73	353	7.372558
14555.60	304	6.349172	16949.61	354	7.393444
14603.48	305	6.370057	16997.49	355	7.414329
14651.36	306	6.390943	17045.37	356	7.435214
14699.24	307	6.411828	17093.25	357	7.456100
14747.12	308	6.432714	17141.13	358	7.476985
14795.00	309	6.453599	17189.01	359	7.497871
14842.88	310	6.474484	17236.89	360	7.518756
14890.76	311	6.495370	17284.77	361	7.539642
14938.64	312	6.516255	17332.65	362	7.560527
14986.52	313	6.537141	17380.53	363	7.581412
15034.40	314	6.558026	17428.41	364	7.602298
15082.28	315	6.578912	17476.29	365	7.623183
15130.16	316	6.599797	17524.18	366	7.644069
15178.04	317	6.620682	17572.06	367	7.664954
15225.92	318	6.641568	17619.94	368	7.685840
15273.80	319	6.662453	17667.82	369	7.706725
15321.68	320	6.683339	17715.70	370	7.727610
15369.56	321	6.704224	17763.58	371	7.748496
15417.44	322	6.725110	17811.46	372	7.769381
15465.32	323	6.745995	17859.34	373	7.790267
15513.20	324	6.766881	17907.22	374	7.811152
15561.08	325	6.787766	17955.10	375	7.832038
15608.96	326	6.808651	18002.98	376	7.852923
15656.84	327	6.829537	18050.86	377	7.873809
15704.73	328	6.850422	18098.74	378	7.894694
15752.61	329	6.871308	18146.62	379	7.915579
15800.49	330	6.892193	18194.50	380	7.936465
15848.37	331	6.913079	18242.38	381	7.957350
15896.25	332	6.933964	18290.26	382	7.978236
15944.13	333	6.954849	18338.14	383	7.999121
15992.01	334	6.975735	18386.02	384	8.020007
16039.89	335	6.996620	18433.90	385	8.040892
16087.77	336	7.017506	18481.78	386	8.061777
16135.65	337	7.038391	18529.66	387	8.082663
16183.53	338	7.059277	18577.54	388	8.103548
16231.41	339	7.080162	18625.42	389	8.124434
16279.29	340	7.101047	18673.30	390	8.145319
16327.17	341	7.121933	18721.18	391	8.166205
16375.05	342	7.142818	18769.06	392	8.187090
16422.93	343	7.163704	18816.94	393	8.207976
16470.81	344	7.184589	18864.82	394	8.228861
16518.69	345	7.205475	18912.70	395	8.249746
16566.57	346	7.226360	18960.58	396	8.270632
16614.45	347	7.247245	19008.46	397	8.291517
16662.33	348	7.268131	19056.34	398	8.312403
16710.21	349	7.289016	19104.22	399	8.333288
16758.09	350	7.309902	19152.10	400	8.354174

kPa ← —3

N/cm² ← —4 — Pa(N/m²)

Pa · s ——— +3 ——→ cP

Pa N/m² kPa kN/m² Pa · s N · s/m²	REF	lbf/ft² kip/ft² lbf–s/ft²	Pa N/m² kPa kN/m² Pa · s N · s/m²	REF	lbf/ft² kip/ft² lbf–s/ft²
19199•98	401	8•375059	21594•00	451	9•419331
19247•86	402	8•395944	21641•88	452	9•440216
19295•74	403	8•416830	21689•76	453	9•461102
19343•62	404	8•437715	21737•64	454	9•481987
19391•51	405	8•458601	21785•52	455	9•502872
19439•39	406	8•479486	21833•40	456	9•523758
19487•27	407	8•500372	21881•28	457	9•544643
19535•15	408	8•521257	21929•16	458	9•565529
19583•03	409	8•542142	21977•04	459	9•586414
19630•91	410	8•563028	22024•92	460	9•607300
19678•79	411	8•583913	22072•80	461	9•628185
19726•67	412	8•604799	22120•68	462	9•649070
19774•55	413	8•625684	22168•56	463	9•669956
19822•43	414	8•646570	22216•44	464	9•690841
19870•31	415	8•667455	22264•32	465	9•711727
19918•19	416	8•688340	22312•20	466	9•732612
19966•07	417	8•709226	22360•08	467	9•753498
20013•95	418	8•730111	22407•96	468	9•774383
20061•83	419	8•750997	22455•84	469	9•795268
20109•71	420	8•771882	22503•72	470	9•816154
20157•59	421	8•792768	22551•60	471	9•837039
20205•47	422	8•813653	22599•48	472	9•857925
20253•35	423	8•834538	22647•36	473	9•878810
20301•23	424	8•855424	22695•24	474	9•899696
20349•11	425	8•876309	22743•12	475	9•920581
20396•99	426	8•897195	22791•00	476	9•941466
20444•87	427	8•918080	22838•88	477	9•962352
20492•75	428	8•938966	22886•76	478	9•983237
20540•63	429	8•959851	22934•64	479	10•00412
20588•51	430	8•980736	22982•52	480	10•02501
20636•39	431	9•001622	23030•41	481	10•04589
20684•27	432	9•022507	23078•29	482	10•06678
20732•15	433	9•043393	23126•17	483	10•08766
20780•03	434	9•064278	23174•05	484	10•10855
20827•91	435	9•085164	23221•93	485	10•12944
20875•79	436	9•106049	23269•81	486	10•15032
20923•67	437	9•126935	23317•69	487	10•17121
20971•55	438	9•147820	23365•57	488	10•19209
21019•43	439	9•168705	23413•45	489	10•21298
21067•31	440	9•189591	23461•33	490	10•23386
21115•19	441	9•210476	23509•21	491	10•25475
21163•07	442	9•231362	23557•09	492	10•27563
21210•96	443	9•252247	23604•97	493	10•29652
21258•84	444	9•273133	23652•85	494	10•31740
21306•72	445	9•294018	23700•73	495	10•33829
21354•60	446	9•314903	23748•61	496	10•35918
21402•48	447	9•335789	23796•49	497	10•38006
21450•36	448	9•356674	23844•37	498	10•40095
21498•24	449	9•377560	23892•25	499	10•42183
21546•12	450	9•398445	23940•13	500	10•44272

Pa N/m² kPa kN/m² Pa · s N · s/m²	REF	lbf/ft² kip/ft² lbf-s/ft²	Pa N/m² kPa kN/m² Pa · s N · s/m²	REF	lbf/ft² kip/ft² lbf-s/ft²
23988·01	501	10·46360	26382·02	551	11·50787
24035·89	502	10·48449	26429·90	552	11·52876
24083·77	503	10·50537	26477·78	553	11·54964
24131·65	504	10·52626	26525·66	554	11·57053
24179·53	505	10·54714	26573·54	555	11·59142
24227·41	506	10·56803	26621·42	556	11·61230
24275·29	507	10·58891	26669·30	557	11·63319
24323·17	508	10·60980	26717·19	558	11·65407
24371·05	509	10·63069	26765·07	559	11·67496
24418·93	510	10·65157	26812·95	560	11·69584
24466·81	511	10·67246	26860·83	561	11·71673
24514·69	512	10·69334	26908·71	562	11·73761
24562·57	513	10·71423	26956·59	563	11·75850
24610·45	514	10·73511	27004·47	564	11·77938
24658·33	515	10·75600	27052·35	565	11·80027
24706·21	516	10·77688	27100·23	566	11·82116
24754·09	517	10·79777	27148·11	567	11·84204
24801·97	518	10·81865	27195·99	568	11·86293
24849·85	519	10·83954	27243·87	569	11·88381
24897·74	520	10·86043	27291·75	570	11·90470
24945·62	521	10·88131	27339·63	571	11·92558
24993·50	522	10·90220	27387·51	572	11·94647
25041·38	523	10·92308	27435·39	573	11·96735
25089·26	524	10·94397	27483·27	574	11·98824
25137·14	525	10·96485	27531·15	575	12·00912
25185·02	526	10·98574	27579·03	576	12·03001
25232·90	527	11·00662	27626·91	577	12·05090
25280·78	528	11·02751	27674·79	578	12·07178
25328·66	529	11·04839	27722·67	579	12·09267
25376·54	530	11·06928	27770·55	580	12·11355
25424·42	531	11·09017	27818·43	581	12·13444
25472·30	532	11·11105	27866·31	582	12·15532
25520·18	533	11·13194	27914·19	583	12·17621
25568·06	534	11·15282	27962·07	584	12·19709
25615·94	535	11·17371	28009·95	585	12·21798
25663·82	536	11·19459	28057·83	586	12·23886
25711·70	537	11·21548	28105·71	587	12·25975
25759·58	538	11·23636	28153·59	588	12·28063
25807·46	539	11·25725	28201·47	589	12·30152
25855·34	540	11·27813	28249·35	590	12·32241
25903·22	541	11·29902	28297·23	591	12·34329
25951·10	542	11·31991	28345·11	592	12·36418
25998·98	543	11·34079	28392·99	593	12·38506
26046·86	544	11·36168	28440·87	594	12·40595
26094·74	545	11·38256	28488·75	595	12·42683
26142·62	546	11·40345	28536·64	596	12·44772
26190·50	547	11·42433	28584·52	597	12·46860
26238·38	548	11·44522	28632·40	598	12·48949
26286·26	549	11·46610	28680·28	599	12·51037
26334·14	550	11·48699	28728·16	600	12·53126

$$\begin{array}{l} \text{kPa} \xleftarrow{\quad -3 \quad} \\ \text{N/cm}^2 \xleftarrow{\quad -4 \quad} \quad \text{Pa(N/m}^2) \\ \text{Pa} \cdot \text{s} \xrightarrow{\quad +3 \quad} \text{cP} \end{array}$$

Pa N/m² / kPa kN/m² / Pa·s N·s/m²	REF	lbf/ft² / kip/ft² / lbf-s/ft²		Pa N/m² / kPa kN/m² / Pa·s N·s/m²	REF	lbf/ft² / kip/ft² / lbf-s/ft²
28776.04	601	12.55215		31170.05	651	13.59642
28823.92	602	12.57303		31217.93	652	13.61730
28871.80	603	12.59392		31265.81	653	13.63819
28919.68	604	12.61480		31313.69	654	13.65907
28967.56	605	12.63569		31361.57	655	13.67996
29015.44	606	12.65657		31409.45	656	13.70084
29063.32	607	12.67746		31457.33	657	13.72173
29111.20	608	12.69834		31505.21	658	13.74262
29159.08	609	12.71923		31553.09	659	13.76350
29206.96	610	12.74011		31600.97	660	13.78439
29254.84	611	12.76100		31648.85	661	13.80527
29302.72	612	12.78189		31696.73	662	13.82616
29350.60	613	12.80277		31744.61	663	13.84704
29398.48	614	12.82366		31792.49	664	13.86793
29446.36	615	12.84454		31840.37	665	13.88881
29494.24	616	12.86543		31888.25	666	13.90970
29542.12	617	12.88631		31936.13	667	13.93058
29590.00	618	12.90720		31984.01	668	13.95147
29637.88	619	12.92808		32031.89	669	13.97236
29685.76	620	12.94897		32079.77	670	13.99324
29733.64	621	12.96985		32127.65	671	14.01413
29781.52	622	12.99074		32175.53	672	14.03501
29829.40	623	13.01163		32223.42	673	14.05590
29877.28	624	13.03251		32271.30	674	14.07678
29925.16	625	13.05340		32319.18	675	14.09767
29973.04	626	13.07428		32367.06	676	14.11855
30020.92	627	13.09517		32414.94	677	14.13944
30068.80	628	13.11605		32462.82	678	14.16032
30116.68	629	13.13694		32510.70	679	14.18121
30164.56	630	13.15782		32558.58	680	14.20209
30212.44	631	13.17871		32606.46	681	14.22298
30260.32	632	13.19959		32654.34	682	14.24387
30308.20	633	13.22048		32702.22	683	14.26475
30356.08	634	13.24136		32750.10	684	14.28564
30403.97	635	13.26225		32797.98	685	14.30652
30451.85	636	13.28314		32845.86	686	14.32741
30499.73	637	13.30402		32893.74	687	14.34829
30547.61	638	13.32491		32941.62	688	14.36918
30595.49	639	13.34579		32989.50	689	14.39006
30643.37	640	13.36668		33037.38	690	14.41095
30691.25	641	13.38756		33085.26	691	14.43183
30739.13	642	13.40845		33133.14	692	14.45272
30787.01	643	13.42933		33181.02	693	14.47361
30834.89	644	13.45022		33228.90	694	14.49449
30882.77	645	13.47110		33276.78	695	14.51538
30930.65	646	13.49199		33324.66	696	14.53626
30978.53	647	13.51288		33372.54	697	14.55715
31026.41	648	13.53376		33420.42	698	14.57803
31074.29	649	13.55465		33468.30	699	14.59892
31122.17	650	13.57553		33516.18	700	14.61980

Pa N/m² kPa kN/m² Pa · s N · s/m²	REF	lbf/ft² kip/ft² lbf-s/ft²	Pa N/m² kPa kN/m² Pa · s N · s/m²	REF	lbf/ft² kip/ft² lbf-s/ft²
33564.06	701	14.64069	35958.08	751	15.68496
33611.94	702	14.66157	36005.96	752	15.70585
33659.82	703	14.68246	36053.84	753	15.72673
33707.70	704	14.70335	36101.72	754	15.74762
33755.58	705	14.72423	36149.60	755	15.76850
33803.46	706	14.74512	36197.48	756	15.78939
33851.34	707	14.76600	36245.36	757	15.81027
33899.22	708	14.78689	36293.24	758	15.83116
33947.10	709	14.80777	36341.12	759	15.85204
33994.98	710	14.82866	36389.00	760	15.87293
34042.86	711	14.84954	36436.88	761	15.89382
34090.75	712	14.87043	36484.76	762	15.91470
34138.63	713	14.89131	36532.64	763	15.93559
34186.51	714	14.91220	36580.52	764	15.95647
34234.39	715	14.93309	36628.40	765	15.97736
34282.27	716	14.95397	36676.28	766	15.99824
34330.15	717	14.97486	36724.16	767	16.01913
34378.03	718	14.99574	36772.04	768	16.04001
34425.91	719	15.01663	36819.92	769	16.06090
34473.79	720	15.03751	36867.80	770	16.08178
34521.67	721	15.05840	36915.68	771	16.10267
34569.55	722	15.07928	36963.56	772	16.12355
34617.43	723	15.10017	37011.44	773	16.14444
34665.31	724	15.12105	37059.32	774	16.16533
34713.19	725	15.14194	37107.20	775	16.18621
34761.07	726	15.16282	37155.08	776	16.20710
34808.95	727	15.18371	37202.96	777	16.22798
34856.83	728	15.20460	37250.84	778	16.24887
34904.71	729	15.22548	37298.72	779	16.26975
34952.59	730	15.24637	37346.60	780	16.29064
35000.47	731	15.26725	37394.48	781	16.31152
35048.35	732	15.28814	37442.36	782	16.33241
35096.23	733	15.30902	37490.24	783	16.35329
35144.11	734	15.32991	37538.12	784	16.37418
35191.99	735	15.35079	37586.00	785	16.39507
35239.87	736	15.37168	37633.88	786	16.41595
35287.75	737	15.39256	37681.76	787	16.43684
35335.63	738	15.41345	37729.65	788	16.45772
35383.51	739	15.43434	37777.52	789	16.47861
35431.39	740	15.45522	37825.41	790	16.49949
35479.27	741	15.47611	37873.29	791	16.52038
35527.15	742	15.49699	37921.17	792	16.54126
35575.03	743	15.51788	37969.05	793	16.56215
35622.91	744	15.53876	38016.93	794	16.58303
35670.79	745	15.55965	38064.81	795	16.60392
35718.67	746	15.58053	38112.69	796	16.62481
35766.55	747	15.60142	38160.57	797	16.64569
35814.43	748	15.62230	38208.45	798	16.66658
35862.31	749	15.64319	38256.33	799	16.68746
35910.19	750	15.66408	38304.21	800	16.70835

```
kPa ◄─────────── -3 ───────────
                                          ┐
N/cm² ◄────────── -4 ──────────── Pa(N/m²)
                                          │
Pa · s ├────────── +3 ──────────► cP
```

Pa N/m² kPa kN/m² Pa · s N · s/m²	REF	lbf/ft² kip/ft² lbf-s/ft²	Pa N/m² kPa kN/m² Pa · s N · s/m²	REF	lbf/ft² lbf-s/ft² lbf-s/ft²
38352.09	801	16.72923	40746.10	851	17.77350
38399.97	802	16.75012	40793.98	852	17.79439
38447.85	803	16.77100	40841.86	853	17.81527
38495.73	804	16.79189	40889.74	854	17.83616
38543.61	805	16.81277	40937.62	855	17.85705
38591.49	806	16.83366	40985.50	856	17.87793
38639.37	807	16.85455	41033.38	857	17.89882
38687.25	808	16.87543	41081.26	858	17.91970
38735.13	809	16.89632	41129.14	859	17.94059
38783.01	810	16.91720	41177.02	860	17.96147
38830.89	811	16.93809	41224.90	861	17.98236
38878.77	812	16.95897	41272.78	862	18.00324
38926.65	813	16.97986	41320.66	863	18.02413
38974.53	814	17.00074	41368.54	864	18.04501
39022.41	815	17.02163	41416.42	865	18.06590
39070.29	816	17.04251	41464.31	866	18.08679
39118.17	817	17.06340	41512.19	867	18.10767
39166.05	818	17.08428	41560.07	868	18.12856
39213.93	819	17.10517	41607.95	869	18.14944
39261.81	820	17.12606	41655.83	870	18.17033
39309.69	821	17.14694	41703.71	871	18.19121
39357.57	822	17.16783	41751.59	872	18.21210
39405.45	823	17.18871	41799.47	873	18.23298
39453.33	824	17.20960	41847.35	874	18.25387
39501.21	825	17.23048	41895.23	875	18.27475
39549.09	826	17.25137	41943.11	876	18.29564
39596.98	827	17.27225	41990.99	877	18.31653
39644.86	828	17.29314	42038.87	878	18.33741
39692.74	829	17.31402	42086.75	879	18.35830
39740.62	830	17.33491	42134.63	880	18.37918
39788.50	831	17.35580	42182.51	881	18.40007
39836.38	832	17.37668	42230.39	882	18.42095
39884.26	833	17.39757	42278.27	883	18.44184
39932.14	834	17.41845	42326.15	884	18.46272
39980.02	835	17.43934	42374.03	885	18.48361
40027.90	836	17.46022	42421.91	886	18.50449
40075.78	837	17.48111	42469.79	887	18.52538
40123.66	838	17.50199	42517.67	888	18.54627
40171.54	839	17.52288	42565.55	889	18.56715
40219.42	840	17.54376	42613.43	890	18.58804
40267.30	841	17.56465	42661.31	891	18.60892
40315.18	842	17.58554	42709.19	892	18.62981
40363.06	843	17.60642	42757.07	893	18.65069
40410.94	844	17.62731	42804.95	894	18.67158
40458.82	845	17.64819	42852.83	895	18.69246
40506.70	846	17.66908	42900.71	896	18.71335
40554.58	847	17.68996	42948.59	897	18.73423
40602.46	848	17.71085	42996.47	898	18.75512
40650.34	849	17.73173	43044.35	899	18.77601
40698.22	850	17.75262	43092.23	900	18.79689

```
        kPa ─────────── +3 ───────────┐
                                        ├── Pa(N/m²)
        N/cm² ────────── +4 ───────────┘

        Pa · s ◄──────── −3 ──────────── cP
```

Pa / N/m² / kPa / kN/m² / Pa·s / N·s/m²	REF	lbf/ft² / kip/ft² / lbf-s/ft²	Pa / N/m² / kPa / kN/m² / Pa·s / N·s/m²	REF	lbf/ft² / kip/ft² / lbf-s/ft²
43140.11	901	18.81778	45534.13	951	19.86205
43187.99	902	18.83866	45582.01	952	19.88293
43235.37	903	18.85955	45629.89	953	19.90382
43283.75	904	18.88043	45677.77	954	19.92470
43331.64	905	18.90132	45725.65	955	19.94559
43379.52	906	18.92220	45773.53	956	19.96647
43427.40	907	18.94309	45821.41	957	19.98736
43475.28	908	18.96397	45869.29	958	20.00825
43523.16	909	18.98486	45917.17	959	20.02913
43571.04	910	19.00574	45965.05	960	20.05002
43618.92	911	19.02663	46012.93	961	20.07090
43666.80	912	19.04752	46060.81	962	20.09179
43714.68	913	19.06840	46108.69	963	20.11267
43762.56	914	19.08929	46156.57	964	20.13356
43810.44	915	19.11017	46204.45	965	20.15444
43858.32	916	19.13106	46252.33	966	20.17533
43906.20	917	19.15194	46300.21	967	20.19621
43954.08	918	19.17283	46348.09	968	20.21710
44001.96	919	19.19371	46395.97	969	20.23799
44049.84	920	19.21460	46443.85	970	20.25887
44097.72	921	19.23548	46491.73	971	20.27976
44145.60	922	19.25637	46539.61	972	20.30064
44193.48	923	19.27726	46587.49	973	20.32153
44241.36	924	19.29814	46635.37	974	20.34241
44289.24	925	19.31903	46683.25	975	20.36330
44337.12	926	19.33991	46731.13	976	20.38418
44385.00	927	19.36080	46779.01	977	20.40507
44432.88	928	19.38168	46826.89	978	20.42595
44480.76	929	19.40257	46874.77	979	20.44684
44528.64	930	19.42345	46922.65	980	20.46773
44576.52	931	19.44434	46970.54	981	20.48861
44624.40	932	19.46522	47018.42	982	20.50950
44672.28	933	19.48611	47066.30	983	20.53038
44720.16	934	19.50700	47114.18	984	20.55127
44768.04	935	19.52788	47162.06	985	20.57215
44815.92	936	19.54877	47209.94	986	20.59304
44863.80	937	19.56965	47257.82	987	20.61392
44911.68	938	19.59054	47305.70	988	20.63481
44959.56	939	19.61142	47353.58	989	20.65569
45007.44	940	19.63231	47401.46	990	20.67658
45055.32	941	19.65319	47449.34	991	20.69746
45103.21	942	19.67408	47497.22	992	20.71835
45151.08	943	19.69496	47545.10	993	20.73924
45198.97	944	19.71585	47592.98	994	20.76012
45246.85	945	19.73673	47640.86	995	20.78101
45294.73	946	19.75762	47688.74	996	20.80189
45342.61	947	19.77851	47736.62	997	20.82278
45390.49	948	19.79939	47784.50	998	20.84366
45438.37	949	19.82028	47832.38	999	20.86455
45486.25	950	19.84116	47880.26	1000	20.88543

HEAT

British Thermal Unit (International Steam Table) — Kilojoule
(Btu — kJ)

British Thermal Unit (International Steam Table) — Joule
(Btu — J)

British Thermal Unit (International Steam Table) — Megajoule
(Btu — MJ)

Therm — Joule
(therm — J)

Therm — Kilojoule
(therm — kJ)

Therm — Megajoule
(therm — MJ)

Therm — Gigajoule
(therm — GJ)

Conversion Factors:
1 IST British Thermal Unit (Btu) = 1.055056 kilojoules (kJ)
(1 kJ = 0.9478170 Btu)

Therm-British Thermal Unit Relation

The "therm" is an English practical unit of heat, such that 1 therm = 100,000 Btu. The following Decimal-shift Diagram 34-A will assist the user of Table 34 in converting between therms and Btu's.

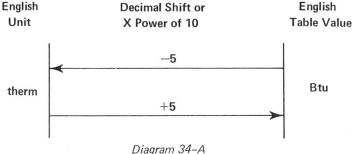

Diagram 34-A

Therefore, from the above conversion factors, 1 therm $= 1.055056 \times 10^5$ kJ. It can be seen that the conversion factor for therm to kJ differs only by a power of 10 from the Btu to kJ relation, from which Table 34 is derived. Therefore, with proper decimal shifting, therms can be converted to SI units through this table.

Example (a): Convert 1.57 therms to kJ.
　From Diagram 34-A: 1.57 therms $= 1.57 \times 10^5$ Btu $= 157 \times 10^3$ Btu
　From table: 157 Btu $= 165.6438$ kJ*
　Therefore: 157×10^3 Btu $= 165.6438 \times 10^3$ kJ $\approx 166 \times 10^3$ kJ

In converting kJ values to J, MJ, and GJ, the Decimal-shift Diagram 34-B, which appears in the headings on alternate table pages, will assist table users. It is derived from: 1 kJ $= (10^3)$ J $= (10^{-3})$ MJ $= (10^{-6})$ GJ.

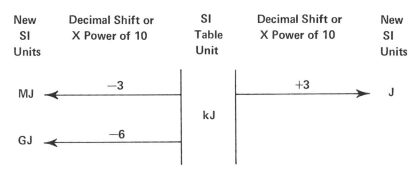

Diagram 34-B

---

*All conversion values obtained can be rounded to desired number of significant figures.

Example (b): Convert 36,500,000 Btu to kJ, J, MJ, and GJ.

36,500,000 Btu = 365 × 10⁵ Btu

From table: 365 Btu = 385.0954 kJ* ≈ 385 kJ

Therefore: 365 × 10⁵ Btu = 385 × 10⁵ kJ

From Diagram 34-B: 1)   385 × 10⁵ kJ = 385 × 10⁵ (10³) J = 385 × 10⁸ J
(kJ to J by multiplying kJ value by 10³.)

2)   385 × 10⁵ kJ = 385 × 10⁵ (10⁻³) MJ
= 385 × 10² MJ = 38,500 MJ
(kJ to MJ by multiplying kJ value by 10⁻³.)

3)   385 × 10⁵ kJ = 385 × 10⁵ (10⁻⁶) GJ
= 385 × 10⁻¹ GJ = 38.5 GJ
(kJ to GJ by multiplying kJ value by 10⁻⁶.)

Diagram 34-C is also given at the top of alternate table pages:

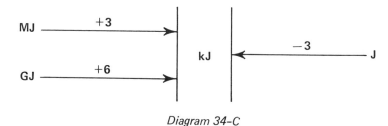

Diagram 34-C

It is the opposite of Diagram 34-B and is used to convert a quantity in J, MJ, GJ, to kJ. The kJ equivalent is then applied to the REF column in the table to obtain Btu.

Example (c): Convert 295 MJ to Btu; and therms.

From Diagram 34-C: 295 MJ = 295 × 10³ kJ

From table: 295 kJ = 279.6060 Btu*

Therefore: 295 × 10³ kJ = 279.6060 × 10³ Btu = 279,606 Btu
≈ 280,000 Btu.

From Diagram 34-A: 280,000 Btu = 280,000 (10⁻⁵) therms = 2.8 therms
(Btu to therms by multiplying Btu value by 10⁻⁵ or by
shifting its decimal 5 places to the left.)

---

*All conversion values obtained can be rounded to desired number of significant figures.

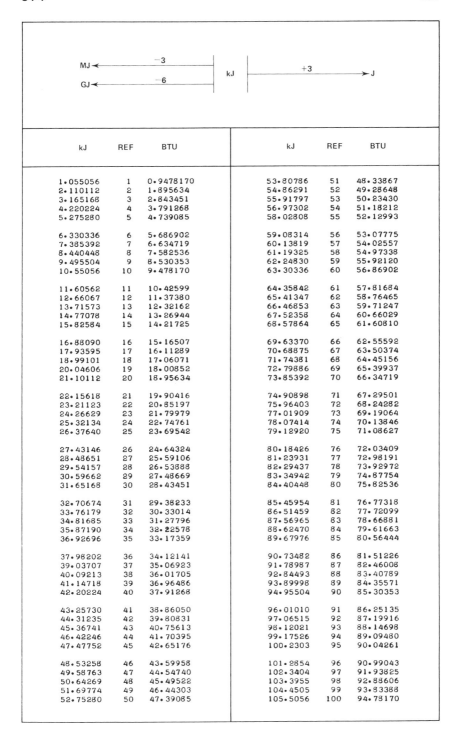

kJ	REF	BTU		kJ	REF	BTU
1.055056	1	0.9478170		53.80786	51	48.33867
2.110112	2	1.895634		54.86291	52	49.28648
3.165168	3	2.843451		55.91797	53	50.23430
4.220224	4	3.791268		56.97302	54	51.18212
5.275280	5	4.739085		58.02808	55	52.12993
6.330336	6	5.686902		59.08314	56	53.07775
7.385392	7	6.634719		60.13819	57	54.02557
8.440448	8	7.582536		61.19325	58	54.97338
9.495504	9	8.530353		62.24830	59	55.92120
10.55056	10	9.478170		63.30336	60	56.86902
11.60562	11	10.42599		64.35842	61	57.81684
12.66067	12	11.37380		65.41347	62	58.76465
13.71573	13	12.32162		66.46853	63	59.71247
14.77078	14	13.26944		67.52358	64	60.66029
15.82584	15	14.21725		68.57864	65	61.60810
16.88090	16	15.16507		69.63370	66	62.55592
17.93595	17	16.11289		70.68875	67	63.50374
18.99101	18	17.06071		71.74381	68	64.45156
20.04606	19	18.00852		72.79886	69	65.39937
21.10112	20	18.95634		73.85392	70	66.34719
22.15618	21	19.90416		74.90898	71	67.29501
23.21123	22	20.85197		75.96403	72	68.24282
24.26629	23	21.79979		77.01909	73	69.19064
25.32134	24	22.74761		78.07414	74	70.13846
26.37640	25	23.69542		79.12920	75	71.08627
27.43146	26	24.64324		80.18426	76	72.03409
28.48651	27	25.59106		81.23931	77	72.98191
29.54157	28	26.53888		82.29437	78	73.92972
30.59662	29	27.48669		83.34942	79	74.87754
31.65168	30	28.43451		84.40448	80	75.82536
32.70674	31	29.38233		85.45954	81	76.77318
33.76179	32	30.33014		86.51459	82	77.72099
34.81685	33	31.27796		87.56965	83	78.66881
35.87190	34	32.22578		88.62470	84	79.61663
36.92696	35	33.17359		89.67976	85	80.56444
37.98202	36	34.12141		90.73482	86	81.51226
39.03707	37	35.06923		91.78987	87	82.46008
40.09213	38	36.01705		92.84493	88	83.40789
41.14718	39	36.96486		93.89998	89	84.35571
42.20224	40	37.91268		94.95504	90	85.30353
43.25730	41	38.86050		96.01010	91	86.25135
44.31235	42	39.80831		97.06515	92	87.19916
45.36741	43	40.75613		98.12021	93	88.14698
46.42246	44	41.70395		99.17526	94	89.09480
47.47752	45	42.65176		100.2303	95	90.04261
48.53258	46	43.59958		101.2854	96	90.99043
49.58763	47	44.54740		102.3404	97	91.93825
50.64269	48	45.49522		103.3955	98	92.88606
51.69774	49	46.44303		104.4505	99	93.83388
52.75280	50	47.39085		105.5056	100	94.78170

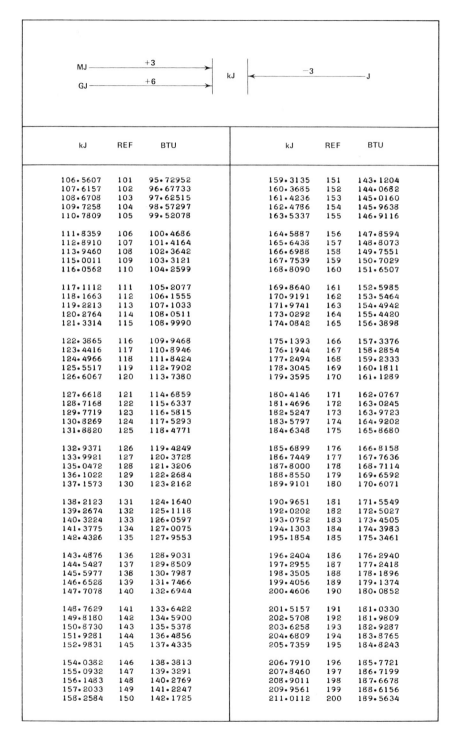

kJ	REF	BTU		kJ	REF	BTU
106.5607	101	95.72952		159.3135	151	143.1204
107.6157	102	96.67733		160.3685	152	144.0682
108.6708	103	97.62515		161.4236	153	145.0160
109.7258	104	98.57297		162.4786	154	145.9638
110.7809	105	99.52078		163.5337	155	146.9116
111.8359	106	100.4686		164.5887	156	147.8594
112.8910	107	101.4164		165.6438	157	148.8073
113.9460	108	102.3642		166.6988	158	149.7551
115.0011	109	103.3121		167.7539	159	150.7029
116.0562	110	104.2599		168.8090	160	151.6507
117.1112	111	105.2077		169.8640	161	152.5985
118.1663	112	106.1555		170.9191	162	153.5464
119.2213	113	107.1033		171.9741	163	154.4942
120.2764	114	108.0511		173.0292	164	155.4420
121.3314	115	108.9990		174.0842	165	156.3898
122.3865	116	109.9468		175.1393	166	157.3376
123.4416	117	110.8946		176.1944	167	158.2854
124.4966	118	111.8424		177.2494	168	159.2333
125.5517	119	112.7902		178.3045	169	160.1811
126.6067	120	113.7380		179.3595	170	161.1289
127.6618	121	114.6859		180.4146	171	162.0767
128.7168	122	115.6337		181.4696	172	163.0245
129.7719	123	116.5815		182.5247	173	163.9723
130.8269	124	117.5293		183.5797	174	164.9202
131.8820	125	118.4771		184.6348	175	165.8680
132.9371	126	119.4249		185.6899	176	166.8158
133.9921	127	120.3728		186.7449	177	167.7636
135.0472	128	121.3206		187.8000	178	168.7114
136.1022	129	122.2684		188.8550	179	169.6592
137.1573	130	123.2162		189.9101	180	170.6071
138.2123	131	124.1640		190.9651	181	171.5549
139.2674	132	125.1118		192.0202	182	172.5027
140.3224	133	126.0597		193.0752	183	173.4505
141.3775	134	127.0075		194.1303	184	174.3983
142.4326	135	127.9553		195.1854	185	175.3461
143.4876	136	128.9031		196.2404	186	176.2940
144.5427	137	129.8509		197.2955	187	177.2418
145.5977	138	130.7987		198.3505	188	178.1896
146.6528	139	131.7466		199.4056	189	179.1374
147.7078	140	132.6944		200.4606	190	180.0852
148.7629	141	133.6422		201.5157	191	181.0330
149.8180	142	134.5900		202.5708	192	181.9809
150.8730	143	135.5378		203.6258	193	182.9287
151.9281	144	136.4856		204.6809	194	183.8765
152.9831	145	137.4335		205.7359	195	184.8243
154.0382	146	138.3813		206.7910	196	185.7721
155.0932	147	139.3291		207.8460	197	186.7199
156.1483	148	140.2769		208.9011	198	187.6678
157.2033	149	141.2247		209.9561	199	188.6156
158.2584	150	142.1725		211.0112	200	189.5634

```
MJ ◄────── -3 ──────┤     ├──────── +3 ────────► J
                    │  kJ │
GJ ◄────── -6 ──────┤     │
```

kJ	REF	BTU		kJ	REF	BTU
212.0663	201	190.5112		264.8191	251	237.9021
213.1213	202	191.4590		265.8741	252	238.8499
214.1764	203	192.4068		266.9292	253	239.7977
215.2314	204	193.3547		267.9842	254	240.7455
216.2865	205	194.3025		269.0393	255	241.6933
217.3415	206	195.2503		270.0943	256	242.6411
218.3966	207	196.1981		271.1494	257	243.5890
219.4516	208	197.1459		272.2044	258	244.5368
220.5067	209	198.0937		273.2595	259	245.4846
221.5618	210	199.0416		274.3146	260	246.4324
222.6168	211	199.9894		275.3696	261	247.3802
223.6719	212	200.9372		276.4247	262	248.3280
224.7269	213	201.8850		277.4797	263	249.2759
225.7820	214	202.8328		278.5348	264	250.2237
226.8370	215	203.7807		279.5898	265	251.1715
227.8921	216	204.7285		280.6449	266	252.1193
228.9472	217	205.6763		281.7000	267	253.0671
230.0022	218	206.6241		282.7550	268	254.0150
231.0573	219	207.5719		283.8101	269	254.9628
232.1123	220	208.5197		284.8651	270	255.9106
233.1674	221	209.4676		285.9202	271	256.8584
234.2224	222	210.4154		286.9752	272	257.8062
235.2775	223	211.3632		288.0303	273	258.7540
236.3325	224	212.3110		289.0853	274	259.7019
237.3876	225	213.2588		290.1404	275	260.6497
238.4427	226	214.2066		291.1955	276	261.5975
239.4977	227	215.1545		292.2505	277	262.5453
240.5528	228	216.1023		293.3056	278	263.4931
241.6078	229	217.0501		294.3606	279	264.4409
242.6629	230	217.9979		295.4157	280	265.3888
243.7179	231	218.9457		296.4707	281	266.3366
244.7730	232	219.8935		297.5258	282	267.2844
245.8280	233	220.8414		298.5808	283	268.2322
246.8831	234	221.7892		299.6359	284	269.1800
247.9382	235	222.7370		300.6910	285	270.1278
248.9932	236	223.6848		301.7460	286	271.0757
250.0483	237	224.6326		302.8011	287	272.0235
251.1033	238	225.5804		303.8561	288	272.9713
252.1584	239	226.5283		304.9112	289	273.9191
253.2134	240	227.4761		305.9662	290	274.8669
254.2685	241	228.4239		307.0213	291	275.8147
255.3236	242	229.3717		308.0764	292	276.7626
256.3786	243	230.3195		309.1314	293	277.7104
257.4337	244	231.2673		310.1865	294	278.6582
258.4887	245	232.2152		311.2415	295	279.6060
259.5438	246	233.1630		312.2966	296	280.5538
260.5988	247	234.1108		313.3516	297	281.5016
261.6539	248	235.0586		314.4067	298	282.4495
262.7089	249	236.0064		315.4617	299	283.3973
263.7640	250	236.9542		316.5168	300	284.3451

MJ —— +3 —→ kJ ←—— −3 —— J
GJ —— +6 —→

kJ	REF	BTU		kJ	REF	BTU
317.5719	301	285.2929		370.3247	351	332.6838
318.6269	302	286.2407		371.3797	352	333.6316
319.6820	303	287.1885		372.4348	353	334.5794
320.7370	304	288.1364		373.4898	354	335.5272
321.7921	305	289.0842		374.5449	355	336.4750
322.8471	306	290.0320		375.5999	356	337.4228
323.9022	307	290.9798		376.6550	357	338.3707
324.9572	308	291.9276		377.7100	358	339.3185
326.0123	309	292.8754		378.7651	359	340.2663
327.0674	310	293.8233		379.8202	360	341.2141
328.1224	311	294.7711		380.8752	361	342.1619
329.1775	312	295.7189		381.9303	362	343.1097
330.2325	313	296.6667		382.9853	363	344.0576
331.2876	314	297.6145		384.0404	364	345.0054
332.3426	315	298.5624		385.0954	365	345.9532
333.3977	316	299.5102		386.1505	366	346.9010
334.4528	317	300.4580		387.2056	367	347.8488
335.5078	318	301.4058		388.2606	368	348.7966
336.5629	319	302.3536		389.3157	369	349.7445
337.6179	320	303.3014		390.3707	370	350.6923
338.6730	321	304.2493		391.4258	371	351.6401
339.7280	322	305.1971		392.4808	372	352.5879
340.7831	323	306.1449		393.5359	373	353.5357
341.8381	324	307.0927		394.5909	374	354.4836
342.8932	325	308.0405		395.6460	375	355.4314
343.9483	326	308.9883		396.7011	376	356.3792
345.0033	327	309.9362		397.7561	377	357.3270
346.0584	328	310.8840		398.8112	378	358.2748
347.1134	329	311.8318		399.8662	379	359.2226
348.1685	330	312.7796		400.9213	380	360.1705
349.2235	331	313.7274		401.9763	381	361.1183
350.2786	332	314.6752		403.0314	382	362.0661
351.3336	333	315.6231		404.0864	383	363.0139
352.3887	334	316.5709		405.1415	384	363.9617
353.4438	335	317.5187		406.1966	385	364.9095
354.4988	336	318.4665		407.2516	386	365.8574
355.5539	337	319.4143		408.3067	387	366.8052
356.6089	338	320.3621		409.3617	388	367.7530
357.6640	339	321.3100		410.4168	389	368.7008
358.7190	340	322.2578		411.4718	390	369.6486
359.7741	341	323.2056		412.5269	391	370.5964
360.8292	342	324.1534		413.5820	392	371.5443
361.8842	343	325.1012		414.6370	393	372.4921
362.9393	344	326.0490		415.6921	394	373.4399
363.9943	345	326.9969		416.7471	395	374.3877
365.0494	346	327.9447		417.8022	396	375.3355
366.1044	347	328.8925		418.8572	397	376.2833
367.1595	348	329.8403		419.9123	398	377.2312
368.2145	349	330.7881		420.9673	399	378.1790
369.2696	350	331.7359		422.0224	400	379.1268

MJ ←————— -3 —————
                          kJ ——————— +3 ———————→ J
GJ ←————— -6 —————

kJ	REF	BTU		kJ	REF	BTU
423.0775	401	380.0746		475.8303	451	427.4655
424.1325	402	381.0224		476.8853	452	428.4133
425.1876	403	381.9702		477.9404	453	429.3611
426.2426	404	382.9181		478.9954	454	430.3089
427.2977	405	383.8659		480.0505	455	431.2567
428.3527	406	384.8137		481.1055	456	432.2045
429.4078	407	385.7615		482.1606	457	433.1524
430.4628	408	386.7093		483.2157	458	434.1002
431.5179	409	387.6571		484.2707	459	435.0480
432.5730	410	388.6050		485.3258	460	435.9958
433.6280	411	389.5528		486.3808	461	436.9436
434.6831	412	390.5006		487.4359	462	437.8914
435.7381	413	391.4484		488.4909	463	438.8393
436.7932	414	392.3962		489.5460	464	439.7871
437.8482	415	393.3440		490.6010	465	440.7349
438.9033	416	394.2919		491.6561	466	441.6827
439.9584	417	395.2397		492.7112	467	442.6305
441.0134	418	396.1875		493.7662	468	443.5783
442.0685	419	397.1353		494.8213	469	444.5262
443.1235	420	398.0831		495.8763	470	445.4740
444.1786	421	399.0309		496.9314	471	446.4218
445.2336	422	399.9788		497.9864	472	447.3696
446.2887	423	400.9266		499.0415	473	448.3174
447.3437	424	401.8744		500.0965	474	449.2653
448.3988	425	402.8222		501.1516	475	450.2131
449.4539	426	403.7700		502.2067	476	451.1609
450.5089	427	404.7179		503.2617	477	452.1087
451.5640	428	405.6657		504.3168	478	453.0565
452.6190	429	406.6135		505.3718	479	454.0043
453.6741	430	407.5613		506.4269	480	454.9522
454.7291	431	408.5091		507.4819	481	455.9000
455.7842	432	409.4569		508.5370	482	456.8478
456.8392	433	410.4048		509.5921	483	457.7956
457.8943	434	411.3526		510.6471	484	458.7434
458.9494	435	412.3004		511.7022	485	459.6912
460.0044	436	413.2482		512.7572	486	460.6391
461.0595	437	414.1960		513.8123	487	461.5869
462.1145	438	415.1438		514.8673	488	462.5347
463.1696	439	416.0917		515.9224	489	463.4825
464.2246	440	417.0395		516.9774	490	464.4303
465.2797	441	417.9873		518.0325	491	465.3781
466.3348	442	418.9351		519.0876	492	466.3260
467.3898	443	419.8829		520.1426	493	467.2738
468.4449	444	420.8307		521.1977	494	468.2216
469.4999	445	421.7786		522.2527	495	469.1694
470.5550	446	422.7264		523.3078	496	470.1172
471.6100	447	423.6742		524.3628	497	471.0650
472.6651	448	424.6220		525.4179	498	472.0129
473.7201	449	425.5698		526.4729	499	472.9607
474.7752	450	426.5176		527.5280	500	473.9085

MJ ———————— +3 ————————→
kJ ←———————— −3 ————————— J
GJ ———————— +6 ————————→

kJ	REF	BTU		kJ	REF	BTU
528.5831	501	474.8563		581.3359	551	522.2472
529.6381	502	475.8041		582.3909	552	523.1950
530.6932	503	476.7519		583.4460	553	524.1428
531.7482	504	477.6998		584.5010	554	525.0906
532.8033	505	478.6476		585.5561	555	526.0384
533.8583	506	479.5954		586.6111	556	526.9862
534.9134	507	480.5432		587.6662	557	527.9341
535.9685	508	481.4910		588.7213	558	528.8819
537.0235	509	482.4388		589.7763	559	529.8297
538.0786	510	483.3867		590.8314	560	530.7775
539.1336	511	484.3345		591.8864	561	531.7253
540.1887	512	485.2823		592.9415	562	532.6731
541.2437	513	486.2301		593.9965	563	533.6210
542.2988	514	487.1779		595.0516	564	534.5688
543.3538	515	488.1257		596.1066	565	535.5166
544.4089	516	489.0736		597.1617	566	536.4644
545.4640	517	490.0214		598.2168	567	537.4122
546.5190	518	490.9692		599.2718	568	538.3600
547.5741	519	491.9170		600.3269	569	539.3079
548.6291	520	492.8648		601.3819	570	540.2557
549.6842	521	493.8126		602.4370	571	541.2035
550.7392	522	494.7605		603.4920	572	542.1513
551.7943	523	495.7083		604.5471	573	543.0991
552.8493	524	496.6561		605.6021	574	544.0470
553.9044	525	497.6039		606.6572	575	544.9948
554.9595	526	498.5517		607.7123	576	545.9426
556.0145	527	499.4995		608.7673	577	546.8904
557.0696	528	500.4474		609.8224	578	547.8382
558.1246	529	501.3952		610.8774	579	548.7860
559.1797	530	502.3430		611.9325	580	549.7338
560.2347	531	503.2908		612.9875	581	550.6817
561.2898	532	504.2386		614.0426	582	551.6295
562.3448	533	505.1865		615.0976	583	552.5773
563.3999	534	506.1343		616.1527	584	553.5251
564.4550	535	507.0821		617.2078	585	554.4729
565.5100	536	508.0299		618.2628	586	555.4208
566.5651	537	508.9777		619.3179	587	556.3686
567.6201	538	509.9255		620.3729	588	557.3164
568.6752	539	510.8734		621.4280	589	558.2642
569.7302	540	511.8212		622.4830	590	559.2120
570.7853	541	512.7690		623.5381	591	560.1598
571.8404	542	513.7168		624.5932	592	561.1077
572.8954	543	514.6646		625.6482	593	562.0555
573.9505	544	515.6124		626.7033	594	563.0033
575.0055	545	516.5603		627.7583	595	563.9511
576.0606	546	517.5081		628.8134	596	564.8989
577.1156	547	518.4559		629.8684	597	565.8467
578.1707	548	519.4037		630.9235	598	566.7946
579.2257	549	520.3515		631.9785	599	567.7424
580.2808	550	521.2993		633.0336	600	568.6902

kJ	REF	BTU		kJ	REF	BTU
634.0887	601	569.6380		686.8415	651	617.0289
635.1437	602	570.5858		687.8965	652	617.9767
636.1988	603	571.5336		688.9516	653	618.9245
637.2538	604	572.4815		690.0066	654	619.8723
638.3089	605	573.4293		691.0617	655	620.8201
639.3639	606	574.3771		692.1167	656	621.7679
640.4190	607	575.3249		693.1718	657	622.7158
641.4741	608	576.2727		694.2269	658	623.6636
642.5291	609	577.2205		695.2819	659	624.6114
643.5842	610	578.1684		696.3370	660	625.5592
644.6392	611	579.1162		697.3920	661	626.5070
645.6943	612	580.0640		698.4471	662	627.4548
646.7493	613	581.0118		699.5021	663	628.4027
647.8044	614	581.9596		700.5572	664	629.3505
648.8594	615	582.9074		701.6122	665	630.2983
649.9145	616	583.8553		702.6673	666	631.2461
650.9696	617	584.8031		703.7224	667	632.1939
652.0246	618	585.7509		704.7774	668	633.1417
653.0797	619	586.6987		705.8325	669	634.0896
654.1347	620	587.6465		706.8875	670	635.0374
655.1898	621	588.5943		707.9426	671	635.9852
656.2448	622	589.5422		708.9976	672	636.9330
657.2999	623	590.4900		710.0527	673	637.8808
658.3549	624	591.4378		711.1077	674	638.8286
659.4100	625	592.3856		712.1628	675	639.7765
660.4651	626	593.3334		713.2179	676	640.7243
661.5201	627	594.2812		714.2729	677	641.6721
662.5752	628	595.2291		715.3280	678	642.6199
663.6302	629	596.1769		716.3830	679	643.5677
664.6853	630	597.1247		717.4381	680	644.5155
665.7403	631	598.0725		718.4931	681	645.4634
666.7954	632	599.0203		719.5482	682	646.4112
667.8504	633	599.9681		720.6032	683	647.3590
668.9055	634	600.9160		721.6583	684	648.3068
669.9606	635	601.8638		722.7134	685	649.2546
671.0156	636	602.8116		723.7684	686	650.2025
672.0707	637	603.7594		724.8235	687	651.1503
673.1257	638	604.7072		725.8785	688	652.0981
674.1808	639	605.6551		726.9336	689	653.0459
675.2358	640	606.6029		727.9886	690	653.9937
676.2909	641	607.5507		729.0437	691	654.9415
677.3460	642	608.4985		730.0988	692	655.8894
678.4010	643	609.4463		731.1538	693	656.8372
679.4561	644	610.3941		732.2089	694	657.7850
680.5111	645	611.3420		733.2639	695	658.7328
681.5662	646	612.2898		734.3190	696	659.6806
682.6212	647	613.2376		735.3740	697	660.6284
683.6763	648	614.1854		736.4291	698	661.5763
684.7313	649	615.1332		737.4841	699	662.5241
685.7864	650	616.0810		738.5392	700	663.4719

```
MJ ———— +3 ———→
                      kJ  ←———— -3 ————— J
GJ ———— +6 ———→
```

kJ	REF	BTU	kJ	REF	BTU
739.5943	701	664.4197	792.3471	751	711.8106
740.6493	702	665.3675	793.4021	752	712.7584
741.7044	703	666.3153	794.4572	753	713.7062
742.7594	704	667.2632	795.5122	754	714.6540
743.8145	705	668.2110	796.5673	755	715.6018
744.8695	706	669.1588	797.6223	756	716.5496
745.9246	707	670.1066	798.6774	757	717.4975
746.9797	708	671.0544	799.7325	758	718.4453
748.0347	709	672.0022	800.7875	759	719.3931
749.0898	710	672.9501	801.8426	760	720.3409
750.1448	711	673.8979	802.8976	761	721.2887
751.1999	712	674.8457	803.9527	762	722.2365
752.2549	713	675.7935	805.0077	763	723.1844
753.3100	714	676.7413	806.0628	764	724.1322
754.3650	715	677.6891	807.1178	765	725.0800
755.4201	716	678.6370	808.1729	766	726.0278
756.4752	717	679.5848	809.2280	767	726.9756
757.5302	718	680.5326	810.2830	768	727.9234
758.5853	719	681.4804	811.3381	769	728.8713
759.6403	720	682.4282	812.3931	770	729.8191
760.6954	721	683.3760	813.4482	771	730.7669
761.7504	722	684.3239	814.5032	772	731.7147
762.8055	723	685.2717	815.5583	773	732.6625
763.8605	724	686.2195	816.6133	774	733.6103
764.9156	725	687.1673	817.6684	775	734.5582
765.9707	726	688.1151	818.7235	776	735.5060
767.0257	727	689.0630	819.7785	777	736.4538
768.0808	728	690.0108	820.8336	778	737.4016
769.1358	729	690.9586	821.8886	779	738.3494
770.1909	730	691.9064	822.9437	780	739.2972
771.2459	731	692.8542	823.9987	781	740.2451
772.3010	732	693.8020	825.0538	782	741.1929
773.3560	733	694.7498	826.1089	783	742.1407
774.4111	734	695.6977	827.1639	784	743.0885
775.4662	735	696.6455	828.2190	785	744.0363
776.5212	736	697.5933	829.2740	786	744.9841
777.5763	737	698.5411	830.3291	787	745.9320
778.6313	738	699.4889	831.3841	788	746.8793
779.6864	739	700.4368	832.4392	789	747.8276
780.7414	740	701.3846	833.4942	790	748.7754
781.7965	741	702.3324	834.5493	791	749.7232
782.8516	742	703.2802	835.6044	792	750.6711
783.9066	743	704.2280	836.6594	793	751.6189
784.9617	744	705.1758	837.7145	794	752.5667
786.0167	745	706.1236	838.7695	795	753.5145
787.0718	746	707.0715	839.8246	796	754.4623
788.1268	747	708.0193	840.8796	797	755.4101
789.1819	748	708.9671	841.9347	798	756.3580
790.2369	749	709.9149	842.9897	799	757.3058
791.2920	750	710.8627	844.0448	800	758.2536

MJ ← — 3
GJ ← —6   kJ   +3 → J

kJ	REF	BTU		kJ	REF	BTU
845.0999	801	759.2014		897.8527	851	806.5923
846.1549	802	760.1492		898.9077	852	807.5401
847.2100	803	761.0970		899.9628	853	808.4879
848.2650	804	762.0449		901.0178	854	809.4357
849.3201	805	762.9927		902.0729	855	810.3835
850.3751	806	763.9405		903.1279	856	811.3313
851.4302	807	764.8883		904.1830	857	812.2792
852.4853	808	765.8361		905.2381	858	813.2270
853.5403	809	766.7839		906.2931	859	814.1748
854.5954	810	767.7318		907.3482	860	815.1226
855.6504	811	768.6796		908.4032	861	816.0704
856.7055	812	769.6274		909.4583	862	817.0182
857.7605	813	770.5752		910.5133	863	817.9661
858.8156	814	771.5230		911.5684	864	818.9139
859.8706	815	772.4708		912.6234	865	819.8617
860.9257	816	773.4187		913.6785	866	820.8095
861.9808	817	774.3665		914.7336	867	821.7573
863.0358	818	775.3143		915.7886	868	822.7051
864.0909	819	776.2621		916.8437	869	823.6530
865.1459	820	777.2099		917.8987	870	824.6008
866.2010	821	778.1577		918.9538	871	825.5486
867.2560	822	779.1056		920.0088	872	826.4964
868.3111	823	780.0534		921.0639	873	827.4442
869.3661	824	781.0012		922.1189	874	828.3920
870.4212	825	781.9490		923.1740	875	829.3399
871.4763	826	782.8968		924.2291	876	830.2877
872.5313	827	783.8446		925.2841	877	831.2355
873.5864	828	784.7925		926.3392	878	832.1833
874.6414	829	785.7403		927.3942	879	833.1311
875.6965	830	786.6881		928.4493	880	834.0789
876.7515	831	787.6359		929.5043	881	835.0268
877.8066	832	788.5837		930.5594	882	835.9746
878.8617	833	789.5315		931.6145	883	836.9224
879.9167	834	790.4794		932.6695	884	837.8702
880.9718	835	791.4272		933.7246	885	838.8180
882.0268	836	792.3750		934.7796	886	839.7658
883.0819	837	793.3228		935.8347	887	840.7137
884.1369	838	794.2706		936.8897	888	841.6615
885.1920	839	795.2184		937.9448	889	842.6093
886.2470	840	796.1663		938.9998	890	843.5571
887.3021	841	797.1141		940.0549	891	844.5049
888.3572	842	798.0619		941.1100	892	845.4528
889.4122	843	799.0097		942.1650	893	846.4006
890.4673	844	799.9575		943.2201	894	847.3484
891.5223	845	800.9053		944.2751	895	848.2962
892.5774	846	801.8532		945.3302	896	849.2440
893.6324	847	802.8010		946.3852	897	850.1918
894.6875	848	803.7488		947.4403	898	851.1396
895.7425	849	804.6966		948.4953	899	852.0875
896.7976	850	805.6444		949.5504	900	853.0353

MJ ———— +3 ————→ kJ ←———— -3 ————→ J
GJ ———— +6 ————→

kJ	REF	BTU		kJ	REF	BTU
950.6055	901	853.9831		1003.358	951	901.3739
951.6605	902	854.9309		1004.413	952	902.3218
952.7156	903	855.8787		1005.468	953	903.2696
953.7706	904	856.8266		1006.523	954	904.2174
954.8257	905	857.7744		1007.578	955	905.1652
955.8807	906	858.7222		1008.634	956	906.1130
956.9358	907	859.6700		1009.689	957	907.0609
957.9909	908	860.6178		1010.744	958	908.0087
959.0459	909	861.5656		1011.799	959	908.9565
960.1010	910	862.5135		1012.854	960	909.9043
961.1560	911	863.4613		1013.909	961	910.8521
962.2111	912	864.4091		1014.964	962	911.7999
963.2661	913	865.3569		1016.019	963	912.7478
964.3212	914	866.3047		1017.074	964	913.6956
965.3762	915	867.2525		1018.129	965	914.6434
966.4313	916	868.2004		1019.184	966	915.5912
967.4864	917	869.1482		1020.239	967	916.5390
968.5414	918	870.0960		1021.294	968	917.4868
969.5965	919	871.0438		1022.349	969	918.4347
970.6515	920	871.9916		1023.404	970	919.3825
971.7066	921	872.9394		1024.459	971	920.3303
972.7616	922	873.8873		1025.514	972	921.2781
973.8167	923	874.8351		1026.569	973	922.2259
974.8717	924	875.7829		1027.625	974	923.1737
975.9268	925	876.7307		1028.680	975	924.1216
976.9819	926	877.6785		1029.735	976	925.0694
978.0369	927	878.6263		1030.790	977	926.0172
979.0920	928	879.5742		1031.845	978	926.9650
980.1470	929	880.5220		1032.900	979	927.9128
981.2021	930	881.4698		1033.955	980	928.8606
982.2571	931	882.4176		1035.010	981	929.8085
983.3122	932	883.3654		1036.065	982	930.7563
984.3673	933	884.3132		1037.120	983	931.7041
985.4223	934	885.2611		1038.175	984	932.6519
986.4774	935	886.2089		1039.230	985	933.5997
987.5324	936	887.1567		1040.285	986	934.5475
988.5875	937	888.1045		1041.340	987	935.4954
989.6425	938	889.0523		1042.395	988	936.4432
990.6976	939	890.0001		1043.450	989	937.3910
991.7526	940	890.9480		1044.505	990	938.3388
992.8077	941	891.8958		1045.561	991	939.2866
993.8628	942	892.8436		1046.616	992	940.2344
994.9178	943	893.7914		1047.671	993	941.1823
995.9729	944	894.7392		1048.726	994	942.1301
997.0279	945	895.6870		1049.781	995	943.0779
998.0830	946	896.6349		1050.836	996	944.0257
999.1380	947	897.5827		1051.891	997	944.9735
1000.193	948	898.5305		1052.946	998	945.9213
1001.248	949	899.4783		1054.001	999	946.8692
1002.303	950	900.4261		1055.056	1000	947.8170

POWER

Horsepower (550 ft-lb/s) — Kilowatt (or Kilojoule/Second)
(hp — kW — kJ/s)
Horsepower — Watt (or Joule/Second)
(hp — W — J/s)
Horsepower — Megawatt (or Megajoule/Second)
(hp — MW — MJ/s)

Conversion Factors:
1 horsepower (hp) = 0.7456999 kilowatts (kW)
(1 kW = 1.341022 hp)

In converting kW values to W and MW, the Decimal-shift Diagram 35-A, which appears in the headings on alternate table pages, will assist table users. It is derived from: $1 \text{ kW} = (10^{3}) \text{ W} = (10^{-3}) \text{ MW}$.

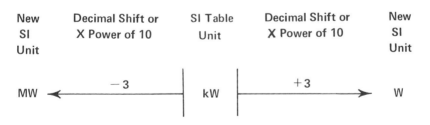

New SI Unit	Decimal Shift or X Power of 10	SI Table Unit	Decimal Shift or X Power of 10	New SI Unit
MW	← −3	kW	+3 →	W

Diagram 35-A

Example (*a*): Convert 3,700 hp to kW, W, and MW.
Since: 3,700 hp = 370 × 10 hp:
From table: 370 hp = 275.9090 kW*
Therefore: 370 × 10 hp = 275.9090 × 10 kW = 2759.090 kW = 2,760 kW.

---

*All conversion values obtained can be rounded to desired number of significant figures.

From Diagram 35-A: 1) 2,760 kW = 2,760 × 10^{-3} MW = 2.76 MW
(kW to MW by multiplying kW value by 10^{-3} or by shifting its decimal 3 places to the left.)

2) 2.760 kW = 2,760 × 10^{3} W = 2,760,000 W
(kW to W by multiplying kW value by 10^{3} or by shifting its decimal 3 places to the right.)

Diagram 35-B is also given at the top of alternate table pages:

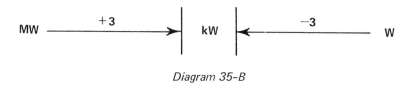

Diagram 35-B

It is the opposite of Diagram 35-A and is used to convert a quantity in W or MW to kW. The kW equivalent is then applied to the REF column in the table to obtain hp.

Example (*b*): Convert 25,500 W to hp.
From Diagram 35-B: 25,500 W = 25,500 × 10^{-3} kW = 255 × 10^{-1} kW
From table: 255 kW = 341.9606 hp*
Therefore: 255 × 10^{-1} kW = 341.9606 × 10^{-1} kW = 34.19606 kW
≈ 34.2 hp.

---

*All conversion values obtained can be rounded to desired number of significant figures.

MW ◄————— -3 ————| kW |————— +3 —————► W

kW kJ/s	REF	hp	kW kJ/s	REF	hp
0.7456999	1	1.341022	38.03069	51	68.39212
1.491400	2	2.682044	38.77639	52	69.73315
2.237100	3	4.023066	39.52209	53	71.07417
2.982800	4	5.364088	40.26779	54	72.41519
3.728500	5	6.705110	41.01349	55	73.75621
4.474199	6	8.046132	41.75919	56	75.09723
5.219899	7	9.387154	42.50489	57	76.43826
5.965599	8	10.72818	43.25059	58	77.77928
6.711299	9	12.06920	43.99629	59	79.12030
7.456999	10	13.41022	44.74199	60	80.46132
8.202699	11	14.75124	45.48769	61	81.80234
8.948399	12	16.09226	46.23339	62	83.14337
9.694099	13	17.43329	46.97909	63	84.48439
10.43980	14	18.77431	47.72479	64	85.82541
11.18550	15	20.11533	48.47049	65	87.16643
11.93120	16	21.45635	49.21619	66	88.50745
12.67690	17	22.79737	49.96189	67	89.84848
13.42260	18	24.13840	50.70759	68	91.18950
14.16830	19	25.47942	51.45329	69	92.53052
14.91400	20	26.82044	52.19899	70	93.87154
15.65970	21	28.16146	52.94469	71	95.21257
16.40540	22	29.50249	53.69039	72	96.55359
17.15110	23	30.84351	54.43609	73	97.89461
17.89680	24	32.18453	55.18179	74	99.23563
18.64250	25	33.52555	55.92749	75	100.5767
19.38820	26	34.86657	56.67319	76	101.9177
20.13390	27	36.20760	57.41889	77	103.2587
20.87960	28	37.54862	58.16459	78	104.5997
21.62530	29	38.88964	58.91029	79	105.9407
22.37100	30	40.23066	59.65599	80	107.2818
23.11670	31	41.57168	60.40169	81	108.6228
23.86240	32	42.91271	61.14739	82	109.9638
24.60810	33	44.25373	61.89309	83	111.3048
25.35380	34	45.59475	62.63879	84	112.6459
26.09950	35	46.93577	63.38449	85	113.9869
26.84520	36	48.27679	64.13019	86	115.3279
27.59090	37	49.61782	64.87589	87	116.6689
28.33660	38	50.95884	65.62159	88	118.0099
29.08230	39	52.29986	66.36729	89	119.3510
29.82800	40	53.64088	67.11299	90	120.6920
30.57370	41	54.98190	67.85869	91	122.0330
31.31940	42	56.32293	68.60439	92	123.3740
32.06510	43	57.66395	69.35009	93	124.7150
32.81080	44	59.00497	70.09579	94	126.0561
33.55650	45	60.34599	70.84149	95	127.3971
34.30220	46	61.68701	71.58719	96	128.7381
35.04789	47	63.02804	72.33289	97	130.0791
35.79360	48	64.36906	73.07859	98	131.4202
36.53930	49	65.71008	73.82429	99	132.7612
37.28500	50	67.05110	74.56999	100	134.1022

```
MW ————————— +3 ————————→  kW  ←————————— −3 ————————— W
```

kW kJ/s	REF	hp		kW kJ/s	REF	hp
75.31569	101	135.4432		112.6007	151	202.4943
76.06139	102	136.7842		113.3464	152	203.8354
76.80709	103	138.1253		114.0921	153	205.1764
77.55279	104	139.4663		114.8378	154	206.5174
78.29849	105	140.8073		115.5835	155	207.8584
79.04419	106	142.1483		116.3292	156	209.1994
79.78989	107	143.4894		117.0749	157	210.5405
80.53559	108	144.8304		117.8206	158	211.8815
81.28129	109	146.1714		118.5663	159	213.2225
82.02699	110	147.5124		119.3120	160	214.5635
82.77269	111	148.8534		120.0577	161	215.9045
83.51839	112	150.1945		120.8034	162	217.2456
84.26409	113	151.5355		121.5491	163	218.5866
85.00979	114	152.8765		122.2948	164	219.9276
85.75549	115	154.2175		123.0405	165	221.2686
86.50119	116	155.5586		123.7862	166	222.6097
87.24689	117	156.8996		124.5319	167	223.9507
87.99259	118	158.2406		125.2776	168	225.2917
88.73829	119	159.5816		126.0233	169	226.6327
89.48399	120	160.9226		126.7690	170	227.9737
90.22969	121	162.2637		127.5147	171	229.3148
90.97539	122	163.6047		128.2604	172	230.6558
91.72109	123	164.9457		129.0061	173	231.9968
92.46679	124	166.2867		129.7518	174	233.3378
93.21249	125	167.6278		130.4975	175	234.6789
93.95819	126	168.9688		131.2432	176	236.0199
94.70389	127	170.3098		131.9889	177	237.3609
95.44959	128	171.6508		132.7346	178	238.7019
96.19529	129	172.9918		133.4803	179	240.0429
96.94099	130	174.3329		134.2260	180	241.3840
97.68669	131	175.6739		134.9717	181	242.7250
98.43239	132	177.0149		135.7174	182	244.0660
99.17809	133	178.3559		136.4631	183	245.4070
99.92379	134	179.6970		137.2088	184	246.7481
100.6695	135	181.0380		137.9545	185	248.0891
101.4152	136	182.3790		138.7002	186	249.4301
102.1609	137	183.7200		139.4459	187	250.7711
102.9066	138	185.0610		140.1916	188	252.1121
103.6523	139	186.4021		140.9373	189	253.4532
104.3980	140	187.7431		141.6830	190	254.7942
105.1437	141	189.0841		142.4287	191	256.1352
105.8894	142	190.4251		143.1744	192	257.4762
106.6351	143	191.7662		143.9201	193	258.8173
107.3808	144	193.1072		144.6658	194	260.1583
108.1265	145	194.4482		145.4115	195	261.4993
108.8722	146	195.7892		146.1572	196	262.8403
109.6179	147	197.1302		146.9029	197	264.1813
110.3636	148	198.4713		147.6486	198	265.5224
111.1093	149	199.8123		148.3943	199	266.8634
111.8550	150	201.1533		149.1400	200	268.2044

MW ←———— −3 ————| kW |———— +3 ————→ W

kW kJ/s	REF	hp	kW kJ/s	REF	hp
149.8857	201	269.5454	187.1707	251	336.5965
150.6314	202	270.8865	187.9164	252	337.9376
151.3771	203	272.2275	188.6621	253	339.2786
152.1228	204	273.5685	189.4078	254	340.6196
152.8685	205	274.9095	190.1535	255	341.9606
153.6142	206	276.2505	190.8992	256	343.3016
154.3599	207	277.5916	191.6449	257	344.6427
155.1056	208	278.9326	192.3906	258	345.9837
155.8513	209	280.2736	193.1363	259	347.3247
156.5970	210	281.6146	193.8820	260	348.6657
157.3427	211	282.9557	194.6277	261	350.0068
158.0884	212	284.2967	195.3734	262	351.3478
158.8341	213	285.6377	196.1191	263	352.6888
159.5798	214	286.9787	196.8648	264	354.0298
160.3255	215	288.3197	197.6105	265	355.3708
161.0712	216	289.6608	198.3562	266	356.7119
161.8169	217	291.0018	199.1019	267	358.0529
162.5626	218	292.3428	199.8476	268	359.3939
163.3083	219	293.6838	200.5933	269	360.7349
164.0540	220	295.0248	201.3390	270	362.0760
164.7997	221	296.3659	202.0847	271	363.4170
165.5454	222	297.7069	202.8304	272	364.7580
166.2911	223	299.0479	203.5761	273	366.0990
167.0368	224	300.3889	204.3218	274	367.4400
167.7825	225	301.7300	205.0675	275	368.7811
168.5282	226	303.0710	205.8132	276	370.1221
169.2739	227	304.4120	206.5589	277	371.4631
170.0196	228	305.7530	207.3046	278	372.8041
170.7653	229	307.0940	208.0503	279	374.1451
171.5110	230	308.4351	208.7960	280	375.4862
172.2567	231	309.7761	209.5417	281	376.8272
173.0024	232	311.1171	210.2874	282	378.1682
173.7481	233	312.4581	211.0331	283	379.5092
174.4938	234	313.7992	211.7788	284	380.8503
175.2395	235	315.1402	212.5245	285	382.1913
175.9852	236	316.4812	213.2702	286	383.5323
176.7309	237	317.8222	214.0159	287	384.8733
177.4766	238	319.1632	214.7616	288	386.2143
178.2223	239	320.5043	215.5073	289	387.5554
178.9680	240	321.8453	216.2530	290	388.8964
179.7137	241	323.1863	216.9987	291	390.2374
180.4594	242	324.5273	217.7444	292	391.5784
181.2051	243	325.8684	218.4901	293	392.9195
181.9508	244	327.2094	219.2358	294	394.2605
182.6965	245	328.5504	219.9815	295	395.6015
183.4422	246	329.8914	220.7272	296	396.9425
184.1879	247	331.2324	221.4729	297	398.2835
184.9336	248	332.5735	222.2186	298	399.6246
185.6793	249	333.9145	222.9643	299	400.9656
186.4250	250	335.2555	223.7100	300	402.3066

MW ──────── +3 ───────→ | kW | ←─────── −3 ──────── W

kW kJ/s	REF	hp	kW kJ/s	REF	hp
224.4557	301	403.6476	261.7407	351	470.6987
225.2014	302	404.9887	262.4864	352	472.0398
225.9471	303	406.3297	263.2321	353	473.3808
226.6928	304	407.6707	263.9778	354	474.7218
227.4385	305	409.0117	264.7235	355	476.0628
228.1842	306	410.3527	265.4692	356	477.4038
228.9299	307	411.6938	266.2149	357	478.7449
229.6756	308	413.0348	266.9606	358	480.0859
230.4213	309	414.3758	267.7063	359	481.4269
231.1670	310	415.7168	268.4520	360	482.7679
231.9127	311	417.0579	269.1977	361	484.1090
232.6584	312	418.3989	269.9434	362	485.4500
233.4041	313	419.7399	270.6891	363	486.7910
234.1498	314	421.0809	271.4348	364	488.1320
234.8955	315	422.4219	272.1805	365	489.4730
235.6412	316	423.7630	272.9262	366	490.8141
236.3869	317	425.1040	273.6719	367	492.1551
237.1326	318	426.4450	274.4176	368	493.4961
237.8783	319	427.7860	275.1633	369	494.8371
238.6240	320	429.1271	275.9090	370	496.1782
239.3697	321	430.4681	276.6547	371	497.5192
240.1154	322	431.8091	277.4004	372	498.8602
240.8611	323	433.1501	278.1461	373	500.2012
241.6068	324	434.4911	278.8918	374	501.5422
242.3525	325	435.8322	279.6375	375	502.8833
243.0982	326	437.1732	280.3832	376	504.2243
243.8439	327	438.5142	281.1289	377	505.5653
244.5896	328	439.8552	281.8746	378	506.9063
245.3353	329	441.1963	282.6203	379	508.2474
246.0810	330	442.5373	283.3660	380	509.5884
246.8267	331	443.8783	284.1117	381	510.9294
247.5724	332	445.2193	284.8574	382	512.2704
248.3181	333	446.5603	285.6031	383	513.6114
249.0638	334	447.9014	286.3488	384	514.9525
249.8095	335	449.2424	287.0945	385	516.2935
250.5552	336	450.5834	287.8402	386	517.6345
251.3009	337	451.9244	288.5859	387	518.9755
252.0466	338	453.2654	289.3316	388	520.3166
252.7923	339	454.6065	290.0773	389	521.6576
253.5380	340	455.9475	290.8230	390	522.9986
254.2837	341	457.2885	291.5687	391	524.3396
255.0294	342	458.6295	292.3144	392	525.6806
255.7751	343	459.9706	293.0601	393	527.0217
256.5208	344	461.3116	293.8058	394	528.3627
257.2665	345	462.6526	294.5515	395	529.7037
258.0122	346	463.9936	295.2972	396	531.0447
258.7579	347	465.3346	296.0429	397	532.3857
259.5036	348	466.6757	296.7886	398	533.7268
260.2493	349	468.0167	297.5343	399	535.0678
260.9950	350	469.3577	298.2800	400	536.4088

MW ←——— −3 ——— | kW | ——— +3 ——→ W

kW kJ/s	REF	hp	kW kJ/s	REF	hp
299.0257	401	537.7498	336.3107	451	604.8009
299.7714	402	539.0909	337.0564	452	606.1420
300.5171	403	540.4319	337.8021	453	607.4830
301.2628	404	541.7729	338.5478	454	608.8240
302.0085	405	543.1139	339.2935	455	610.1650
302.7542	406	544.4549	340.0392	456	611.5061
303.4999	407	545.7960	340.7849	457	612.8471
304.2456	408	547.1370	341.5306	458	614.1881
304.9913	409	548.4780	342.2763	459	615.5291
305.7370	410	549.8190	343.0220	460	616.8701
306.4827	411	551.1601	343.7677	461	618.2112
307.2284	412	552.5011	344.5134	462	619.5522
307.9741	413	553.8421	345.2591	463	620.8932
308.7198	414	555.1831	346.0048	464	622.2342
309.4655	415	556.5241	346.7505	465	623.5752
310.2112	416	557.8652	347.4962	466	624.9163
310.9569	417	559.2062	348.2419	467	626.2573
311.7026	418	560.5472	348.9876	468	627.5983
312.4483	419	561.8882	349.7333	469	628.9393
313.1940	420	563.2293	350.4790	470	630.2804
313.9397	421	564.5703	351.2247	471	631.6214
314.6854	422	565.9113	351.9704	472	632.9624
315.4311	423	567.2523	352.7161	473	634.3034
316.1768	424	568.5933	353.4618	474	635.6444
316.9225	425	569.9344	354.2075	475	636.9855
317.6682	426	571.2754	354.9532	476	638.3265
318.4139	427	572.6164	355.6989	477	639.6675
319.1596	428	573.9574	356.4445	478	641.0085
319.9053	429	575.2985	357.1903	479	642.3496
320.6510	430	576.6395	357.9360	480	643.6906
321.3967	431	577.9805	358.6817	481	645.0316
322.1424	432	579.3215	359.4273	482	646.3726
322.8881	433	580.6625	360.1730	483	647.7136
323.6338	434	582.0036	360.9188	484	649.0547
324.3795	435	583.3446	361.6645	485	650.3957
325.1252	436	584.6856	362.4101	486	651.7367
325.8709	437	586.0266	363.1558	487	653.0777
326.6166	438	587.3677	363.9016	488	654.4188
327.3623	439	588.7087	364.6473	489	655.7598
328.1080	440	590.0497	365.3929	490	657.1008
328.8537	441	591.3907	366.1386	491	658.4418
329.5994	442	592.7317	366.8843	492	659.7828
330.3451	443	594.0728	367.6301	493	661.1239
331.0908	444	595.4138	368.3757	494	662.4649
331.8365	445	596.7548	369.1214	495	663.8059
332.5822	446	598.0958	369.8671	496	665.1469
333.3279	447	599.4369	370.6129	497	666.4880
334.0736	448	600.7779	371.3585	498	667.8290
334.8193	449	602.1189	372.1042	499	669.1700
335.5650	450	603.4599	372.8499	500	670.5110

MW ——————— +3 ——————→ kW ←———— −3 ———— W

kW kJ/s	REF	hp		kW kJ/s	REF	hp
373.5956	501	671.8520		410.8806	551	738.9031
374.3413	502	673.1931		411.6263	552	740.2442
375.0870	503	674.5341		412.3720	553	741.5852
375.8327	504	675.8751		413.1177	554	742.9262
376.5784	505	677.2161		413.8634	555	744.2672
377.3241	506	678.5572		414.6091	556	745.6083
378.0698	507	679.8982		415.3548	557	746.9493
378.8155	508	681.2392		416.1005	558	748.2903
379.5612	509	682.5802		416.8462	559	749.6313
380.3069	510	683.9212		417.5919	560	750.9723
381.0526	511	685.2623		418.3376	561	752.3134
381.7983	512	686.6033		419.0833	562	753.6544
382.5440	513	687.9443		419.8290	563	754.9954
383.2897	514	689.2853		420.5747	564	756.3364
384.0354	515	690.6264		421.3204	565	757.6775
384.7811	516	691.9674		422.0661	566	759.0185
385.5268	517	693.3084		422.8118	567	760.3595
386.2725	518	694.6494		423.5575	568	761.7005
387.0182	519	695.9904		424.3032	569	763.0415
387.7639	520	697.3315		425.0489	570	764.3826
388.5096	521	698.6725		425.7946	571	765.7236
389.2553	522	700.0135		426.5403	572	767.0646
390.0010	523	701.3545		427.2860	573	768.4056
390.7467	524	702.6955		428.0317	574	769.7467
391.4924	525	704.0366		428.7774	575	771.0877
392.2381	526	705.3776		429.5231	576	772.4287
392.9838	527	706.7186		430.2688	577	773.7697
393.7295	528	708.0596		431.0145	578	775.1107
394.4752	529	709.4007		431.7602	579	776.4518
395.2209	530	710.7417		432.5059	580	777.7928
395.9666	531	712.0827		433.2516	581	779.1338
396.7123	532	713.4237		433.9973	582	780.4748
397.4580	533	714.7647		434.7430	583	781.8158
398.2037	534	716.1058		435.4887	584	783.1569
398.9494	535	717.4468		436.2344	585	784.4979
399.6951	536	718.7878		436.9801	586	785.8389
400.4408	537	720.1288		437.7258	587	787.1799
401.1865	538	721.4699		438.4715	588	788.5210
401.9322	539	722.8109		439.2172	589	789.8620
402.6779	540	724.1519		439.9629	590	791.2030
403.4236	541	725.4929		440.7086	591	792.5440
404.1693	542	726.8339		441.4543	592	793.8850
404.9150	543	728.1750		442.2000	593	795.2261
405.6607	544	729.5160		442.9457	594	796.5671
406.4064	545	730.8570		443.6914	595	797.9081
407.1521	546	732.1980		444.4371	596	799.2491
407.8978	547	733.5391		445.1828	597	800.5902
408.6435	548	734.8801		445.9285	598	801.9312
409.3892	549	736.2211		446.6742	599	803.2722
410.1349	550	737.5621		447.4199	600	804.6132

MW ←	−3		kW		+3	→ W

kW kJ/s	REF	hp	kW kJ/s	REF	hp
448.1656	601	805.9542	485.4506	651	873.0053
448.9113	602	807.2953	486.1963	652	874.3464
449.6570	603	808.6363	486.9420	653	875.6874
450.4027	604	809.9773	487.6877	654	877.0284
451.1484	605	811.3183	488.4334	655	878.3694
451.8941	606	812.6594	489.1791	656	879.7105
452.6398	607	814.0004	489.9248	657	881.0515
453.3855	608	815.3414	490.6705	658	882.3925
454.1312	609	816.6824	491.4162	659	883.7335
454.8769	610	818.0234	492.1619	660	885.0745
455.6226	611	819.3645	492.9076	661	886.4156
456.3683	612	820.7055	493.6533	662	887.7566
457.1140	613	822.0465	494.3990	663	889.0976
457.8597	614	823.3875	495.1447	664	890.4386
458.6054	615	824.7286	495.8904	665	891.7797
459.3511	616	826.0696	496.6361	666	893.1207
460.0968	617	827.4106	497.3818	667	894.4617
460.8425	618	828.7516	498.1275	668	895.8027
461.5882	619	830.0926	498.8732	669	897.1437
462.3339	620	831.4337	499.6189	670	898.4848
463.0796	621	832.7747	500.3646	671	899.8258
463.8253	622	834.1157	501.1103	672	901.1668
464.5710	623	835.4567	501.8560	673	902.5078
465.3167	624	836.7978	502.6017	674	903.8489
466.0624	625	838.1388	503.3474	675	905.1899
466.8081	626	839.4798	504.0931	676	906.5309
467.5538	627	840.8208	504.8388	677	907.8719
468.2995	628	842.1618	505.5845	678	909.2129
469.0452	629	843.5029	506.3302	679	910.5540
469.7909	630	844.8439	507.0759	680	911.8950
470.5366	631	846.1849	507.8216	681	913.2360
471.2823	632	847.5259	508.5673	682	914.5770
472.0280	633	848.8670	509.3130	683	915.9181
472.7737	634	850.2080	510.0587	684	917.2591
473.5194	635	851.5490	510.8044	685	918.6001
474.2651	636	852.8900	511.5501	686	919.9411
475.0108	637	854.2310	512.2958	687	921.2821
475.7565	638	855.5721	513.0415	688	922.6232
476.5022	639	856.9131	513.7872	689	923.9642
477.2479	640	858.2541	514.5329	690	925.3052
477.9936	641	859.5951	515.2786	691	926.6462
478.7393	642	860.9361	516.0243	692	927.9873
479.4850	643	862.2772	516.7700	693	929.3283
480.2307	644	863.6182	517.5157	694	930.6693
480.9764	645	864.9592	518.2614	695	932.0103
481.7221	646	866.3002	519.0071	696	933.3513
482.4678	647	867.6413	519.7528	697	934.6924
483.2135	648	868.9823	520.4985	698	936.0334
483.9592	649	870.3233	521.2442	699	937.3744
484.7049	650	871.6643	521.9899	700	938.7154

MW ——————— +3 ——————→ | kW | ←—————— −3 ——————— W

kW kJ/s	REF	hp	kW kJ/s	REF	hp
522.7356	701	940.0564	560.0206	751	1007.108
523.4813	702	941.3975	560.7663	752	1008.449
524.2270	703	942.7385	561.5120	753	1009.790
524.9727	704	944.0795	562.2577	754	1011.131
525.7184	705	945.4205	563.0034	755	1012.472
526.4641	706	946.7616	563.7491	756	1013.813
527.2098	707	948.1026	564.4948	757	1015.154
527.9555	708	949.4436	565.2405	758	1016.495
528.7012	709	950.7846	565.9862	759	1017.836
529.4469	710	952.1256	566.7319	760	1019.177
530.1926	711	953.4667	567.4776	761	1020.518
530.9383	712	954.8077	568.2233	762	1021.859
531.6840	713	956.1487	568.9690	763	1023.200
532.4297	714	957.4897	569.7147	764	1024.541
533.1754	715	958.8308	570.4604	765	1025.882
533.9211	716	960.1718	571.2061	766	1027.223
534.6668	717	961.5128	571.9518	767	1028.564
535.4125	718	962.8538	572.6975	768	1029.905
536.1582	719	964.1948	573.4432	769	1031.246
536.9039	720	965.5359	574.1889	770	1032.587
537.6496	721	966.8769	574.9346	771	1033.928
538.3953	722	968.2179	575.6803	772	1035.269
539.1410	723	969.5589	576.4260	773	1036.610
539.8867	724	970.9000	577.1717	774	1037.951
540.6324	725	972.2410	577.9174	775	1039.292
541.3781	726	973.5820	578.6631	776	1040.633
542.1238	727	974.9230	579.4088	777	1041.974
542.8695	728	976.2640	580.1545	778	1043.315
543.6152	729	977.6051	580.9002	779	1044.656
544.3609	730	978.9461	581.6459	780	1045.997
545.1066	731	980.2871	582.3916	781	1047.338
545.8523	732	981.6281	583.1373	782	1048.679
546.5980	733	982.9692	583.8830	783	1050.020
547.3437	734	984.3102	584.6287	784	1051.361
548.0894	735	985.6512	585.3744	785	1052.702
548.8351	736	986.9922	586.1201	786	1054.043
549.5808	737	988.3332	586.8658	787	1055.384
550.3265	738	989.6743	587.6115	788	1056.725
551.0722	739	991.0153	588.3572	789	1058.066
551.8179	740	992.3563	589.1029	790	1059.407
552.5636	741	993.6973	589.8486	791	1060.748
553.3093	742	995.0384	590.5943	792	1062.089
554.0550	743	996.3794	591.3400	793	1063.430
554.8007	744	997.7204	592.0857	794	1064.771
555.5464	745	999.0614	592.8314	795	1066.113
556.2921	746	1000.402	593.5771	796	1067.454
557.0378	747	1001.743	594.3228	797	1068.795
557.7835	748	1003.084	595.0685	798	1070.136
558.5292	749	1004.426	595.8142	799	1071.477
559.2749	750	1005.767	596.5599	800	1072.818

MW ←———— −3 ————| kW |———— +3 ————→ W

kW kJ/s	REF	hp	kW kJ/s	REF	hp
597.3056	801	1074.159	634.5906	851	1141.210
598.0513	802	1075.500	635.3363	852	1142.551
598.7970	803	1076.841	636.0820	853	1143.892
599.5427	804	1078.182	636.8277	854	1145.233
600.2884	805	1079.523	637.5734	855	1146.574
601.0341	806	1080.864	638.3191	856	1147.915
601.7798	807	1082.205	639.0648	857	1149.256
602.5255	808	1083.546	639.8105	858	1150.597
603.2712	809	1084.887	640.5562	859	1151.938
604.0169	810	1086.228	641.3019	860	1153.279
604.7626	811	1087.569	642.0476	861	1154.620
605.5083	812	1088.910	642.7933	862	1155.961
606.2540	813	1090.251	643.5390	863	1157.302
606.9997	814	1091.592	644.2847	864	1158.643
607.7454	815	1092.933	645.0304	865	1159.984
608.4911	816	1094.274	645.7761	866	1161.325
609.2368	817	1095.615	646.5218	867	1162.666
609.9825	818	1096.956	647.2675	868	1164.007
610.7282	819	1098.297	648.0132	869	1165.348
611.4739	820	1099.638	648.7589	870	1166.689
612.2196	821	1100.979	649.5046	871	1168.030
612.9653	822	1102.320	650.2503	872	1169.371
613.7110	823	1103.661	650.9960	873	1170.712
614.4567	824	1105.002	651.7417	874	1172.053
615.2024	825	1106.343	652.4874	875	1173.394
615.9481	826	1107.684	653.2331	876	1174.735
616.6938	827	1109.025	653.9788	877	1176.076
617.4395	828	1110.366	654.7245	878	1177.417
618.1852	829	1111.707	655.4702	879	1178.758
618.9309	830	1113.048	656.2159	880	1180.099
619.6766	831	1114.389	656.9616	881	1181.440
620.4223	832	1115.730	657.7073	882	1182.781
621.1680	833	1117.071	658.4530	883	1184.122
621.9137	834	1118.412	659.1987	884	1185.463
622.6594	835	1119.753	659.9444	885	1186.805
623.4051	836	1121.094	660.6901	886	1188.146
624.1508	837	1122.435	661.4358	887	1189.487
624.8965	838	1123.776	662.1815	888	1190.828
625.6422	839	1125.117	662.9272	889	1192.169
626.3879	840	1126.459	663.6729	890	1193.510
627.1336	841	1127.800	664.4186	891	1194.851
627.8793	842	1129.141	665.1643	892	1196.192
628.6250	843	1130.482	665.9100	893	1197.533
629.3707	844	1131.823	666.6557	894	1198.874
630.1164	845	1133.164	667.4014	895	1200.215
630.8621	846	1134.505	668.1471	896	1201.556
631.6078	847	1135.846	668.8928	897	1202.897
632.3535	848	1137.187	669.6385	898	1204.238
633.0992	849	1138.528	670.3842	899	1205.579
633.8449	850	1139.869	671.1299	900	1206.920

MW ———— +3 ————→ kW ←———— − 3 ———— W

kW kJ/s	REF	hp	kW kJ/s	REF	hp
671.8756	901	1208.261	709.1606	951	1275.312
672.6213	902	1209.602	709.9063	952	1276.653
673.3670	903	1210.943	710.6520	953	1277.994
674.1127	904	1212.284	711.3977	954	1279.335
674.8584	905	1213.625	712.1434	955	1280.676
675.6041	906	1214.966	712.8891	956	1282.017
676.3498	907	1216.307	713.6348	957	1283.358
677.0955	908	1217.648	714.3805	958	1284.699
677.8412	909	1218.989	715.1262	959	1286.040
678.5869	910	1220.330	715.8719	960	1287.381
679.3326	911	1221.671	716.6176	961	1288.722
680.0783	912	1223.012	717.3633	962	1290.063
680.8240	913	1224.353	718.1090	963	1291.404
681.5697	914	1225.694	718.8547	964	1292.745
682.3154	915	1227.035	719.6004	965	1294.086
683.0611	916	1228.376	720.3461	966	1295.427
683.8068	917	1229.717	721.0918	967	1296.768
684.5525	918	1231.058	721.8375	968	1298.109
685.2982	919	1232.399	722.5832	969	1299.450
686.0439	920	1233.740	723.3289	970	1300.791
686.7896	921	1235.081	724.0746	971	1302.132
687.5353	922	1236.422	724.8203	972	1303.473
688.2810	923	1237.763	725.5660	973	1304.814
689.0267	924	1239.104	726.3117	974	1306.155
689.7724	925	1240.445	727.0574	975	1307.496
690.5181	926	1241.786	727.8031	976	1308.838
691.2638	927	1243.127	728.5488	977	1310.179
692.0095	928	1244.468	729.2945	978	1311.520
692.7552	929	1245.809	730.0402	979	1312.861
693.5009	930	1247.150	730.7859	980	1314.202
694.2466	931	1248.492	731.5316	981	1315.543
694.9923	932	1249.833	732.2773	982	1316.884
695.7380	933	1251.174	733.0230	983	1318.225
696.4837	934	1252.515	733.7687	984	1319.566
697.2294	935	1253.856	734.5144	985	1320.907
697.9751	936	1255.197	735.2601	986	1322.248
698.7208	937	1256.538	736.0058	987	1323.589
699.4665	938	1257.879	736.7515	988	1324.930
700.2122	939	1259.220	737.4972	989	1326.271
700.9579	940	1260.561	738.2429	990	1327.612
701.7036	941	1261.902	738.9886	991	1328.953
702.4493	942	1263.243	739.7343	992	1330.294
703.1950	943	1264.584	740.4800	993	1331.635
703.9407	944	1265.925	741.2257	994	1332.976
704.6864	945	1267.266	741.9714	995	1334.317
705.4321	946	1268.607	742.7171	996	1335.658
706.1778	947	1269.948	743.4628	997	1336.999
706.9235	948	1271.289	744.2085	998	1338.340
707.6692	949	1272.630	744.9542	999	1339.681
708.4149	950	1273.971	745.6999	1000	1341.022

36

HEAT FLOW

<u>British Thermal Unit (International Steam Table)/Hour — Watt (or Joule/Second)</u>
$$(Btu/h - W - J/s)$$
British Thermal Unit (International Steam Table)/Hour — Kilowatt (or Kilojoule/Second)
$$(Btu/h - kW - kJ/s)$$

Conversion Factors:
  1 IST British Thermal Unit/hour (Btu/h) = **0.2930711 Watt (W)**
  (1 W = **3.412141 Btu/h**)

In converting W values to kW or kW to W, the Decimal-shift Diagram 36-A, which appears at the top of alternate table pages, will assist table users. It is derived from: 1 kW = 10^3 W.

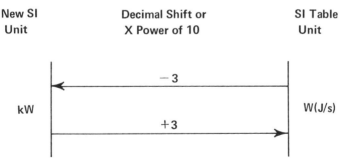

New SI Unit	Decimal Shift or X Power of 10	SI Table Unit

Diagram 36-A

Example (*a*): Convert 4,170 Btu/h to W, kW.
  Since: 4,170 Btu/h = 417 X 10 Btu/h:
  From table: 417 Btu/h = 122.2106 W* ≈ 122 W

---

*All conversion values obtained can be rounded to desired number of significant figures.

421

Therefore: 417×10 Btu/h $\approx 122 \times 10$ W $\approx 1{,}220$ W

From Diagram 36-A: $1{,}220$ W $= 1{,}220 \times 10^{-3}$ kW $= 1.22$ kW

(W to kW by multiplying W value by 10^{-3} or by shifting its decimal 3 places to the left.)

Example (*b*): Convert 81,150 kW to Btu/h.

From Diagram 36-A: $81{,}150$ kW $= 81{,}150 \times 10^{3}$ W $= 811.5 \times 10^{5}$ W

Since: 811.5 W $= (811.0 + 0.5)$ W:

From table: 811 W $= 2767.247$ Btu/h*

$\qquad\qquad\quad 5$ W $= 17.06071$ Btu/h*

Therefore: with proper decimal shifting and addition:

$$811.0 \text{ W} = 2{,}767.247 \qquad \text{Btu/h}$$
$$\underline{0.5 \text{ W} = \qquad\ 1.706071 \text{ Btu/h}}$$
$$811.5 \text{ W} = 2{,}768.953 \qquad \text{Btu/h} \approx 2{,}769 \text{ Btu/h}$$

From above: 811.5×10^{5} W $\approx 2{,}769 \times 10^{5}$ Btu/h $\approx 276.9 \times 10^{6}$ Btu/h.

---

*All conversion values obtained can be rounded to desired number of significant figures.

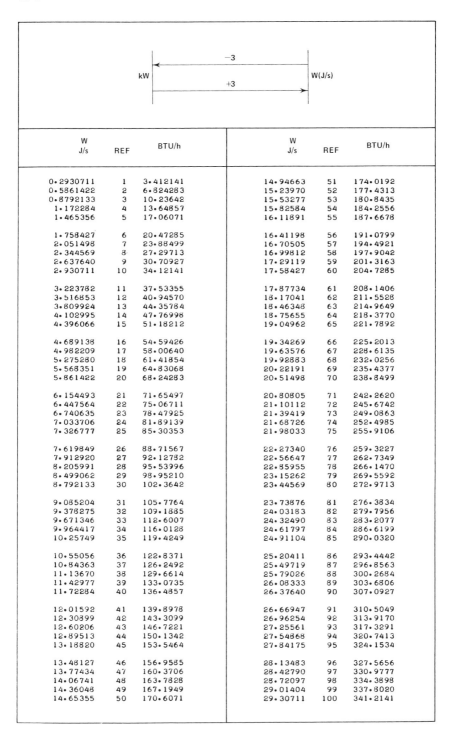

W J/s	REF	BTU/h	W J/s	REF	BTU/h
0.2930711	1	3.412141	14.94663	51	174.0192
0.5861422	2	6.824283	15.23970	52	177.4313
0.8792133	3	10.23642	15.53277	53	180.8435
1.172284	4	13.64857	15.82584	54	184.2556
1.465356	5	17.06071	16.11891	55	187.6678
1.758427	6	20.47285	16.41198	56	191.0799
2.051498	7	23.88499	16.70505	57	194.4921
2.344569	8	27.29713	16.99812	58	197.9042
2.637640	9	30.70927	17.29119	59	201.3163
2.930711	10	34.12141	17.58427	60	204.7285
3.223782	11	37.53355	17.87734	61	208.1406
3.516853	12	40.94570	18.17041	62	211.5528
3.809924	13	44.35784	18.46348	63	214.9649
4.102995	14	47.76998	18.75655	64	218.3770
4.396066	15	51.18212	19.04962	65	221.7892
4.689138	16	54.59426	19.34269	66	225.2013
4.982209	17	58.00640	19.63576	67	228.6135
5.275280	18	61.41854	19.92883	68	232.0256
5.568351	19	64.83068	20.22191	69	235.4377
5.861422	20	68.24283	20.51498	70	238.8499
6.154493	21	71.65497	20.80805	71	242.2620
6.447564	22	75.06711	21.10112	72	245.6742
6.740635	23	78.47925	21.39419	73	249.0863
7.033706	24	81.89139	21.68726	74	252.4985
7.326777	25	85.30353	21.98033	75	255.9106
7.619849	26	88.71567	22.27340	76	259.3227
7.912920	27	92.12782	22.56647	77	262.7349
8.205991	28	95.53996	22.85955	78	266.1470
8.499062	29	98.95210	23.15262	79	269.5592
8.792133	30	102.3642	23.44569	80	272.9713
9.085204	31	105.7764	23.73876	81	276.3834
9.378275	32	109.1885	24.03183	82	279.7956
9.671346	33	112.6007	24.32490	83	283.2077
9.964417	34	116.0128	24.61797	84	286.6199
10.25749	35	119.4249	24.91104	85	290.0320
10.55056	36	122.8371	25.20411	86	293.4442
10.84363	37	126.2492	25.49719	87	296.8563
11.13670	38	129.6614	25.79026	88	300.2684
11.42977	39	133.0735	26.08333	89	303.6806
11.72284	40	136.4857	26.37640	90	307.0927
12.01592	41	139.8978	26.66947	91	310.5049
12.30399	42	143.3099	26.96254	92	313.9170
12.60206	43	146.7221	27.25561	93	317.3291
12.89513	44	150.1342	27.54868	94	320.7413
13.18820	45	153.5464	27.84175	95	324.1534
13.48127	46	156.9585	28.13483	96	327.5656
13.77434	47	160.3706	28.42790	97	330.9777
14.06741	48	163.7828	28.72097	98	334.3898
14.36048	49	167.1949	29.01404	99	337.8020
14.65355	50	170.6071	29.30711	100	341.2141

W J/s	REF	BTU/h	W J/s	REF	BTU/h
29·60018	101	344·6263	44·25374	151	515·2333
29·89325	102	348·0384	44·54681	152	518·6455
30·18632	103	351·4506	44·83988	153	522·0576
30·47939	104	354·8627	45·13295	154	525·4698
30·77247	105	358·2748	45·42602	155	528·8819
31·06554	106	361·6870	45·71909	156	532·2940
31·35861	107	365·0991	46·01216	157	535·7062
31·65168	108	368·5113	46·30523	158	539·1183
31·94475	109	371·9234	46·59830	159	542·5305
32·23782	110	375·3355	46·89138	160	545·9426
32·53089	111	378·7477	47·18445	161	549·3548
32·82396	112	382·1598	47·47752	162	552·7669
33·11703	113	385·5720	47·77059	163	556·1790
33·41011	114	388·9841	48·06366	164	559·5912
33·70318	115	392·3962	48·35673	165	563·0033
33·99625	116	395·8084	48·64980	166	566·4155
34·28932	117	399·2205	48·94287	167	569·8276
34·58239	118	402·6327	49·23594	168	573·2397
34·87546	119	406·0448	49·52902	169	576·6519
35·16853	120	409·4570	49·82209	170	580·0640
35·46160	121	412·8691	50·11516	171	583·4762
35·75467	122	416·2812	50·40823	172	586·8883
36·04775	123	419·6934	50·70130	173	590·3004
36·34082	124	423·1055	50·99437	174	593·7126
36·63389	125	426·5177	51·28744	175	597·1247
36·92696	126	429·9298	51·58051	176	600·5369
37·22003	127	433·3419	51·87358	177	603·9490
37·51310	128	436·7541	52·16666	178	607·3612
37·80617	129	440·1662	52·45973	179	610·7733
38·09924	130	443·5784	52·75280	180	614·1854
38·39231	131	446·9905	53·04587	181	617·5976
38·68539	132	450·4026	53·33894	182	621·0097
38·97846	133	453·8148	53·63201	183	624·4219
39·27153	134	457·2269	53·92508	184	627·8340
39·56460	135	460·6391	54·21815	185	631·2461
39·85767	136	464·0512	54·51122	186	634·6583
40·15074	137	467·4634	54·80430	187	638·0704
40·44381	138	470·8755	55·09737	188	641·4826
40·73688	139	474·2876	55·39044	189	644·8947
41·02995	140	477·6998	55·68351	190	648·3068
41·32302	141	481·1119	55·97658	191	651·7190
41·61610	142	484·5241	56·26965	192	655·1311
41·90917	143	487·9362	56·56272	193	658·5433
42·20224	144	491·3483	56·85579	194	661·9554
42·49531	145	494·7605	57·14886	195	665·3676
42·78838	146	498·1726	57·44194	196	668·7797
43·08145	147	501·5848	57·73501	197	672·1918
43·37452	148	504·9969	58·02808	198	675·6040
43·66759	149	508·4091	58·32115	199	679·0161
43·96066	150	511·8212	58·61422	200	682·4283

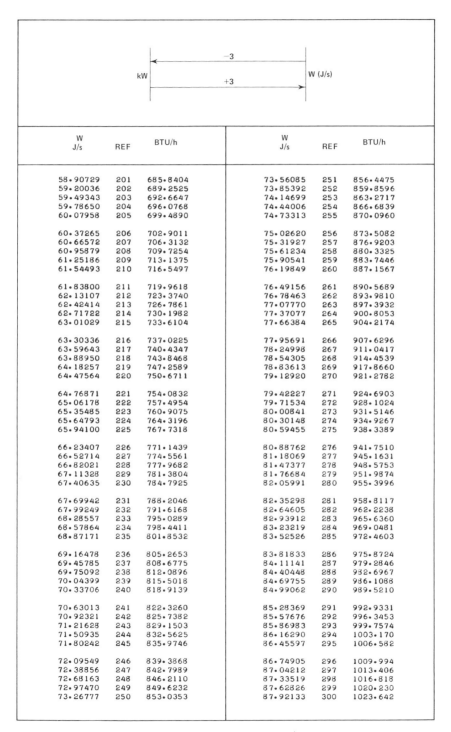

W J/s	REF	BTU/h	W J/s	REF	BTU/h
58.90729	201	685.8404	73.56085	251	856.4475
59.20036	202	689.2525	73.85392	252	859.8596
59.49343	203	692.6647	74.14699	253	863.2717
59.78650	204	696.0768	74.44006	254	866.6839
60.07958	205	699.4890	74.73313	255	870.0960
60.37265	206	702.9011	75.02620	256	873.5082
60.66572	207	706.3132	75.31927	257	876.9203
60.95879	208	709.7254	75.61234	258	880.3325
61.25186	209	713.1375	75.90541	259	883.7446
61.54493	210	716.5497	76.19849	260	887.1567
61.83800	211	719.9618	76.49156	261	890.5689
62.13107	212	723.3740	76.78463	262	893.9810
62.42414	213	726.7861	77.07770	263	897.3932
62.71722	214	730.1982	77.37077	264	900.8053
63.01029	215	733.6104	77.66384	265	904.2174
63.30336	216	737.0225	77.95691	266	907.6296
63.59643	217	740.4347	78.24998	267	911.0417
63.88950	218	743.8468	78.54305	268	914.4539
64.18257	219	747.2589	78.83613	269	917.8660
64.47564	220	750.6711	79.12920	270	921.2782
64.76871	221	754.0832	79.42227	271	924.6903
65.06178	222	757.4954	79.71534	272	928.1024
65.35485	223	760.9075	80.00841	273	931.5146
65.64793	224	764.3196	80.30148	274	934.9267
65.94100	225	767.7318	80.59455	275	938.3389
66.23407	226	771.1439	80.88762	276	941.7510
66.52714	227	774.5561	81.18069	277	945.1631
66.82021	228	777.9682	81.47377	278	948.5753
67.11328	229	781.3804	81.76684	279	951.9874
67.40635	230	784.7925	82.05991	280	955.3996
67.69942	231	788.2046	82.35298	281	958.8117
67.99249	232	791.6168	82.64605	282	962.2238
68.28557	233	795.0289	82.93912	283	965.6360
68.57864	234	798.4411	83.23219	284	969.0481
68.87171	235	801.8532	83.52526	285	972.4603
69.16478	236	805.2653	83.81833	286	975.8724
69.45785	237	808.6775	84.11141	287	979.2846
69.75092	238	812.0896	84.40448	288	982.6967
70.04399	239	815.5018	84.69755	289	986.1088
70.33706	240	818.9139	84.99062	290	989.5210
70.63013	241	822.3260	85.28369	291	992.9331
70.92321	242	825.7382	85.57676	292	996.3453
71.21628	243	829.1503	85.86983	293	999.7574
71.50935	244	832.5625	86.16290	294	1003.170
71.80242	245	835.9746	86.45597	295	1006.582
72.09549	246	839.3868	86.74905	296	1009.994
72.38856	247	842.7989	87.04212	297	1013.406
72.68163	248	846.2110	87.33519	298	1016.818
72.97470	249	849.6232	87.62826	299	1020.230
73.26777	250	853.0353	87.92133	300	1023.642

W J/s	REF	BTU/h	W J/s	REF	BTU/h
88.21440	301	1027.055	102.8680	351	1197.662
88.50747	302	1030.467	103.1610	352	1201.074
88.80054	303	1033.879	103.4541	353	1204.486
89.09361	304	1037.291	103.7472	354	1207.898
89.38669	305	1040.703	104.0402	355	1211.310
89.67976	306	1044.115	104.3333	356	1214.722
89.97283	307	1047.527	104.6264	357	1218.134
90.26590	308	1050.940	104.9195	358	1221.547
90.55897	309	1054.352	105.2125	359	1224.959
90.85204	310	1057.764	105.5056	360	1228.371
91.14511	311	1061.176	105.7987	361	1231.783
91.43818	312	1064.588	106.0917	362	1235.195
91.73125	313	1068.000	106.3848	363	1238.607
92.02433	314	1071.412	106.6779	364	1242.019
92.31740	315	1074.825	106.9710	365	1245.432
92.61047	316	1078.237	107.2640	366	1248.844
92.90354	317	1081.649	107.5571	367	1252.256
93.19661	318	1085.061	107.8502	368	1255.668
93.48968	319	1088.473	108.1432	369	1259.080
93.78275	320	1091.885	108.4363	370	1262.492
94.07582	321	1095.297	108.7294	371	1265.904
94.36889	322	1098.710	109.0224	372	1269.317
94.66196	323	1102.122	109.3155	373	1272.729
94.95504	324	1105.534	109.6086	374	1276.141
95.24811	325	1108.946	109.9017	375	1279.553
95.54118	326	1112.358	110.1947	376	1282.965
95.83425	327	1115.770	110.4878	377	1286.377
96.12732	328	1119.182	110.7809	378	1289.789
96.42039	329	1122.594	111.0739	379	1293.202
96.71346	330	1126.007	111.3670	380	1296.614
97.00653	331	1129.419	111.6601	381	1300.026
97.29960	332	1132.831	111.9532	382	1303.438
97.59268	333	1136.243	112.2462	383	1306.850
97.88575	334	1139.655	112.5393	384	1310.262
98.17882	335	1143.067	112.8324	385	1313.674
98.47189	336	1146.479	113.1254	386	1317.087
98.76496	337	1149.892	113.4185	387	1320.499
99.05803	338	1153.304	113.7116	388	1323.911
99.35110	339	1156.716	114.0047	389	1327.323
99.64417	340	1160.128	114.2977	390	1330.735
99.93724	341	1163.540	114.5908	391	1334.147
100.2303	342	1166.952	114.8839	392	1337.559
100.5234	343	1170.364	115.1769	393	1340.972
100.8165	344	1173.777	115.4700	394	1344.384
101.1095	345	1177.189	115.7631	395	1347.796
101.4026	346	1180.601	116.0562	396	1351.208
101.6957	347	1184.013	116.3492	397	1354.620
101.9887	348	1187.425	116.6423	398	1358.032
102.2818	349	1190.837	116.9354	399	1361.444
102.5749	350	1194.249	117.2284	400	1364.857

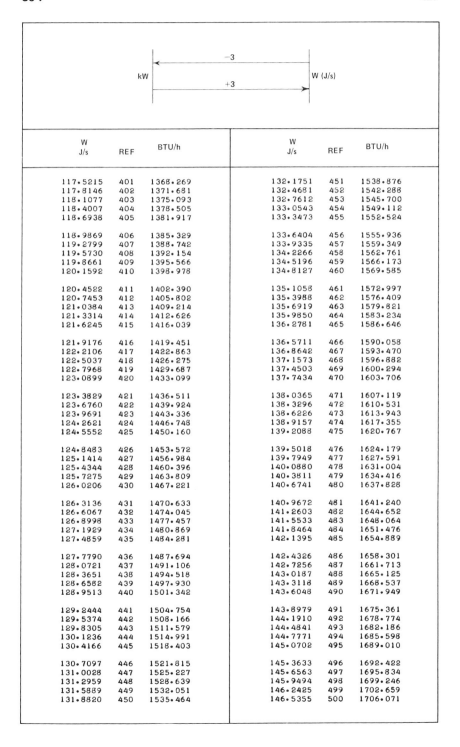

W J/s	REF	BTU/h		W J/s	REF	BTU/h
117.5215	401	1368.269		132.1751	451	1538.876
117.8146	402	1371.681		132.4681	452	1542.288
118.1077	403	1375.093		132.7612	453	1545.700
118.4007	404	1378.505		133.0543	454	1549.112
118.6938	405	1381.917		133.3473	455	1552.524
118.9869	406	1385.329		133.6404	456	1555.936
119.2799	407	1388.742		133.9335	457	1559.349
119.5730	408	1392.154		134.2266	458	1562.761
119.8661	409	1395.566		134.5196	459	1566.173
120.1592	410	1398.978		134.8127	460	1569.585
120.4522	411	1402.390		135.1058	461	1572.997
120.7453	412	1405.802		135.3988	462	1576.409
121.0384	413	1409.214		135.6919	463	1579.821
121.3314	414	1412.626		135.9850	464	1583.234
121.6245	415	1416.039		136.2781	465	1586.646
121.9176	416	1419.451		136.5711	466	1590.058
122.2106	417	1422.863		136.8642	467	1593.470
122.5037	418	1426.275		137.1573	468	1596.882
122.7968	419	1429.687		137.4503	469	1600.294
123.0899	420	1433.099		137.7434	470	1603.706
123.3829	421	1436.511		138.0365	471	1607.119
123.6760	422	1439.924		138.3296	472	1610.531
123.9691	423	1443.336		138.6226	473	1613.943
124.2621	424	1446.748		138.9157	474	1617.355
124.5552	425	1450.160		139.2088	475	1620.767
124.8483	426	1453.572		139.5018	476	1624.179
125.1414	427	1456.984		139.7949	477	1627.591
125.4344	428	1460.396		140.0880	478	1631.004
125.7275	429	1463.809		140.3811	479	1634.416
126.0206	430	1467.221		140.6741	480	1637.828
126.3136	431	1470.633		140.9672	481	1641.240
126.6067	432	1474.045		141.2603	482	1644.652
126.8998	433	1477.457		141.5533	483	1648.064
127.1929	434	1480.869		141.8464	484	1651.476
127.4859	435	1484.281		142.1395	485	1654.889
127.7790	436	1487.694		142.4326	486	1658.301
128.0721	437	1491.106		142.7256	487	1661.713
128.3651	438	1494.518		143.0187	488	1665.125
128.6582	439	1497.930		143.3118	489	1668.537
128.9513	440	1501.342		143.6048	490	1671.949
129.2444	441	1504.754		143.8979	491	1675.361
129.5374	442	1508.166		144.1910	492	1678.774
129.8305	443	1511.579		144.4841	493	1682.186
130.1236	444	1514.991		144.7771	494	1685.598
130.4166	445	1518.403		145.0702	495	1689.010
130.7097	446	1521.815		145.3633	496	1692.422
131.0028	447	1525.227		145.6563	497	1695.834
131.2959	448	1528.639		145.9494	498	1699.246
131.5889	449	1532.051		146.2425	499	1702.659
131.8820	450	1535.464		146.5355	500	1706.071

W J/s	REF	BTU/h	W J/s	REF	BTU/h
146.8286	501	1709.483	161.4822	551	1880.090
147.1217	502	1712.895	161.7752	552	1883.502
147.4148	503	1716.307	162.0683	553	1886.914
147.7078	504	1719.719	162.3614	554	1890.326
148.0009	505	1723.131	162.6545	555	1893.738
148.2940	506	1726.543	162.9475	556	1897.151
148.5870	507	1729.956	163.2406	557	1900.563
148.8801	508	1733.368	163.5337	558	1903.975
149.1732	509	1736.780	163.8267	559	1907.387
149.4663	510	1740.192	164.1198	560	1910.799
149.7593	511	1743.604	164.4129	561	1914.211
150.0524	512	1747.016	164.7060	562	1917.623
150.3455	513	1750.428	164.9990	563	1921.036
150.6385	514	1753.841	165.2921	564	1924.448
150.9316	515	1757.253	165.5852	565	1927.860
151.2247	516	1760.665	165.8782	566	1931.272
151.5178	517	1764.077	166.1713	567	1934.684
151.8108	518	1767.489	166.4644	568	1938.096
152.1039	519	1770.901	166.7575	569	1941.508
152.3970	520	1774.313	167.0505	570	1944.921
152.6900	521	1777.726	167.3436	571	1948.333
152.9831	522	1781.138	167.6367	572	1951.745
153.2762	523	1784.550	167.9297	573	1955.157
153.5693	524	1787.962	168.2228	574	1958.569
153.8623	525	1791.374	168.5159	575	1961.981
154.1554	526	1794.786	168.8090	576	1965.393
154.4485	527	1798.198	169.1020	577	1968.806
154.7415	528	1801.611	169.3951	578	1972.218
155.0346	529	1805.023	169.6882	579	1975.630
155.3277	530	1808.435	169.9812	580	1979.042
155.6208	531	1811.847	170.2743	581	1982.454
155.9138	532	1815.259	170.5674	582	1985.866
156.2069	533	1818.671	170.8605	583	1989.278
156.5000	534	1822.083	171.1535	584	1992.691
156.7930	535	1825.496	171.4466	585	1996.103
157.0861	536	1828.908	171.7397	586	1999.515
157.3792	537	1832.320	172.0327	587	2002.927
157.6723	538	1835.732	172.3258	588	2006.339
157.9653	539	1839.144	172.6189	589	2009.751
158.2584	540	1842.556	172.9119	590	2013.163
158.5515	541	1845.968	173.2050	591	2016.576
158.8445	542	1849.381	173.4981	592	2019.988
159.1376	543	1852.793	173.7912	593	2023.400
159.4307	544	1856.205	174.0842	594	2026.812
159.7237	545	1859.617	174.3773	595	2030.224
160.0168	546	1863.029	174.6704	596	2033.636
160.3099	547	1866.441	174.9634	597	2037.048
160.6030	548	1869.853	175.2565	598	2040.460
160.8960	549	1873.266	175.5496	599	2043.873
161.1891	550	1876.678	175.8427	600	2047.285

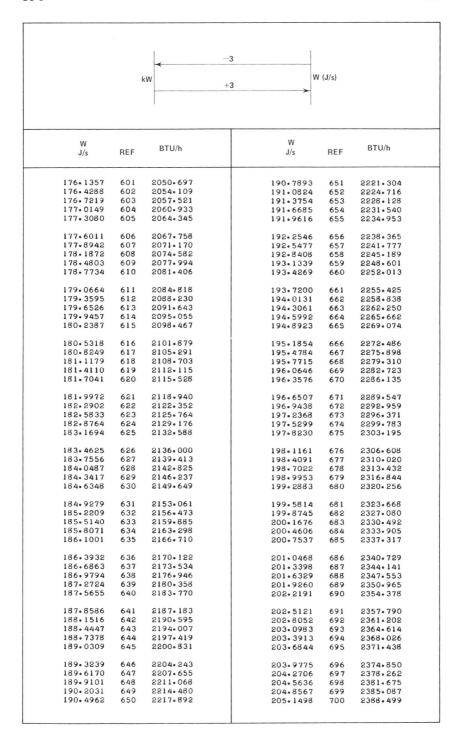

W J/s	REF	BTU/h	W J/s	REF	BTU/h
176.1357	601	2050.697	190.7893	651	2221.304
176.4288	602	2054.109	191.0824	652	2224.716
176.7219	603	2057.521	191.3754	653	2228.128
177.0149	604	2060.933	191.6685	654	2231.540
177.3080	605	2064.345	191.9616	655	2234.953
177.6011	606	2067.758	192.2546	656	2238.365
177.8942	607	2071.170	192.5477	657	2241.777
178.1872	608	2074.582	192.8408	658	2245.189
178.4803	609	2077.994	193.1339	659	2248.601
178.7734	610	2081.406	193.4269	660	2252.013
179.0664	611	2084.818	193.7200	661	2255.425
179.3595	612	2088.230	194.0131	662	2258.838
179.6526	613	2091.643	194.3061	663	2262.250
179.9457	614	2095.055	194.5992	664	2265.662
180.2387	615	2098.467	194.8923	665	2269.074
180.5318	616	2101.879	195.1854	666	2272.486
180.8249	617	2105.291	195.4784	667	2275.898
181.1179	618	2108.703	195.7715	668	2279.310
181.4110	619	2112.115	196.0646	669	2282.723
181.7041	620	2115.528	196.3576	670	2286.135
181.9972	621	2118.940	196.6507	671	2289.547
182.2902	622	2122.352	196.9438	672	2292.959
182.5833	623	2125.764	197.2368	673	2296.371
182.8764	624	2129.176	197.5299	674	2299.783
183.1694	625	2132.588	197.8230	675	2303.195
183.4625	626	2136.000	198.1161	676	2306.608
183.7556	627	2139.413	198.4091	677	2310.020
184.0487	628	2142.825	198.7022	678	2313.432
184.3417	629	2146.237	198.9953	679	2316.844
184.6348	630	2149.649	199.2883	680	2320.256
184.9279	631	2153.061	199.5814	681	2323.668
185.2209	632	2156.473	199.8745	682	2327.080
185.5140	633	2159.885	200.1676	683	2330.492
185.8071	634	2163.298	200.4606	684	2333.905
186.1001	635	2166.710	200.7537	685	2337.317
186.3932	636	2170.122	201.0468	686	2340.729
186.6863	637	2173.534	201.3398	687	2344.141
186.9794	638	2176.946	201.6329	688	2347.553
187.2724	639	2180.358	201.9260	689	2350.965
187.5655	640	2183.770	202.2191	690	2354.378
187.8586	641	2187.183	202.5121	691	2357.790
188.1516	642	2190.595	202.8052	692	2361.202
188.4447	643	2194.007	203.0983	693	2364.614
188.7378	644	2197.419	203.3913	694	2368.026
189.0309	645	2200.831	203.6844	695	2371.438
189.3239	646	2204.243	203.9775	696	2374.850
189.6170	647	2207.655	204.2706	697	2378.262
189.9101	648	2211.068	204.5636	698	2381.675
190.2031	649	2214.480	204.8567	699	2385.087
190.4962	650	2217.892	205.1498	700	2388.499

W J/s	REF	BTU/h	W J/s	REF	BTU/h
205.4428	701	2391.911	220.0964	751	2562.518
205.7359	702	2395.323	220.3895	752	2565.930
206.0290	703	2398.735	220.6825	753	2569.342
206.3221	704	2402.147	220.9756	754	2572.755
206.6151	705	2405.560	221.2687	755	2576.167
206.9082	706	2408.972	221.5618	756	2579.579
207.2013	707	2412.384	221.8548	757	2582.991
207.4943	708	2415.796	222.1479	758	2586.403
207.7874	709	2419.208	222.4410	759	2589.815
208.0805	710	2422.620	222.7340	760	2593.227
208.3736	711	2426.032	223.0271	761	2596.640
208.6666	712	2429.445	223.3202	762	2600.052
208.9597	713	2432.857	223.6132	763	2603.464
209.2528	714	2436.269	223.9063	764	2606.876
209.5458	715	2439.681	224.1994	765	2610.288
209.8389	716	2443.093	224.4925	766	2613.700
210.1320	717	2446.505	224.7855	767	2617.112
210.4250	718	2449.917	225.0786	768	2620.525
210.7181	719	2453.330	225.3717	769	2623.937
211.0112	720	2456.742	225.6647	770	2627.349
211.3043	721	2460.154	225.9578	771	2630.761
211.5973	722	2463.566	226.2509	772	2634.173
211.8904	723	2466.978	226.5440	773	2637.585
212.1835	724	2470.390	226.8370	774	2640.997
212.4765	725	2473.802	227.1301	775	2644.410
212.7696	726	2477.215	227.4232	776	2647.822
213.0627	727	2480.627	227.7162	777	2651.234
213.3558	728	2484.039	228.0093	778	2654.646
213.6488	729	2487.451	228.3024	779	2658.058
213.9419	730	2490.863	228.5955	780	2661.470
214.2350	731	2494.275	228.8885	781	2664.882
214.5280	732	2497.687	229.1816	782	2668.294
214.8211	733	2501.100	229.4747	783	2671.707
215.1142	734	2504.512	229.7677	784	2675.119
215.4073	735	2507.924	230.0608	785	2678.531
215.7003	736	2511.336	230.3539	786	2681.943
215.9934	737	2514.748	230.6470	787	2685.355
216.2865	738	2518.160	230.9400	788	2688.767
216.5795	739	2521.572	231.2331	789	2692.179
216.8726	740	2524.985	231.5262	790	2695.592
217.1657	741	2528.397	231.8192	791	2699.004
217.4588	742	2531.809	232.1123	792	2702.416
217.7518	743	2535.221	232.4054	793	2705.828
218.0449	744	2538.633	232.6985	794	2709.240
218.3380	745	2542.045	232.9915	795	2712.652
218.6310	746	2545.457	233.2846	796	2716.064
218.9241	747	2548.870	233.5777	797	2719.477
219.2172	748	2552.282	233.8707	798	2722.889
219.5103	749	2555.694	234.1638	799	2726.301
219.8033	750	2559.106	234.4569	800	2729.713

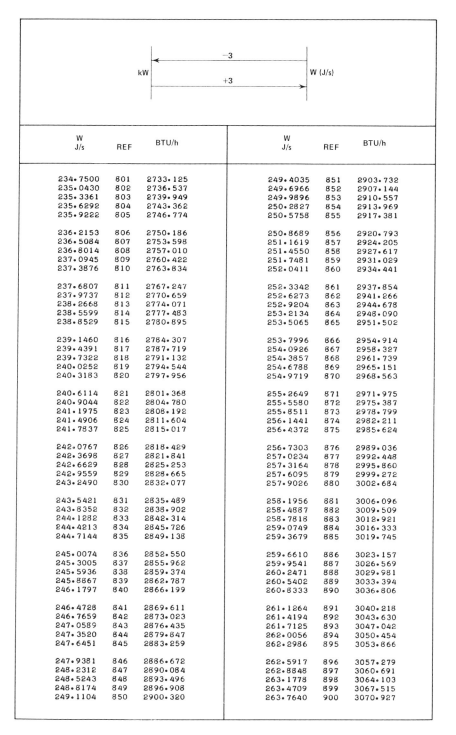

W J/s	REF	BTU/h	W J/s	REF	BTU/h
234.7500	801	2733.125	249.4035	851	2903.732
235.0430	802	2736.537	249.6966	852	2907.144
235.3361	803	2739.949	249.9896	853	2910.557
235.6292	804	2743.362	250.2827	854	2913.969
235.9222	805	2746.774	250.5758	855	2917.381
236.2153	806	2750.186	250.8689	856	2920.793
236.5084	807	2753.598	251.1619	857	2924.205
236.8014	808	2757.010	251.4550	858	2927.617
237.0945	809	2760.422	251.7481	859	2931.029
237.3876	810	2763.834	252.0411	860	2934.441
237.6807	811	2767.247	252.3342	861	2937.854
237.9737	812	2770.659	252.6273	862	2941.266
238.2668	813	2774.071	252.9204	863	2944.678
238.5599	814	2777.483	253.2134	864	2948.090
238.8529	815	2780.895	253.5065	865	2951.502
239.1460	816	2784.307	253.7996	866	2954.914
239.4391	817	2787.719	254.0926	867	2958.327
239.7322	818	2791.132	254.3857	868	2961.739
240.0252	819	2794.544	254.6788	869	2965.151
240.3183	820	2797.956	254.9719	870	2968.563
240.6114	821	2801.368	255.2649	871	2971.975
240.9044	822	2804.780	255.5580	872	2975.387
241.1975	823	2808.192	255.8511	873	2978.799
241.4906	824	2811.604	256.1441	874	2982.211
241.7837	825	2815.017	256.4372	875	2985.624
242.0767	826	2818.429	256.7303	876	2989.036
242.3698	827	2821.841	257.0234	877	2992.448
242.6629	828	2825.253	257.3164	878	2995.860
242.9559	829	2828.665	257.6095	879	2999.272
243.2490	830	2832.077	257.9026	880	3002.684
243.5421	831	2835.489	258.1956	881	3006.096
243.8352	832	2838.902	258.4887	882	3009.509
244.1282	833	2842.314	258.7818	883	3012.921
244.4213	834	2845.726	259.0749	884	3016.333
244.7144	835	2849.138	259.3679	885	3019.745
245.0074	836	2852.550	259.6610	886	3023.157
245.3005	837	2855.962	259.9541	887	3026.569
245.5936	838	2859.374	260.2471	888	3029.981
245.8867	839	2862.787	260.5402	889	3033.394
246.1797	840	2866.199	260.8333	890	3036.806
246.4728	841	2869.611	261.1264	891	3040.218
246.7659	842	2873.023	261.4194	892	3043.630
247.0589	843	2876.435	261.7125	893	3047.042
247.3520	844	2879.847	262.0056	894	3050.454
247.6451	845	2883.259	262.2986	895	3053.866
247.9381	846	2886.672	262.5917	896	3057.279
248.2312	847	2890.084	262.8848	897	3060.691
248.5243	848	2893.496	263.1778	898	3064.103
248.8174	849	2896.908	263.4709	899	3067.515
249.1104	850	2900.320	263.7640	900	3070.927

W J/s	REF	BTU/h		W J/s	REF	BTU/h
264.0571	901	3074.339		278.7106	951	3244.946
264.3501	902	3077.751		279.0037	952	3248.359
264.6432	903	3081.164		279.2968	953	3251.771
264.9363	904	3084.576		279.5898	954	3255.183
265.2293	905	3087.988		279.8829	955	3258.595
265.5224	906	3091.400		280.1760	956	3262.007
265.8155	907	3094.812		280.4690	957	3265.419
266.1086	908	3098.224		280.7621	958	3268.831
266.4016	909	3101.636		281.0552	959	3272.243
266.6947	910	3105.049		281.3483	960	3275.656
266.9878	911	3108.461		281.6413	961	3279.068
267.2808	912	3111.873		281.9344	962	3282.480
267.5739	913	3115.285		282.2275	963	3285.892
267.8670	914	3118.697		282.5205	964	3289.304
268.1601	915	3122.109		282.8136	965	3292.716
268.4531	916	3125.521		283.1067	966	3296.128
268.7462	917	3128.934		283.3998	967	3299.541
269.0393	918	3132.346		283.6928	968	3302.953
269.3323	919	3135.758		283.9859	969	3306.365
269.6254	920	3139.170		284.2790	970	3309.777
269.9185	921	3142.582		284.5720	971	3313.189
270.2116	922	3145.994		284.8651	972	3316.601
270.5046	923	3149.406		285.1582	973	3320.013
270.7977	924	3152.819		285.4512	974	3323.426
271.0908	925	3156.231		285.7443	975	3326.838
271.3838	926	3159.643		286.0374	976	3330.250
271.6769	927	3163.055		286.3305	977	3333.662
271.9700	928	3166.467		286.6235	978	3337.074
272.2631	929	3169.879		286.9166	979	3340.486
272.5561	930	3173.291		287.2097	980	3343.898
272.8492	931	3176.704		287.5027	981	3347.311
273.1423	932	3180.116		287.7958	982	3350.723
273.4353	933	3183.528		288.0889	983	3354.135
273.7284	934	3186.940		288.3820	984	3357.547
274.0215	935	3190.352		288.6750	985	3360.959
274.3145	936	3193.764		288.9681	986	3364.371
274.6076	937	3197.176		289.2612	987	3367.783
274.9007	938	3200.589		289.5542	988	3371.196
275.1938	939	3204.001		289.8473	989	3374.608
275.4868	940	3207.413		290.1404	990	3378.020
275.7799	941	3210.825		290.4335	991	3381.432
276.0730	942	3214.237		290.7265	992	3384.844
276.3660	943	3217.649		291.0196	993	3388.256
276.6591	944	3221.061		291.3127	994	3391.668
276.9522	945	3224.474		291.6057	995	3395.081
277.2453	946	3227.886		291.8988	996	3398.493
277.5383	947	3231.298		292.1919	997	3401.905
277.8314	948	3234.710		292.4850	998	3405.317
278.1245	949	3238.122		292.7780	999	3408.729
278.4175	950	3241.534		293.0711	1000	3412.141

37

TEMPERATURE AND TEMPERATURE DIFFERENCE

Section 37 is composed of two separate temperature tables. The first, Table 37A, permits the direct conversion of thermodynamic or absolute temperature between the English (Rankine Scale) and SI (Kelvin Scale) temperature units. Values in Table 37A can also be used to obtain conversions of temperature difference (Δt) between Rankine or Fahrenheit, and Kelvin or Celsius (formerly centigrade) scales. By use of conversion formulas given in the special introduction to Table 37A, the thermodynamic temperature values obtained from the table can be quickly changed into their Fahrenheit or Celsius equivalents.

Also, temperature differences provided by Table 37A can be used with values from Table 37B to obtain intermediate temperatures not given in this table. The introduction and examples to Table 37B explain this procedure.

For the table user who wishes to work directly with Fahrenheit and Celsius values, the editors have included Table 37B in this section. It is based on conveniently chosen REF temperature points and intervals.

433

TEMPERATURE (t)

Rankine — Kelvin
  (°R − K)

Fahrenheit — Celsius
  (°F − °C)

TEMPERATURE DIFFERENCE (Δt)

Change in Rankine	Change in Kelvin
(Δt_R)	(Δt_K)
or	or
Change in Fahrenheit	Change in Celsius
(Δt_F)	(Δt_C)

Conversion Factors: temperature difference (Δt_R) in Rankine (°R)
  = 5/9 temperature difference (Δt_K) in Kelvin (K)

English and SI Temperature Scales

The temperature scale for SI is the International Thermodynamic Tempera-ture Scale, having the kelvin (K) as its basic unit. However, because of its wide-spread usage, the Celsius (°C) (formerly Centigrade) or International Practical Temperature Scale is permitted in SI.

The thermodynamic or absolute temperature in the English system of measure-ment is the Rankine scale. It has as its unit the degree Rankine (°R). However, the Fahrenheit (°F) Scale is more commonly used in customary temperature measure.

Use of Table 37A

Only thermodynamic or absolute temperatures (°R or K) are converted through Table 37A. The following Relations Chart which appears at the top of alternate table pages, will assist the user in transforming $^\circ$R and K values to and from their "practical" forms ($^\circ$F and $^\circ$C).

$$K = {}^\circ C + 273.15$$
$$^\circ R = {}^\circ F + 459.67$$
$$^\circ C = K - 273.15$$
$$^\circ F = {}^\circ R - 459.67$$

Example (a): Convert 584°R to K; and $^\circ$C.
   From table: 584°R = 324.4444 K* $\approx$ 324.44 K
   From Relations Chart: $^\circ$C = K $-$ 273.15
   Therefore: $^\circ$C = 324.44 $-$ 273.15 = 51.29° $\approx$ 51°.

Example (b): Convert 174.65 K to $^\circ$R; and $^\circ$F.
   Since: 174.65 K = (174.00 + 0.65) K:
   From table: 174 K = 313.2000° R
                65 K = 117.0000° R
   Therefore: with proper decimal shifting and addition:
      174.00 K = 313.2000°R
       0.65 K =   1.1700°R
      174.65 K = 314.37°R
   From Relations Chart: $^\circ$F = $^\circ$R $-$ 459.67
   $^\circ$F = 314.37 $-$ 459.67 = $-$145.30

Temperature Difference (Δt)

   Table 37A can also be used for the conversion of temperature changes or differentials (Δt) between the customary and SI systems. For Δt values, however, the values in the table must be considered as quantitive changes in temperature. The table user must determine independently the proper algebraic signs. For a rise in temperature, the unsigned table value will suffice or, depending on the users requirement, a + notation can be used. A drop in temperature would then use a minus ($-$) sign.
   Temperature differences on the Kelvin scale are the same as on the Celsius scale, and can be given in either K or $^\circ$C. Also, the temperature differences on the Rankine scale are the same as on the Fahrenheit scale.

Example (c): Convert the drop in temperature from 73°F to $-$15°F into SI units.
   Since: $\Delta t_F = (-15 - 73)^\circ$ F = $-$88° F (or $^\circ$R):
   From table: 88° F = 48.88889° C*
   Therefore: temperature change $\approx$ 49° C (or K).

---

*All conversion values obtained can be rounded to desired number of significant figures.

Example (*d*): Convert the temperature rise of 284 K to 413 K into English units.
Since: $\Delta t_K = (413 - 284)\,K = +129\,K$:
From table: $129\,K = 232.2000°\,R^*$
Therefore: temperature change $\approx 232°\,R$ (or $°F$).

---

*All conversion values obtained can be rounded to desired number of significant figures.

K = °C + 273.15					
°R = °F + 459.67					
°C = K − 273.15					
°F = °R − 459.67					

°R Δt_R Δt_F	REF	°K Δt_k Δt_c	°R Δt_R Δt_F	REF	°K Δt_k Δt_c
0.5555556	1	1.800000	28.33333	51	91.80000
1.111111	2	3.600000	28.88889	52	93.60000
1.666667	3	5.400000	29.44444	53	95.40000
2.222222	4	7.200000	30.00000	54	97.20000
2.777778	5	9.000000	30.55556	55	99.00000
3.333333	6	10.80000	31.11111	56	100.8000
3.888889	7	12.60000	31.66667	57	102.6000
4.444444	8	14.40000	32.22222	58	104.4000
5.000000	9	16.20000	32.77778	59	106.2000
5.555556	10	18.00000	33.33333	60	108.0000
6.111111	11	19.80000	33.88889	61	109.8000
6.666667	12	21.60000	34.44444	62	111.6000
7.222222	13	23.40000	35.00000	63	113.4000
7.777778	14	25.20000	35.55556	64	115.2000
8.333333	15	27.00000	36.11111	65	117.0000
8.888889	16	28.80000	36.66667	66	118.8000
9.444444	17	30.60000	37.22222	67	120.6000
0.000000	18	32.40000	37.77778	68	122.4000
10.55556	19	34.20000	38.33333	69	124.2000
11.11111	20	36.00000	38.88889	70	126.0000
11.66667	21	37.80000	39.44444	71	127.8000
12.22222	22	39.60000	40.00000	72	129.6000
12.77778	23	41.40000	40.55556	73	131.4000
13.33333	24	43.20000	41.11111	74	133.2000
13.88889	25	45.00000	41.66667	75	135.0000
14.44444	26	46.80000	42.22222	76	136.8000
15.00000	27	48.60000	42.77778	77	138.6000
15.55556	28	50.40000	43.33333	78	140.4000
16.11111	29	52.20000	43.88889	79	142.2000
16.66667	30	54.00000	44.44444	80	144.0000
17.22222	31	55.80000	45.00000	81	145.8000
17.77778	32	57.60000	45.55556	82	147.6000
18.33333	33	59.40000	46.11111	83	149.4000
18.88889	34	61.20000	46.66667	84	151.2000
19.44444	35	63.00000	47.22222	85	153.0000
20.00000	36	64.80000	47.77778	86	154.8000
20.55556	37	66.60000	48.33333	87	156.6000
21.11111	38	68.40000	48.88889	88	158.4000
21.66667	39	70.20000	49.44444	89	160.2000
22.22222	40	72.00000	50.00000	90	162.0000
22.77778	41	73.80000	50.55556	91	163.8000
23.33333	42	75.60000	51.11111	92	165.6000
23.88889	43	77.40000	51.66667	93	167.4000
24.44444	44	79.20000	52.22222	94	169.2000
25.00000	45	81.00000	52.77778	95	171.0000
25.55556	46	82.80000	53.33333	96	172.8000
26.11111	47	84.60000	53.88889	97	174.6000
26.66667	48	86.40000	54.44444	98	176.4000
27.22222	49	88.20000	55.00000	99	178.2000
27.77778	50	90.00000	55.55556	100	180.0000

°R Δt_R Δt_F	REF	°K Δt_k Δt_c	°R Δt_R Δt_F	REF	°K Δt_k Δt_c
56.11111	101	181.8000	83.88889	151	271.8000
56.66667	102	183.6000	84.44444	152	273.6000
57.22222	103	185.4000	85.00000	153	275.4000
57.77778	104	187.2000	85.55556	154	277.2000
58.33333	105	189.0000	86.11111	155	279.0000
58.88889	106	190.8000	86.66667	156	280.8000
59.44444	107	192.6000	87.22222	157	282.6000
60.00000	108	194.4000	87.77778	158	284.4000
60.55556	109	196.2000	88.33333	159	286.2000
61.11111	110	198.0000	88.88889	160	288.0000
61.66667	111	199.8000	89.44444	161	289.8000
62.22222	112	201.6000	90.00000	162	291.6000
62.77778	113	203.4000	90.55556	163	293.4000
63.33333	114	205.2000	91.11111	164	295.2000
63.88889	115	207.0000	91.66667	165	297.0000
64.44444	116	208.8000	92.22222	166	298.8000
65.00000	117	210.6000	92.77778	167	300.6000
65.55556	118	212.4000	93.33333	168	302.4000
66.11111	119	214.2000	93.88889	169	304.2000
66.66667	120	216.0000	94.44444	170	306.0000
67.22222	121	217.8000	95.00000	171	307.8000
67.77778	122	219.6000	95.55556	172	309.6000
68.33333	123	221.4000	96.11111	173	311.4000
68.88889	124	223.2000	96.66667	174	313.2000
69.44444	125	225.0000	97.22222	175	315.0000
70.00000	126	226.8000	97.77778	176	316.8000
70.55556	127	228.6000	98.33333	177	318.6000
71.11111	128	230.4000	98.88889	178	320.4000
71.66667	129	232.2000	99.44444	179	322.2000
72.22222	130	234.0000	00.00000	180	324.0000
72.77778	131	235.8000	100.5556	181	325.8000
73.33333	132	237.6000	101.1111	182	327.6000
73.88889	133	239.4000	101.6667	183	329.4000
74.44444	134	241.2000	102.2222	184	331.2000
75.00000	135	243.0000	102.7778	185	333.0000
75.55556	136	244.8000	103.3333	186	334.8000
76.11111	137	246.6000	103.8889	187	336.6000
76.66667	138	248.4000	104.4444	188	338.4000
77.22222	139	250.2000	105.0000	189	340.2000
77.77778	140	252.0000	105.5556	190	342.0000
78.33333	141	253.8000	106.1111	191	343.8000
78.88889	142	255.6000	106.6667	192	345.6000
79.44444	143	257.4000	107.2222	193	347.4000
80.00000	144	259.2000	107.7778	194	349.2000
80.55556	145	261.0000	108.3333	195	351.0000
81.11111	146	262.8000	108.8889	196	352.8000
81.66667	147	264.6000	109.4444	197	354.6000
82.22222	148	266.4000	110.0000	198	356.4000
82.77778	149	268.2000	110.5556	199	358.2000
83.33333	150	270.0000	111.1111	200	360.0000

$$K = °C + 273.15$$

$$°R = °F + 459.67$$

$$°C = K - 273.15$$

$$°F = °R - 459.67$$

°R Δt_R Δt_F	REF	°K Δt_k Δt_C	°R Δt_R Δt_F	REF	°K Δt_k Δt_C
111.6667	201	361.8000	139.4444	251	451.8000
112.2222	202	363.6000	140.0000	252	453.6000
112.7778	203	365.4000	140.5556	253	455.4000
113.3333	204	367.2000	141.1111	254	457.2000
113.8889	205	369.0000	141.6667	255	459.0000
114.4444	206	370.8000	142.2222	256	460.8000
115.0000	207	372.6000	142.7778	257	462.6000
115.5556	208	374.4000	143.3333	258	464.4000
116.1111	209	376.2000	143.8889	259	466.2000
116.6667	210	378.0000	144.4444	260	468.0000
117.2222	211	379.8000	145.0000	261	469.8000
117.7778	212	381.6000	145.5556	262	471.6000
118.3333	213	383.4000	146.1111	263	473.4000
118.8889	214	385.2000	146.6667	264	475.2000
119.4444	215	387.0000	147.2222	265	477.0000
120.0000	216	388.8000	147.7778	266	478.8000
120.5556	217	390.6000	148.3333	267	480.6000
121.1111	218	392.4000	148.8889	268	482.4000
121.6667	219	394.2000	149.4444	269	484.2000
122.2222	220	396.0000	150.0000	270	486.0000
122.7778	221	397.8000	150.5556	271	487.8000
123.3333	222	399.6000	151.1111	272	489.6000
123.8889	223	401.4000	151.6667	273	491.4000
124.4444	224	403.2000	152.2222	274	493.2000
125.0000	225	405.0000	152.7778	275	495.0000
125.5556	226	406.8000	153.3333	276	496.8000
126.1111	227	408.6000	153.8889	277	498.6000
126.6667	228	410.4000	154.4444	278	500.4000
127.2222	229	412.2000	155.0000	279	502.2000
127.7778	230	414.0000	155.5556	280	504.0000
128.3333	231	415.8000	156.1111	281	505.8000
128.8889	232	417.6000	156.6667	282	507.6000
129.4444	233	419.4000	157.2222	283	509.4000
130.0000	234	421.2000	157.7778	284	511.2000
130.5556	235	423.0000	158.3333	285	513.0000
131.1111	236	424.8000	158.8889	286	514.8000
131.6667	237	426.6000	159.4444	287	516.6000
132.2222	238	428.4000	160.0000	288	518.4000
132.7778	239	430.2000	160.5556	289	520.2000
133.3333	240	432.0000	161.1111	290	522.0000
133.8889	241	433.8000	161.6667	291	523.8000
134.4444	242	435.6000	162.2222	292	525.6000
135.0000	243	437.4000	162.7778	293	527.4000
135.5556	244	439.2000	163.3333	294	529.2000
136.1111	245	441.0000	163.8889	295	531.0000
136.6667	246	442.8000	164.4444	296	532.8000
137.2222	247	444.6000	165.0000	297	534.6000
137.7778	248	446.4000	165.5556	298	536.4000
138.3333	249	448.2000	166.1111	299	538.2000
138.8889	250	450.0000	166.6667	300	540.0000

°R Δt_R Δt_F		REF	°K Δt_k Δt_c	°R Δt_R Δt_F		REF	°K Δt_k Δt_c
167.2222		301	541.8000	195.0000		351	631.8000
167.7778		302	543.6000	195.5556		352	633.6000
168.3333		303	545.4000	196.1111		353	635.4000
168.8889		304	547.2000	196.6667		354	637.2000
169.4444		305	549.0000	197.2222		355	639.0000
170.0000		306	550.8000	197.7778		356	640.8000
170.5556		307	552.6000	198.3333		357	642.6000
171.1111		308	554.4000	198.8889		358	644.4000
171.6667		309	556.2000	199.4444		359	646.2000
172.2222		310	558.0000	200.0000		360	648.0000
172.7778		311	559.8000	200.5556		361	649.8000
173.3333		312	561.6000	201.1111		362	651.6000
173.8889		313	563.4000	201.6667		363	653.4000
174.4444		314	565.2000	202.2222		364	655.2000
175.0000		315	567.0000	202.7778		365	657.0000
175.5556		316	568.8000	203.3333		366	658.8000
176.1111		317	570.6000	203.8889		367	660.6000
176.6667		318	572.4000	204.4444		368	662.4000
177.2222		319	574.2000	205.0000		369	664.2000
177.7778		320	576.0000	205.5556		370	666.0000
178.3333		321	577.8000	206.1111		371	667.8000
178.8889		322	579.6000	206.6667		372	669.6000
179.4444		323	581.4000	207.2222		373	671.4000
180.0000		324	583.2000	207.7778		374	673.2000
180.5556		325	585.0000	208.3333		375	675.0000
181.1111		326	586.8000	208.8889		376	676.8000
181.6667		327	588.6000	209.4444		377	678.6000
182.2222		328	590.4000	210.0000		378	680.4000
182.7778		329	592.2000	210.5556		379	682.2000
183.3333		330	594.0000	211.1111		380	684.0000
183.8889		331	595.8000	211.6667		381	685.8000
184.4444		332	597.6000	212.2222		382	687.6000
185.0000		333	599.4000	212.7778		383	689.4000
185.5556		334	601.2000	213.3333		384	691.2000
186.1111		335	603.0000	213.8889		385	693.0000
186.6667		336	604.8000	214.4444		386	694.8000
187.2222		337	606.6000	215.0000		387	696.6000
187.7778		338	608.4000	215.5556		388	698.4000
188.3333		339	610.2000	216.1111		389	700.2000
188.8889		340	612.0000	216.6667		390	702.0000
189.4444		341	613.8000	217.2222		391	703.8000
190.0000		342	615.6000	217.7778		392	705.6000
190.5556		343	617.4000	218.3333		393	707.4000
191.1111		344	619.2000	218.8889		394	709.2000
191.6667		345	621.0000	219.4444		395	711.0000
192.2222		346	622.8000	220.0000		396	712.8000
192.7778		347	624.6000	220.5556		397	714.6000
193.3333		348	626.4000	221.1111		398	716.4000
193.8889		349	628.2000	221.6667		399	718.2000
194.4444		350	630.0000	222.2222		400	720.0000

$$K = °C + 273.15$$

$$°R = °F + 459.67$$

$$°C = K - 273.15$$

$$°F = °R - 459.67$$

$°R$ Δt_R Δt_F	REF	$°K$ Δt_k Δt_c	$°R$ Δt_R Δt_F	REF	$°K$ Δt_k Δt_c
222.7778	401	721.8000	250.5556	451	811.8000
223.3333	402	723.6000	251.1111	452	813.6000
223.8889	403	725.4000	251.6667	453	815.4000
224.4444	404	727.2000	252.2222	454	817.2000
225.0000	405	729.0000	252.7778	455	819.0000
225.5556	406	730.8000	253.3333	456	820.8000
226.1111	407	732.6000	253.8889	457	822.6000
226.6667	408	734.4000	254.4444	458	824.4000
227.2222	409	736.2000	255.0000	459	826.2000
227.7778	410	738.0000	255.5556	460	828.0000
228.3333	411	739.8000	256.1111	461	829.8000
228.8889	412	741.6000	256.6667	462	831.6000
229.4444	413	743.4000	257.2222	463	833.4000
230.0000	414	745.2000	257.7778	464	835.2000
230.5556	415	747.0000	258.3333	465	837.0000
231.1111	416	748.8000	258.8889	466	838.8000
231.6667	417	750.6000	259.4444	467	840.6000
232.2222	418	752.4000	260.0000	468	842.4000
232.7778	419	754.2000	260.5556	469	844.2000
233.3333	420	756.0000	261.1111	470	846.0000
233.8889	421	757.8000	261.6667	471	847.8000
234.4444	422	759.6000	262.2222	472	849.6000
235.0000	423	761.4000	262.7778	473	851.4000
235.5556	424	763.2000	263.3333	474	853.2000
236.1111	425	765.0000	263.8889	475	855.0000
236.6667	426	766.8000	264.4444	476	856.8000
237.2222	427	768.6000	265.0000	477	858.6000
237.7778	428	770.4000	265.5556	478	860.4000
238.3333	429	772.2000	266.1111	479	862.2000
238.8889	430	774.0000	266.6667	480	864.0000
239.4444	431	775.8000	267.2222	481	865.8000
240.0000	432	777.6000	267.7778	482	867.6000
240.5556	433	779.4000	268.3333	483	869.4000
241.1111	434	781.2000	268.8889	484	871.2000
241.6667	435	783.0000	269.4444	485	873.0000
242.2222	436	784.8000	270.0000	486	874.8000
242.7778	437	786.6000	270.5556	487	876.6000
243.3333	438	788.4000	271.1111	488	878.4000
243.8889	439	790.2000	271.6667	489	880.2000
244.4444	440	792.0000	272.2222	490	882.0000
245.0000	441	793.8000	272.7778	491	883.8000
245.5556	442	795.6000	273.3333	492	885.6000
246.1111	443	797.4000	273.8889	493	887.4000
246.6667	444	799.2000	274.4444	494	889.2000
247.2222	445	801.0000	275.0000	495	891.0000
247.7778	446	802.8000	275.5556	496	892.8000
248.3333	447	804.6000	276.1111	497	894.6000
248.8889	448	806.4000	276.6667	498	896.4000
249.4444	449	808.2000	277.2222	499	898.2000
250.0000	450	810.0000	277.7778	500	900.0000

°R Δt_R Δt_F	REF	°K Δt_k Δt_C	°R Δt_R Δt_F	REF	°K Δt_k Δt_C
278.3333	501	901.8000	306.1111	551	991.8000
278.8889	502	903.6000	306.6667	552	993.6000
279.4444	503	905.4000	307.2222	553	995.4000
280.0000	504	907.2000	307.7778	554	997.2000
280.5556	505	909.0000	308.3333	555	999.0000
281.1111	506	910.8000	308.8889	556	1000.800
281.6667	507	912.6000	309.4444	557	1002.600
282.2222	508	914.4000	310.0000	558	1004.400
282.7778	509	916.2000	310.5556	559	1006.200
283.3333	510	918.0000	311.1111	560	1008.000
283.8889	511	919.8000	311.6667	561	1009.800
284.4444	512	921.6000	312.2222	562	1011.600
285.0000	513	923.4000	312.7778	563	1013.400
285.5556	514	925.2000	313.3333	564	1015.200
286.1111	515	927.0000	313.8889	565	1017.000
286.6667	516	928.8000	314.4444	566	1018.800
287.2222	517	930.6000	315.0000	567	1020.600
287.7778	518	932.4000	315.5556	568	1022.400
288.3333	519	934.2000	316.1111	569	1024.200
288.8889	520	936.0000	316.6667	570	1026.000
289.4444	521	937.8000	317.2222	571	1027.800
290.0000	522	939.6000	317.7778	572	1029.600
290.5556	523	941.4000	318.3333	573	1031.400
291.1111	524	943.2000	318.8889	574	1033.200
291.6667	525	945.0000	319.4444	575	1035.000
292.2222	526	946.8000	320.0000	576	1036.800
292.7778	527	948.6000	320.5556	577	1038.600
293.3333	528	950.4000	321.1111	578	1040.400
293.8889	529	952.2000	321.6667	579	1042.200
294.4444	530	954.0000	322.2222	580	1044.000
295.0000	531	955.8000	322.7778	581	1045.800
295.5556	532	957.6000	323.3333	582	1047.600
296.1111	533	959.4000	323.8889	583	1049.400
296.6667	534	961.2000	324.4444	584	1051.200
297.2222	535	963.0000	325.0000	585	1053.000
297.7778	536	964.8000	325.5556	586	1054.800
298.3333	537	966.6000	326.1111	587	1056.600
298.8889	538	968.4000	326.6667	588	1058.400
299.4444	539	970.2000	327.2222	589	1060.200
300.0000	540	972.0000	327.7778	590	1062.000
300.5556	541	973.8000	328.3333	591	1063.800
301.1111	542	975.6000	328.8889	592	1065.600
301.6667	543	977.4000	329.4444	593	1067.400
302.2222	544	979.2000	330.0000	594	1069.200
302.7778	545	981.0000	330.5556	595	1071.000
303.3333	546	982.8000	331.1111	596	1072.800
303.8889	547	984.6000	331.6667	597	1074.600
304.4444	548	986.4000	332.2222	598	1076.400
305.0000	549	988.2000	332.7778	599	1078.200
305.5556	550	990.0000	333.3333	600	1080.000

$$K = °C + 273.15$$

$$°R = °F + 459.67$$

$$°C = K - 273.15$$

$$°F = °R - 459.67$$

°R Δt_R Δt_F	REF	°K Δt_k Δt_c	°R Δt_R Δt_F	REF	°K Δt_k Δt_c
333.8889	601	1081.800	361.6667	651	1171.800
334.4444	602	1083.600	362.2222	652	1173.600
335.0000	603	1085.400	362.7778	653	1175.400
335.5556	604	1087.200	363.3333	654	1177.200
336.1111	605	1089.000	363.8889	655	1179.000
336.6667	606	1090.800	364.4444	656	1180.800
337.2222	607	1092.600	365.0000	657	1182.600
337.7778	608	1094.400	365.5556	658	1184.400
338.3333	609	1096.200	366.1111	659	1186.200
338.8889	610	1098.000	366.6667	660	1188.000
339.4444	611	1099.800	367.2222	661	1189.800
340.0000	612	1101.600	367.7778	662	1191.600
340.5556	613	1103.400	368.3333	663	1193.400
341.1111	614	1105.200	368.8889	664	1195.200
341.6667	615	1107.000	369.4444	665	1197.000
342.2222	616	1108.800	370.0000	666	1198.800
342.7778	617	1110.600	370.5556	667	1200.600
343.3333	618	1112.400	371.1111	668	1202.400
343.8889	619	1114.200	371.6667	669	1204.200
344.4444	620	1116.000	372.2222	670	1206.000
345.0000	621	1117.800	372.7778	671	1207.800
345.5556	622	1119.600	373.3333	672	1209.600
346.1111	623	1121.400	373.8889	673	1211.400
346.6667	624	1123.200	374.4444	674	1213.200
347.2222	625	1125.000	375.0000	675	1215.000
347.7778	626	1126.800	375.5556	676	1216.800
348.3333	627	1128.600	376.1111	677	1218.600
348.8889	628	1130.400	376.6667	678	1220.400
349.4444	629	1132.200	377.2222	679	1222.200
350.0000	630	1134.000	377.7778	680	1224.000
350.5556	631	1135.800	378.3333	681	1225.800
351.1111	632	1137.600	378.8889	682	1227.600
351.6667	633	1139.400	379.4444	683	1229.400
352.2222	634	1141.200	380.0000	684	1231.200
352.7778	635	1143.000	380.5556	685	1233.000
353.3333	636	1144.800	381.1111	686	1234.800
353.8889	637	1146.600	381.6667	687	1236.600
354.4444	638	1148.400	382.2222	688	1238.400
355.0000	639	1150.200	382.7778	689	1240.200
355.5556	640	1152.000	383.3333	690	1242.000
356.1111	641	1153.800	383.8889	691	1243.800
356.6667	642	1155.600	384.4444	692	1245.600
357.2222	643	1157.400	385.0000	693	1247.400
357.7778	644	1159.200	385.5556	694	1249.200
358.3333	645	1161.000	386.1111	695	1251.000
358.8889	646	1162.800	386.6667	696	1252.800
359.4444	647	1164.600	387.2222	697	1254.600
360.0000	648	1166.400	387.7778	698	1256.400
360.5556	649	1168.200	388.3333	699	1258.200
361.1111	650	1170.000	388.8889	700	1260.000

°R Δt_R Δt_F	REF	°K Δt_k Δt_c	°R Δt_R Δt_F	REF	°K Δt_k Δt_c
389.4444	701	1261.800	417.2222	751	1351.800
390.0000	702	1263.600	417.7778	752	1353.600
390.5556	703	1265.400	418.3333	753	1355.400
391.1111	704	1267.200	418.8889	754	1357.200
391.6667	705	1269.000	419.4444	755	1359.000
392.2222	706	1270.800	420.0000	756	1360.800
392.7778	707	1272.600	420.5556	757	1362.600
393.3333	708	1274.400	421.1111	758	1364.400
393.8889	709	1276.200	421.6667	759	1366.200
394.4444	710	1278.000	422.2222	760	1368.000
395.0000	711	1279.800	422.7778	761	1369.800
395.5556	712	1281.600	423.3333	762	1371.600
396.1111	713	1283.400	423.8889	763	1373.400
396.6667	714	1285.200	424.4444	764	1375.200
397.2222	715	1287.000	425.0000	765	1377.000
397.7778	716	1288.800	425.5556	766	1378.800
398.3333	717	1290.600	426.1111	767	1380.600
398.8889	718	1292.400	426.6667	768	1382.400
399.4444	719	1294.200	427.2222	769	1384.200
400.0000	720	1296.000	427.7778	770	1386.000
400.5556	721	1297.800	428.3333	771	1387.800
401.1111	722	1299.600	428.8889	772	1389.600
401.6667	723	1301.400	429.4444	773	1391.400
402.2222	724	1303.200	430.0000	774	1393.200
402.7778	725	1305.000	430.5556	775	1395.000
403.3333	726	1306.800	431.1111	776	1396.800
403.8889	727	1308.600	431.6667	777	1398.600
404.4444	728	1310.400	432.2222	778	1400.400
405.0000	729	1312.200	432.7778	779	1402.200
405.5556	730	1314.000	433.3333	780	1404.000
406.1111	731	1315.800	433.8889	781	1405.800
406.6667	732	1317.600	434.4444	782	1407.600
407.2222	733	1319.400	435.0000	783	1409.400
407.7778	734	1321.200	435.5556	784	1411.200
408.3333	735	1323.000	436.1111	785	1413.000
408.8889	736	1324.800	436.6667	786	1414.800
409.4444	737	1326.600	437.2222	787	1416.600
410.0000	738	1328.400	437.7778	788	1418.400
410.5556	739	1330.200	438.3333	789	1420.200
411.1111	740	1332.000	438.8889	790	1422.000
411.6667	741	1333.800	439.4444	791	1423.800
412.2222	742	1335.600	440.0000	792	1425.600
412.7778	743	1337.400	440.5556	793	1427.400
413.3333	744	1339.200	441.1111	794	1429.200
413.8889	745	1341.000	441.6667	795	1431.000
414.4444	746	1342.800	442.2222	796	1432.800
415.0000	747	1344.600	442.7778	797	1434.600
415.5556	748	1346.400	443.3333	798	1436.400
416.1111	749	1348.200	443.8889	799	1438.200
416.6667	750	1350.000	444.4444	800	1440.000

$$K = {}^\circ C + 273.15$$

$$^\circ R = {}^\circ F + 459.67$$

$$^\circ C = K - 273.15$$

$$^\circ F = {}^\circ R - 459.67$$

°R Δt_R Δt_F	REF	°K Δt_k Δt_c	°R Δt_R Δt_F	REF	°K Δt_k Δt_c
445.0000	801	1441.800	472.7778	851	1531.800
445.5556	802	1443.600	473.3333	852	1533.600
446.1111	803	1445.400	473.8889	853	1535.400
446.6667	804	1447.200	474.4444	854	1537.200
447.2222	805	1449.000	475.0000	855	1539.000
447.7778	806	1450.800	475.5556	856	1540.800
448.3333	807	1452.600	476.1111	857	1542.600
448.8889	808	1454.400	476.6667	858	1544.400
449.4444	809	1456.200	477.2222	859	1546.200
450.0000	810	1458.000	477.7778	860	1548.000
450.5556	811	1459.800	478.3333	861	1549.800
451.1111	812	1461.600	478.8889	862	1551.600
451.6667	813	1463.400	479.4444	863	1553.400
452.2222	814	1465.200	480.0000	864	1555.200
452.7778	815	1467.000	480.5556	865	1557.000
453.3333	816	1468.800	481.1111	866	1558.800
453.8889	817	1470.600	481.6667	867	1560.600
454.4444	818	1472.400	482.2222	868	1562.400
455.0000	819	1474.200	482.7778	869	1564.200
455.5556	820	1476.000	483.3333	870	1566.000
456.1111	821	1477.800	483.8889	871	1567.800
456.6667	822	1479.600	484.4444	872	1569.600
457.2222	823	1481.400	485.0000	873	1571.400
457.7778	824	1483.200	485.5556	874	1573.200
458.3333	825	1485.000	486.1111	875	1575.000
458.8889	826	1486.800	486.6667	876	1576.800
459.4444	827	1488.600	487.2222	877	1578.600
460.0000	828	1490.400	487.7778	878	1580.400
460.5556	829	1492.200	488.3333	879	1582.200
461.1111	830	1494.000	488.8889	880	1584.000
461.6667	831	1495.800	489.4444	881	1585.800
462.2222	832	1497.600	490.0000	882	1587.600
462.7778	833	1499.400	490.5556	883	1589.400
463.3333	834	1501.200	491.1111	884	1591.200
463.8889	835	1503.000	491.6667	885	1593.000
464.4444	836	1504.800	492.2222	886	1594.800
465.0000	837	1506.600	492.7778	887	1596.600
465.5556	838	1508.400	493.3333	888	1598.400
466.1111	839	1510.200	493.8889	889	1600.200
466.6667	840	1512.000	494.4444	890	1602.000
467.2222	841	1513.800	495.0000	891	1603.800
467.7778	842	1515.600	495.5556	892	1605.600
468.3333	843	1517.400	496.1111	893	1607.400
468.8889	844	1519.200	496.6667	894	1609.200
469.4444	845	1521.000	497.2222	895	1611.000
470.0000	846	1522.800	497.7778	896	1612.800
470.5556	847	1524.600	498.3333	897	1614.600
471.1111	848	1526.400	498.8889	898	1616.400
471.6667	849	1528.200	499.4444	899	1618.200
472.2222	850	1530.000	500.0000	900	1620.000

°R Δt_R / Δt_R	REF	°K Δt_k / Δt_c	°R Δt_R / Δt_F	REF	°K Δt_k / Δt_c
500.5556	901	1621.800	528.3333	951	1711.800
501.1111	902	1623.600	528.8889	952	1713.600
501.6667	903	1625.400	529.4444	953	1715.400
502.2222	904	1627.200	530.0000	954	1717.200
502.7778	905	1629.000	530.5555	955	1719.000
503.3333	906	1630.800	531.1111	956	1720.800
503.8889	907	1632.600	531.6667	957	1722.600
504.4444	908	1634.400	532.2222	958	1724.400
505.0000	909	1636.200	532.7778	959	1726.200
505.5556	910	1638.000	533.3333	960	1728.000
506.1111	911	1639.800	533.8889	961	1729.800
506.6667	912	1641.600	534.4444	962	1731.600
507.2222	913	1643.400	535.0000	963	1733.400
507.7778	914	1645.200	535.5555	964	1735.200
508.3333	915	1647.000	536.1111	965	1737.000
508.8889	916	1648.800	536.6667	966	1738.800
509.4444	917	1650.600	537.2222	967	1740.600
510.0000	918	1652.400	537.7778	968	1742.400
510.5556	919	1654.200	538.3333	969	1744.200
511.1111	920	1656.000	538.8889	970	1746.000
511.6667	921	1657.800	539.4444	971	1747.800
512.2222	922	1659.600	540.0000	972	1749.600
512.7778	923	1661.400	540.5555	973	1751.400
513.3333	924	1663.200	541.1111	974	1753.200
513.8889	925	1665.000	541.6667	975	1755.000
514.4444	926	1666.800	542.2222	976	1756.800
515.0000	927	1668.600	542.7778	977	1758.600
515.5555	928	1670.400	543.3333	978	1760.400
516.1111	929	1672.200	543.8889	979	1762.200
516.6667	930	1674.000	544.4444	980	1764.000
517.2222	931	1675.800	545.0000	981	1765.800
517.7778	932	1677.600	545.5555	982	1767.600
518.3333	933	1679.400	546.1111	983	1769.400
518.8889	934	1681.200	546.6667	984	1771.200
519.4444	935	1683.000	547.2222	985	1773.000
520.0000	936	1684.800	547.7778	986	1774.800
520.5555	937	1686.600	548.3333	987	1776.600
521.1111	938	1688.400	548.8889	988	1778.400
521.6667	939	1690.200	549.4444	989	1780.200
522.2222	940	1692.000	550.0000	990	1782.000
522.7778	941	1693.800	550.5555	991	1783.800
523.3333	942	1695.600	551.1111	992	1785.600
523.8889	943	1697.400	551.6667	993	1787.400
524.4444	944	1699.200	552.2222	994	1789.200
525.0000	945	1701.000	552.7778	995	1791.000
525.5555	946	1702.800	553.3333	996	1792.800
526.1111	947	1704.600	553.8889	997	1794.600
526.6667	948	1706.400	554.4444	998	1796.400
527.2222	949	1708.200	555.0000	999	1798.200
527.7778	950	1710.000	555.5555	1000	1800.000

37B

TEMPERATURE

Fahrenheit − Celsius
  (°F − °C)
Fahrenheit to Kelvin
  (°F to K)
Celsius to Rankine
  (°C to °R)

Conversion Relations:

$$\text{Fahrenheit Temperature } (°F) = \frac{9 \times \text{Celsius Temperature } (°C)}{5} + 32$$

$$(°C = 5/9 \ [°F - 32])$$

As noted in the special introduction to this table section, Table 37B permits the direct conversion of temperature values between Fahrenheit and Celsius (formerly Centigrade). The range of REF values advance from −459.4 to 5000 in varying intervals. These intervals narrow in the most commonly used temperature ranges.

Intermediate temperatures that lie within the interval between table values can be obtained by the proper addition or subtraction of °F or °C temperature changes (Δt) from Table 37A.

The following relations will assist the table user in working with thermodynamic temperatures in Kelvin (K) and Rankine (°R). However, only °F and °C equivalents are to be used in Table 37B.

$$K = °C + 273.2$$
$$°R = °F + 459.7$$
$$°C = K - 273.2$$
$$°F = °R - 459.7$$

Example (a): Convert 166° to °C; and K.
  From table: $166°F = 74.4°C$
  From Relations Chart: $K = °C + 273.2$
  Therefore: $K = 74.4 + 273.2 = 347.6$.

Example (b): Convert 230°C to °F; and °R.
  From table: 230°C = 466.0°F
  From Relations Chart: °R = °F + 459.7
  Therefore: °R = 446.0 + 459.7 = 905.7
Example (c): Convert 1750°F to °C.
  From Table 37B: 1500°F = 815.6°C
  Since: temperature difference (Δt) = 1750 − 1500 = 250°F
  From Table 37A: 250°F = 138.8889°C ≈ 138.9°C
  Therefore: 1500°F = 815.6°C

$$\frac{250°F = 138.9°C}{1750°F = 954.5°C}$$

Example (d): Convert − 83 °C to °F.
  From Table 37B: − 80 °C = −112 °F
  Since: temperature difference (Δt) = −83° − (−80°) = −3°C
  From Table 37A: 3°C = 5.4°F ≈ 5°F or −3°C ≈ −5°F

  Therefore: − 80°C = −112°F

$$\frac{-3°C =   -5°F}{-83°C = -117°F}$$

Fahrenheit — Celsius (Centigrade) Conversion Table

Deg.C.	Ref	Deg.F.	Deg.C.	Ref	Deg.F.	Deg.C.	Ref.	Deg.F.	Deg.C.	Ref	Deg.F.
−273	−459.4	...	−101	−150	−238	− 8.3	17	62.6	9.4	49	120.2
−268	−450	...	− 96	−140	−220	− 7.8	18	64.4	10.0	50	122.0
−262	−440	...	− 90	−130	−202	− 7.2	19	66.2	10.6	51	123.8
−257	−430	...	− 84	−120	−184	− 6.7	20	68.0	11.1	52	125.6
−251	−420	...	− 79	−110	−166	− 6.1	21	69.8	11.7	53	127.4
−246	−410	...	− 73	−100	−148	− 5.6	22	71.6	12.2	54	129.2
−240	−400	...	− 68	− 90	−130	− 5.0	23	73.4	12.8	55	131.0
−234	−390	...	− 62	− 80	−112	− 4.4	24	75.2	13.3	56	132.8
−229	−380	...	− 57	− 70	− 94	− 3.9	25	77.0	13.9	57	134.6
−223	−370	...	− 51	− 60	− 76	− 3.3	26	78.8	14.4	58	136.4
−218	−360	...	−46	−50	−58	− 2.8	27	80.6	15.0	59	138.2
−212	−350	...	−40	−40	−40	− 2.2	28	82.4	15.6	60	140.0
−207	−340	...	−34	−30	−22	− 1.7	29	84.2	16.1	61	141.8
−201	−330	...	−29	−20	− 4	− 1.1	30	86.0	16.7	62	143.6
−196	−320	...	−23	−10	14	− 0.6	31	87.8	17.2	63	145.4
−190	−310	...	−17.8	0	32—	0—	32	89.6	17.8	64	147.2
−184	−300	...	−17.2	1	33.8	0.6	33	91.4	18.3	65	149.0
−179	−290	...	−16.7	2	35.6	1.1	34	93.2	18.9	66	150.8
−173	−280	...	−16.1	3	37.4	1.7	35	95.0	19.4	67	152.6
−169	−273	−459.4	−15.6	4	39.2	2.2	36	96.8	20.0	68	154.4
−168	−270	−454	−15.0	5	41.0	2.7	37	98.6	20.6	69	156.2
−162	−260	−436	−14.4	6	42.8	3.3	38	100.4	21.1	70	158.0
−157	−250	−418	−13.9	7	44.6	3.9	39	102.2	21.7	71	159.8
−151	−240	−400	−13.3	8	46.4	4.4	40	104.0	22.2	72	161.6
−146	−230	−382	−12.8	9	48.2	5.0	41	105.8	22.8	73	163.4
−140	−220	−364	−12.2	10	50.0	5.6	42	107.6	23.3	74	165.2
−134	−210	−346	−11.7	11	51.8	6.1	43	109.4	23.9	75	167.0
−129	−200	−328	−11.1	12	53.6	6.7	44	111.2	24.4	76	168.8
−123	−190	−310	−10.6	13	55.4	7.2	45	113.0	25.0	77	170.6
−118	−180	−292	−10.0	14	57.2	7.8	46	114.8	25.6	78	172.4
−112	−170	−274	− 9.4	15	59.0	8.3	47	116.6	26.1	79	174.2
−107	−160	−256	− 8.9	16	60.8	8.9	48	118.4	26.7	80	176.0

$$K = {}^\circ C + 273.2 \qquad\qquad {}^\circ C = K - 273.2$$

$$^\circ R = {}^\circ F + 459.7 \qquad\qquad {}^\circ F = {}^\circ R - 459.7$$

Fahrenheit — Celsius (Centigrade) Conversion Table (*Continued*)

Deg.C.	Ref.	Deg.F.	Deg.C.	Ref.	Deg.F.	Deg.C.	Ref.	Deg.F.	Deg.C.	Ref.	Deg.F.
27.2	81	177.8	58.3	137	278.6	89.4	193	379.4	304.4	580	1076
27.8	82	179.6	58.9	138	280.4	90.0	194	381.2	310.0	590	1094
28.3	83	181.4	59.4	139	282.2	90.6	195	383.0	315.6	600	1112
28.9	84	183.2	60.0	140	284.0	91.1	196	384.8	321.1	610	1130
29.4	85	185.0	60.6	141	285.8	91.7	197	386.6	326.7	620	1148
30.0	86	186.8	61.1	142	287.6	92.2	198	388.4	332.2	630	1166
30.6	87	188.6	61.7	143	289.4	92.8	199	390.2	337.8	640	1184
31.1	88	190.4	62.2	144	291.2	93.3	200	392.0	343.3	650	1202
31.7	89	192.2	62.8	145	293.0	93.9	201	393.8	348.9	660	1220
32.2	90	194.0	63.3	146	294.8	94.4	202	395.6	354.4	670	1238
32.8	91	195.8	63.9	147	296.6	95.0	203	397.4	360.0	680	1256
33.3	92	197.6	64.4	148	298.4	95.6	204	399.2	365.6	690	1274
33.9	93	199.4	65.0	149	300.2	96.1	205	401.0	371.1	700	1292
34.4	94	201.2	65.6	150	302.0	96.7	206	402.8	376.7	710	1310
35.0	95	203.0	66.1	151	303.8	97.2	207	404.6	382.2	720	1328
35.6	96	204.8	66.7	152	305.6	97.8	208	406.4	387.8	730	1346
36.1	97	206.6	67.2	153	307.4	98.3	209	408.2	393.3	740	1364
36.7	98	208.4	67.8	154	309.2	98.9	210	410.0	398.9	750	1382
37.2	99	210.2	68.3	155	311.0	99.4	211	411.8	404.4	760	1400
37.8	100	212.0	68.9	156	312.8	100.0	212	413.6	410.0	770	1418
38.3	101	213.8	69.4	157	314.6	104.4	220	428.0	415.6	780	1436
38.9	102	215.6	70.0	158	316.4	110.0	230	446.0	421.1	790	1454
39.4	103	217.4	70.6	159	318.2	115.6	240	464.0	426.7	800	1472
40.0	104	219.2	71.1	160	320.0	121.1	250	482.0	432.2	810	1490
40.6	105	221.0	71.7	161	321.8	126.7	260	500.0	437.8	820	1508
41.1	106	222.8	72.2	162	323.6	132.2	270	518.0	443.3	830	1526
41.7	107	224.6	72.8	163	325.4	137.8	280	536.0	448.9	840	1544
42.2	108	226.4	73.3	164	327.2	143.3	290	554.0	454.4	850	1562
42.8	109	228.2	73.9	165	329.0	148.9	300	572.0	460.0	860	1580
43.3	110	230.0	74.4	166	330.8	154.4	310	590.0	465.6	870	1598
43.9	111	231.8	75.0	167	332.6	160.0	320	608.0	471.1	880	1616
44.4	112	233.6	75.6	168	334.4	165.6	330	626.0	476.7	890	1634
45.0	113	235.4	76.1	169	336.2	171.1	340	644.0	482.2	900	1652
45.6	114	237.2	76.7	170	338.0	176.7	350	662.0	487.8	910	1670
46.1	115	239.0	77.2	171	339.8	182.2	360	680.0	493.3	920	1688
46.7	116	240.8	77.8	172	341.6	187.8	370	698.0	498.9	930	1706
47.2	117	242.6	78.3	173	343.4	193.3	380	716.0	504.4	940	1724
47.8	118	244.4	78.9	174	345.2	198.9	390	734.0	510.0	950	1742
48.3	119	246.2	79.4	175	347.0	204.4	400	752.0	515.6	960	1760
48.9	120	248.0	80.0	176	348.8	210	410	770.0	521.1	970	1778
49.4	121	249.8	80.6	177	350.6	215.6	420	788	526.7	980	1796
50.0	122	251.6	81.1	178	352.4	221.1	430	806	532.2	990	1814
50.6	123	253.4	81.7	179	354.2	226.7	440	824	537.8	1000	1832
51.1	124	255.2	82.2	180	356.0	232.2	450	842	565.6	1050	1922
51.7	125	257.0	82.8	181	357.8	237.8	460	860	593.3	1100	2012
52.2	126	258.8	83.3	182	359.6	243.3	470	878	621.1	1150	2102
52.8	127	260.6	83.9	183	361.4	248.9	480	896	648.9	1200	2192
53.3	128	262.4	84.4	184	363.2	254.4	490	914	676.7	1250	2282
53.9	129	264.2	85.0	185	365.0	260.0	500	932	704.4	1300	2372
54.4	130	266.0	85.6	186	366.8	265.6	510	950	732.2	1350	2462
55.0	131	267.8	86.1	187	368.6	271.1	520	968	760.0	1400	2552
55.6	132	269.6	86.7	188	370.4	276.7	530	986	787.8	1450	2642
56.1	133	271.4	87.2	189	372.2	282.2	540	1004	815.6	1500	2732
56.7	134	273.2	87.8	190	374.0	287.8	550	1022	1093.9	2000	3632
57.2	135	275.0	88.3	191	375.8	293.3	560	1040	1648.9	3000	5432
57.8	136	276.8	88.9	192	377.6	298.9	570	1058	2760.0	5000	9032

SECTION D

Using Powers of Ten

Using Powers of Ten

Since the modernized metric system, SI, simplifies the varying of physical quantity units by only shifting the decimal points or multiplying by powers of ten, the following introduction to the Powers of Ten Notation may provide a helpful review to many table users.

Powers of Ten Notation

In this system of notation every number is expressed by two factors, one of which is some integer from 1 to 9 followed by a decimal and the other is some power of 10.

Thus, 10,000 is expressed as 1.0000×10^4 and 10,463 as 1.0463×10^4. The number 43 is expressed 4.3×10 and 568 is expressed 5.68×10^2.

In the case of decimals, the number 0.0001 which as a fraction is $\frac{1}{10,000}$ is expressed as 1×10^{-4} and 0.0001463 is expressed as 1.463×10^{-4}. The decimal 0.498 is expressed as 4.98×10^{-1} and 0.03146 is expressed as 3.146×10^{-2}.

Rules for Converting any Number to Powers of Ten Notation

Any number can be converted to the powers of ten notation by means of one of two rules.

Rule 1. If the number is a whole number or a whole number and a decimal so that it has digits to the left of the decimal point, the decimal point is moved a sufficient number of places to the *left* to bring it to the immediate right of the first digit. With the decimal point shifted to this position, the number so written comprises the *first* factor when written in powers of ten notation.

The number of places that the decimal point is moved to the left, to bring it immediately to the right of the first digit is the *positive* index or power of 10 that comprises the *second* factor when written in powers of ten notation.

Thus, to write 4639 in this notation, the decimal point is moved three places to the left giving the two factors: 4.639×10^3. Similarly,

$$431.412 = 4.31412 \times 10^2$$
$$986388 = 9.86388 \times 10^5$$
$$7006 = 7.006 \times 10^3$$

Rule 2. If the number is a decimal, i.e., it has digits entirely to the right of the decimal point, then the decimal point is moved a sufficient number of places to the *right* to bring it immediately to the right of the first digit. With the decimal point shifted to this position, the number so written comprises the *first* factor when written in powers of ten notation.

The number of places that the decimal point is moved to the *right* to bring it immediately to the right of the first digit is the *negative* index or power of 10 that follows the number when written in powers of ten notation.

Thus, to bring the decimal point in 0.005721 to the immediate right of the first digit which is 5, it must be moved *three* places to the right, giving the two factors: 5.721×10^{-3}. Similarly,

$$0.469 = 4.69 \times 10^{-1}$$

$$0.0000516 = 5.16 \times 10^{-5}$$

Multiplying Numbers Written in Powers of Ten Notation

When multiplying two numbers written in the powers of ten notation together, the procedure is as follows:

1. Multiply the first factor of one number by the first factor of the other to obtain the first factor of the product.

2. Add the index of the second factor (which is some power of 10) of one number to the index of the second factor of the other number to obtain the index of the second factor (which is some power of 10) in the product. Thus,

$$(4.31 \times 10^{-2}) \times (9.0125 \times 10) =$$

$$(4.31 \times 9.0125) \times 10^{-2+1} = 38.844 \times 10^{-1}$$

$$(5.986 \times 10^4) \times (4.375 \times 10^3) =$$

$$(5.986 \times 4.375) \times 10^{4+3} = 26.189 \times 10^7$$

in each case rounding the first factor off to three decimal places.

When multiplying several numbers written in this notation together, the procedure is the same. All of the first factors are multiplied together to get the first factor of the product and all of the indices of the respective powers of ten are added together, taking into account their respective signs, to get the index of the second factor of the product. Thus, $(4.02 \times 10^{-3}) \times (3.987 \times 10) \times (4.863 \times 10^5) = (4.02 \times 3.987 \times 4.863) \times (10^{-3+1+5}) = 77.94 \times 10^3$ rounding off the first factor to two decimal places.

Dividing Numbers Written in Powers of Ten Notation

When dividing one number by another when both are written in this notation, the procedure is as follows:

1. Divide the first factor of the dividend by the first factor of the divisor to get the first factor of the quotient.

2. Subtract the index of the second factor of the divisor from the index of the second factor of the dividend, taking into account their respective signs, to get the index of the second factor of the quotient. Thus,

$$(4.31 \times 10^{-2}) \div (9.0125 \times 10) =$$

$$(4.31 \div 9.0125) \times (10^{-2-1}) = 0.4782 \times 10^{-3}$$

It can be seen, then, that where several numbers of different magnitudes are to be multiplied and divided this system of notation is helpful.

Example: Find the quotient of $\dfrac{250 \times 4698 \times 0.00039}{43678 \times 0.002 \times 0.0147}$

Solution: Changing all of these numbers to powers of ten notation and performing the operations indicated:

$$\frac{(2.5 \times 10^2) \times (4.698 \times 10^3) \times (3.9 \times 10^{-4})}{(4.3678 \times 10^4) \times (2 \times 10^{-3}) \times (1.47 \times 10^{-2})}$$

$$= \frac{(2.5 \times 4.698 \times 3.9)\,(10^{2+3-4})}{(4.3678 \times 2 \times 1.47)\,(10^{4-3-2})} = \frac{45.806 \times 10}{12.841 \times 10^{-1}}$$

$$= 3.5672 \times 10^{1\,-(-1)}$$

$$= 3.5672 \times 10^2$$

$$= 356.72$$

Index

For the convenience of the table user, this index is composed entirely of unit symbols. It is divided into two sections. The first, beginning below, gives entries for the conversion of English to metric equivalents. The second section, beginning on page 457, gives entries for metric to English equivalents.

To find general categories, such as Area, Volume Flow Rate, or Kinematic Viscosity, the reader should refer to the table of contents beginning on page V. From it, under category headings in boldface type, he can choose appropriate metric or English units to fit his need.

ENGLISH TO METRIC

METRIC TO ENGLISH